The New
Chemistry

The New Chemistry

Editor-in-chief **Nina Hall**

CAMBRIDGE
UNIVERSITY PRESS

PUBLISHED BY THE PRESS SYNDICATE OF THE UNIVERSITY OF CAMBRIDGE
The Pitt Building, Trumpington Street, Cambridge, United Kingdom

CAMBRIDGE UNIVERSITY PRESS
The Edinburgh Building, Cambridge CB2 2RU, UK
40 West 20th Street, New York, NY 10011–4211, USA
10 Stamford Road, Oakleigh, VIC 3166, Australia
Ruiz de Alarcón 13, 28014 Madrid, Spain
Dock House, The Waterfront, Cape Town 8001, South Africa

http://www.cambridge.org

© Cambridge University Press 2000

First published 2000

Printed in the United Kingdom at the University Press, Cambridge

Typeface Adobe Rotis 9.5/12 pt   *System* QuarkXPress™   [SE]

A catalogue record for this book is available from the British Library

ISBN 0 521 45224 4 hardback

Contents

Contributors

Jim Baggott (Independent management consultant)
David R. Baghurst (Imperial College of Science, Technology & Medicine)
Philip Ball (Nature)
Richard H. Bromilow (IACR – Rothamsted, Harpenden)
Paul Calvert (University of Arizona, Tucson)
Malcolm Chisholm (Indiana University)
Paul Christensen (University of Newcastle upon Tyne)
Guy Dewel (Université Libre de Bruxelles)
Peter P. Edwards (University of Birmingham)
Nina Hall (Independent science writer)
Andrew Hamnett (University of Newcastle upon Tyne)
Roald Hoffman (Cornell University)
Bhupinder P. S. Khambay (IACR – Rothamsted, Harpenden)
Dilip Kondepudi (Wake Forest University, Winston-Salem)
Jean-Marie Lehn* (Université Louis Pasteur)
Paul D. Lickiss (Imperial College of Science, Technology & Medicine)
Walter D. Loveland (Oregon State University)
D. Michael P. Mingos (University of Oxford)
Bob Munn (UMIST)
John N. Murrell (University of Sussex)
K. C. Nicolaou (The Scripps Research Institute, La Jolla)
T. Oshima (The Scripps Research Institute, La Jolla)
Ilya Prigogine* (University of Texas, Austin)
Stanley Roberts (University of Liverpool)
Colin Russell (Open University, UK)
Glenn T. Seaborg* (Lawrence Berkeley Laboratory)
Tatsuya Shono (Kyoto University)
Gabor A. Somorjai (University of California, Berkeley)
Jim Staunton (University of Cambridge)
Kira Weissman (University of Cambridge)
Robert J. P. Williams (University of Oxford)
E. W. Yue (The Scripps Research Institute, La Jolla)

* Nobel Prize winner

Preface

The subject of chemistry is vast. It covers virtually all aspects of the behaviour of atoms and molecules – from the creation of the elements in the stars to the complex molecules of life. Chemistry, however, is much more than just about investigating the Universe at the molecular level; its central remit (which is quite different from those of other scientific disciplines) is to synthesize new forms of matter, many of which are extremely useful, for example, pharmaceuticals. *The New Chemistry* aims to illustrate the ingenuity and imaginative breadth of modern molecular science by presenting a showcase of research over the past 30 years. It is, of course, an impossible task to be comprehensive; were we to have covered every area, we would have easily filled several volumes. Nevertheless, we have collected here a broad range of reviews from some of the world's most renowned chemists (including several Nobel Prize winners) in areas where there have been recent exciting advances.

Most general books on chemistry published in the past have concentrated on applications and on the material benefits of chemical discoveries. We have taken a slightly different tack. This book looks at the underpinning science, such as the strategies behind making complex molecules, the intricate chemistry of metals, the chemistry that happens at surfaces, and the study of chemical bonding and reactions. We also wanted to illustrate the interdisciplinary nature of chemistry, in particular its central role in the life sciences and increasingly in the development of novel electronics and energy sources. We have also included areas less often covered in the standard chemistry book, such as nuclear chemistry and chemistry far from equilibrium.

One huge area that we have not been able to include is analytical chemistry. This subject is so large that it would easily fill a volume on its own. We do, of course, fully recognize that analytical techniques designed to solve the structures of molecules, such as nuclear magnetic resonance and X-ray crystallography, have revolutionized research both in pure chemistry and in molecular biology – as have chromatographic techniques that allow minute concentrations of molecules to be separated. For similar reasons of space we have not been able to review the improvements to the environment being made through the design of 'greener' industrial processes.

Despite these inevitable omissions, we hope that you will enjoy reading about developments in the core of the subject presented here. It is worth remembering that chemistry is about investigating and creating ultimate complexity in nature, and as such presents a wonderful intellectual challenge, which *The New Chemistry* aptly illustrates.

Nina Hall

Acknowledgements

Chapter 6

Malcolm Chisholm thanks the Alexander von Humboldt Foundation for a fellowship and his good friends and colleagues Professors Gottfried Huttner and Walter Siebert for their kindness and hospitality at the Anorganisch-chemisches Institut der Universität Heidelberg, where this chapter was prepared.

Chapter 7

This work was supported by the Director, Office of Energy Research, Office of Basic Energy Sciences, Materials Sciences Division of the USA's Department of Energy under contract DE-AC03-76SF00098.

Chapter 8

K. C. Nicolaou, E. W. Yue and T. Oshima are pleased to acknowledge the contributions of their collaborators. They also wish to acknowledge with many thanks financial support for their research programmes from the National Institutes of Health (USA), The Skaggs Institute of Chemical Biology, The Scripps Research Institute and their friends in the pharmaceutical industry.

Chapter 10

Stanley Roberts thanks Professor Roger Newton (Maybridge Chemicals, Tintagel), Dr. Graham Sewell (University of Plymouth) and Dr. Sheila Buxton (Royal Society of Chemistry, Cambridge) for helpful comments and advice during the preparation of the manuscript.

Chapter 16

Dilip Kondepudi thanks the National Science Foundation of the USA for supporting this work through grant CHE-9527095. Guy Dewel is a Research Associate with the FNRS (Belgium). The authors would like to acknowledge the following sources of funding: the USA's Department of Energy (grant DE-FGO3-94ER14465); the Robert A. Welch Foundation (grant F-0364); La Communauté Française de Belgique and La Loterie Nationale Belge.

Introduction

Roald Hoffmann

Before the word for it (in any language) came to be, there was chemistry. For one defining aspect of human beings has always been the meld of mind and hands in transforming matter. Look at a wall painting in the tomb of Rekhmire at Thebes and you see men sweating at the bellows as they separate gold from dross. There, kohl, a dark cosmetic mixture, is being made. There, people are cooking.

People want to change the natural into the useful unnatural. Or, if you prefer, into the improved man- or human-made. To add value is to profit. So from the beginning transformation was essential, whether in the making of metals and alloys, in medicinal preparations, in cooking, in dyeing and colouring, in tanning leather or in cosmetics. It took some time for this most fundamental of chemical activities to be called by its contemporary name – synthesis.

Human beings are curious as well as practical. So questions are asked about matter, be it natural or transformed. They are really very simple questions: 'What have I got?', 'How did that transformation happen?', 'Why that change or structure, rather than others?' . . . The first of these questions was central to the work that marked the definitive emergence of chemistry as a science – Antoine Laurent Lavoisier's 1789 *Elementary Treatise on Chemistry*. Analysis, the weighty struggle with the question 'What is it?', remained absolutely central for a long time. So Goethe, the prime minister of a grand duchy in what is now Germany, would take two weeks off to follow a course of analysis with Johann Wolfgang Döbereiner in nearby Jena.

If analysis, at least at the elemental level, was at the heart of chemistry 200 years ago, in 1901 the questions 'How?' and 'Why?' were the ones I think most often asked and answered. At a certain level – that of the behaviour of matter on the average, at the macroscopic level – the understanding came from thermodynamics, a beautifully worked-through blend of practicality (steam engines at its origins) and sublime mathematics (the Gibbs phase rule). For the first time, industrial processes could be rationally optimized – it was around 1910 that the Haber–Bosch process for nitrogen fixation was elaborated, a synthesis of ammonia so successful that it now competes on a global scale with the natural processing of the atmosphere's nitrogen. Answering a different call for 'How?', we see the first explorations of the wondrously evolved biochemical pathways in just this period, but also the beginnings of the deplorable split between chemistry and biochemistry.

A new century opens before us. Where are we now? In what way is chemistry today different from the chemistry we admired 100 years ago? I see fantastic progress in the ways we have looked before at matter – synthesis ever more proficient and prominent; analysis answering (and asking) still more detailed questions – 'What have I got?' being replaced by 'How little have I got of what?' I see the study of the mechanisms of chemical reactions (the answer to the 'How?' question), the atomistic tracing of the transformation, pursued down to the nearly unimaginably small time interval of a femtosecond. I see theory (dealing with the 'Why' question) finally entering into real predictive competition with experiment, even as it, theory, rushes pell-mell into simulation rather than understanding. These are wonders and marvels, all in an industrial setting that is a positive contributor to the trade balance of many an industrialized country.

However, when I look at the riches of the new chemistry, I see some leitmotifs that were just not there a century ago. Let me call them 'the molecular vision', being 'in control', and 'taking care'.

Regarding the molecular vision, leaf through this book – how many pictures of chemical reactions in a flask or coloured crystals can you see? How many representations of molecules are there? Molecular structures have been with us since the middle of the nineteenth century, but in the study of molecules both quiescent and in motion there has been a sea change in the second half of the twentieth century. Without waiting for microscopes (we're still waiting, hype to the contrary), using cool thought, hot hands and our ingenious instruments, we have probed inside recalcitrant matter. The diffraction of X-rays, modern computers and nuclear magnetic resonance (before it was denuclearized in a fit of euphemism) have taught us the arrangement of atoms in space in substantive metric detail. Today, when a chemist thinks of a reaction, he or she sees it in the light of a double flame (to adopt Octavio Paz' felicitous phrase) – the macroscopic transformation, as of old, and the microscopic, molecular change.

No, those molecules are not hard balls held together by sticks or springs. They are quantum objects that demand a dual vision. As our instruments interrogate them with light, molecules behave quantum mechanically, with the sometimes mysterious logic of such objects . . . and yet, and yet, for most practical purposes molecules may be assembled (in a hands-off manner) in very architectural, ball-and-stick ways. The twinned perspective is that these molecules are eminently manipulable geometrical objects *and* they are wave-packets of matter.

In thinking about molecules as they change (for that remains the essence of chemistry) we have invented incredibly fast strobe lights to freeze their motion even as they career around at the speed of sound on a very crowded dance floor. We have also developed very specific ways to put energy into parts of a molecule, by using a new type of light – the intense, monochromatic laser beam. The chemist of 2001 makes the new surely, with a double vision, that of substance and molecule.

Regarding being in control; 100 years ago there was so much to discover – How does inheritance work?, 'What is inside diamond and graphite?' – that chemists were just entranced by the intricacy of the world as it is. True, industrial chemists were not mesemerized, simply proceeded to make in novel, unnatural ways what there was profit in making. It there were no cheap pure aluminium in the world of 1884, young Héroult and young Hall told us how to make it,

A cursory look at modern chemistry reveals it compulsively making new molecules with exquisite control. True, the makers are human and so bound to puff serendipitous creation up with *ex-post-facto* motive; and yes, obsession with control seems inordinately, sometimes humorously, male – in a science in which women are approaching parity in numbers – but these are just my carpings in contemplating a marvellous achievement: chemists, men and women, have learned to assemble incredibly complex structures with fantastic control.

The working out of that creative obsession for control is essential in a world in which as small a difference between molecules as that between a left and a right hand can spell the difference between toxicity and pharmaceutical efficacy. A catalyst that allows one company's process to be 10% more efficient can put a competitor out of business. Materials with exquisitely controlled surfaces and properties are shaping the information revolution. The genetically manipulated feedstocks (that's also chemistry) that confer resistance against a common pest have transformed agriculture.

Yes, we are taking a leaf from nature's wondrous ways of self-assembly and evolutionary strategies; and yes, by engineering properties way beyond what nature provides, we, masters of cultural evolution, make the new.

Regarding taking care; we look at nature not only as servant muse, but also with true love. For, talented as we are, we have transformed many of the grand cycles of this planet. More than half of the sulfur atoms in your amino acids have seen the inside of a sulfuric-acid factory. Some essential renewable resources, such as the petroleum feedstocks of our chemical industry, will be depleted within the next couple of centuries. The moral tenor of this age, not just the Greens, demands that whatever we transform be done with social justice and with respect for nature.

The signal change in the chemical industry is that, when a new process is introduced, ecological and safety considerations are now paramount. This did not come about willingly, mind you – I recall the screaming of the automotive industry about how they could never, absolutely never, reduce emissions of CO, hydrocarbons and NO_x by a factor of 20. Forced to do so by public pressure and government regulation, the industry, in one of the major scientific and engineering advances of the twentieth century, devised the 'three-way' catalyst to do just what it had said could never be done.

The ecological imperative has crept down much more slowly to inventive yet unconcerned academia. However, I see its formative events there – in the interest in atmospheric chemistry and in the ingenious construction of novel organic processes to avoid the use of organic solvents. A government carrot in the form of new research funds for green chemistry would, in my unpopular opinion, be just what is needed to channel the ingenuity of my colleagues, who love to say that they just want to do what is interesting, but . . . The same nicely obsessive penchant for control as that which is used to make molecules do acrobatics can and is being turned to the attainment of a necessary balance between our given imperative to create and our love for the world. Chemists should and do care.

The Search for New Elements

Glenn T. Seaborg and Walter D. Loveland

Introduction

The chemical elements are the building blocks of nature. All substances are combinations of these elements. There are 112 known chemical elements, the heaviest naturally occurring element being uranium ($Z=92$). Twenty of the heaviest chemical elements, the transuranium elements, are man-made. The story of their synthesis, their properties, their impact on chemistry and physics and their importance to society is as fascinating as is the chance of a further expansion in their number.

In this chapter, we will discuss how to make these elements and their chemical properties and importance to chemistry. We will discuss the chances of making new elements. We will conclude by discussing the practical applications of the transuranium elements, including concerns about their presence in the environment. We show the cast of characters for our story in Table 1 that lists the names of the transuranium elements and Figure 1 that depicts their places in the modern Periodic Table.

The History of the Chemical Elements

The story of the chemical elements begins in ancient times. At the time of Christ, people knew about nine chemical elements (C, S, Fe, Cu, Ag, Sn, Au, Hg and Pb). These elements had been isolated in pure form and put to use in various ways. From the charcoal of fires, one had carbon (C). One knew the elements that were found free in nature – gold (Au), silver (Ag) and copper (Cu). Sulfur (S) was known, as was how to extract the pure metals from ores of copper (Cu), mercury (Hg), lead (Pb) and tin (Sn).

Figure 1
The modern Periodic Table showing the transuranium elements.

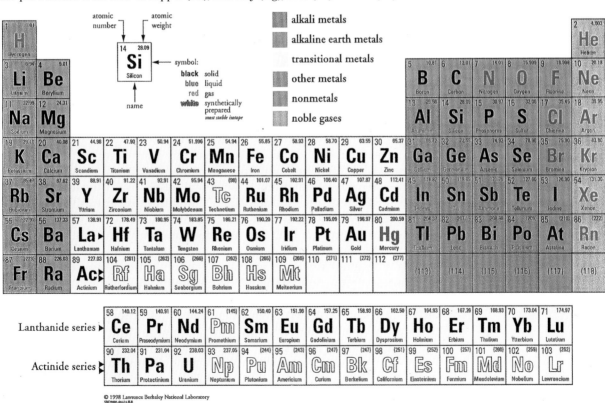

© 1998 Lawrence Berkeley National Laboratory

Table 1	The transuranium elements	
Atomic number	Element	Symbol
93	Neptunium	Np
94	Plutonium	Pu
95	Americium	Am
96	Curium	Cm
97	Berkelium	Bk
98	Californium	Cf
99	Einsteinium	Es
100	Fermium	Fm
101	Mendelevium	Md
102	Nobelium	No
103	Lawrencium	Lr
104	Rutherfordium	Rf
105	Hahnium[a]	Ha
106	Seaborgium	Sg
107	Bohrium	Bh
108	Hassium	Hs
109	Meitnerium	Mt
110		
111		
112		

Note:

[a] The IUPAC has recommended the name dubnium (symbol Db) for element 105.

During the Middle Ages, alchemists discovered arsenic (As), antimony (Sb) and bismuth (Bi). The first element discovered by one person and recognized as such was phosphorus (P). The German alchemist Hennig Brand discovered phosphorus ('bearer of light') in 1669. Brand made phosphorus from dried urine and discovered that it had the remarkable property of glowing in the dark when exposed to air.

Cobalt was discovered in 1737 and nickel about 14 years later. Cobalt and nickel ores previously had been mistaken for copper ore. Since they refused to yield copper, these ores obviously were possessed by evil spirits and thus had gained the names *Kobold* (goblin) and *Kupfernickel* (Old Nick's copper). These names persisted.

The pace of the discovery of elements quickened during the latter part of the eighteenth century through the work of Henry Cavendish, Daniel Rutherford, Joseph Priestley, Karl Scheele and Antoine Lavoiser. The number of elements known to man by the middle 1770s was about twenty. During the next 25 years, eleven more elements were discovered. Between 1800 and 1869, the number of elements known to man nearly doubled.

In 1869, the great Russian chemist Dmitri Mendeleev and the German chemist Lothar Meyer independently uncovered the principle of the Periodic Table. First, they arranged the known elements (about two thirds of the naturally occurring elements) in order of increasing atomic weight. Hydrogen didn't fit in very well, so they started lining up the elements beginning with lithium and beryllium. They found that, if they finished a row and then started a second row below it, they got elements with similar chemical properties falling below one another across the table. As they extended the table, they found several groups of elements that did not fit their seven columns. These elements were incorporated into the table later.

Figure 2
The Periodic Table of Mendeleev.

REIHEN	GRUPPE I. R^2O	GRUPPE II. RO	GRUPPE III. R^2O^3	GRUPPE IV. RH^4 RO^2	GRUPPE V. RH^3 R^2O^5	GRUPPE VI. RH^2 RO^3	GRUPPE VII. RH R^2O^7	GRUPPE VIII. RO^4
1	H=1							
2	Li = 7	Be = 9,4	B = 11	C = 12	N = 14	O = 16	F = 19	
3	Na = 23	Mg = 24	Al = 27,3	Si = 28	P = 31	S = 32	Cl = 35,5	
4	K = 39	Ca = 40	— = 44	Ti = 48	V = 51	Cr = 52	Mn = 55	Fe = 56, Co = 59, Ni = 59, Cu = 63.
5	(Cu = 63)	Zn = 65	— = 68	— = 72	As = 75	Se = 78	Br = 80	
6	Rb = 85	Sr = 87	?Yt = 88	Zr = 90	Nb = 94	Mo = 96	— = 100	Ru = 104, Rh = 104, Pd = 106, Ag = 108.
7	(Ag = 108)	Cd = 112	In = 113	Sn = 118	Sb = 122	Te = 125	J = 127	
8	Cs = 133	Ba = 137	?Di = 138	?Ce = 140	—	—	—	— — — —
9	(—)	—	—	—	—	—	—	
10	—	—	?Er = 178	?La = 180	Ta = 182	W = 184	—	Os = 195, Ir = 197, Pt = 198, Au = 199.
11	(Au = 199)	Hg = 200	Tl = 204	Pb = 207	Bi = 208	—	—	
12	—	—	—	Th = 231	—	U = 240	—	— — — —

Mendeleev, in particular, noticed that he had to skip several places to maintain vertical columns in which all the elements listed had similar properties (see Figure 2). Mendeleev's contribution was that he recognized the gaps in his Periodic Table and that they should be filled by elements that had not yet been discovered.

He went further than that. He had the courage to predict what three of the undiscovered elements would look like, how much they would weigh and how they would react chemically. Since he expected the three elements to have properties similar to those of boron, aluminium and silicon, respectively, he tentatively named them eka-boron, eka-aluminium, and eka-silicon (from the Sanskrit *eka*, meaning 'next'). These missing elements were discovered in Scandinavia (scandium), France (gallium) and Germany (germanium) a few years later.

By the end of the first third of the twentieth century, the total number of chemical elements had increased to 88. This included the rare-gas family that had to be fitted into the scheme of Figure 2 by adding another column and a series of elements – the rare earths or lanthanides – located in the place of a single element, lanthanum. Only four of the first 92 elements were missing, i.e. those with atomic numbers 43, 61, 85 and 87. We show in Figure 3 the Periodic Table as it was known in 1940.

The first man-made element was that with atomic number 43. Carlo Perrier and Emilio Segre identified it in 1937. A foil of molybdenum metal was irradiated with 8 MeV deuterons in the Berkeley cyclotron and sent to Perrier and Segre in Italy. They identified two radioactivities (with half-lives of 62 and 90 days) as chemically different from all the other 88 elements and thus due to element 43. These two radioactivities are known now to be $^{95}Tc^m$ and $^{97}Tc^m$, where the 'm' designation represents a long-lived excited state. They later (in 1947) named the new element 'technetium' (Tc) after the Greek word $\tau\epsilon\chi\nu\epsilon\tau o\varsigma$ (artificial) because this was the first element to be produced by artificial means.

This discovery raised the possibility of filling in the other missing elements (61, 85 and 87) and possibly even extending the Periodic Table to elements beyond uranium. It also raised some general questions about the limits of the Periodic Table, methods of synthesis of elements and so on.

Element 61 was first identified by Jacob Marinsky, Lawrence Glendenin and Charles Coryell during the Manhattan Project in 1945. These investigators made a chemical identification, using ion exchange, of two isotopes of element 61, ^{147}Pm and ^{149}Pm, formed by the neutron-induced fission of uranium. The name promethium was suggested for this element by Grace Mary Coryell in honour of the figure in Greek mythology who stole fire from the gods for human use, an apt analogy for the Manhattan Project.

Element 85, astatine (At), was first identified in 1940 as the result of production of the radioactive isotope ^{211}At by the irradiation of bismuth with 32-MeV alpha-particles

Figure 3
The Periodic Table of the 1930s; atomic numbers of undiscovered elements are in shaded squares.

1 H																	2 He
3 Li	4 Be											5 B	6 C	7 N	8 O	9 F	10 Ne
11 Na	12 Mg											13 Al	14 Si	15 P	16 S	17 Cl	18 Ar
19 K	20 Ca	21 Sc	22 Ti	23 V	24 Cr	25 Mn	26 Fe	27 Co	28 Ni	29 Cu	30 Zn	31 Ga	32 Ge	33 As	34 Se	35 Br	36 Kr
37 Rb	38 Sr	39 Y	40 Zr	41 Nb	42 Mo	43	44 Ru	45 Rh	46 Pd	47 Ag	48 Cd	49 In	50 Sn	51 Sb	52 Te	53 I	54 Xe
55 Cs	56 Ba	57-71* La–Lu*	72 Hf	73 Ta	74 W	75 Re	76 Os	77 Ir	78 Pt	79 Au	80 Hg	81 Tl	82 Pb	83 Bi	84 Po	85	86 Rn
87	88 Ra	89 Ac	90 Th	91 Pa	92 U	93	94	95	96	97	98	99	100	101	102	103	104

*Lanthanides	57 La	58 Ce	59 Pr	60 Nd	61	62 Sm	63 Eu	64 Gd	65 Tb	66 Dy	67 Ho	68 Er	69 Tm	70 Yb	71 Lu

in the 60-inch cyclotron at Berkeley. In a series of experiments involving radiotracers, Dale Corson, Kenneth MacKenzie and Emilio Segre were able to show that this alpha-particle-emitting isotope (^{211}At) was a new chemical element. The name astatine is derived from a Greek word meaning 'unstable'. Later it was found that radioactive isotopes of astatine occur as part of the natural decay series.

Element 87 (francium (Fr)) was discovered by Marguerite Perey in 1939. She showed that the alpha-decay of ^{227}Ac led to a new beta-particle-emitting substance with a half-life of 21 min (^{223}Fr). This new substance behaved chemically like an alkali metal. Perey named the new element francium in honour of her native country.

The Problems of Element Synthesis

The extent to which man can extend the 'natural' number of chemical elements is limited by the characteristics of the fundamental forces of nature. A commonly used model of the gross properties of nuclei predicts that a nucleus will instantaneously undergo fission when

$$E_c = 2E_s$$

where E_c and E_s are the Coulomb energy and surface energy, respectively, of the nucleus that is represented as a uniformly charged liquid drop. The quantities E_c and E_s are given as

$$E_c = \frac{3}{5}\frac{(Ze)^2}{R} = k_c \frac{Z^2}{A^{1/3}}$$

$$E_s = 4\pi R^2 \gamma = k_s A^{2/3}$$

where γ is the nuclear surface tension ($\simeq$1 MeV fm^{-2}), Z is the atomic number and R is the nuclear radius (which is proportional to $A^{1/3}$, where A is the mass number). The limiting value of the atomic number, Z_{LIMIT}, is then

$$Z^2_{\mathrm{LIMIT}} = 2(k_s/k_c)A_{\mathrm{LIMIT}}$$

If we remember that the neutron/proton ratio in heavy nuclei is about 1.5/1, then

$$Z_{\mathrm{LIMIT}} = 5(k_s/k_c)$$

Thus the upper bound to the Periodic Table is proportional to the ratio of two fundamental constants related to the strengths of the nuclear (surface) and electro-magnetic forces. The ratio k_s/k_c is about 20–25 and thus we expect 100–125 chem-ical elements. Only a moderate extension of the 'natural' Periodic Table by man would be expected to be possible.

The synthesis of a new element involves more than just colliding two nuclei whose atomic numbers are such that they sum to a value corresponding to an unknown element. Heavy nuclei are, in general, quite fissionable. If they are made with significant excitation, they will decay by fission, leaving no identifiable heavy residue of their formation. So one must balance the factors governing the 'production' of a new nucleus carefully with those factors governing its 'survival'. The 'production factors' determine the yield of the primary reaction products whereas the 'survival factors' determine which primary product nuclei de-excite by particle emission, which allows them to survive, and which nuclei de-excite by fission, which destroys them. Amongst the 'production factors' are items such as the 'starting material', the target nuclei, which must be available in sufficient quantity and suitable form. We must have enough transmuting projectile nuclei also. The transmutation reaction must occur with adequate probability to ensure a good yield of the product nucleus in a form suitable for further study. Equally important is that the product nuclei be produced with dis-tributions of excitation energy and angular momentum such that the product nuclei will de-excite by particle or photon emission rather than the disastrous fission process. The competition between particle emission and fission as de-excitation paths depends on the excitation energy, angular momentum and intrinsic stability of the product nucleus, which is related to the atomic and mass numbers of the product (see Box 1). For a description of the detection of product nuclei, see Box 2.

Nuclear synthesis is similar in some ways to inorganic and organic chemical syn-theses in that the synthetic chemist or physicist has to understand the reactions involved and the structures and stabilities of the intermediate species. Although, in principle, the outcome of any synthesis reaction is calculable, in practice such calcula-tions are, for the most part, very difficult. Instead, the cleverness of the scientists involved, their manipulative skills and the instrumentation available for their use determine the success of many synthetic efforts.

The synthesis reactions used to 'discover' the man-made elements are given in Table 2. All these reactions are complete fusion reactions in which the reacting nuclei fuse, equilibrate and de-excite in a manner that is independent of their mode of formation. The nucleus is said to have 'amnesia' about its mode of formation. Other production reactions involving partial capture of the projectile nucleus are also possible.

The reactions in Table 2 can be divided into four classes: the neutron-induced reac-tions ($Z=61, 93, 95, 99$ and 100); the light-charged-particle-induced reactions ($Z=43, 85, 94, 96$–98 and 101); the 'hot fusion' reactions ($Z=102$–106) and the 'cold fusion' reactions ($Z=107$–112). In the neutron-induced reactions to make the trans-uranium nuclei, the capture of a neutron does not create a new element, but the sub-sequent β^- decays do. Reactions of light charged particles with exotic actinide target nuclei allow one to increase the atomic number of the product by one or two units from that of the target nucleus. To make the heaviest elements, one needs to add several protons to the target nucleus by a reaction with a heavy ion. Such 'hot fusion' reactions with actinide target nuclei lead to highly excited intermediate species that decay mostly by fission but occasionally by emitting neutrons, thus producing new nuclei. However, as the atomic number of the product nuclei increases, so does the probability of fission leading to very poor probabilities of survival for the putative new species. The Russian nuclear physicist Yuri Oganessian pointed out that a way around this problem was to fuse heavier projectile nuclei with nuclei in the

Box 1 Element-synthesis calculations

The reactions used to synthesize heavy nuclei are, quite often, very improbable reactions, representing minor branches relative to the main reaction. Their probabilities of occurrence with respect to the main synthesis reaction are frequently less than 10^{-6}. Hence it is intrinsically difficult to describe these reactions accurately from a theoretical point of view. Instead, workers in this field have frequently resorted to semi-empirical prescriptions to guide their efforts.

The German physicist Peter Armbruster constructed a systematic diagram of the probability of fusion of two heavy nuclei at energies near the reaction barrier. This is shown in Figure 4. To use this graph, one picks values of the atomic numbers of projectile and target nuclei and reads off the expected value for the cross section for producing a completely fused species. The excitation energy of the completely fused species can then be read from Figure 5, which is based upon the nuclear masses of Peter Möller, J. Rayford Nix,

Władysław Świątecki and William Myers. Taking as a rough rule of thumb that, for each 10 MeV of excitation energy, the probability of survival of the fused system drops by a factor of 10^2, one can then compute the cross section for producing a given species.

For example, the successful synthesis of ^{265}Hs (^{265}108) involved the reaction

$$^{58}\text{Fe} + {}^{208}\text{Pb} \rightarrow {}^{265}\text{Hs} + n$$

From Figure 4, one predicts the fusion cross section to be 10^{-32} cm^2, while Figure 5 would suggest an excitation energy of about 20 MeV. Thus one would roughly estimate the overall cross section for producing ^{265}Hs to be

$$10^{-32} \times (10^{-2})^2 \simeq 10^{-36} \text{ cm}^2$$

(The measured cross section was 2×10^{-35} cm^2.)

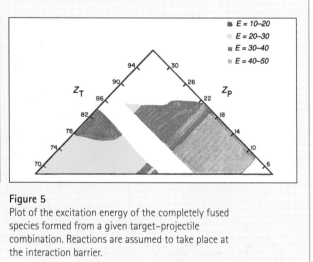

Figure 4
A plot of the contours of $\log_{10} \sigma_{\text{fus}}$ (where σ_{fus} is the s-wave fusion cross section at the interaction barrier).

Figure 5
Plot of the excitation energy of the completely fused species formed from a given target–projectile combination. Reactions are assumed to take place at the interaction barrier.

lead–bismuth region. Because of the special stability of the lead-bismuth nuclei, the resulting fused species would be formed 'cold' and could, with some reasonable probability, decay by emitting only a single neutron.

The History of the Discovery of Transuranium Elements

The first scientific attempts to prepare the elements beyond uranium were made by Enrico Fermi, Emilio Segre and co-workers in Rome in 1934, shortly after the existence of the neutron had been discovered. This group of investigators irradiated uranium with slow neutrons and found several radioactive products, which were thought to be due to new elements. However, chemical studies by Otto Hahn and Fritz Strassman in Berlin showed that these species were isotopes of the known elements formed by the fission of uranium into two approximately equal parts. This discovery of nuclear

Box 2 How to detect heavy-element atoms

The detection of atoms of a new element has always focused on measuring the atomic number of the new species and showing that it is different from all known values of *Z*. Unambiguous methods for establishing the atomic number include chemical separations, measurement of the X-ray spectrum accompanying a nuclear decay process and establishment of a genetic relationship between the unknown new nucleus and some known nuclide. As the quest for new elements focuses on still heavier species, the probability of producing the new elements has decreased and one has had to devote more attention to the problem of detecting a few atoms of a new species amidst a background of many orders of magnitude more atoms of other elements. Thus, modern attempts to make new heavy-element atoms usually involve some kind of 'separator'.

An example of a modern separator is the velocity filter SHIP (Figure 6) at the GSI in Darmstadt, Germany. In this separator, nuclear reaction products (from the target wheel) undergo different electrostatic deflections (in a crossed magnetic field) depending on whether they are fission fragments, scattered beam particles or the desired heavy-element residues. The efficiency of the separator is about 50% for heavy-element residues, while transfer products and scattered beam nuclei are rejected by factors of 10^{14} and 10^{11}, respectively. The heavy recoil atoms are implanted in the silicon detectors. Their implantation energies and positions are correlated to any subsequent decays of the nuclei to establish genetic relationships to known nuclei.

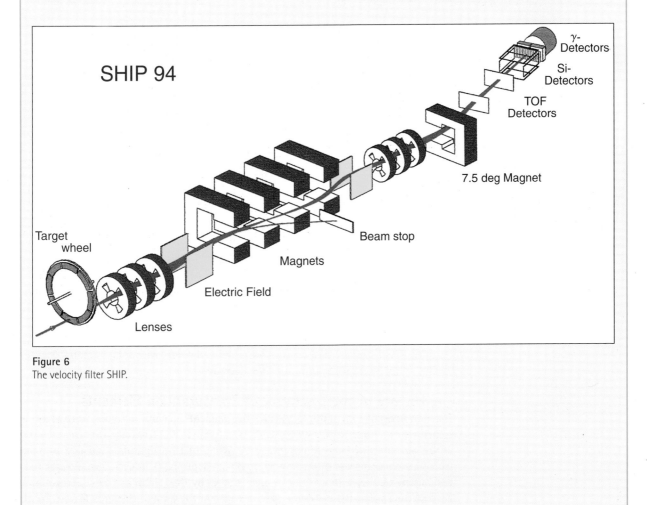

Figure 6
The velocity filter SHIP.

Table 2 A summary of syntheses of man-made elements

Atomic number	Name and symbol	Synthesis reaction	Half-life
43	Technetium (Tc)	$^{94}Mo + {}^{2}H \rightarrow {}^{95}Tc^{m} + n$	61 days
		$^{96}Mo + {}^{2}H \rightarrow {}^{97}Tc^{m} + n$	90.5 days
61	Promethium (Pm)	$^{235}U + n \rightarrow {}^{147}Pm, {}^{149}Pm$	2.6 years
			53.1 h
85	Astatine (At)	$^{209}Bi + {}^{4}He \rightarrow {}^{211}At + 2n$	7.2 h
87	Francium (Fr)	$^{227}Ac \rightarrow {}^{223}Fr + \alpha$	21.8 min
93	Neptunium (Np)	$^{238}U + n \rightarrow {}^{239}U +$ $^{239}U \xrightarrow{\beta^-} {}^{239}Np$	2.35 days
94	Plutonium (Pu)	$^{238}U + {}^{2}H \rightarrow {}^{238}Np + 2n$ $^{238}Np \xrightarrow{\beta^-} {}^{238}Pu$	86.4 years
95	Americium (Am)	$^{239}Pu + n \rightarrow {}^{240}Pu +$ $^{240}Pu + n \rightarrow {}^{241}Pu +$ $^{241}Pu \xrightarrow{\beta^-} {}^{241}Am$	433 years
96	Curium (Cm)	$^{239}Pu + {}^{4}He \rightarrow {}^{242}Cm + n$	162.5 days
97	Berkelium (Bk)	$^{241}Am + {}^{4}He \rightarrow {}^{243}Bk + 2n$	4.5 h
98	Californium (Cf)	$^{242}Cm + {}^{4}He \rightarrow {}^{245}Cf + n$	44 min
99	Einsteinium (Es)	'Mike' thermonuclear explosion (leading to ^{253}Es)	20 days
100	Fermium (Fm)	'Mike' thermonuclear explosion (leading to ^{255}Fm)	20 h
101	Mendelevium (Md)	$^{253}Es + {}^{4}He \rightarrow {}^{256}Md + n$	76 min
102	Nobelium (No)	$^{244}Cm + {}^{12}C \rightarrow {}^{252}No + 4n$	2.3 s
103	Lawrencium (Lr)	$\left.\begin{array}{l}^{250}Cf \\ {}^{251}Cf \\ {}^{252}Cf\end{array}\right\} + {}^{11}B \rightarrow {}^{258}Lr + \left\{\begin{array}{l}3n \\ 4n \\ 5n\end{array}\right.$ $\left.\begin{array}{l}^{250}Cf \\ {}^{251}Cf \\ {}^{252}Cf\end{array}\right\} + {}^{10}B \rightarrow {}^{258}Lr + \left\{\begin{array}{l}2n \\ 3n \\ 4n\end{array}\right.$	4.3 s
104	Rutherfordium (Rf)	$^{249}Cf + {}^{12}C \rightarrow {}^{257}Rf + 4n$	3.4 s
		$^{249}Cf + {}^{13}C \rightarrow {}^{259}Rf + 3n$	3.8 s
105	Hahnium (Ha)	$^{249}Cf + {}^{15}N \rightarrow {}^{260}Ha + 4n$	1.5 s
106	Seaborgium (Sg)	$^{249}Cf + {}^{18}O \rightarrow {}^{263}106 + 4n$	0.9 s
107	Bohrium (Bh)	$^{209}Bi + {}^{54}Cr \rightarrow {}^{262}107 + n$	102 ms
108	Hassium (Hs)	$^{208}Pb + {}^{58}Fe \rightarrow {}^{265}108 + n$	1.8 ms
109	Meitnerium (Mt)	$^{209}Bi + {}^{58}Fe \rightarrow {}^{266}109 + n$	3.4 ms
110		$^{209}Bi + {}^{59}Co \rightarrow {}^{267}110 + n$	4 μs
		$^{208}Pb + {}^{62}Ni \rightarrow {}^{269}110 + n$	170 μs
		$^{208}Pb + {}^{64}Ni \rightarrow {}^{271}110 + n$	56 ms
		$^{244}Pu + {}^{34}S \rightarrow {}^{273}110 + 5n$	118 ms
111		$^{209}Bi + {}^{64}Ni \rightarrow {}^{272}111 + n$	1.5 ms
112		$^{208}Pb + {}^{70}Zn \rightarrow {}^{277}112 + n$	240 μs

fission in December of 1938 was thus a by-product of man's quest for the transuranium elements.

With poetic justice, the actual discovery of the first transuranium element came as part of an experiment performed to study the nuclear fission process. Edwin McMillan, working at the University of California at Berkeley in the spring of 1939, was trying to measure the energies of the two recoiling fragments from the neutron-induced fission of uranium. He placed a thin layer of uranium oxide on one piece of paper. Next to this he stacked very thin sheets of cigarette paper to stop and collect the fission fragments of uranium. During his studies he found that there was another radioactive product of the reaction – one that did not recoil enough to escape the uranium layer as did the fission products. He suspected that this product was formed by the capture of a neutron by the more abundant isotope of uranium, $^{238}_{92}U$. McMillan and Philip Abelson (Figure 7), who joined him in this research, showed in 1940, by chemical means, that this product is an isotope of element 93, $^{239}_{93}Np$, formed in the following sequence:

$$^{238}_{92}U + n \rightarrow \; ^{239}U + \gamma$$

$$^{239}_{92}U \xrightarrow[t_{1/2}=23.5 \text{ min}]{\beta^-} \; ^{239}_{93}Np \; (t_{1/2}=2.36 \text{ days})$$

Neptunium, the element beyond uranium, was named after the planet Neptune because this planet is beyond the planet Uranus, after which uranium had been named.

Plutonium was the second transuranium element to be discovered. By bombarding uranium with charged particles, in particular, deuterons (^2H), using the 60-inch cyclotron at the University of California at Berkeley, Glenn T. Seaborg, McMillan, Joseph W. Kennedy and Arthur C. Wahl (Figure 8) succeeded in preparing a new isotope of neptunium, ^{238}Np, which decayed by β^- emission to ^{238}Pu, that is,

$$^{238}_{92}U + \; ^2H \rightarrow \; ^{238}_{93}Np + 2n$$

$$^{239}_{93}Np \xrightarrow[t_{1/2}=2.12 \text{ days}]{\beta^-} \; ^{238}_{94}Pu \; (t_{1/2}=87.7 \text{ years})$$

Early in 1941, ^{239}Pu, the most important isotope of plutonium, was discovered by Kennedy, Segre, Wahl and Seaborg. ^{239}Pu was produced by the decay of ^{239}Np, which had been produced by the irradiation of ^{238}U by neutrons, using the reaction

$$^{239}_{92}U + \; ^1_0n \rightarrow \; ^{239}_{92}U + \gamma$$

$$^{239}_{92} \xrightarrow[t_{1/2}=23.5 \text{ min}]{\beta^-} \; ^{239}_{93}Np \xrightarrow[t_{1/2}=2.35 \text{ days}]{\beta^-} \; ^{239}_{94}Pu \; (t_{1/2}=24\,110 \text{ years})$$

This isotope, ^{239}Pu, was shown to have a cross section for thermal neutron-induced fission that exceeded that of ^{235}U, a property that made it important for nuclear weapons, considering that no isotope separation was necessary for its preparation, unlike for ^{235}U. Plutonium was named after the planet Pluto, following the pattern used in naming neptunium.

The next transuranium elements to be discovered, americium and curium (Am and Cm; $Z = 95$ and 96, respectively) represent an important milestone in chemistry, namely the recognition of a new group of elements in the Periodic Table, the actinides. According to the Periodic Table of Figure 3, one expected americium and curium to be eka-iridium and eka-platinum, i.e. to have chemical properties similar to those of iridium and platinum. In 1944, Seaborg conceived the idea that all the known elements heavier than actinium ($Z = 89$) had been misplaced in the Periodic Table. He postulated that the elements heavier than actinium form a second series similar to the lanthanide elements (Figure 1), called the actinide series. This series would end in element 103 (Lr) and, analogously to the lanthanides, these elements would have a common oxidation state of $+3$.

Once this redox property had been understood, the use of a proper chemical procedure led quickly to the identification of an isotope of a new element. Thus, a new alpha-particle-emitting nuclide, now known to be $^{242}_{94}Cm$ (half-life 162.9 days), was produced by Seaborg, Albert Ghiorso and Ralph James in the summer of 1944 by the bombardment of $^{239}_{94}Pu$ with 32–MeV helium ions:

Figure 7
The discoverers of neptunium, Edwin M. McMillan (top) and Philip H. Abelson (bottom).

$$^{239}_{94}\text{Pu} + ^{4}_{2}\text{He} \rightarrow ^{242}_{96}\text{Cm} + ^{1}_{0}\text{n}$$

The bombardment took place in the Berkeley 60-inch cyclotron, after which the material was shipped to the Metallurgical Laboratory at Chicago for chemical separation and identification. A crucial step in the identification of the alpha-particle-emitting nuclide as an isotope of element 96, $^{242}_{96}\text{Cm}$, was the identification of the known $^{238}_{94}\text{Pu}$ as the alpha-decay daughter of the new nuclide.

The identification of an isotope of element 95, by Seaborg, Ghiorso, James and Leon Morgan in late 1944 and early 1945, followed the identification of this isotope of element 96 (^{242}Cm) as a result of the bombardment of $^{239}_{94}\text{Pu}$ with neutrons in a nuclear reactor. The production reactions, involving multiple neutron capture by plutonium, are

$$^{239}_{94}\text{Pu} + ^{1}_{0}\text{n} \rightarrow ^{240}_{94}\text{Pu} + \gamma$$

$$^{240}_{94}\text{Pu} + ^{1}_{0}\text{n} \rightarrow ^{241}_{94}\text{Pu} + \gamma$$

$$^{241}_{94}\text{Pu} \xrightarrow[t_{1/2} = 14.4 \text{ years}]{\beta^-} ^{241}_{95}\text{Am} \ (t_{1/2} = 432.7 \text{ years})$$

$$^{241}_{95}\text{Am} + ^{1}_{0}\text{n} \rightarrow ^{242}_{95}\text{Am} + \gamma$$

$$^{242}_{95}\text{Am} \xrightarrow[t_{1/2} = 16.0 \text{ h}]{\beta^-} ^{242}_{96}\text{Cm}$$

Figure 8
The co-discoverers of plutonium, Joseph W. Kennedy (25 December 1940), Arthur C. Wahl and Glenn T. Seaborg. Seaborg and Wahl are shown (in February 1966) with the sample of ^{239}Pu in which fission was demonstrated in 1941 (the cigar box was that of G. N. Lewis).

The years after World War II led to the discovery of elements 97–103 and the completion of the actinide series. Although the story of the discovery of each of these elements is fascinating, we shall, in the interests of brevity, refer the reader elsewhere (see the references) for detailed accounts of most of these discoveries. As an example of the techniques involved, we shall discuss the discovery of element 101 (mendelevium).

The discovery of mendelevium was one of the most dramatic in the sequence of syntheses of transuranium elements. It was the first case in which a new element was produced and identified one atom at a time. By 1955, scientists at Berkeley had prepared an equilibrium amount of about 10^9 atoms of $^{253}_{99}\text{Es}$ by neutron irradiation of plutonium in the Materials Testing Reactor in Idaho. From a 'back of the envelope' calculation done by Ghiorso during an airplane flight, they thought that it might be possible to prepare element 101 using the reaction

$$^{253}_{99}\text{Es} + ^{4}_{2}\text{He} \rightarrow ^{256}_{101}\text{Md} + ^{1}_{0}\text{n}$$

The amount of element 101 expected to be produced in an experiment can be calculated using the formula

$$N_{101} = \frac{N_{\text{Es}} \sigma \varphi (1 - e^{-\lambda t})}{\lambda}$$

where N_{101} and N_{Es} are the numbers of atoms of element 101 produced and of $^{253}_{99}\text{Es}$ target atoms, respectively, σ is the reaction cross section (estimated to be of the order of 10^{-27} cm^2), φ is the flux of helium ions ($\approx 10^{14}$ particles s^{-1}), λ is the decay constant of $^{256}_{101}\text{Md}$ (estimated to be $\approx 10^{-4}$ s^{-1}) and t is the duration of each bombardment ($\approx 10^4$ s):

$$N_{101} \approx \frac{10^9 \times 10^{-27} \times 10^{14}(1 - e^{-10^{-4} \times 10^{+4}})}{10^{-4}} \approx 1 \text{ atom}$$

Thus the production of only one atom of element 101 per experiment could be expected!

Adding immensely to the complexity of the experiment was the absolute necessity for the chemical separation of the one atom of element 101 from the 10^9 atoms of einsteinium in the target and its ultimate, complete chemical identification by separation with the ion-exchange method (see Box 3). This separation and identification would have to take place within a period of hours, or perhaps even 1 h or less. Furthermore,

---

Box 3 The ion-exchange technique in heavy-element research

Ion exchange is a very important method of chemical separation for the transuranium elements. It is fast, efficient and has the unique ability of permitting determination of the atomic numbers of the elements being separated even for samples containing a few atoms. In ion exchange, anions or cations are apportioned between a mobile aqueous phase and a stationary solid phase. Most commonly the solid phase will be an organic polymer containing sulfonic acid groups (a cation-exchange resin) or quaternary ammonium groups (an anion-exchange resin). Actinide ions with oxidation states ranging from $+3$ to $+6$ can be adsorbed by cation-exchange resins and eluted with anions such as citrate, lactate and α-hydroxyisobutyrate. In anion exchange, the actinides are complexed with ligands like Cl^- to form anionic complexes that can then be adsorbed on anion-exchange resins.

The order of elution of the actinide ions from a cation-exchange column is generally in order of the radii of the hydrated ions, the hydrated ions of the heaviest elements eluting first; thus, lawrencium is eluted first and americium last among the tripositive ions (see Figure 9), although this ordering need not be maintained in all situations. The exact position at which a given element will elute can be predicted ahead of time by careful comparison of the given column for the elution of actinides and lanthanides (Figure 9).

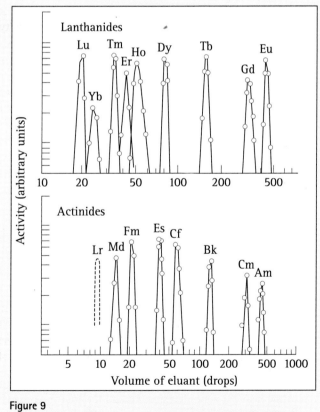

Figure 9
Elution of tripositive lanthanide and actinide ions on Dowex 50 cation-exchange resin using ammonium α-hydroxyiosbutyrate as an eluant.

the target material had a 20-day half-life and one needed a non-destructive technique allowing re-use of the target material.

The definitive experiments were performed during a memorable, all-night session, on 18 February 1955. To increase the number of events that might be observed at one time, three successive 3-h bombardments were performed and, in turn, their transmutation products were quickly and completely separated by the ion-exchange method. Some of the nuclide $^{253}_{99}$Es was present in each case so, together with the $^{246}_{98}$Cf produced from $^{244}_{96}$Cm that was also present in the target (via the $^{244}_{96}$Cm(^{4}He, 2n) reaction), it was possible to define the positions from which the elements came off the column used to contain the ion-exchange resin. Five spontaneous fission counters then were used to count simultaneously the corresponding drops of solution from the three runs.

A total of five spontaneous fission counts was observed at the position for element 101, whereas eight spontaneous fission counts were also observed at the position for element 100. No such counts were observed for any other position. The original data are presented in Figure 10.

Figure 10

Original elution data corresponding to the discovery of mendelevium, 18 February 1955. The curves for einsteinium-253 (given the old symbol E^{253}) and californium-246 are for alpha-particle emission. (Dowex 50 ion-exchange resin was used and the eluting agent was ammonium α-hydroxyisobutyrate.)

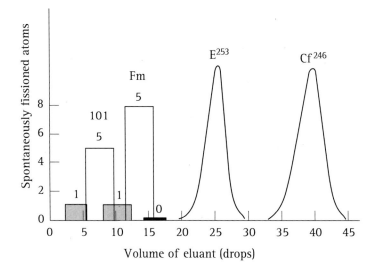

The synthesis of the transactinides is noteworthy from chemical, nuclear and sociological viewpoints. From the chemical point of view, rutherfordium ($Z = 104$) is important because it was the first transactinide element. From Figure 1, we would expect rutherfordium to behave as a group IVB element, such as hafnium or zirconium, but not like the heavy actinides. Its solution chemistry, deduced from chromatography experiments, is different from that of the actinides and resembles that of zirconium and hafnium. Detailed gas chromatography has revealed important deviations from expected periodic-table trends and relativistic quantum-chemical calculations.

From the nuclear point of view, one finds the previously mentioned shift in synthesis techniques from actinide-based, 'hot fusion' reactions for the lighter transactinides to lead-based, 'cold fusion' reactions for the heaviest elements.

From the sociological viewpoint, one notes a change in scientific leadership in this field during the period in which these elements were first synthesized. In the synthesis of the heaviest actinide elements and the first three transactinides, the group headed by Albert Ghiorso (see Figure 11) in Berkeley, California dominated the field, with important contributions coming from the Joint Institute for Nuclear Research at Dubna. Starting with element 107, bohrium, Bh, leadership in this field passed to the group headed by Peter Armbruster (Figure 11) at the GSI in Darmstadt, with important contributions from the Dubna laboratory. Current experimental work at the GSI is being done by the same group, which is now directed by Sigurd Hofmann.

An Aside from the Junior Author

We would be remiss not to comment on another of the transactinide elements, element 106. Element 106 has been named seaborgium (symbol Sg) after one of the authors of this chapter. Glenn, the co-discoverer of plutonium and nine other transuranium elements, said upon this occasion (Figure 12) 'It is the greatest honour ever bestowed upon me – even better, I think, than winning the Nobel Prize'. Seaborgium was first synthesized in 1974 using the ^{249}Cf(^{18}O,n)^{263}Sg reaction by a team led by Ken Hulet and Albert Ghiorso. The discovery was confirmed in 1993 in an experiment at Berkeley by a team led by Ken Gregorich.

Superheavy Elements

Until 1970, we thought that the practical limit of the Periodic Table would be reached at about element 108 (Figure 13). By extrapolating the experimental data on half-lives of heavy elements, we concluded that the half-lives of the longest-lived isotopes of the heavy elements beyond about element 108 would be so short ($\lesssim 10^{-6}$ s), due to spontaneous fission decay, that we could not produce and study them. However, during the late 1960s and early 1970s, nuclear theorists, using techniques developed by Vilen Strutinsky and Władysław Świątecki, predicted that special stability against fission would be associated with proton number $Z = 114$ and neutron number $N = 184$. These 'superheavy elements' were predicted to have half-lives of the order of the age of the universe. They were predicted to form an 'island' of stability separated from the 'peninsula' of known nuclei (Figure 14(*a*)).

We now know that these predictions were wrong, in part. Although we believe that there is a group of 'superheavy' nuclei whose half-lives are relatively long compared with those of lower Z elements, we do not believe that they form an 'island' of stability. Rather, we picture them as a continuation of the peninsula of known nuclei (Figure 14(*b*)). We also believe that their half-lives are short on geological time scales. Therefore, they do not exist in nature. The most stable of the superheavy nuclei, those with $Z = 112$ and $N \simeq 184$, are predicted to decay by alpha-particle emission with half-lives of about 20 days.

The principal problem associated with the superheavy nuclei is not whether they could exist (which is considered relatively certain) but how to make them. Literally hundreds of synthesis reactions have been proposed and several have been tried. All have failed because either the formation of the fused system is too improbable (Figure 4) or its excitation energy is too large (Figure 5), resulting in too small a probability for formation of the product nuclei. For example, the most widely studied synthesis reaction is the ^{48}Ca + ^{248}Cm reaction. Using Figures 4 and 5 as a guide, we roughly estimate the fusion cross section to be 3×10^{-32} cm^2, the excitation energy to be about 30 MeV (survival probability 10^{-6}) and predict a formation cross section of 10^{-38} cm^2. With modern technology, this would correspond to one atom per month. This production rate is at the limit of modern technology.

Some questions concerning the extra stability associated with $Z \simeq 114$ have recently been raised. (The special stability associated with $N = 184$ seems to be certain.) Some theorists believe that the next 'magic' number of protons will be 126, not 114. In this case, it would be very difficult to make such nuclei.

Others have taken the approach that, without the special stability associated with nuclear shell structure, elements as light as $Z = 106$–108 would have negligibly short half-lives. The existence of these nuclei with millisecond half-lives is a demonstration

Figure 11
Albert Ghiorso (left) and Peter Armbruster (right) as they participated in a collaborative search for superheavy elements. (Photograph from the Laurence Berkeley National Laboratory.)

Figure 12
Glenn Seaborg points out seaborgium in the Periodic Table.

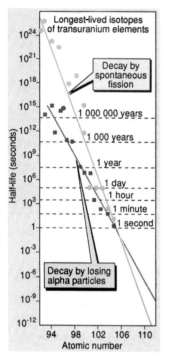

Figure 13

The half-lives of the longest-lived isotope of each element versus atomic number Z, circa 1970.

that we have already made superheavy nuclei, according to this view. The shell stabilization of these nuclei, which are deformed, is due to the special stability of the $N = 162$ configuration in deformed nuclei. (The 'traditional' superheavy nuclei with $Z \approx 114$ and $N = 184$ are believed to have spherical shapes.)

Prospects for the Future – The Synthesis of New Heavy Nuclei

The known transnobelium nuclei are shown in Figure 15. One sees a general decrease in the half-lives of the nuclei and a tendency towards nuclei becoming more proton-rich with increasing Z. Given the short half-lives shown in Figure 15, one might wonder about the likelihood of being able to synthesize and study new heavy nuclei or elements. However, as shown in Figures 16 and 17, there are interesting possibilities of new, longer-lived heavy nuclei. The first possible region of interest (Figure 16) is the neutron-rich isotopes of elements 104–108. These nuclei are predicted to have half-lives that are substantially longer (by a factor of 10–1000) than those of the known longest-lived isotopes of these elements. Instead of having half-lives of seconds (or a fraction of a second), one predicts that the neutron-rich isotopes of elements 104–108 will have half-lives of days to years. The impact of having these longer-lived species available for studies of the chemistry and atomic physics of these elements could be tremendous. Detailed studies that are not possible now could become routine. Thus there is substantial interest in making new neutron-rich isotopes of the known heavy elements.

Of course there is also the old quest to make new chemical elements. In Figure 17 (taken from the work of Robert Smolanczuk and co-workers), we show the predicted half-lives for decay by spontaneous fission and alpha-decay for the isotopes of elements 104–120. Although the overall half-lives for the accessible isotopes are substantially shorter than those of the heaviest known elements (microseconds instead of milliseconds), these species are readily detectable with modern technology (Box 2). For elements 114 and 116, α-decay is a favoured mode of decay for many accessible isotopes. Because the detection of a primary α-decay (and subsequent time-correlated daughter decays) is a well-established method of determining the atomic number, the dominance of α-decay is favourable to the experimenter. Such is also true for a limited range of very neutron-rich isotopes of elements 118 and 120. Some of the most neutron-rich isotopes of the elements of $Z \leq 114$ are predicted to have total half-lives (including α-decay) that are long enough to allow studies of their atomic physics and chemistry.

Having explained what the motivations behind the synthesis of new elements might be, we need to examine the prospects of synthesizing these species. The authors have developed a semi-empirical formalism based, in part, on the considerations leading to Figures 4 and 5. (The probabilities of survival of the nuclei are calculated using more detailed prescriptions for decay by fission or neutron emission, as are the probabilities of their formation.) We have calculated the probability of every possible combination of every possible radioactive or stable target nucleus and stable projectile nucleus to make new nuclei. We have focused our attention on two classes of nuclear reactions, complete fusion and multi-nucleon transfer. The calculations qualitatively reproduce all the known data on heavy-element synthesis using these reactions (Figure 18).

Assuming experimental conditions (accelerator beam intensities and target thicknesses) that are typical for the best modern laboratories, we have calculated the rates of nuclide formation that might be expected for some of the most promising reactions to synthesize new elements (Table 3). For elements 113 and 114, synthesis by 'hot fusion' reactions appears to be more promising than using the 'cold fusion' reactions, although the uncertainties in the estimates are large.

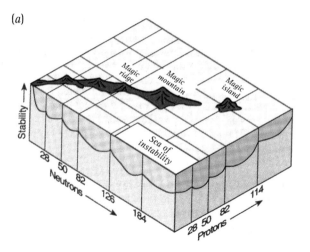

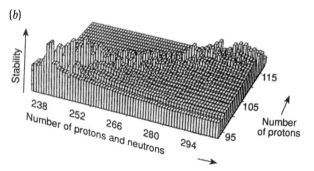

Figure 14
(*a*) An allegorical representation of the stability of heavy nuclides, *circa* 1975, showing the 'superheavy island'. (*b*) A modern plot of the predicted half-lives of the heaviest nuclei. Note that there is an isthmus connecting the known and superheavy nuclei.

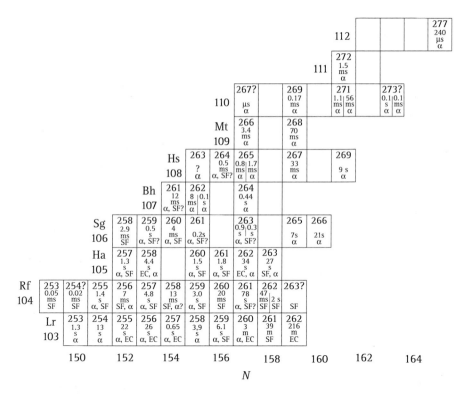

Figure 15
A chart of the transnobelium nuclides. SF, spontaneous fission.

An alternative approach to estimating heavy-element-production cross sections is by judicious extrapolation of current data. This approach is shown in Figure 19, taken from the work of Sigurd Hofmann. One can estimate production cross sections for element 113 (via the ^{209}Bi(^{70}Zn,n)^{278}113 reaction) of about one picobarn (1 pb) and for element 114 (via the ^{208}Pb(^{76}Ge,n)^{283}114 reaction) of about 1 pb also.

The synthesis of elements 113 and 114 appears very challenging. It should be noted that a recent attempt to synthesize element 113 by the GSI group was not successful. An upper-limit cross section for the production of element 113 by the ^{209}Bi(^{70}Zn, n)^{278}113 reaction was estimated to be about 0.6 pb, consistently with current understanding of the difficulty of this synthesis.

Figure 16

Predicted half-lives of the isotopes of elements 104–108.

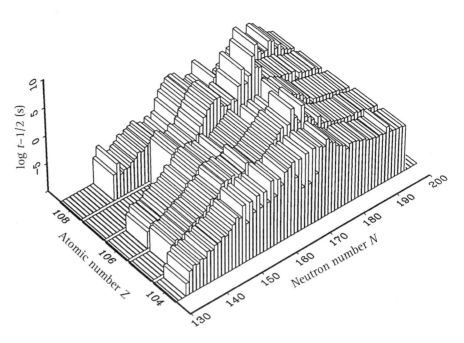

Figure 17

Predicted half-lives for decay by spontaneous fission (sf) and α–particle emission for the heaviest elements.

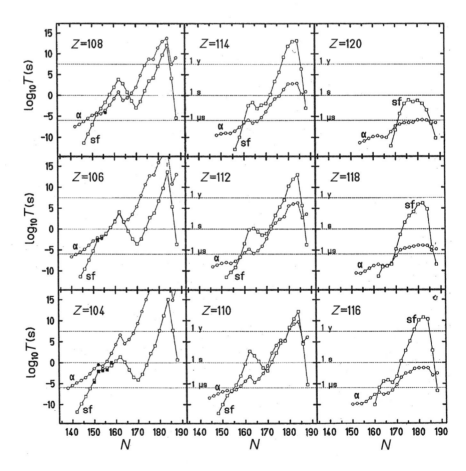

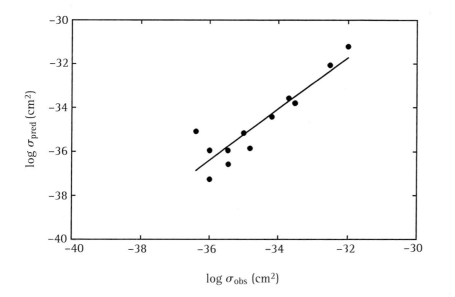

Figure 18
A plot of the predicted versus observed cross sections for the formation of new elements, using the authors' formalism.

Table 3	Promising element-synthesis reactions	
Element	Reaction	Predicted cross section
113	^{209}Bi(^{70}Zn, n)^{278}113	15 fb
	^{208}Pb(^{74}Ga, n)^{281}113	20 fb
	^{237}Np(^{48}Ca, 3n)^{282}113	0.6 pb
114	^{244}Pu(^{48}Ca, 3n)^{289}114	2.0 pb
	^{208}Pb(^{76}Ge, n)^{283}114	10 fb
	^{207}Pb(^{76}Ge, n)^{282}114	7 fb
115	^{209}Bi(^{76}Ge, n)^{284}115	0.7 fb
	^{208}Pb(^{83}As, γ)^{291}115	4 pb
	^{243}Am(^{48}Ca, 3n)^{298}115	70 fb
116	^{208}Pb(^{82}Se, γ)^{290}116	1.3 pb
	^{208}Pb(^{80}Se, n)^{287}116	0.4 fb
117	^{209}Bi(^{82}Se, n)^{284}117	35 ab

Note:

The uncertainty in the predicted cross sections is one to two orders of magnitude. The radiative-capture reactions are speculative insofar as attempts to perform them have not proven successful. The cross sections are reported in units of barns (10^{-24} cm$^2 = 1$ b) and subdivisions (picobarns, 1 pb $= 10^{-36}$ cm^2; femtobarns, 1 fb $= 10^{-39}$ cm^2; and attobarns, 1 ab $= 10^{-42}$ cm^2).

Margareta Magda (now at SUNY, Stony Brook) and co-workers have estimated that a modest extension of the known nuclei to more neutron-rich isotopes of elements 104–106 by using multi-nucleon transfer reactions should be feasible. They calculate that, by using the heaviest available nuclide, ^{254}Es, as a target and bombarding it with the neutron-rich nuclide, ^{48}Ca, one can make 1–100 atoms per day of species like ^{265}Rf, ^{265}Ha and ^{267}106. Similar calculations by the authors concerning the more readily available ^{253}Es also indicate that there are promising situations.

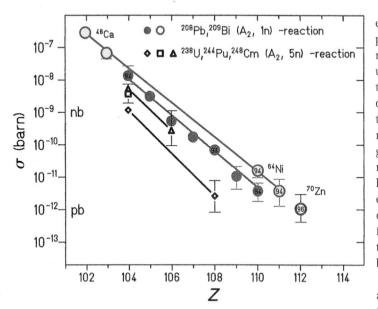

Figure 19
Empirical predictions of cross sections for the production of heavy elements.

However promising these calculations are, experimenters have become excited by another possible avenue ·for synthesizing new heavy nuclei. This· possibility involves the use of unstable radioactive nuclei to induce the synthesis reactions. The motivation for this development comes, in part, from the realization that the more neutron-rich heavy nuclei are more stable (i.e. have longer half-lives), in general, than the known proton-rich heavy nuclei (Figure 16). Another important source of hope is the prediction that the reaction of extremely neutron-rich, radioactive nuclei with other nuclei will lead to a greater probability of formation of the fused species and, most importantly, a lower excitation energy, leading to better chances of survival.

Although there are limited facilities available at present to allow such experiments to be performed, the nuclear science community is very excited about the new physics opportunities such facilities would afford. The potential application of radioactive nuclear beams to make new heavy nuclei is but one small part of many new avenues of research, including nuclear astrophysics and the synthesis and study of other exotic nuclei, that await the users of such facilities.

At present one is limited to using stable nuclei or nuclei with relatively long half-lives as the 'starting material' for nuclear synthesis. By producing, accelerating and reacting short-lived radioactive nuclei ($t_{1/2} > 0.01$ s) with an accelerator, one can extend the number of available 'starting projectile nuclei' by a factor of more than ten. Figure 20 shows the projected output of radioactive nuclei from a large, proposed North American radioactive beam facility designated the IsoSpin Laboratory (ISL). When one compares the relative number of stable nuclei with the range of possible radioactive beam nuclei, one has to be impressed with how our synthetic possibilities would expand.

Two possible designs for such facilities seem to be favoured. The first possibility, typified by the currently operating facility at Michigan State University, involves fragmenting a high-energy heavy-ion beam ($E^{beam} \simeq 50$–100 MeV/nucleon) and then magnetically separating and transporting the high-energy radioactive projectile fragments to an experimental area where they can be studied or used to induce reactions, etc. The second type of design, the ISOL design, is typified by the ISOLDE facility at the CERN. In this type of facility, a nuclear reaction such as fission or spallation is induced in a large mass of material. This material serves as the ion source for a mass separator, with the radioactive reaction products recoiling or diffusing out of the material into the separator. The result is low-energy ($\simeq 60$ keV) mass-separated beams of radioactive nuclei. These nuclei can be further accelerated and used for study, synthesis reactions and so on.

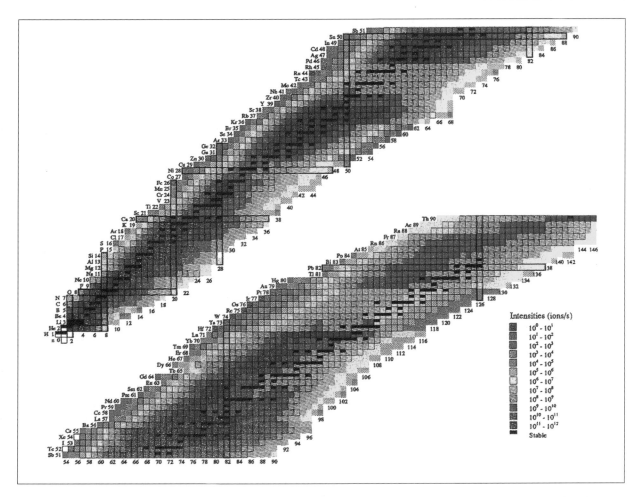

Intensities (ions/s)

$10^0 - 10^1$
$10^1 - 10^2$
$10^2 - 10^3$
$10^3 - 10^4$
$10^4 - 10^5$
$10^5 - 10^6$
$10^6 - 10^7$
$10^7 - 10^8$
$10^8 - 10^9$
$10^9 - 10^{10}$
$10^{10} - 10^{11}$
$10^{11} - 10^{12}$
Stable

With the plethora of such nuclei available (Figure 20), one can write down an almost uncountable array of possible synthesis reactions. However, writing down such reactions doesn't mean that they will occur with sufficient probability or that the radioactive nuclei are available in sufficient quantity to allow practical synthesis of new nuclei. With this problem in mind, the authors have made a 'brute force,' numerical evaluation of every possible reaction between every available species (shown in Figure 20) and all available target nuclei. Both complete fusion and multi-nucleon transfer reactions were considered, but no special enhancements of fusion due to the neutron-rich projectiles were considered. Some typical results are shown in Table 4. The estimates of Table 4 should be regarded as lower limits on the production rates because they do not include the effect of any unknown enhancement in the fusion cross section with more neutron-rich nuclei.

We can summarize the prospects for synthesis of new heavy nuclei by stating that there appear to be many promising opportunities to make new, neutron-rich isotopes of elements 104–108 using stable or radioactive beams. There also appears to be a good chance of synthesizing new heavy elements using stable projectiles.

The Chemistry of the Transuranium Elements

The chemical behaviour of the transuranium elements is interesting because of its complexity and the insights offered into the chemistry of the other elements. The placing of these man-made elements into the Periodic Table (Figure 1) is one of the few significant alterations of the Periodic Table of Mendeleev. Since so little is known

Figure 20
Typical radioactive beams that can be expected from the North American radioactive beam facility, the ISL.

Table 4	Predicted rates of synthesis for heavy neutron-rich nuclei using radioactive beams

Reaction	Rate (atoms/day)
^{209}Bi(^{53}Sc,2n)^{260}Rf	7
^{252}Cf(^{14}C,3n)^{263}Rf	70 000
^{249}Cf(^{14}C,3n)^{260}Rf	25 000
^{209}Bi(^{53}Ti,2n)^{260}Ha	30
^{253}Es(^{14}C,3n)^{264}Ha	23 000
^{252}Cf(^{20}O,3n)^{269}Sg	20
^{209}Bi(^{58}Cr,2n)^{265}Bh	0.4
^{253}Es(^{20}O,3n)^{270}Bh	7
^{208}Pb(^{58}Mn,2n)^{264}Bh	3
^{208}Pb(^{63}Fe,2n)^{269}Hs	0.1
^{226}Ra(^{48}Ca,3n)^{272}Hs	20

about the chemistry of the transactinide elements, one has the unique opportunity of being able to test periodic-table predictions of chemical behaviour before the relevant experiments are done.

The first thing that strikes one as one contemplates the chemistry of the actinide elements is their complex chemistry. One graphic demonstration of this is shown in Figure 21, in which the colours of the oxidation states of plutonium in solution are shown. One is struck by the number of accessible oxidation states (III–VII) and how their properties (colours) change with the complexing agent. Such diversity in solution chemistry is certainly not typical of the lanthanides and indicates the special character of the actinides.

To understand the differences in chemical behaviour between the lanthanides and the actinides, which both have partially filled d and f orbitals, one needs to understand a bit about their electron configurations. As the atomic number Z of the nucleus increases, the electrons become more tightly bound. As their binding energy increases, so does their velocity. For electrons in the 1s shell, the average velocity is roughly Z au. (The speed of light, c, is 137.035 au.) Thus, for $Z = 90$, the velocity is $90c/137$ or $0.51c$. The Schrödinger equation is no longer appropriate and one must use the fully relativistic treatment of Dirac.

The solution of the Dirac equation for the hydrogen-like atoms leads to wave functions that are products of radial and angular factors. The angular factors are shown in Figure 22. Each state is specified by four quantum numbers whose meaning is different from that for the Schrödinger equation. They are n, the principal quantum number with values of 1, 2, 3, . . .; ℓ, the azimuthal quantum number with values 0, 1, 2, . . ., $n-1$, denoted by s, p, d, f and g; j, the angular momentum quantum number with values of $\ell \pm \frac{1}{2}$ (usually denoted as a subscript to ℓ) and m, the magnetic quantum number, taking on half-integer values from $-j$ to $+j$. Thus the three p, the five d and the seven f orbitals are no longer degenerate and split into one $p_{1/2}$, two $p_{3/2}$, two $d_{3/2}$, three $d_{5/2}$, three $f_{5/2}$ and four $f_{7/2}$ levels, the occupancy of each of them being $2j + 1$. This is called the 'spin–orbit' splitting. The orbital shapes are given by j and m, orbitals of the same j and m having the same shape. What is surprising to the traditional chemist in Figure 22 is that the p orbitals are shaped like a sphere ($p_{1/2}$), a doughnut ($p_{3/2}$, $m = \frac{3}{2}$) and a dog-bone ($p_{3/2}$, $m = \frac{1}{2}$). One notes that the state with the highest m value for a given j value always has a doughnut-shaped distribution whereas the lowest

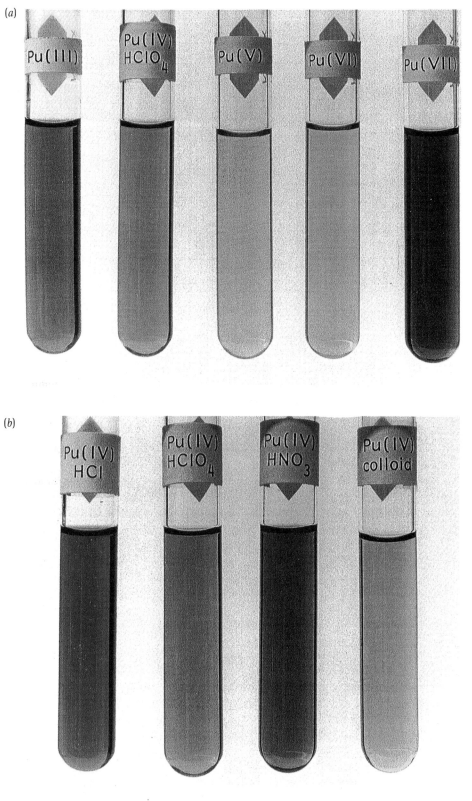

Figure 21
(*a*) Colours of the oxidation states of Pu in HClO$_4$, except for Pu(VII) which is in strong base. (*b*) Colours of Pu(IV) in various media.

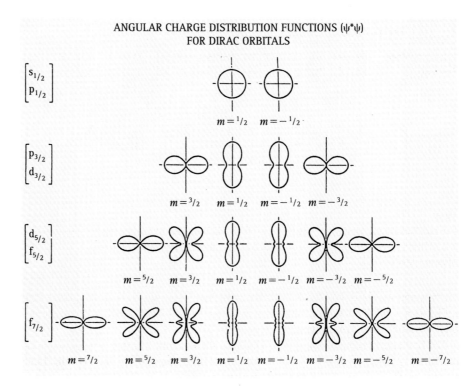

ANGULAR CHARGE DISTRIBUTION FUNCTIONS ($\psi^*\psi$)
FOR DIRAC ORBITALS

value of m corresponds to a distribution stretched along the z-axis with no nodes. States of intermediate m are multi-lobed toroids.

The effect of using relativistic rather than non-relativistic quantum mechanics on the predicted atomic orbitals is threefold: (a) a contraction and stabilization of the $s_{1/2}$ and $p_{1/2}$ shells, (b) the splitting of the energy levels due to the spin–orbit coupling and (c) an expansion (and destabilization) of the outer d and all f shells. These effects are of approximately equal magnitude and increase as Z^2. The chemical consequences of these effects have been documented for elements such as gold, for which, for example, the yellow colour is due to relativistic effects. (Non-relativistic quantum mechanics would predict gold and silver to have similar colours.)

The known oxidation states of the actinide elements are shown in Table 5. The lower oxidation states are stabilized by acid whereas the higher oxidation states are more stable in basic solutions. In solution, the 2+, 3+ and 4+ species are present as metal cations, whereas the higher oxidation states are present as oxo-cations, MO_2^+ and MO_2^{2+}. Relativistic quantum mechanics predicts the ground state of lawrencium to be $5f^{14}7s^27p_{1/2}$, not $5f^{14}7s^26d^1$. This might lead to a stable +1 oxidation state for lawrencium (Lr), but scientists who have done experiments designed to look for this state have not observed it. An upper limit for the reduction potential of $E^o \leq 0.44$ V for the Lr^{3+}/Lr^{1+} half-reaction has been determined.

It should be noted that the study of the chemistry of the elements with $Z > 100$ is very difficult. These elements have short half-lives and the typical rates of production are about one atom per experiment. The experiments must be carried out hundreds of times and the results summed to produce statistically meaningful results.

The elements Lr–112 are expected (non-relativistically) to be d-block elements because they are expected to involve the filling of the 6d orbital. However, relativistic calculations have shown that rutherfordium prefers a 6d7p electron configuration rather than the $6d^2$ configuration expected non-relativistically and expected from simple extrapolation of periodic-table trends. This prediction also implies that $RfCl_4$ should be more covalently bonded than its homologues $HfCl_4$ and $ZrCl_4$. In particular,

Table 5 The oxidation states of the actinide elements

Atomic number	89	90	91	92	93	94	95	96	97	98	99	100	101	102	103
Element	Ac	Th	Pa	U	Np	Pu	Am	Cm	Bk	Cf	Es	Fm	Md	No	Lr
													1?		
						(2)	(2)		(2)	(2)	2	2	2		
	3	(3)	(3)	3	3	3	3	3	3	3	3	3	3	3	3
		4	4	4	4	4	4	4	4	4	(4)	4?			
			5	5	5	5	5	5?			5?				
				6	6	6	6	6?							
					7	(7)	7?								

Note:
The most common oxidation states are underlined, unstable oxidation states are shown in parentheses. Question marks indicate species that have been claimed but not substantiated.

the calculations show $RfCl_4$ to be more volatile than $HfCl_4$, which is more volatile than $ZrCl_4$, with bond dissociation energies in the order $RfCl_4 > ZrCl_4 > HfCl_4$. (The periodic-table extrapolations would predict the volatility sequence $ZrCl_4 > HfCl_4 > RfCl_4$.)

The first aqueous chemistry of rutherfordium showed that it eluted from liquid chromatography columns as a 4+ ion, consistently with its position in the Periodic Table as a d-block element rather than a trivalent actinide. Gas chromatography of the rutherfordium halides has shown the volatility sequence to be $ZrCl_4 \approx RfCl_4 > HfCl_4$, with a similar sequence for the tetrabromides. Thus rutherfordium neither follows the expected periodic-table trend nor is its behaviour in accord with relativistic calculations.

The aqueous chemistry of hahnium has also shown unexpected trends. Hahnium does not behave like its homologue tantalum in aqueous solutions but is similar to niobium or the pseudo-group VB element, protactinium, under certain conditions. For example, hahnium did not extract from methylisobutylketone under conditions for which tantalum is extracted but niobium is not. The extraction of hahnium, niobium, tantalum and protactinium from VIB M HCl solutions by amines agreed with relativistic calculations.

Gas-phase thermochromatography of $NbBr_5$, $TaBr_5$ and $HaBr_5$ shows $NbBr_5$ and $TaBr_5$ to behave similarly whereas $HaBr_5$ is less volatile. Just the opposite trend was predicted by relativistic calculations.

Thus the chemistry of hahnium and rutherfordium deviates significantly from periodic-table trends, a fact that is partly explained by relativistic calculations.

The study of the chemistry of seaborgium is remarkable for its technical difficulty as well as the insight offered. In an experiment carried out over a 2-year period, fifteen atoms of seaborgium were identified in a thermal chromatography experiment. From this experiment, one concluded that the volatility sequence $MoO_2Cl_2 > WO_2Cl_2 > SgO_2Cl_2$ was followed. This observation agreed both with the extrapolations of periodic-table trends and with relativistic calculations. In an aqueous chemistry experiment, three atoms of seaborgium were detected, showing seaborgium to have the hexavalent character expected of a group VIB element. The most stable oxidation state of seaborgium is +6 and, like its homologues molybdenum and tungsten, seaborgium forms neutral or anionic oxo and oxohalide compounds.

To study the chemistry of elements 107 (Bh) and 108 (Hs) one must be able to produce isotopes of these elements with half-lives long enough for chemical studies. ^{269}Hs is reported to have $t_{1/2} \approx 9\,s$ and ^{267}Bh is expected to have a similar half-life. Because of the small probability of producing these nuclei, methods for chemical study must be very sensitive. Among the projected methods of study, liquid–liquid extraction and gas-phase thermochemistry are thought to be the most viable. Of particular interest is the predicted formation of the very volatile tetraoxide, HsO_4. New methods focusing on this compound may give the desired speed of separation and high efficiency.

Further detailed studies of the chemistry of rutherfordium, hahnium and seaborgium are planned. One possible goal is to make observations such that the observed quantity (redox potential, deflection in a magnetic field) can be directly related to relativistic quantum mechanical calculations rather than becoming involved in situations in which the theory is used to interpret gross thermodynamic properties.

Practical Applications

Nuclear Fission Power

One might think naively that the man-made transuranium elements have little practical importance, considering that the rates of production for many of them are measured in atoms per day. Nothing could be further from the truth. One might even say that one of the most significant focal points of human activity and concern during the last fifty years has been with the practical applications of man-made plutonium.

One of the most important practical applications of uranium and the transuranium elements is the production of electrical power from fission. Although a detailed discussion of nuclear power is beyond the scope of this chapter, a few remarks are in order.

A typical nuclear reactor system is shown in Figure 23. The reactor consists of a core of fissionable material (usually UO_2, enriched to 3.3% ^{235}U) in which the chain reaction takes place. The energy released by the fission process, which is primarily in the form of the kinetic energy of the fission fragments, is absorbed as heat in the core. Also present in the core is the water moderator which slows the fission neutrons to thermal energies at which their probability of inducing another fission is greatest. A reflector helps to prevent neutrons from escaping from the core. The heat is removed from the core by a coolant and the chain reaction is controlled by rods of neutron-absorbing material inserted into the core. In the pressurized water reactor (PWR), the heat energy produces steam for the turbine through the use of a heat exchanger, whereas in a boiling water reactor (BWR), the steam is produced for direct use in the turbine.

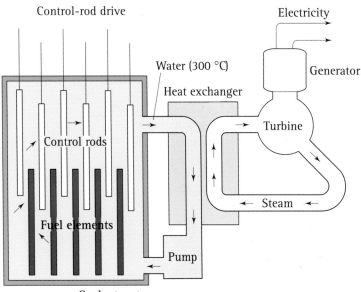

Figure 23
A schematic diagram of a pressurized-water nuclear power plant.

Reactors are classified as thermal, fast or intermediate according to the energy of the neutrons inducing nuclear fission in the core. A thermal reactor is one in which the neutrons are in thermal equilibrium with the reactor's materials. In a typical thermal power reactor, the neutron speeds are about 3000 m s^{-1}. At these speeds, the neutrons are readily absorbed by many materials. The reactor fuel in thermal reactors is primarily ^{235}U, although ^{239}Pu is also used following its production by neutron absorption by ^{238}U. Ordinary water can act both as a moderator and as the coolant in thermal reactors. It is also possible to moderate thermal reactors using graphite or heavy water and cool them by the flow of a gas such as carbon dioxide or helium. In a fast reactor the average speed of the neutrons is about 15×10^6 m s^{-1}. At these high speeds the probability of a neutron being absorbed or causing fission is lower. One fission event results in the release of about 200 MeV of energy, or about 3.20×10^{-4} erg, which corresponds to 3.20×10^{-11} W s. Thus 3.1×10^{10} fissions s^{-1} produces 1 W of power as heat. The fission of 1 g of uranium or plutonium per day liberates 0.96×10^3 kW, or about 1 MW. This is the energy equivalent of 3 tons of coal or about 600 gallons of fuel oil per day.

One remarkable fact about nuclear reactors is that they can produce their own fuel. If the total nuclear capacity installed worldwide is 3.5×10^5 MW, one can estimate that about 110 tons of ^{239}Pu are produced each year. In such converter reactors the primary power source is the fission of ^{235}U. The more abundant isotope of uranium, ^{238}U, which is always present, even in enriched uranium fuel, is converted to ^{239}Pu, which could, in principle, be used to fuel another reactor. In the plutonium-production reactors, such as those that were operated at Hanford, Washington, the ratio of ^{239}Pu produced to ^{235}U burned was about 0.8.

As of 1997, more than 444 nuclear power plants were operating around the world, generating more than 2300 TW h of electricity in 33 countries. Some countries depend vitally on the electricity generated by nuclear power. In 1996, France generated 77% of its electricity from nuclear power plants, Lithuania 83%, Belgium 57%, Sweden 52%, the Slovak Republic 45%, Switzerland 45%, the Ukraine 43%, Bulgaria 42% and Hungary 41%. Japan and Korea generated 34% of their electricity from nuclear power, Germany 31%, the UK 28%, the USA 20% and Canada and the Russian Federation 14%. Furthermore, although the USA was not a leader in percentage, it had the largest total output of electricity from nuclear power: 629 000 GW h generated in 1997 from 107 plants, generating 20% of the USA's electrical power.

The important actinides in irradiated uranium fuel are uranium, neptunium, plutonium, americium and curium. The paths for the production of these radionuclides in a nuclear reactor are shown in Figure 24. Neutron capture by ^{235}U leads to the formation of ^{237}Np, which, in turn, can, by (n, γ) and (n, 2n) reactions, form ^{238}Pu and ^{236}Pu. ^{237}Np is an important, long-lived component of radioactive waste. ^{238}Pu is the largest contributor to the total alpha activity of plutonium in irradiated fuel. The daughters of ^{232}U produced by the decay of ^{236}Pu emit high-energy gamma rays and are of concern in shielding during recycling plutonium. The largest amount of the plutonium present in irradiated fuel is as ^{239}Pu, formed by neutron capture by ^{238}U followed by beta-decay of ^{239}U and ^{239}Np. Neutron capture by ^{239}Pu leads to ^{240}Pu, ^{241}Pu, ^{242}Pu and ^{243}Pu. Decays of ^{241}Pu and ^{243}Pu lead to the formation of ^{241}Am and ^{243}Am. Neutron capture by ^{241}Am ultimately leads to the production of ^{242}Cm, which is the most intense source of alpha activity in spent fuel. Some heavier curium isotopes are also produced, as shown in Figure 24. The curium isotopes present in spent fuel decay back to plutonium, thus acting as the greatest source of the α-active ^{238}Pu in the high-level waste from fuel reprocessing.

In a yearly operating cycle of a typical (1000 MW) pressurized water reactor, the spent fuel at discharge contains about 25 MT of uranium and about 250 kg of plutonium (1 MT (one metric ton) = 1000 kg). Some 40% of the energy produced in the course of a nuclear fuel cycle comes from ^{239}Pu. (Since about 20% of the electricity generated in the USA in 1997 came from nuclear power plants, about three times as much electricity was generated from the synthetic element plutonium (251×10^9 kW h in 1997) as was generated from oil-fired electrical generating plants (78×10^9 kW h in 1997).

One way to express the toxicity of the spent fuel is as the 'water-dilution volume,' the volume of water needed to dilute these radionuclides so that drinking that water

Figure 24

Nuclide chains producing
plutonium, neptunium, americium
and curium in nuclear power
reactors.

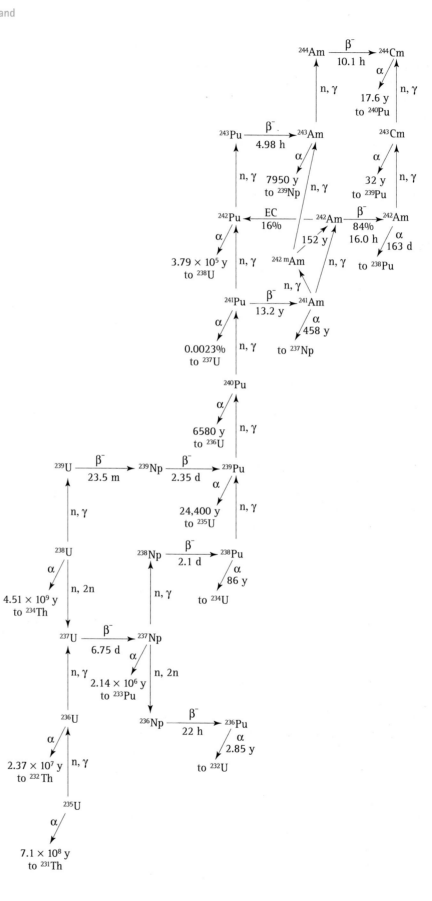

will result in a small, generally considered acceptable, risk (an exposure of about 0.5 rem per year). In Figure 25, we show the water-dilution volumes needed for spent reactor fuel. One can see that, during the first ten years after production, the toxicity is governed by the fission products, primarily ^{90}Sr and ^{137}Cs. After that, the toxicity is determined by ^{241}Am, ^{239,240}Pu and then ^{237}Np. The very-long-term toxicity is governed by the ^{238}U decay products, ^{226}Ra and ^{210}Pb. In comparison with the original uranium ore used to make the reactor fuel, one concludes that the spent fuel is the primary factor governing the toxicity of the nuclear fuel cycle for millions of years.

Radionuclide Power Sources

One special practical application of the transuranium elements is their use in radio-nuclide power sources. In a radionuclide power source, the decay energy of a radio-nuclide is absorbed in an appropriate material, giving rise to heat. Using thermo-electrical devices without moving parts, this heat can be converted into electricity. The requirements for the radionuclides used in such sources are that they be easily shielded, emit weakly penetrating radiation, have rea-sonably long half-lives, have a specific power of 0.2 W g^{-1} or more, have good resis-tance to corrosion, be insoluble in water, be of low cost and be reasonably available. The alpha-particle-emitting nuclides ^{238}Pu ($t_{1/2} = $ 88 years), ^{244}Cm ($t_{1/2} = 18.1$ years) and ^{245}Cm ($t_{1/2} = 8500$ years) satisfy these criteria. A few grams to kilograms of such nuclides, in appropriately shielded containers, provide sources with power levels up to hundreds of watts. Such sources are of light weight, rugged and portable.

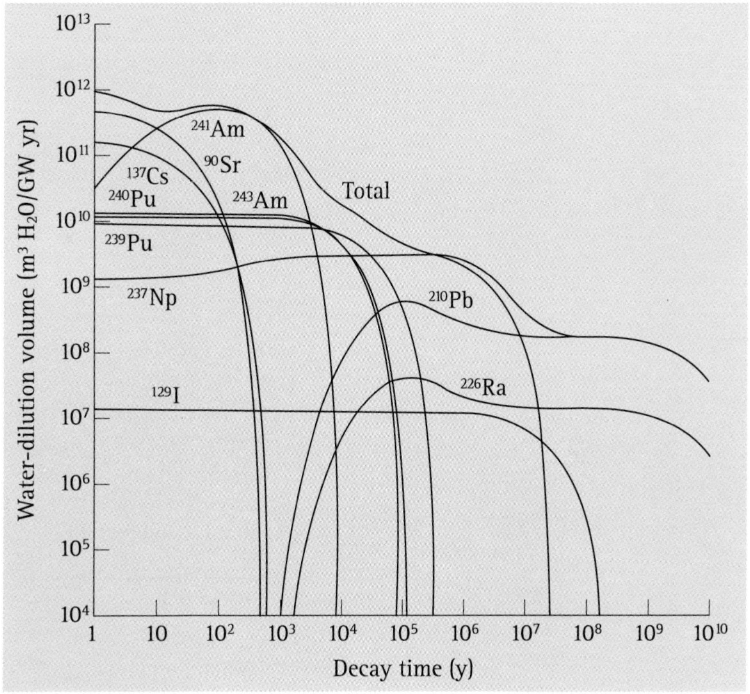

Figure 25
Principal contributions to the long-term ingestion toxicity of spent fuel.

Sources fuelled by ^{238}Pu have become important in space applications. Several satel-lites and remote-sensing instrument packages have employed such power sources, termed SNAP (systems for nuclear auxiliary power) units. SNAP sources were used as the power sources for instrument packages on the five Apollo missions, the Viking unmanned Mars lander and the Pioneer and Voyager probes to Jupiter, Saturn, Uranus, Neptune, Pluto and beyond. Early versions of the SNAP sources were designed to burn up upon re-entry into the earth's atmosphere. Following the controversy concerning the burn-up of such a source in 1963, subsequent sources were designed to remain intact during re-entry and after impact. Such sources have been recovered, demon-strating the viability of the design.

The Environmental Chemistry of the Transuranium Elements

With the large annual production of neptunium, plutonium and the higher actinides in the nuclear power industry, there has been increasing concern about the possible release of these elements to the environment. This concern has been heightened by the nuclear reactor accidents at Three Mile Island and Chernobyl. Coupled with the prospect of cleaning up the detritus of the nuclear weapons programmes of the major nations and the general lack of a publically acceptable method of long-term disposal of nuclear waste, there is considerable interest in the environmental chemistry of the transuranium elements.

Table 6 Events leading to large injections of radionuclides into the atmosphere

Source	Country	Time	Radioactivity (Bq)[a]	Important nuclides
Hiroshima and Nagasaki	Japan	1945	4×10^{16}	Fission products of actinides
Atmospheric weapons tests	USA USSR	1945–1963	2×10^{20}	Fission products of actinides
Windscale	UK	1957	10^{15}	^{131}I
Chelyabinsk (Kysthym)	USSR	1957	8×10^{16}	Fission products of ^{90}Sr and ^{137}Cs
Three Mile Island	USA	1979	10^{12}	Noble gases and ^{131}I
Chernobyl	USSR	1986	2×10^{18}	^{137}Cs

Note:

[a] 1 Becquerel (Bq) is one disintegration per second.

Source: From G. Choppin, J. O. Liljinzin and J. Rydberg, *Radiochemistry and Nuclear Chemistry* (Pergamon, London, 1994).

Plutonium is clearly the most significant transuranium element in the environment. The plutonium in the environment is due primarily to atmospheric testing of nuclear weapons, secondarily to the disintegration upon re-entry of satellites equipped with ^{238}Pu power sources and, lastly, to the processing of irradiated fuel and fabrication of fuel in the nuclear power industry and the plutonium-production programme. Some major releases of radionuclides are summarized in Table 6. During the period 1950–1963, about 4.2 tons of plutonium (mostly a mixture of ^{239}Pu and ^{240}Pu) was injected into the atmosphere as a result of nuclear weapons testing. Because of the high temperatures involved, most of this plutonium was thought to be in the form of a refractory oxide. Most of this plutonium has been re-deposited on the earth, with concentrations highest at the mid-latitudes. Of the 350 000 curies (Ci) of ^{238}Pu and ^{239}Pu originally present in the atmosphere, about 1000 Ci remained in 1989. Approximately 9.7×10^{6} Ci of ^{241}Pu were also injected into the atmosphere during weapons testing. When this completely decays ($t_{1/2}$ of ^{241}Pu is 14.4 years), a total of about 3.4×10^{5} Ci of ^{241}Am will be formed. About an additional 1.4 tons of plutonium has been deposited in the ground (as of 1989) due to surface and sub-surface nuclear weapons testing. Approximately 16 000 Ci of ^{238}Pu were injected into the atmosphere when a satellite containing an isotopic power source disintegrated over the Indian Ocean in 1964. The Chernobyl accident caused the release of about 800 Ci ^{238}Pu, about 700 Ci of ^{239}Pu and about 1000 Ci of ^{240}Pu, representing about 3% of the reactor core's inventory. This activity was dispersed over large areas of the USSR and other countries in Europe. The amount of plutonium in the environment due to fuel reprocessing is small.

Over 99% of the plutonium released into the environment ends up in the soil and in sediments. The global average concentration of plutonium in soils is $5 \times 10^{-4} - 2 \times 10^{-2}$ pCi g^{-1} dry weight, most of the plutonium being near the soil's surface. The concentrations of plutonium in natural waters are quite low, an average concentration being around 10^{-4} pCi l^{-1}, i.e. about 10^{-18} M. (More than 96% of any plutonium released into an aquatic ecosystem ends up in the sediments. In these sediments, there is some translocation of the plutonium to the sediment's surface due to the activities of benthic biota.) Less than 1% (and perhaps closer to 0.1%) of all the plutonium in the environment ends up in the biota. The concentrations of plutonium in vegetation range from 10^{-5} to 2%, with concentrations in litter and animals ranging from 10^{-4} to 3% and 10^{-8} to 1%, respectively. *None of these concentrations has been observed to cause any discernible effect.*

Despite the extremely low concentrations of the transuranium elements in water, most of the environmental chemistry of these elements has been focused on their behaviour in the aquatic environment. One notes that the neutrality of natural water (pH = 5–9) results in extensive hydrolysis of the highly charged ions except for Pu(V) and a very low solubility. In addition, natural waters contain organics as well as microscopic and macroscopic concentrations of various inorganic species such as metals and anions that can compete, complex or react with the transuranium species. The final concentrations of the actinide elements in the environment are thus the result of a complex set of competing chemical reactions such as hydrolysis, complexation, redox reactions and colloid formation. As a consequence, the aqueous environmental chemistry of the transuranium elements is significantly different from their ordinary solution chemistry in the laboratory.

In natural waters, hydrolysis is the primary factor determining concentration. The tendency to hydrolyse follows the relative effective charge of the ions. This is known to be

$$An^{4+} > AnO_2^{2+} > An^{3+} > AnO_2^+$$

(where An represents an actinide element). The hydrolysis reaction can be written as

$$xAn^{m+} + yOH^- \rightleftharpoons An_x(OH)_y$$

The hydrolysis products can be monomeric, polynuclear or colloidal.

Some strongly complexing inorganic anions are present in natural waters, such as HCO_3^-/CO_3^{2-}, Cl^-, SO_4^{2-} and PO_4^{3-}, etc. The order of complexation of these anions is

$$CO_3^{2-} > SO_4^{2-} > PO_4^{3-} > Cl^- \cdots$$

Also present in many natural waters are humic/fulvic acid, citric acid, etc. These organics also can complex actinides. In Figure 26, we show the relative stability constants for the first complexation reaction of various ligands with actinides of various oxidation states. Clearly the carbonate and humate ions, together with hydrolysis, dominate the chemistry. The tetravalent actinide ions will tend towards undergoing hydrolysis reactions or complexation with carbonate rather than formation of the humate/fulvate.

The aquatic solution chemistry of the actinides is also influenced by pH and redox potential (E_h). The approximate ranges of pH and E_h for natural waters are shown in Figure 27. The pH varies from 4 to 9.5 and E_h from −300 mV to +500 mV. In these pH and E_h ranges, neptunium and plutonium can be present in several oxidation states whereas americium and curium will be trivalent. In the oxidizing environment of surface waters, Np(V), Pu(IV), Pu(V) and Am(III) will be the dominant species whereas in the reducing environment of deep groundwater, other species may be present.

Colloids are always present in natural waters containing the transuranium elements. (Colloids are defined as particles with sizes in the range 1–450 nm. These particles form stable suspensions in natural waters.) Colloids of the transuranium elements can be

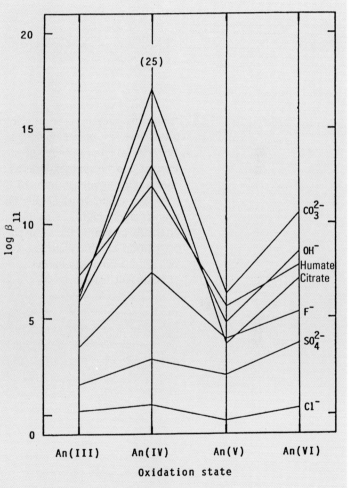

Figure 26
A comparison of complexation stability constants for the interaction of several ligands with various actinide oxidation states.

Figure 27

A Pourbaix diagram showing the ranges of pH and E_h values in natural waters.

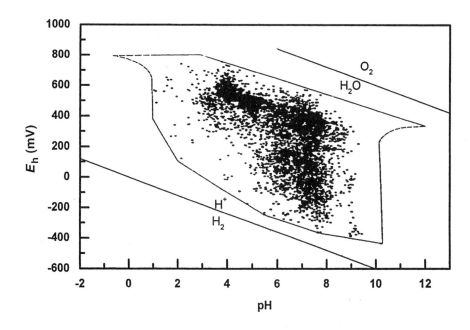

formed by hydrolysis of transuranium ions or by the sorption of transuranium elements onto the 'naturally occurring colloids.' The naturally occurring colloids include such species as metal hydroxides, silicate polymers and organics (such as humates). In Figure 21, we show colloidal Pu(IV), which is a unique green colour. The mobility of the transuranium elements in an aquifer is determined largely by the mobility of its pseudo-colloids, i.e. those colloidal species formed by the adsorption of the transuranium ions upon the naturally occurring colloids.

The speciation of the transuranium elements in waters is thus a complex function of hydrolysis, colloid formation, redox reactions and complexation with available ligands. The solubility (mobility) is hence highly dependent on the particular aquatic environment and its characteristics.

However, bearing in mind these caveats, we can make certain generalizations about the behaviour of the actinide elements in natural waters. Americium and curium remain in the $+3$ oxidation state over the natural range of environmental conditions. For plutonium, Pu(III) is unstable with respect to oxidation at environmental acidities and so the other three states are observed, the dominant oxidation state in natural waters being Pu(V). (Humic materials cause a slow reduction of Pu(V) to Pu(IV), so that Pu(IV) is found to be important in waters containing significant amounts of organic material.) Under reducing conditions, neptunium should be present as Np(IV) and behave like Pu(IV); under oxidizing conditions, NpO_2^+ will be the stable species. In marine waters, Pu(IV) and the transplutonium elements will tend to undergo hydrolysis to form insoluble hydroxides and oxides. However, these elements can also form strong complexes with inorganic anions (OH^-, CO_3^{2-}, HPO_4^{2-}, F^- and SO_4^{2-}) and organic complexing agents that may be present. The speciation and solubility of these elements are largely determined by hydrolysis and formation of carbonate, fluoride and phosphate complexes. Stable soluble species include Pu(V,VI) and Np(V), although under most conditions the actinides will form insoluble species that concentrate in the sediments.

Pu(IV), which forms highly charged polymers, is strongly sorbed onto soils and sediments. Other actinide III and IV oxidation states also bind by ion exchange to clays. The uptake of these species by solids is in the same sequence as the order of hydrolysis: Pu > Am(III) > U(VI) > Np(V). The uptake of these actinides by plants appears to be in the reverse order to that of hydrolysis: Np(V) > U(VI) > Am(III) > Pu(IV), plants exhibiting little ability to assimilate the immobile hydrolysed species. The further concentration of these species in the food chain with subsequent deposition in man appears to be minor. Of the roughly 4 tons of plutonium released into the environment in atmospheric testing of nuclear weapons, the total amount fixed in the world population is less than 1 g (of this amount, most (99.9%) was inhaled rather than ingested).

Further Reading

1. J. I. Kim, Chemical behaviour of transuranic elements in natural aquatic systems in *Handbook on the Physics and Chemistry of the Actinides*, Vol. 4, edited by A. J. Freeman and C. Keller (North-Holland, Amsterdam, 1986) pp. 413–456.
2. *The Transactinide Elements, Proceedings of the 41st Robert Welch Conference on Chemical Research* (Welch, Houston, 1998).
3. G. T. Seaborg and W. Loveland, *The Elements Beyond Uranium* (Wiley, New York, 1990) and the many references contained therein.
4. G. T. Seaborg and E. G. Valens, *Elements of the Universe* (Dutton, New York, 1958).

Glenn T. Seaborg

In February, 1999, the senior author of this chapter, Glenn T. Seaborg, passed away after a long illness. Since Prof. Seaborg had very actively participated in all stages of the writing and revision of this chapter and the chapter reflected his ideas, the junior author (WDL) decided to leave the manuscript as Seaborg last saw it in the summer of 1998 rather than revise it to include a discussion of the exciting developments that occurred in 1999 and 2000.

2 Bonding and the Theory of Atoms and Molecules

John N. Murrell

Introduction

Theoretical chemistry, computational chemistry and mathematical chemistry are all terms used to describe methods for quantitatively interpreting and predicting chemical (molecular) properties. The terms have some overlap but are not synonymous. Computational chemistry is clearly associated with the development of computers from the mid-1950s onwards; numerical agreement between the outcomes of a computer 'experiment' and a 'laboratory' experiment is an indication that the underlying model is satisfactory. Mathematical chemistry typically uses branches of pure mathematics such as group theory and graph theory to classify chemical systems, classification being an important step on the road to understanding. The theorems will also lead to what I would call, yes/no answers – is a spectroscopic transition allowed (yes) or forbidden (no). Theoretical chemistry can be said to cover any model that interprets chemical systems in a quantitative or semi-quantitative fashion. In this sense it started with such nineteenth-century ideas as valence and periodicity, but the subject developed to its current level of maturity only after the quantum-mechanical revolution of the 1920s.

The first application of quantum mechanics to chemistry was to the chemical bond. Heitler and London in 1927 put earlier ideas of G. N. Lewis on electron pairing into quantitative form by showing, for the first time, that H_2 was more stable than two separate hydrogen atoms. These first calculations were very crude, giving a binding energy of H_2 only 66% of its experimental value, but by 1933 the calculations had reached 99% of the exact result and by 1960 essentially exact agreement between theory and experiment had been achieved. Anyone now reading accounts of calculations before the computer age would be amazed at the achievements made with the simple computational tools (at best using mechanical calculators) then available.

There are essentially two strands of theoretical chemistry. In one of these there is an attempt to solve the exact equations of classical and quantum mechanics to ever greater accuracy; clearly the ability to do this parallels advances in computer power. In the other strand one solves simplified equations (models) exactly and likewise the models can be made more sophisticated with time. However, it is certainly no accident that the crudest calculations and simplest models have produced the most important concepts which underlie chemical processes. Much of our current chemical language pre-dates the computer age, or, to put it more starkly, not much of our language clearly post-dates the computer age.

When we look at theoretical chemistry at the beginning of the twenty-first century we are therefore looking mainly at its predictive capabilities for a subject that is already mature. Chemists are currently not expecting dramatic surprises – which is not to say that they will not occur. Computation is expected to point the way to the structure and properties of molecules that have not yet been prepared; also of materials that are to be chemically synthesized. In this they are seen as an important aid to the experimentalist who has certain targets in mind.

Chemical systems obey the basic laws of physics encompassed in classical and quantum mechanics; it is rarely necessary to go further and bring in relativity or quantum electrodynamics. It might be thought that quantum theory is always necessary since, by the de Broglie relation $\lambda = h/p$, the wavelength λ of a nucleus or electron moving with a momentum p is usually large compared with the distance over which the forces acting on the particle change significantly (a sure sign that quantum effects are operating). However, the characteristic features of quantum systems, such as discrete energies, interference patterns and tunnelling, are often not resolved in an experiment, particularly one involving an assembly of particles with a distribution of energies or velocities.

For electrons in individual atoms or molecules, quantum mechanics is always needed. However, for nuclei, particularly when there are many of them, one can often get all one wants from classical mechanics. This is an important point because the equations of classical mechanics for n particles are much easier to solve by numerical methods than are the quantum-mechanical equations for the same system; this is generally true for $n > 2$.

The Chemical Bond

As has already been mentioned, our understanding of why atoms combine together to form molecules and why the most stable molecules satisfy the rules of valence was unclear until the development of quantum mechanics. Theoretical chemists now have available many computer packages (prepared after huge efforts at writing software) that allow them to obtain approximate solutions to the Schrödinger equation for the electrons and to deduce the structure, stability and other properties of a molecule. It has, however, been very difficult to obtain results that provide confident predictions for the experimentalist, except for rather simple (few-electron) molecules. Even today there are many calculations for which theoretical chemists will happily call on the maximum computer power available; so, it is necessary to explain why the calculations are so difficult.

The energy of a molecule is a balance between the attractive forces between electrons and nuclei and the repulsive forces acting between nuclei or between electrons. Multiply charged molecules, either positive or negative, are rarely stable with respect to dissociation in the gas phase because the repulsive forces win (in solution there are other factors that affect the stability).

Because nuclei are so much heavier than electrons, one can obtain an energy for electrons moving in a potential from fixed nuclei and then use this to determine the nuclear motion; this is called the Born–Oppenheimer approximation. It is the first of these tasks, that of calculating the energy of the electrons, that is particularly difficult, because the repulsion between the electrons is an important part of the chemical bond energy and, for many molecules, unless it is calculated with high accuracy the results are poor.

The role of computers for chemical bond calculations was mapped out in a landmark paper in *Nature*, published in 1956 by Boys, Cook, Reeves and Shavitt, under the title 'Automatic fundamental calculations of molecular structure'. They not only proposed a method that is still the basis of many current calculations but also performed calculations on what was then quite a complicated system, namely the water molecule. They used the first University of Cambridge computer, EDSAC, which carried out 1000 operations per second and had a 1000-word store. A single calculation took 40 h (how did they keep it running?). Such a calculation today might take 4 s.

The most general approach to the electronic-structure problem is to take as a first approximation a model in which the electrons move in the potential of the nuclei and the *average* potential of the other electrons. This is called the self-consistent-field (SCF) approximation and until, say, the mid-1970s was as far as most calculations went. Indeed, a simpler approach in which inter-electron repulsion was ignored, or treated as a quantity for which one could make allowance by an appropriate choice of parameters, was widely used. This was the independent-electron or Hückel approach, which is analogous to the tight-binding approach in solid-state physics. However, stopping at the SCF level can give errors in the energy of about 1 eV per electron pair (a typical chemical bond energy is only a few electronvolts), so there are many molecules for which the SCF level gives a very poor answer. Thus, today, if one is aiming for accurate predictions regarding small polyatomic molecules, some method that reliably treats the so-called electron-correlation problem has to be used, either starting from the SCF level as an approximation or, which is even more difficult, treating interelectron repulsion explicitly right from the beginning.

Let me give two examples of problems that are amenable to accurate treatment by *ab initio* methods (accurate means that theoretical predictions are on a par with, or better than, experimental measurements). In 1967 an unstable species was isolated at low temperatures in an inert gas matrix and, from some features of its vibrational spectrum, it was predicted to be HNC (isocyanic acid), an unstable isomer of HCN (hydrogen cyanide). In 1971 a microwave emission line at 90.665 GHz from an astronomical source was attributed to the same species, this tentative assignment being based on the geometry assumed in the earlier work. The microwave spectrum of HNC was not observed in the laboratory until 1976.

The first high-quality calculations on the bond lengths of HNC (beyond the SCF level) were in 1973. From these the microwave frequency was predicted to be 90.86 GHz; this was certainly good enough to confirm the assignment. By 1975 calculations for bond lengths and vibration frequencies were considered accurate and the calculated energy barrier for isomerization to the stable species HCN was predicted to be 1.52 eV.

My second example concerns the structure of methylene, CH_2. This is a molecule that has been a challenge to calculations for many years, but, interestingly, the first very crude calculations were sufficient to change the structural interpretation of the spectroscopic data. Carbon is tetravalent in nearly all of its stable compounds. However, many unstable molecules, in which it exhibits other valencies, have been studied by spectroscopy; they may also be, as CH_2 is, important transients in organic reactions.

Methylene has two low-energy states; the electrons couple together to give either a singlet spin or a triplet spin. The structure of the singlet is fairly straightforward to explain through the formation of electron-pair bonds between the unpaired 2p electrons of the carbon and the hydrogen 1s electrons. On this basis we would expect a bond angle of 90°, which is the angle between the two 2p orbitals, whereas it is actually 102°. However, this is the higher of the two energy states by about 30 kJ mol^{-1}.

In 1960, Boys predicted that the triplet state would be the ground state and he calculated a bond angle of 129°. However, Herzberg, the most eminent spectroscopist of his time, showed in 1964 that the triplet CH_2 spectrum was typical of that of a linear molecule, in conflict with theory. This threw down the gauntlet to theoreticians and several calculations to support Boys' result followed; in particular, much better calculations by Schaeffer in 1970 gave the angle as 135°. It was one of the minor triumphs of theory that a later re-analysis of the spectrum by Herzberg concluded that the angle was 136°.

The message to be learnt from these calculations, which today one could do equally well on molecules with, say, three times as many atoms, is that theoretical chemistry in its strict computational mode is a powerful tool for the identification of molecules from their spectroscopic properties and for the elucidation of their structures. There are, however, many more difficult problems that stretch even the current computer resources. Firstly, excited electronic states are generally more difficult to calculate than are ground states. Secondly, there are many systems (unlike HNC and CH_2) for which the SCF approximation is not a good starting point; this is generally true for transition metal complexes, for example, and for these many of the existing and heavily used computer programmes will fail. Lastly, probably most importantly, chemists and biologists are also interested in the structures of very large molecules and for these the standard 'small-molecule' *ab initio* programmes can still not be applied.

Many of the approximate or semi-empirical methods for determining electronic structures which were developed for small molecules, particularly organic molecules, during the period leading up to about 1970 are now being used to treat very large molecules. Even today new variants and new packages appear.

We have become more knowledgeable about the potentials which act within molecules, at least those which are made up of standard bonds; we know in energetic terms how bonds stretch, angles bend, distant atoms repel or attract, etc. and all of this knowledge can be expressed by writing simple algebraic functions of the internal co-ordinates of the molecule. This leads to what are called 'molecular mechanics' potentials. When these are applied to a molecule that we expect to have a standard structure (not the triplet states of CH_2, for example), the lowest point on this potential will correspond to the equilibrium structure. Finding this minimum from some arbitrary structure may, of course, involve quite a lot of computation because potentials of this type, involving a large number of variables, have a large number of minima, so any computer programme need not necessarily find the lowest.

The whole procedure for handling very large molecules and seeing how they pack together is of great interest to protein and polymer chemists and is considered by many to be an essential tool for designing drugs. Some commercial firms have been enthusiastic about these computational techniques only to be disapppointed by the results. However, there is no doubt that, with a continuing increase in computational power and more sophisticated programmes, the role of computational chemistry in such fields is bound to become more important (Figures 1 and 2).

Let me return briefly to the matter of inter-electron interaction. This manifests itself in two ways. Firstly, just like in classical electrostatics, it manifests itself by the repulsion between like charges; we need know only electron densities in order to calculate this, not wave functions. The second way is through an energy associated with the specific pair-correlation of electron positions (if electron one is at r_1, what is the probability of finding electron two at r_2?); this is usually called the exchange-correlation energy and it is the difficult and time-consuming part to calculate because it can involve difficult integrals and it tends to be slowly convergent in wave-function expansions.

A breakthrough came in 1964, in papers by Hohenberg, Kohn and Sham, in which they showed that the total electron energy was a functional of the electron density; the energy is uniquely determined by the density and we do not need to know the wave function. The snag is that we do not know the functional, although we know some aspects of its general form.

Knowledge about the density functional came from a variety of sources: the wave functions of the free-electron gas and examination of accurate wave functions for simple atoms and molecules. The usual attack on it is to perform an expansion in the gradient of the density, the first term being a functional first noted by Dirac. By the mid-1980s the subject was sufficiently well advanced that programmes were available, first for solids and then for molecules, which provided results of comparable accuracy to the best 'standard' approach, but with a fraction of the computing effort. The saving in computing time obtained by using density-functional methods increases rapidly with the number of electrons.

Density-functional methods still have some way to go because we know that all the current models fail in some cases. For example, they generally fail to give the correct

John N. Murrell

Figure 1

Structures of clusters with He_3^+ cores. (Reprinted from *Molecular Physics*, vol. 87, No. 4 (1996), Knowles and Murrell *The Structures and Stabilities of Helium Cluster Ions*, with permission from Taylor and Francis http://www.tandf.co.uk/journals)

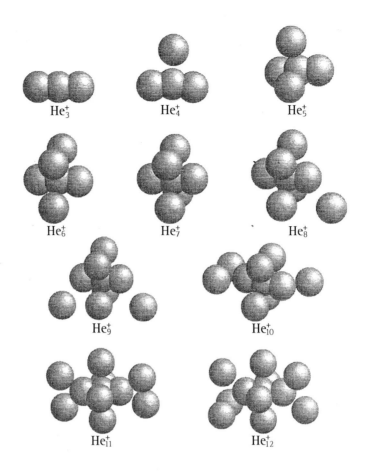

form for the long-range interaction between atoms. Many are of the opinion that these methods provide the way forwards for the future, but there is still a healthy number of sceptics.

A final point on computing: with the increasing sophistication both of hardware and of software, in particular, instantly available graphics, one can now purchase what I consider to be amazing all-singing, all-dancing, at-a-touch-of-the-mouse, packages to do electronic-structure calculations. They are a great temptation, but unfortunately you cannot get instant knowledge about which of the variety of procedures on offer have scientific validity for the task you have in mind. You can no more do useful science with these packages without a good understanding of the underlying theory than you can do good organic synthesis by randomly mixing bottles.

Electronic-structure calculations give more than geometry and energy; from the computed wave functions one can deduce electron densities and properties relating to the interaction of molecules with electrical, magnetic and electromagnetic fields. We can certainly expect to be predicting more new materials for, say, optoelectronics or superconductivity in the future. However, the most important current role for electronic-structure calculations is probably in the field of reaction dynamics.

Reaction Dynamics

The traditional reaction in chemistry is one carried out in a bulk phase, in gases, liquids or, less commonly, solids. The rate constants of such reactions generally follow a simple law for variation with temperature, deduced in the nineteenth century by Arrhénius, which involves only two parameters. One of these is an exponent that is

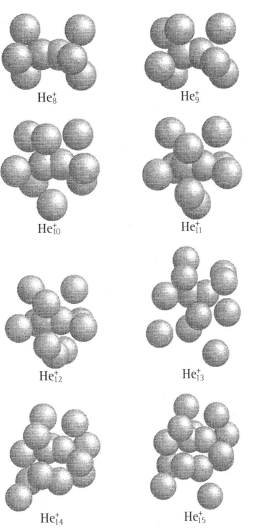

Figure 2
Structures of clusters with He_2^+ cores. (Reprinted from *Molecular Physics*, vol. 87, No. 4 (1996), Knowles and Murrell *The Structures and Stabilities of Helium Cluster Ions*, with permission from Taylor and Francis.)

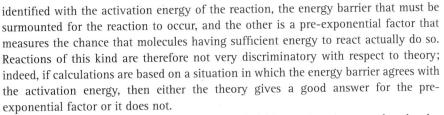

identified with the activation energy of the reaction, the energy barrier that must be surmounted for the reaction to occur, and the other is a pre-exponential factor that measures the chance that molecules having sufficient energy to react actually do so. Reactions of this kind are therefore not very discriminatory with respect to theory; indeed, if calculations are based on a situation in which the energy barrier agrees with the activation energy, then either the theory gives a good answer for the pre-exponential factor or it does not.

A much more severe test of theory is provided by reactions in crossed molecular beams. These are experiments in which two collimated beams of atoms or molecules are crossed in a reaction chamber and the products detected at various angles to the incident beams. If the reactants are prepared in specific energy states with discrete velocities and the scattered reactants analysed for their energies and their angles of scattering, then a very large amount of data with which to test reaction-rate theories is available. However, not all crossed-beam experiments are 'ideal'. In practice there is never perfect energy selection or angular resolution and both of these failings reduce the amount of data to challenge theory; in particular, some of the subtleties associated with the quantum mechanics of the process can be lost.

Central to the interpretation of chemical dynamics is the concept of the potential energy surface (PES) (Figure 3). This is the electronic energy of the molecule as a function of the internal coordinates (geometrical variables) of the molecule. To determine a PES it is necessary to perform a sufficient number of electronic-structure calculations to map out all configurations that could feasibly be involved as the nuclei pass from reactants to products. Moreover, insofar as all electronic-structure calculations are inherently approximate, the errors must be similar in magnitude over the whole PES; this is quite difficult to achieve and certainly requires a procedure that handles electron correlation accurately. It is probably true to say that even today the only accurate PESs we have available are those for a few intensively studied small systems such as $H + H_2$ and $F + H_2$.

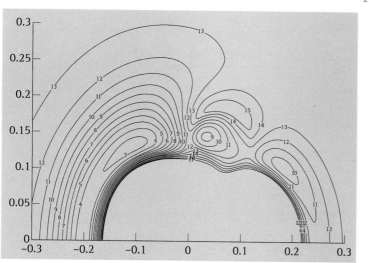

Figure 3

Contours for the movement of O around SO with a fixed bond length of 0.166 nm. The SO moiety lies along the *x* axis with its centre of mass at the origin and O in the positive direction. Contour 3 corresponds to −10 eV; the contour interval is 0.5 eV. Axes are labelled in nanometres. (Reprinted from *Molecular Physics*, vol. 56, No. 4 (1985), J. N. Murrell *et al. Potential Energy Functions for the Ground State Surfaces of SO$_2$ and S$_2$O*, with permission form Taylor and Francis.)

To interpret a crossed-beam reaction one must calculate the nuclear dynamics on the appropriate PES. A first step, which seems easy but proves to be quite difficult, is to fit the calculated discrete set of points on the PES to analytical functions that cover the surface wholly or in parts. The next step is to use these functions either to calculate classical trajectories that pass from reactants to products or to calculate quantum wave functions that pass between these limits. Computer packages for performing both of these calculations are available. Calculating classical trajectories is fairly routine but one has to do a sufficient number so that the appropriate averaging of the initial conditions can be performed; 1000 trajectories is none too many to cover all angles and distances of approach in a typical bimolecular collision.

In quantum scattering we start with an initial wave function representing the reactants approaching with specified momenta and internal energy states and integrate this through the scattering process to find the components of the wave function on exit from the reaction; this leads by standard processes to a reaction cross section. However, such calculations are currently limited to handling just a few atoms; the collision of two diatomic molecules is a heavy computational problem. One is saved by the fact that the quantal aspects of the problem are often difficult to resolve experimentally for larger systems, so the classical trajectory method is normally sufficient to analyse experiments. Nevertheless, there are some long-standing problems, such as the exchange of isotopes in hydrogen-atom–hydrogen-molecule collisions, that have only recently been solved by very good experiments and very good three-dimensional quantum-scattering calculations.

It is largely the interplay of theory and experiment for crossed-beam reactions that has advanced our understanding of chemical reactions, at least for those in the gas phase. Even the very simple classical trajectory calculations carried out by Polanyi and co-workers in the 1960s, on collinear atom–diatomic-molecule collisions, explained for the first time how the energy released in a chemical reaction could appear as vibrational energy or as translational kinetic energy, depending on the nature of the PES, particularly the position of the saddle point.

For bulk reactions of the Arrhénius type statistical (pseudo-thermodynamic) theories, which were first developed in the 1930s, are available. Central to these is the idea that there is some region of the PES that, if it is reached, leads on to the reaction, the probability of entering this region being based on the principles of statistical thermodynamics. The most advanced forms of these theories, namely variational transition-state theory for bimolecular reactions and the so-called RRKM theory for unimolecular processes, are quite successful and not too computationally demanding. Nevertheless, it has to be admitted that to calculate the rate of any chemical process entirely from first principles is a task that is far from routine at present.

The Computer Simulation of Liquids

The computer simulation of liquid-state properties typifies a number of topics in which some equilibrium property of bulk matter is calculated and interpreted from the starting point of the potential acting between the components. For example, we may wish to examine the properties of liquid benzene and see how they are determined by the PES for two interacting benzene molecules. The situation may be more complicated because the total potential for the bulk need not be the sum of all pair potentials; this is certainly true for liquid water, for example.

Of course, molecules also interact in the gas phase and this is what makes gases 'real' instead of 'ideal' (i.e. those obeying Boyle's law). However, for gases there is a wide body of theoretical techniques available under the umbrella of analytical statistical mechanics by which non-ideality can be calculated and this was a heavily researched field before the computer age. The reason that gases are amenable to analytical treatment is not that the potentials are simpler but rather that gas molecules by and large interact only two at a time. The so-called cluster expansions of statistical mechanics converge for gases but do not converge for liquids; there are also some simplifications for solids because it is adequate to adopt a model in which molecules move about fixed lattice sites and only rarely exchange sites.

The computer simulation of liquids does not meet the same difficulties as analytical statistical mechanics because classic trajectories can be calculated numerically for large numbers of particles. Of course, one cannot handle 10^{23} particles (a typical bulk number) but the way around this was found in the 1960s by Alder and Wainwright. In a landmark computer experiment they computed the dynamics of 32 hard spheres in a cubic box and modelled the whole structure by translating the positions and momenta to an infinite three-dimensional lattice of this box. Today one would be computing the motion of 10^4 particles interacting with physically realistic potentials. Of course, the true liquid does not have the artificial long-range order possessed by the model, but the short-range properties of the liquid, which can be linked to experimental observations, should not be affected by this; any uncertainty on this point can be resolved by increasing the size of the cube and showing that the properties remain unchanged.

Molecular dynamics simulations of the type described can give time-dependent properties of the liquid but most commonly are taken to produce a set of configurations and forces that, after averaging, give equilibrium properties. For example, an important property of liquids, which can be deduced experimentally by X-ray diffraction, is the radial distribution function $g(r)$. This is a measure of the probability of finding two atoms separated by a distance r. In a perfect crystal $g(r)$ oscillates with a series of peaks representing the distance to nearest neighbours, second-nearest neighbours, etc.; the oscillations would extend to infinity. In a liquid $g(r)$ will have a broad oscillation representing nearest neighbours and perhaps a slight second peak for second-nearest neighbours, but these oscillations decay rapidly to zero. Examination of how $g(r)$ varies with temperature is a clear way of establishing a melting point for the system; the classic and unexpected result of Alder and Wainwright was to show that a system of hard spheres had both a solid and a liquid form (Figure 4).

If one is interested only in the equilibrium properties of liquids then there is an alternative approach to molecular dynamics that can be used which is computationally simpler, called the Monte Carlo method. Equilibrium statistical mechanics tells us that

Figure 4
Radial distribution functions for the bulk at 0, 800, 1050 and 1150 K. (Reprinted from *Surface Science* 373 (1997), H. Cox *et al. Modelling of Surface Relaxation and Melting of Aluminium*, p. 80, with permission from Elsevier Science.)

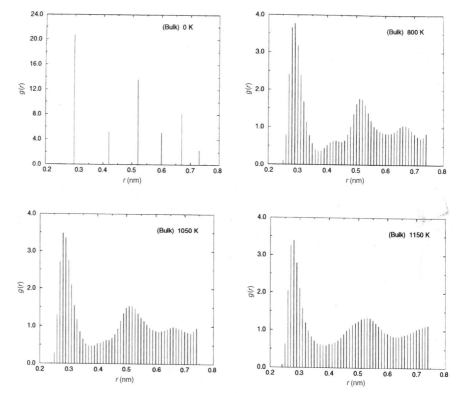

a particular arrangement of atoms of nuclei will appear with a probability given by the Boltzmann factor $\exp[-U/(kT)]$, where U is its energy, T is the temperature and k is Boltzmann's constant. If a random arrangement is laid down in the computer (by choosing random numbers for the coordinates) and its energy calculated, then the Boltzmann factor tells us how it is to be weighted. The simplistic approach of generating a succession of independent random configurations is very inefficient because the majority of these come with very high energies so that their contributions to equilibrium properties are negligible. The efficient approach, devised by Metropolis, is to generate each configuration from an earlier one in such a way that there is a trend towards lower energies and therefore more important configurations; and to do this in such a way that each configuration generated has an *equal* probability for inclusion in the average. The algorithm for doing this by generating random numbers and selecting configurations is well documented.

The computer simulation of liquids has led to a great improvement in our understanding of structure and processes going on in the liquid state, particularly melting. The methods have also been used on the very interesting phases called liquid crystals. The latter are molecules with very anisotropic (non-spherical) potentials and these potentials lead to a loss of translational order but a retention of orientational order. It is the collective orientation of molecules, stimulated by external fields or surfaces, that makes liquid crystals so important for display purposes.

The computer simulation of liquids and other macroscopic phases would clearly fall within the category of computational chemistry and likewise the current emphases in chemical bond theory and reaction dynamics are also computational; the future is bound to see a continuation of growth in this work. However, that does not mean that some of the important analytical techniques developed in the pre-computer age are not as valid and useful as they ever were. To make this point I conclude this chapter with some brief remarks on two areas of mathematical chemistry.

2 Bonding, Atoms and Molecules

Mathematical Chemistry

I take my examples of this topic to be group theory and graph theory, for these have probably been the most important areas of pure mathematics used in chemistry. Group theory is extremely important because many key molecules have high symmetry; typical are methane, benzene, octahedral metal complexes and buckminsterfullerene (C_{60}). This should not be taken to imply that all molecules have the maximum symmetry that could be envisaged; S_2O, for example, has two different types of sulfur atom having the bonding structure S–S–O rather than S–O–S. Indeed, an important theorem deduced from group theory is that high symmetries are always broken for some types of electronic state of molecules of high symmetry; this is the Jahn–Teller theorem.

The first application of group theory to spectroscopy was by Brester in 1924 for the vibrations of crystals. The widespread application to molecular problems has been through point groups, which are the set of operations (rotations, reflections etc.) that leave a solid body unchanged. The most important use of group theory is to give the possible symmetry species (more properly, representations), of molecular states, particularly electronic and vibrational, and to give selection rules for transitions between these. To take a simple example, H_2O belongs to the point group C_{2v} and the O–H stretching vibrational modes belong to the symmetry types labelled A_1 and B_1 rather than to the other possible types for this group, A_2 and B_2. The infrared spectra associated with these two modes of vibration have distinguishable characteristics, giving rise to what are known as parallel and perpendicular bands, respectively. For the spectroscopy of high-symmetry molecules, group theory is invaluable in distinguishing between allowed and forbidden transitions; forbidden transitions give bands that are, at most, very weak in the infrared spectrum. Conversely, the appearance in a spectrum of a pattern of strong and weak bands is an indication of the molecular symmetry. For example, from the vibrational spectrum of H_2O, infrared and Raman, one can deduce that the molecule is bent rather than linear; this is a trivial example, but there are many non-trivial examples, such as the distinction between, say, tetrahedral and square-planar transition-metal complexes.

Graph theory is not as important in chemistry as group theory but it does have applications to molecular structure. If we associate atoms with vertices and bonds with edges (using graph-theory terms) then a graph can be a direct representation of structure insofar as the connections are concerned – but not in respect of specific bond lengths and bond angles. Thus the carbon atoms in aromatic hydrocarbons form a framework with linked bonds, which can be defined by a graph, and, in the Hückel model of electronic structure in which interactions occur only across bonds and are all of the same type, certain features of the pattern of electronic energy states (molecular orbitals) can be deduced from graph theory.

One of the earliest uses of graph theory in chemistry was for the enumeration of isomers; these are chemically distinguishable molecules with the same set of atoms but different bonding arrangements. Thus the general formula C_4H_{10} has two isomers, which are called normal butane and isobutane. The number of isomers for this class of molecules (the alkanes) was determined by Cayley in 1875 by counting the number of 'trees' having vertices of degree four or less.

A much more recent example is the newly discovered family of fullerenes, which are closed carbon cages, C_n, made up of hexagonal and pentagonal rings. To close the cage it is necessary to have twelve pentagons and the smallest molecule without adjacent pentagons that can be found is C_{60}. It was proved by graph theory that the next molecule that has this property is C_{70}; even-numbered structures between 60 and 70 cannot be constructed unless they have adjacent pentagons. The structural instability associated with adjacent pentagons nicely explains why C_{60} and C_{70} were the dominant large carbon molecules produced by laser ablation of graphite in a helium atmosphere.

Summary

In some sense all chemists do theoretical chemistry because no experimentalist fails to interpret results. However, thinking of theory as a professional sub-set of chemistry, there is a wide range of skills based on physics, mathematics and computation that can be used to provide quantitative explanations and predictions of chemical processes. The subject is undoubtedly dominated by computation today, but I would still put the high point of the subject in the post-quantum-mechanics, pre-computer period of the 1930s to 1960s.

Further Reading

1. F. Jensen, *Introduction to Computational Chemistry* (John Wiley and Sons, 1999).
2. J. N. Murrell and S. D. Bosanac, *Introduction to the Theory of Atomic and Molecular Collisions* (John Wiley and Sons, 1989).
3. M. P. Allen and D. J. Tildesley, *Computer Simulation of Liquids* (Clarendon Press, 1987).

Chemistry in a New Light

Jim Baggott

Introduction

When quantum theory was discovered and developed into a consistent framework during the 1920s and 1930s, it completely changed our understanding of the internal physics of the atom. Chemists were not slow to recognize that a revolution had occurred and the first applications of the new physics to the world of chemistry that soon followed provided new insights into chemical bonding, structure and spectroscopy. Amongst the greatest of theoretical challenges was then (and remains so today) the application of quantum principles to the elementary act of chemical reaction.

In the years following the quantum revolution, it seemed that the task of the chemist was simply to work out how the new theory supplied by the physicist could be most successfully applied to problems of chemical interest. Chemists hung on to some sense of self-respect by pointing to the difficulty – if not the sheer impossibility – of applying quantum theory's complex mathematical structure to most systems of relevance to chemistry. Principles are one thing, they said, practice another. The reality is even more heartening. Over the last few decades, it has been experimental chemists who have done most to expose the quantum nature of chemical change. Experimental chemists have defined the physical boundaries of the new chemistry.

The success of the experimentalists can be ascribed to many factors, not least their ingenuity in devising elaborate experiments. However, it is clear that none of this could have been possible without the futuristic technology of lasers. Laser light has made it happen.

Three Questions in Search of Some Answers

On the surface, a chemical reaction is simplicity itself. You put energy in some form or another into a collection of reactant molecules and the molecules react together to form new products. This is what drives every chemical reaction, from the complex biochemical transformations taking place inside our own bodies to the large-scale production processes of the chemicals industry. The level of complexity to be found in even the simplest of chemical reactions depends on how far beneath the surface you wish to look. Chemists, like all scientists, are blessed with a child-like curiosity about how things work that makes surface impressions much less than satisfactory.

Put yourself in the position of an experimental chemist specializing in the intricacies of chemical reaction dynamics (the *how?* of chemical reactions at the detailed level of forces and motion) and kinetics (the *how fast?* of chemical reactions). You are offered fulfilment of three wishes. Perhaps you put aside all notions of phenomenal cosmic power and ask for answers to three questions instead (see Figure 1):

'How is the energy I put into a reactant molecule redistributed amongst the reactant's various quantum states and how quickly does this happen?'
'What are the detailed nuclear motions that link the reactant to the product through its 'transition state' and how quickly do these motions occur?'
'In which of the various possible quantum states are the products formed and what are the speeds of the chemical changes connecting each individual quantum state in the reactant to each of the quantum states of the product?'

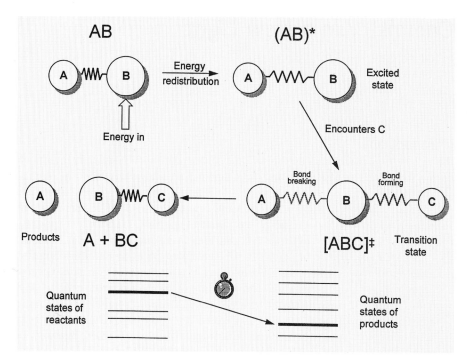

You ask these three questions because, in the context of the current quantum paradigm which provides the basis on which we are attempting to build an understanding of the larger world, these are *the* fundamental questions. Knowing the answers to these questions takes us to the very limit of what there is to know about the nature of a chemical reaction.

The entire history of chemical reaction dynamics and kinetics has been about providing some approximate answers to these three questions. If the redistribution of energy can be assumed to occur much faster than the reaction itself, then the energy can be assumed to be statistically distributed amongst the reactant's various quantum states and hence calculated in a relatively straightforward manner. With the application of a fertile imagination, knowing the structures of the reactants and the products allows us to construct plausible (but entirely theoretical) structures for the transition states, the 'in-between' states in which chemical bonds are assumed to be partly broken and new bonds partly formed. If we cannot measure the speed of the reaction on a quantum-state-to-quantum-state level, then our theoretical picture of the transition state at least allows us to make a stab at calculating a classical (non-quantum) reaction rate.

This is fine as far as it goes, which is not very far. We know the start and end points reasonably well, but the middle is a rather mysterious construct of the mind based more on educated guesswork than on science. One chemist has compared this situation to the challenge of trying to understand what happens in *Hamlet* from witnessing only the opening and closing scenes – from the introduction of our cast of characters to a stage strewn with dead bodies.

Chemists still have some way to go before they will be able to provide full answers to the three questions as a matter of routine. However, by using lasers and some associated optical magic, they have successfully investigated the processes of chemical change at the level of individual molecular quantum states and at the nuclear level of motion on breathtaking timescales measured in millionths of a billionth of a second. By using lasers, they are defying nature and attempting to wrest control of the very instant of chemical change.

Engineering with Light

Whether they produce beautiful, pencil-thin beams of brilliantly coloured light or high-energy pulses of invisible infrared or ultraviolet radiation, all lasers operate on the same basic principles. To make a laser we need three things: a source of energy, a material (gas, liquid or solid) capable of being stimulated to emit radiation under the right circumstances and an optical cavity to amplify the stimulated emission.

At normal temperatures and pressures, molecules present in a gas (such as the air) or a liquid (in a glass of water) will be in thermal, or so-called Boltzmann, equilibrium. Collisions between the molecules will lead to a few of them acquiring energy and being promoted to quantum states higher up the 'energy ladder'. However, the numbers of molecules in quantum states of lower energy will, on average, be larger than the numbers of molecules in higher energy states. Using an energy source such as a high-voltage discharge, spark discharge, intense flashlamp or laser, it is possible in some chemical systems to create a situation in which there are more molecules in some higher energy states than there are in lower energy states. This is called a population inversion and is illustrated in Figure 2. As molecules fall from the higher to the lower states, they may emit radiation with a frequency characteristic of the difference in energy between the states. This radiation can then stimulate emission from other excited molecules. The result can be a cascade of stimulated emission at a characteristic frequency. Instead of the radiation diminishing as it passes through the material, it actually increases in brightness.

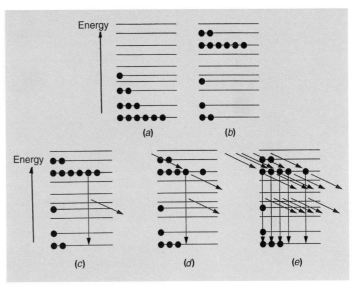

An optical cavity is simply a system of mirrors designed to reflect the radiation back through the material (Figure 3). With each pass, the radiation stimulates more emission, increases in brightness and so is amplified for as long as the population inversion is maintained. By making one of the mirrors partially transmitting (like a two-way mirror), it is possible to extract some of the radiation as useful laser output.

This output has special characteristics that make lasers extremely powerful tools for experimentation. Because the frequency of the laser radiation is dictated by the quantum states of the laser material, this may already be sharply defined and highly monochromatic (a 'single' colour or wavelength). By introducing additional optical devices inside the cavity, it is possible to reduce the spread of frequencies further and to make the output continuously variable (or 'tunable') over a relatively broad frequency range. With this kind of radiation, selectively exciting and/or monitoring individual quantum states in molecules becomes possible.

The radiation circulating inside the cavity is also spatially and temporally coherent, which means that the crests and the troughs of the electromagnetic waves are aligned both in space and in time. This and other properties of the radiation inside the cavity can be exploited by using a variety of optical devices to manipulate the characteristics of the laser output. Thus, in a colliding pulse mode-locked laser, an otherwise continuous laser beam is turned into a series of light pulses lasting less than 100 fs (a femtosecond is 10^{-15} s).

Perhaps the most notable property of many lasers is their reputed ability to deliver pulses of radiation of high power. The key advantage that this confers on the experimental chemist is the ability to use lasers to excite molecules in sufficient numbers to allow their properties and behaviours to be measured and monitored in the laboratory.

Figure 2

In this picture each line represents a specific quantum state of a chemical substance in the gas, liquid or solid phase. Each dot indicates a molecule in the substance that is in one of the quantum states. Under normal conditions (a), there are more molecules in the lowest-energy quantum states. After pumping with an intense source of energy, it is possible to create the situation in (b), in which there are more molecules present in some excited quantum state than there are in the lowest energy states. Spontaneous emission by one of the excited molecules releases light into the optical cavity with a wavelength characteristic of the difference in energy between the quantum states (c). That light can go on to stimulate emission from other excited molecules (d), ultimately resulting in a cascade of stimulated emission (e). Some of this emission can be extracted from the optical cavity as useful laser output.

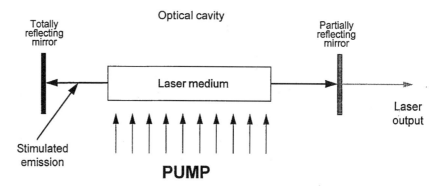

Careful tailoring of the laser pulse's duration, polarization and frequency can expose the intricate internal workings of molecules in the process of reacting.

Making Light of Reaction Dynamics

The study of photodissociation – using light to break molecules apart – has been a source of tremendous insight into the deliciously complex world of chemical reaction dynamics. In a general bimolecular reaction of the type $A + BC \rightarrow AB + C$, the atom or molecule A collides with the molecule BC and they combine to produce a transition state, usually denoted $[ABC]^{\ddagger}$, which then breaks apart to form the products. In their search for answers to the fundamental questions of chemical reaction dynamics, chemists have settled for a half-way house. If they can somehow create the transition state $[ABC]^{\ddagger}$, then observing how this goes on to produce $AB + C$ at least gives them half the story. This is done in practice by exciting a stable ABC molecule, photodissociating the AB–C bond and monitoring the quantum states of the product AB.

The simplest (yet amongst the most highly revealing) type of information that can be gained from this kind of experiment is the basic distribution of product AB molecules over all the quantum states of AB that can potentially be populated, including electronic, vibrational and rotational states. For example, photodissociation of water (H_2O) molecules in the gas phase using ultraviolet radiation with a wavelength of around 157 nm produces OH fragments that are found to be rotationally 'cold', meaning that the population of rotational quantum states of OH is heavily biased towards the lowest states. However, when H_2O is dissociated using higher energy 121.6-nm radiation, the resulting OH fragments are both electronically excited and rotationally 'hot', despite the fact that the amount of energy 'left over' to go into rotation is considerably smaller in the second case than it is in the first. It is impossible to understand these results without reference to the detailed contortions of the excited H_2O molecules in the process of breaking apart.

In the first case, excitation at 157 nm promotes the H_2O molecule from its ground state to its first electronically excited state. This state is said to be dissociative or repulsive: the H–OH bond breaks 'instantly' and the two fragments rush away from each other without imparting any kind of 'kick' to the OH fragment that would be necessary in order to set it spinning end over end. The OH-radical products are therefore formed with little or no rotational excitation. In contrast, excitation at 121.6 nm generates the second electronically excited state of H_2O, which has distinctly different properties. This excited state prefers to be linear, unlike the ground state, which prefers a bent structure with an equilibrium HÔH angle of 104°. As soon as it is excited, the H_2O molecule opens up and a strong torque is exerted. When the H–OH bond breaks, that torque is translated into significant rotational motion of the OH fragment. This is illustrated in Figure 4.

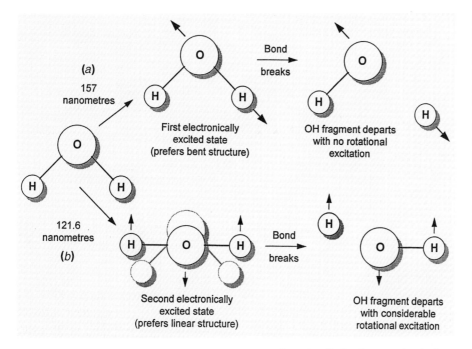

Figure 4
Excitation of the H_2O molecule with 157-nm radiation breaks one of the O—H bonds 'instantly' (a). The fragments rush away from each other and there is no opportunity for the departing OH radical to acquire any rotational excitation. In contrast, excitation with 121.6-nm radiation generates an electronically excited state of the H_2O molecule that prefers to adopt a linear structure (b). The molecule contorts to a linear shape and, on breaking of an O—H bond, the torque exerted on the molecule imparts considerable rotational excitation to the departing OH fragment. This occurs despite the fact that the total energy 'available' after breaking the bond is three times larger in the case of dissociation at 157 nm.

Within figure:

(a) 157 nanometres

First electronically excited state (prefers bent structure)

Bond breaks

OH fragment departs with no rotational excitation

(b) 121.6 nanometres

Second electronically excited state (prefers linear structure)

Bond breaks

OH fragment departs with considerable rotational excitation

Although these explanations may appear somewhat simplistic, they are backed up by detailed theoretical calculations of the structures of the first and second excited states of H_2O. The results demonstrate the utility of measuring products' quantum-state distributions and the explanation shows how such measurements provide insights into the dynamics of the dissociation process.

Other kinds of information are available from photodissociation experiments by virtue of the simple fact that laser light is usually linearly polarized – the light waves oscillate in one spatial dimension only. The probability of a molecule absorbing light is greatest when the amplitude of the light wave is at its peak, so the directionality provided by the laser beam's polarization can be used to impose a reference direction on the subsequent dissociation. This means that certain vector quantities associated with the fragments (such as 'recoil' velocities and rotational angular momenta) are correlated to each other and to other vector quantities associated with the absorption of light. These correlations can be measured in the laboratory using lasers and their interpretation adds further to our understanding of the detailed dynamics of the transition state [ABC]‡.

The photodissociation of hydrogen peroxide (H_2O_2) provides one of the best examples of the value of measuring vector correlations. The structure of the hydrogen peroxide molecule in its lowest energy state can be imagined by reference to this book as you are currently reading it. With the book open, the O—O bond can be imagined to lie along the spine, in the crease between the two facing pages. One O—H bond then lies in the plane of the left-hand page and the other O—H bond lies in the plane of the right-hand page. When H_2O_2 is photodissociated, the O—O bond breaks and there are two possible sources of rotational excitation of the resulting OH fragments, as shown in Figure 5. One source is the bending of the O—H bonds in their respective planes (the hydrogen atoms 'waggling' backwards and forwards in the planes of the left-hand and right-hand pages). When the molecule splits apart, this waggling becomes end-over-end rotation of the OH fragments as they move away from each other.

The second source of rotational excitation is the torsional motion of the H_2O_2 molecule, which can be envisaged as an opening and closing of the book itself. When the molecule splits apart, the OH fragments rotate end over end as before, but now the

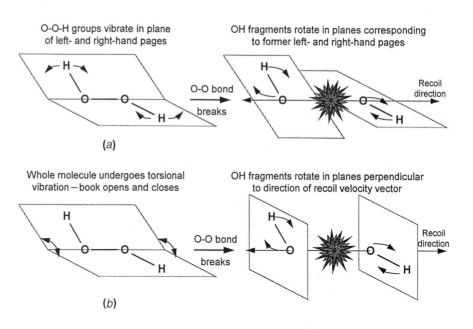

Figure 5
The hydrogen peroxide molecule has a structure that resembles an open book. The O—O bond lies along the spine, one O—H bond lies in the plane of the left-hand page and the other O—H bond lies in the plane of the right-hand page. When H_2O_2 is photodissociated, two potential causes of rotation of OH fragments can be identified. In the first, the rotation is due to in-plane bending vibrations of the O—O—H groups (a). In the second, it is due to torsional motion of the whole molecule – opening and closing the book (b). In (b), the OH fragments rotate in planes that lie perpendicular to the direction of recoil. Only an experimental measurement of the correlation between the recoil and angular momentum vectors can reveal the actual cause of the rotation. Experiments show that (b) is the major source.

direction of rotation is perpendicular to the direction of the O—O bond and hence perpendicular to the recoil-velocity vector. Only a measurement of the correlation between the recoil velocity and the rotational angular momentum vectors can discriminate between these two possibilities. Experiments reveal that the major source of rotational excitation is the torsional motions of the contorting H_2O_2 molecule.

In some further developments of these techniques, Richard Zare and his colleagues at Stanford University in California are hoping to employ lasers to measure all the parameters of a chemical reaction that would normally require techniques involving crossed molecular beams. The basic idea is to force a high-pressure mixture of gases containing AX and BC molecules through a narrow aperture into a vacuum. As the gases expand, they rapidly cool to extremely low temperatures and move through the vacuum at the same velocity. By using a laser to photodissociate a proportion of the AX molecules, the Stanford chemists will produce reactant A atoms with well-defined velocities relative to those of the BC molecules. The internal rotational quantum states of BC will not be completely known, but, because of the extremely low temperatures, relatively few rotational quantum states will be likely to be populated. Thus, A reacts with BC under very well-defined (and known) conditions.

By doing some laser spectroscopy, the chemists will measure the electronic, vibrational and rotational quantum states of the AB products and will also determine the distribution of velocities of the AB products. Simple vector equations and some trigonometry will thus allow them to calculate the distribution of angles over which the AB products are recoiling from the transition state [ABC]‡. Using polarized light from the laser will further allow them to measure vector correlations, providing some of the most complete quantum-state-to-quantum-state information ever obtained for a bimolecular reaction.

Femtochemistry

The molecule cyanogen iodide (ICN) is irradiated with ultraviolet light of sufficient energy to break the I–CN bond. After it has absorbed the radiation, how long on average does it take an individual bond to break, creating completely free iodine atoms (I) and cyanide radicals (CN)? The answer, we now know, is 200 fs. In making this measurement in 1987 (Figure 6), Ahmed Zewail and his colleagues Marcos Dantus and Mark

Figure 6
Zewail's experiment.

Rosker at the California Insitute of Technology in Pasadena launched us into a new era of experimental reaction dynamics. They launched us into the world of *femtochemistry*.

Zewail's experiments on ICN were classic, so-called 'pump–probe' experiments. They fired pulses of ultraviolet light of wavelength 308 nm into a sample of ICN gas. Each pulse had a duration of about 60 fs and 'pumped' the reaction, effectively setting the laboratory clock ticking. At intervals of 10 fs, the chemists fired in 'probe' laser pulses with longer wavelengths and duration 60 fs in order to monitor the progress of the bond-breaking event. What they actually observed was fluorescence from CN radicals excited by the probe laser pulses (Figure 7).

By probing with a wavelength of 388.0 nm, a wavelength that is absorbed by CN radicals only when they are completely free of the influence of their erstwhile iodine-atom partners, Zewail and his colleagues could determine how long it took on average for the iodine atoms and cyanide radicals to separate completely. They found that it took 200 fs. What's more, the Caltech scientists found that, by tuning the probing laser wavelength slightly to the red, to 388.9 nm, they could excite fluorescence from CN

Figure 7

The femtosecond-dynamics studies of the photodissociation of ICN carried out by Zewail and his colleagues launched us into a new era of femtochemistry. The dissociation is initiated (or 'pumped') using a 60-fs laser pulse with a wavelength of 308 nm. As the bond in the transition state [I–CN]⁺ lengthens, it enters the 0.06-nm-wide window provided by the probing laser pulses (*a*). The position of the window along the length of the bond is set by tuning the wavelength of the probe laser. When molecules that are in the transition state enter the window, some of them are excited by the probing laser pulses and break up to produce free CN radicals. The free CN radicals are detected by their resulting fluorescence signal. As the remaining molecules pass through the transition-state region, they pass outside the window and the fluorescence signal falls. The dissociation is completed within about 200 fs. Varying the wavelength of the probe laser effectively moves the window around, as can be seen in the corresponding signals shown in (*b*).

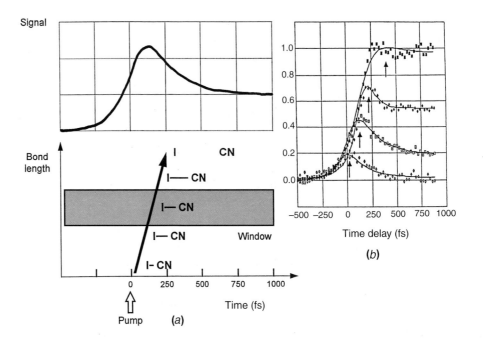

radicals under different degrees of influence of their I atom partners, namely at different moments during the bond-breaking process.

In essence, the probing laser wavelength acts like a movable 'window'. By moving the window to look only at completely free CN radicals, the scientists could determine how long it took for free CN radicals to appear there. However, the window can also be moved to look at the reaction at any point along the lengthening I–CN bond, namely at the various structures that constitute the transition state for the reaction. As the ICN molecules respond to the ultraviolet pumping radiation, the bonds extend into the 'viewing area' defined by the window. Some of the [I–CN]‡ transition states caught in the window absorb the probe radiation and break apart. Fluorescence from the resulting free CN radicals is detected. Because the bonds of the remaining [I–CN]‡ transition states extend beyond the window, the probe radiation is no longer absorbed and the fluorescence from free CN radicals falls in intensity. Probe pulses of duration 60 fs provide windows about 0.06 nm wide.

These experiments are pretty extraordinary in their ability to reveal the motions of molecules in the act of falling apart. However, subsequent experiments on sodium iodide (NaI) reported by Zewail, Todd Rose and Mark Tasker became the benchmark for the new science of femtochemistry. These experiments really showed what the new techniques could do.

Sodium iodide is a relative of sodium chloride, which is common table salt. In contrast to the situation in ICN, the first excited state of NaI is not dissociative. Exciting ICN with ultraviolet light of sufficient energy effectively breaks the I–CN bond almost 'instantly', with no turning back. However, exciting NaI with ultraviolet light creates an excited state that has the opportunity to vibrate a few times before falling apart. This happens because the excited state is actually a mixture of covalent (Na–I) and ionic (Na⁺I⁻) structures and, as the bond vibrates, the structure oscillates back and forth between predominantly covalent and predominantly ionic forms. Only the covalent component of the structure can dissociate to form free sodium and iodine atoms.

Zewail and his colleagues used pulses of 310-nm radiation of duration 60 fs to pump the reaction. Using probing laser pulses of 589 nm to excite free sodium atoms allowed them to measure how quickly the Na–I bond breaks by monitoring the familiar yellow sodium D-line emission. On average, the bond breaks in 4 ps (4×10^{-12} s).

It became apparent that the build-up of the emission signals from free sodium atoms exhibited reproducible wiggles. By changing the probing laser wavelength to 580 nm, the chemists moved the observation window onto the Na–I transition state and the wiggles became pronounced oscillations in the emission signals (see Figure 8).

The explanation for this is as follows. When the bond in the excited Na–I molecule begins to extend following its excitation by 310-nm radiation, its structure takes on an increasingly ionic character. By the time the 'bond' has reached its maximum extension, the structure is almost completely ionic. The motion turns around and the bond begins to compress. As it compresses, it takes on a covalent character once more. The structure oscillates between covalent and ionic forms with a period of 1.3 ps.

However, when the bond extends to a certain critical distance, it is possible for the small covalent component of the structure to 'escape', breaking the bond and forming free sodium and iodine atoms. On average, about 10% of the molecules in the transition state [Na–I]‡ will break apart on reaching this critical distance. The probe radiation of 580 nm monitors only the covalent part of the structure, so the measured emission signal falls rapidly beyond time zero and oscillates in intensity in time with the bond vibration. The signal gradually diminishes, because, with each vibration, about 10% of the Na–I molecules fall apart.

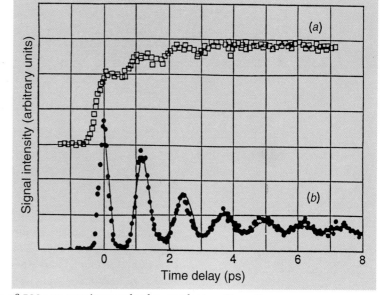

The interpretation of these results is truly remarkable. The signals captured in Zewail's apparatus correspond to the real-time observation of an excited molecule vibrating, 10% of molecules in the transition state that vibrate past some critical bond distance breaking apart to form free sodium and iodine atoms. In the early 1980s, the exercise of a little imagination could probably have provided a picture of the dynamics of the dissociation of sodium iodide. Now we can see all the intermediate acts for ourselves, directly.

Zewail has continued to develop the science of femtochemistry, applying it to all manner of chemical systems in which enhancement of temporal resolution can provide new insights into how the chemistry works. These systems include the dissociation of HgI_2, the molecular motions of I_2 and ICl, bimolecular reactions between H atoms and CO_2 and between HBr and I_2, isomerization reactions, H-atom-transfer and proton-transfer reactions, reactions taking place in a solvent of noble gas atoms and at high pressures and diradical intermediates involved in the dimerization of ethene molecules to produce cyclobutane.

It is worth noting here that the use of laser pulses of durations measured in femtoseconds to prepare excited molecules begs questions concerning what might be called the 'energy resolution' of such experiments. According to Werner Heisenberg's famous uncertainty principle, the uncertainty in the energy of a laser pulse increases as its duration is reduced. The uncertainty in energy associated with a 60-fs laser pulse means that it is not possible to excite individual quantum states in a molecule. Instead, several states are coherently excited to produce what is known as a quantum

Figure 8

The photodissociation of sodium iodide really demonstrated the potential of femtochemistry to reveal an incredibly detailed picture of the dynamics of the transition state. The upper signal (a) was obtained by tuning the probe laser to monitor free sodium atoms and shows that the dissociation is effectively complete within 4 ps. The lower signal (b) is the result of tuning the probe laser (or moving the window) to catch the transition state [Na–I]‡ in the process of dissociating. The oscillations in the signal correspond to the vibrations of the Na–I bond and have a period of about 1.3 ps. When the bond length in the transition state reaches a certain critical distance, about 10% of the molecules may dissociate, which is why the signal diminishes in intensity with each oscillation.

'wave-packet' state. All the results described here can be given a strictly quantum inter-pretation based on the dynamics of wave-packet states.

However, the uncertainty principle does set a limit on the ultimate temporal resolu-tion that can be considered relevant to chemical systems. Reducing the duration of a laser pulse much below the current record of 6 fs is unlikely to be of much value to chemists. For one thing, as was amply demonstrated in the I–CN example, the fastest of chemical reactions are unlikely to take place in less time than it takes a single bond to vibrate once. Furthermore, a sub-femtosecond temporal resolution implies an energy resolution broader than the energy required to break some chemical bonds directly. By using the sophisticated technology of lasers chemists are, in the words of Nobel lau-reate George Porter, 'near the end of the race against time'.

Chemists in Control

From the moment lasers were first introduced as a means of initiating chemical reac-tions, chemists began to dream of almost unlimited power and control. The precision with which lasers could deliver energy to molecules seemed to suggest that control-ling the speeds and directions of chemical reactions was going to be straightforward. After all, to break a chemical bond in a reaction one need only put enough energy into it. Unfortunately, chemists reckoned without the sheer difficulty of selectively excit-ing individual chemical bonds and the rapid redistribution of energy within molecules that accompanied any such attempt. More than 20 years of experimentation with high-power pulsed infrared lasers and with continuous-wave visible lasers produced results that were no more selective than could have been obtained with a Bunsen burner.

However, a major breakthrough was achieved in 1989, when Fleming Crim and his colleagues Amit Sinha and Mark Hsaio at the University of Wisconsin at Madison studied the reaction between hydrogen atoms and vibrationally excited HDO, partially deuterated water. In HDO, the heavier D atom causes the O–D bond to vibrate with a distinctly different frequency from that for the O–H bond. Thus, when the O–H bond is vibrating, the movements of the central O atom do not match the corresponding movements required to excite the O–D bond vibration, with the result that the two bond vibrations are effectively isolated from one another. Contrast this situation to that of ordinary water, H_2O. In H_2O, any attempt to excite one O–H bond would lead to rapid excitation of the other and the nuclear motions would quickly settle down into a collective vibration of the whole molecule. Such collective motions are called normal modes.

The reaction between H atoms and vibrationally excited HDO can proceed in two different ways, as shown in Figure 9. Either the O–H bond breaks, producing H_2 and leaving an OD radical, or the O–D bond breaks, forming HD and leaving an OH radical. Crim and his colleagues found that, on selectively exciting the O–H bond in HDO using laser radiation of wavelength 720 nm, the OD fragment was formed 100 times more frequently than was OH. By putting energy into the O–H bond, the chemists had shown that they could cause it to break preferentially in the subsequent reaction with a H atom. Further studies by Zare and his colleagues at Stanford University demonstrated the selectivity that is possible in this reaction. By exciting the O–H bond in HDO using an infrared laser, the Stanford chemists obtained *only* H_2 and OD products in the sub-sequent reaction with H atoms. By exciting the O–D bond, they obtained only HD and OH.

These experiments were crucial in that they demonstrated clearly for the first time that some form of bond selectivity in a chemical reaction is possible using lasers. Both Crim and Zare and their colleagues have gone on to find more examples. Crim has reported the results of bond-selectivity in the reactions of HDO with chlorine atoms and is currently looking at the reaction between hydrogen cyanide (HCN) and chlorine

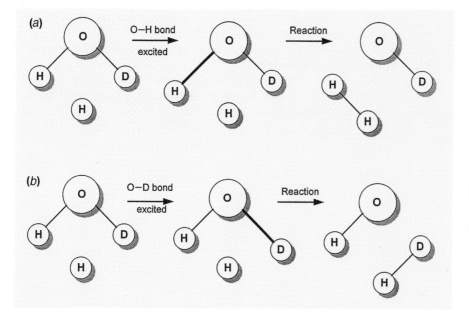

Figure 9
The reaction between hydrogen atoms and vibrationally excited HDO molecules can proceed in one of two different ways, producing $H_2 + OD$ or $HD + OH$. When the O—H bond in HDO is vibrationally excited, the products $H_2 + OD$ are formed preferentially (*a*). Similarly, when the O—D bond is excited, the products $HD + OH$ are formed preferentially. (*b*). This was the first clearly demonstrated example of bond-selectivity in a chemical reaction – by choosing which bond to excite, the chemists could dictate the course of the subsequent reaction.

atoms. Zare, Robert Guettler and Glenn Jones have recently shown how the selective excitation of a specific vibrational mode in positively charged ammonia ions (NH_3^+) leads to the enhancement of the reaction between NH_3^+ and ND_3 to produce NH_2 radicals and NHD_3^+. Scott Anderson and his colleagues at the University of New York at Stony Brook have also demonstrated selectivity in the reaction between positively charged ethyne ions ($C_2H_2^+$) and methane.

These last examples demonstrate mode-selectivity rather than bond-selectivity. However, given that the individual bond-stretching vibrations in HDO are just special kinds of vibrational modes, it would seem that controlling chemical reactions is all about exciting such modes so that they more closely resemble the transition-state structures involved in the reactions. This means relying on the target molecules' natural vibrations and being able to push these further towards the kinds of distorted structures they need to adopt in their respective transition states.

A much more active form of laser control relies on generating these distorted structures without having to start from the molecules' natural vibrations. The proposal is that one could use one laser pulse with a delicately shaped temporal profile, or a series of carefully controlled and timed laser pulses, to push a target molecule more and more towards a specific structure, so that it is much more likely to react in this way rather than that. These are theoretical proposals only, but they are proposals based on more than 20 years' experience (mostly frustrating) in the use of lasers to control and direct chemical reactions along desired pathways. The progress secured during the last 10 years suggests that active laser control of chemical reactions lies not too far in the future.

Further Reading

1. R. D. Levine and R. B. Bernstein, *Molecular Reaction Dynamics and Chemical Reactivity* (Oxford University Press, Oxford, 1987).
2. M. N. R. Ashfold and J. E. Baggott (editors), *Molecular Photodissociation Dynamics* (Royal Society of Chemistry, London, 1987).
3. R. Schinke, *Photodissociation Dynamics* (Cambridge University Press, Cambridge, 1993).

4. J. Manz and L. Wöste (editors), *Femtosecond Chemistry* (VCH Publishers, Inc., New York, 1994).
5. A. H. Zewail, *Femtochemistry: Ultrafast Dynamics of the Chemical Bond* (Parts I and II) (World Scientific, Singapore, 1994).
6. *Lasers: Invention to Application* (National Academy Press, Washington, 1987).
7. D. C. O'Shea, W. R. Callen and W. T. Rhodes, *An Introduction to Lasers and their Applications* (Addison-Wesley, Reading, MA, 1977).

Novel Energy Sources for Reactions

4

Most chemical reactions at normal temperatures and pressures require the input of energy. Traditionally this has been heat. However, in recent decades, chemists have exploited other forms of energy to trigger chemical reactions – electrical energy in electrochemical cells, electromagnetic energy in the form of microwaves and irradiation by ultrasound.

Electrosynthesis

Tatsuya Shono

Introduction

The donation of thermal energy is a minimal pre-requisite for the promotion of an organic reaction. In addition to the thermal energy, the donation of electronic energy to the reaction system leads to the development of a new world of organic synthesis.

Discussion

Chemical reactions do not take place without the donation of thermal energy from outside the reaction system, but, on the other hand, thermal energy is usually the only form of energy which is required to promote the chemical reaction. In some limited cases, photochemical energy is also used, together with thermal energy, to promote the chemical reaction. It has recently been found, however, that the donation of electronic energy to the reaction system results in the development of a new world of organic reactions, that is, electro-organic reactions, which had never been known hitherto. Organic synthesis using the electro-organic reaction as the key reaction is called electrosynthesis.

The first organic reaction which was promoted by the donation of electronic energy to the reaction system is the famous Kolbe electrolysis (Box 1) found by Helmann Kolbe at Leipzig University in 1849, though the mechanism of this reaction was not known at that time.

In 1964, that is, 115 years after the discovery of the Kolbe electrolysis, Manuel M. Baizer of the Monsanto Company found the reductive dimerization of acrylonitrile to adiponitrile:

Box 1 The Kolbe electrolysis

When an electrical current is passed through a solution of a salt of a carboxylic acid (RCOOM, M is an alkali metal) with a sufficiently high anodic potential, the carboxylic acid anion is oxidized at the anode and decomposed to a radical intermediate R• and carbon dioxide as shown in the following equation (1). In the next step, the radical R• dimerizes to the dimer R—R (equation (2)):

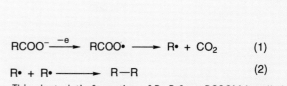

$$RCOO^- \xrightarrow{-e} RCOO\bullet \longrightarrow R\bullet + CO_2 \qquad (1)$$

$$R\bullet + R\bullet \longrightarrow R\text{—}R \qquad (2)$$

This electrolytic formation of R—R from RCOOM is called the Kolbe electrolysis and it is one of the most promising methods for the synthesis of R—R.

$$2\ CH_2{=}CHCN \xrightarrow{\ +\,e\ } NC(CH_2)_4CN \tag{3}$$

This dimerization takes place at the cathode when an electrical current is passed through a solution of acrylonitrile.

This reaction seems to have been the first example of electrosynthesis to be successfully commercialized. Although the details of the mechanisms of this dimerization are not always clear, its initiation step is undoubtedly the formation of an anion radical intermediate by donation of an electron to the unsaturated bond of acrylonitrile. As this dimerization clearly shows, the typical character of the electro-organic reaction is that the reactant is activated by the donation of an electron to the molecular orbital of the reactant (reduction type) or the removal of an electron from the orbital (oxidation type).

The differences among the patterns of activation of the reactant for the usual thermally activated reaction, the photochemical reaction and the electro-organic reaction are shown schematically in Figure 1.

In the ground state, each of the orbitals of the reactant is occupied by two electrons. In the case of thermal activation, the arrangement of electrons does not change, whereas one electron belonging to the highest occupied molecular orbital (HOMO) is promoted to the lowest unoccupied molecular orbital (LUMO) in the case of photochemical activation. There are two types, namely, oxidation and reduction, of electrolytic activation. In the case of reduction, one electron is donated to the LUMO of the reactant from the cathode, whereas one electron is removed from the HOMO of the reactant by the anode in the case of oxidation.

In the case of thermal or photochemical activation, the total number of electrons is not changed in the activation, whereas in electrolytic activation, the number of electrons is increased by one in the case of reduction and decreased by one in the case of oxidation.

Hence, the main differences between electrolytic activation and the other two types

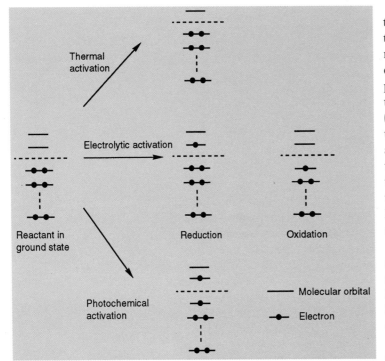

Figure 1

The difference among the patterns of activation of the reactant for the usual thermally activated reaction, photochemical reaction and electro-organic reaction.

of activation may be summarized as follows: the arrangement of electrons in the molecular orbitals is different and the total numbers of electrons are different. Therefore, it is quite obvious that the reactivity of the electrolytically activated reactant is dramatically different from that of the thermally or photochemically activated reactant.

I have classified the electro-organic reactions into four categories, that is, (1) direct reactions, (2) indirect reactions (reactions promoted by mediators), (3) electrochemical generation of active reagents and (4) reactions promoted by chemically reactive electrodes, as shown in Figure 2.

(1) In the direct reaction, the reactant is activated by the direct transfer of electrons between the reactant and the electrode. Although the direct reaction is the most typical pattern of the electro-organic reaction, it is a requisite of this type of reaction that the oxidation and reduction potentials of the reactant be in the range which is accessible to the method of direct electron transfer.

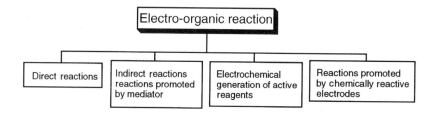

Figure 2
Types of electro-organic reactions.

The oxidation and reduction potentials of organic compounds are, however, often beyond that range and then the direct electron transfer becomes impossible.

(2) In contrast to the direct reaction, the reactant may be oxidized or reduced when a suitable third substance which has been added to the reaction system mediates the electron transfer. The third substance is named a mediator and the reaction promoted by the mediator is called the indirect reaction.

(3) The activated reactant generated in the electro-organic reaction is generally highly active and hence it cannot be called a reagent. Sometimes, however, the activated reactant possesses a moderate stability and it can survive for a reasonably long time. This type of activated reactant is called an electrogenerated reagent and it may promote some unique reactions that have never been known in the usual organic chemistry.

(4) In the electro-organic reaction, the role of the electrode is generally just as a donor or acceptor of an electron and it is not involved in the reaction as something like a reagent. It has been found recently, however, that some electrodes made of chemically reactive materials promote unique reactions and the electrode positively contributes to the reaction in this case.

The electro-organic reaction has some unique characteristics in comparison with the usual chemical reaction. In the case of the usual chemical reaction in a homogeneous solution, the distribution of reactants in the solution is just uniform; hence, the reaction takes place uniformly in the solution. In the case of the electro-organic reaction, however, the activation of the reactant does not take place in the bulk of solution, but rather the formation of the activated reactant and subsequently its reaction take place in the vicinity of the electrode's surface, since the activation is achieved by the transfer of electrons between the reactant and the electrode.

The fact that the reaction takes place very close to the electrode's surface gives very unique characteristics to the electro-organic reaction. Some examples are shown below.

One of the characteristics is the stereochemical selectivity in the reaction. When an electrical current is passed through a solution of 3-methylcyclohexene in ethanoic acid, for instance, acetoxylation of the allylic position of cyclohexene takes place at the anode. This reaction is called the anodic allylic substitution and a remarkably high *cis*-selectivity is found compared with that of the usual organic reaction in a homogeneous solution (equation (4)). This high *cis*-selectivity is explained by considering the stereoselective adsorption of cyclohexene onto the anode (Figure 3).

$$(4)$$

	cis/trans
Anodic oxidation	2.8–3.4
Reaction in a homogeneous solution	0.23–0.27

Figure 3
The stereoselective adsorption of
cyclohexene onto the anode.

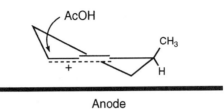

Anode

When 3-methylcyclohexene is adsorbed onto the anode, the adsorption takes place at the side opposite to that on which the methyl group is located, since the methyl group is big enough to obstruct the adsorption. After 3-methylcyclohexene has been activated at the anode, ethanoic acid attacks the activated reactant from the same side as that of the methyl group (the *cis* side) since it is impossible for ethanoic acid to get into the tight space between the reactant and the anode.

The hydroxylation of a steroidal compound takes place at the anode with high stereoselectivity ($\beta/\alpha = 10-14$) and the selectivity is similar to that of microsomal hydroxylation ($\beta/\alpha = 8-14$), whereas it is much better than that of the usual chemical hydroxylation with peracid in a homogeneous solution ($\beta/\alpha \approx 3$) (equation (5)).

The fact that the stereoselectivities observed in the anodic and microsomal hydroxylations are very similar clearly shows that the anodic oxidation is a very effective method for simulating the metabolic oxidation. Indeed, anodic oxidation is an effective means for analysing the metabolic mechanism and also for the large-scale preparation of the metabolite.

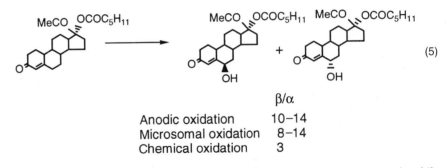

$$\begin{array}{ll} & \beta/\alpha \\ \text{Anodic oxidation} & 10\text{--}14 \\ \text{Microsomal oxidation} & 8\text{--}14 \\ \text{Chemical oxidation} & 3 \end{array}$$

It is well known that the organic reaction generally takes place between a nucleophile (Nu) and an electrophile E (equation (6)). The change of the polarity of a reactant from E to Nu and vice versa is called *Umpolung* and it is one of the most important characteristics of the electro-organic reaction. Thus, the donation of an electron to a compound E, for instance, changes its polarity to Nu. Then, the original compound E reacts as if it were a nucleophile Nu. The polarity of a compound Nu is also changed from Nu to E in the same way by the removal of an electron (equation (7)). Thus, a variety of new types of synthetic reaction has been promoted in the electro-organic reaction by this *Umpolung*.

$$E \ + \ Nu \ = \ E\text{--}Nu \tag{6}$$

$$\left. \begin{array}{l} E \xrightarrow{+e} [\,E^- = Nu\,] \xrightarrow{E} E\text{--}E \\ Nu \xrightarrow{-e} [\,Nu^+ = E\,] \xrightarrow{Nu} Nu\text{--}Nu \end{array} \right\} \tag{7}$$

$$R^1R^2C\overset{+}{=}NR^3R^4 \xrightarrow{+2e} R^1R^2\overset{-}{C}\text{--}NR^3R^4 \xrightarrow{R^5X} R^1R^2R^5C\text{--}NR^3R^4 \tag{8}$$

The reaction shown in equation (8) is a typical example of the *Umpolung* achieved in the electro-organic reaction. Namely, an iminium cation (E) is changed to a carban-

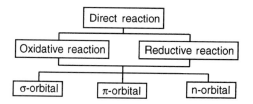

Figure 4

The electron transfer between the
reactant and the electrode.

ion (Nu) by donation of two electrons to the iminium cation. The carbanion (Nu) is
easily alkylated with an alkyl halide.

It is completely impossible in the usual reaction for acid and base to exist at the
same time in a homogeneous solution. In the electro-organic reaction, however, acid
is able to exist in the vicinity of the anode regardless of whether the bulk of the solution is acidic or basic; likewise base can exist in the vicinity of the cathode. Thus, the
fact that both acid and base exist independently of each other in a homogeneous solution gives the electro-organic reaction a unique characteristic. The alkylation shown
in equation (8) is a typical example. Although this alkylation is carried out in the presence of a strong acid, the carbanion (Nu, a base) is not protonated by the acid (H^+)
before it is alkylated with an alkyl halide, because the carbanion (Nu) is formed in the
vicinity of the cathode where the base and a neutral reactant (an alkyl halide) exist
safely whereas acid (H^+) is not stable and is easily discharged.

As I have already mentioned above, the distribution of reactants in the usual homogeneous chemical reaction is just uniform. In the case of the electro-organic reaction,
however, a certain reactant is selectively adsorbed onto an electrode so that the concentration of the reactant in the vicinity of electrode is remarkably greater than that
elsewhere. This selective adsorption of a certain reactant onto an electrode gives a
unique characteristic to the electro-organic reaction. The Kolbe electrolysis (equations
(1) and (2)) is the typical reaction. The formation of R−R in the Kolbe electrolysis is
rationally explained by considering the selective adsorption. The key intermediate in
the Kolbe electrolysis, that is, $R^\bullet$ is very unstable and reactive so that it tends to react
with a nearby compound such as solvent molecule and hence the dimerization of $R^\bullet$ to
R−R is rather difficult unless its concentration is very high. Although $RCOO^\bullet$ and $R^\bullet$ are
also formed by thermal decomposition of diacyl peroxide (RCOO−OOCR) in solution,
the yield of R−R is usually low since the formation of $R^\bullet$ at a high concentration is
extremely difficult in this case. On the other hand, in the case of the Kolbe electrolysis,
$RCOO^-$ is selectively adsorbed onto the anode so that the concentration of $R^\bullet$ is sufficiently high to form the dimer R−R in the vicinity of the anode in a reasonable yield.

I have shown above four categories for the classification of the electro-organic reactions. The first category is the direct reaction in which the reactant is activated by the
direct transfer of electrons between the reactant and the electrode. The direct reaction is
further classified into two types of reaction, that is, reductive and oxidative reactions.
Each type of reaction is divided into three groups according to the type of molecular
orbital (σ and π orbitals and the orbital of a non-bonding electron pair (an n orbital) which
participates in the transfer of electrons between the reactant and electrode (Figure 4).

In the oxidative reaction, however, the removal of an electron from a σ orbital is generally difficult owing to the high oxidation potential of a σ bond. Therefore, the oxidation of a σ bond is generally impossible unless the oxidation potential of the σ bond is
lowered by some factor such as bond strain. On the other hand, the removal of an electron from a π orbital, that is, the oxidation of a π bond, is much easier than the oxidation of a σ bond, since the oxidation potential of a π bond is generally sufficiently low.

The simplest π-bond system is a mono-alkene and one of the most typical reactions
is anodic allylic substitution. I have already mentioned above the high stereoselectively
found in this reaction.

In the case of a conjugated diene, the removal of an electron from the diene at the anode is much easier than that for a mono-alkene due to the fact that the oxidation potential is lower. Mainly 1,4-addition to the diene takes place (equation (9)).

$$R \diagdown \diagup R' \xrightarrow{-e} R \overset{+\bullet}{\diagdown\diagup} R' \xrightarrow{2\ Nu,\ -e} R \diagdown \underset{Nu}{\diagup} \overset{Nu}{\diagup} R' \tag{9}$$

Nu = nucleophile

Although the removal of an electron from a conjugated triene is expected to take place much easier than that from a diene, the oxidation of a straight-chain triene seems practically useless owing to the possibility of the formation of a complex mixture of products.

On the contrary, the removal of an electron from a cyclic triene such as cycloheptatriene (CHT) is much easier due to the formation of a tropylium cation as the intermediate and also because trapping of the intermediate with MeOH forms 7-MeO-CHT. The electrosynthesis of 7-MeO-CHT is characterized not only by the high yield but also by its being very simple and easy to perform. Since 7-MeO-CHT is an equivalent compound to the tropylium salt, this easy synthesis of 7-MeO-CHT is very important for making progress in the synthetic chemistry of seven-membered aromatic compounds. As shown in equation (10), 7-MeO-CHT is easily transformed into tropone acetal through thermal rearrangement followed by oxidation at the anode:

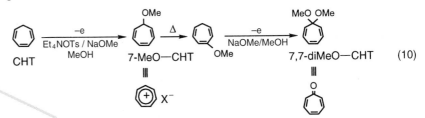

$$(10)$$

The reactions shown above are promoted by the removal of an electron from a π orbital of an unsaturated system where no functional group is located. The oxidation of an unsaturated system substituted with an acetoxyl group, for instance, has also been found to be important as the electrosynthetic reaction

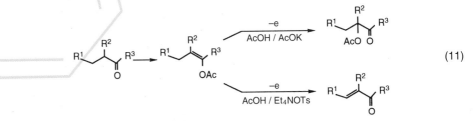

$$(11)$$

The reactions shown in equation (11), which are characterized by the introduction of a substituent at the α-position of a ketone and also the formation of an enone, are very useful for the electrosynthesis of a variety of compounds. The reactions are, for instance, very effective for achieving transposition of a carbonyl group of a ketone from its original position (the 1-position) to the 2-, 3- or 4-position:

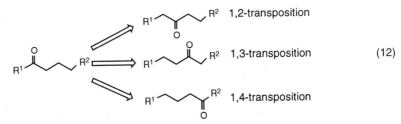

$$(12)$$

Although the removal of an electron from the aromatic ring system is not always impossible, the oxidation of benzenoid compounds is rather useless in electrosynthesis. On the other hand, the removal of an electron from hetero-aromatic compounds such as furan takes place rather easily at the anode and this oxidation has been applied extensively to the synthesis of compounds containing the skeletons of cyclopentenone, pyrone and pyridine. The most typical and successful result is the synthesis of a flavouring agent called Maltol:

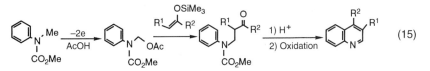

(13)

An electron can be removed from the orbital of the non-bonding electron pair located on the oxygen atom, though the direct oxidation of aliphatic alcohols or ethers does not lead to the development of a useful electrosynthesis. On the other hand, glycols are easily oxidized to give the corresponding carbonyl compounds.

Removal of an electron from the non-bonding electron pair located on the nitrogen atom results in the development of a variety of new reactions that had never been known in organic chemistry. The general pattern of the newly found reaction is the substitution reaction at the α-position of an amine and it is a powerful method in electrosynthesis.

The principle of the new method is that the amine is stabilized by its transformation to the corresponding carbamate and it is oxidized at the anode in methanol. The intermediate formed by the removal of an electron from the nitrogen atom of the carbamate is successfully trapped by methanol and the product is isolated as the carbamate methoxylated at the position α to the nitrogen atom:

(14)

Reaction of the α-methoxylated carbamate with a Lewis-acid catalyst forms a cation intermediate and its reaction with a variety of nucleophiles such as silyl enol ether, allyl silane, isonitrile, lithium acetylide, the Grignard reagent, furan and trialkyl phosphite is a powerful means for synthesizing a variety of compounds containing a nitrogen atom. Examples are summarized in Figure 5.

Although the oxidation of an amine itself is not useful as a synthetic reaction, the electro-organic methoxylation of N,N-dimethylaniline at its methyl group is an exceptionally useful reaction in electrosynthesis. For example, this method has been applied successfully to the synthesis of the quinoline skeleton:

(15)

In contrast to the oxidative reaction, the donation of an electron from the cathode to the σ orbital of a single bond is not always difficult. One of the typical reactions is the reductive cleavage of a σ bond between a carbon atom and a hetero-atom followed by the formation of a new carbon–carbon bond. The formation of a double bond promoted by the reductive 1,2-elimination of two functional groups is shown below as a

Figure 5
Reactions of the α-methoxylated carbamate with a variety of nucleophiles such as silyl enol ether, allyl silane, isonitrile, lithium acetylide, the Grignard reagent, furan and trialkyl phosphite are powerful methods for the synthesis of a variety of compounds containing a nitrogen atom. (Z = CO$_2$Me)

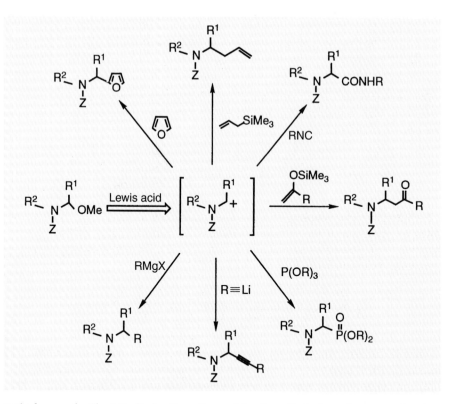

typical example. The 1,2-elimination of a combination of a hydroxyl group and a phenylthio group, for instance, results in the elongation of an aldehyde to a higher homologue of the aldehyde:

$$RCHO \xrightarrow{-\overset{OMe}{\underset{SPh}{<}}} R\overset{OH}{\underset{SPh}{|}}{OMe} \xrightarrow{+e} R\diagdown \overset{OMe}{} \xrightarrow{H_3O^+} RCH_2CHO \tag{16}$$

The donation of an electron from the cathode to a π orbital, that is, the reduction of a π bond, is usually possible, provided that the π bond is not an isolated unsaturated carbon–carbon bond. The most typical pattern of reaction belonging to this group is the formation of a new bond promoted by the reduction of a π bond. The reductive coupling of a carbonyl group with a variety of unsaturated systems is the typical example. The coupling with an intramolecular double bond, for instance, takes place at high yield with remarkable regioselectivity and stereoselectivity (equation (17)). This type of coupling is unique to electrosynthesis.

$$R\overset{O}{\overset{\|}{\diagup}}(CH_2)_n\diagup\diagdown \xrightarrow{+2e} \quad (17)$$

It is rather surprising that, besides a double bond, a triple bond, a cyclic triene system and even an aromatic ring are able to couple with the carbonyl group.

In contrast to the intramolecular coupling shown above, the intermolecular coupling of a carbonyl group with an unsaturated bond needs a special device for the reaction conditions. Namely, the intermolecular coupling of a carbonyl group with a mono-alkene, diene, cyclic triene or vinyl silane is successfully promoted by using carbon fibre as the material of the cathode:

$$R^1 \diagdown + R^2 \overset{O}{\underset{}{\diagup}} R^3 \xrightarrow[\text{carbon} \atop \text{fibre cathode}]{+2e} R^1 \diagup\diagdown\diagup\overset{OH}{\underset{R^2}{\diagup}} R^3 \qquad (18)$$

The intermolecular coupling of an allylic alcohol with a ketone takes place with remarkably high regioselectivity and stereoselectivity and this coupling results in a new electrosynthesis of optically active 1,4-diols:

$$R^1 \overset{O}{\underset{}{\diagup}} R^2 + R^3 \diagdown\diagup\overset{R^4}{\underset{OH}{\diagup}} \xrightarrow[\text{DMF / Bu}_4\text{NBF}_4]{+e} R^1 \overset{R^2}{\underset{HO\ \ R^3\ \ OH}{\diagup\diagdown\diagup}} R^4 \qquad (19)$$

The intramolecular coupling of a carbonyl group with a cyano group is also promoted by reduction at the cathode and yields the corrresponding cyclized ketone (equation (20)). This coupling reaction is very effective for synthesis of compounds containing cyclopentanone skeletons.

$$R^1 \overset{O}{\underset{R^2}{\diagup}}\diagdown\diagup CN \xrightarrow[i\text{-PrOH}]{+e} \overset{O}{\diagup}\overset{OH\ R^1}{\underset{R^2}{}} + \overset{O}{\diagup}\overset{R^1}{\underset{R^2}{}} \qquad (20)$$

When the non-bonding electron pair on a hetero-atom forms an onium salt, the electron-deficient hetero-atom becomes reducible at the cathode. The formation of a phosphonium ylid by the reduction of a phosphonium salt is a typical example. The formation of a phosphonium ylid requires a strong base in the usual chemical method and this strongly basic condition often causes severe difficulties in organic synthesis, whereas the electroreductive formation of a phosphonium ylid does not require the presence of any base:

$$\overset{+}{Ph_3PCH_2R^1} \xrightarrow{+e} \overset{+\ -}{Ph_3PCHR^1} \xrightarrow{R^2 \overset{O}{\underset{}{\diagup}} R^3} \overset{CH-R}{\underset{R^2 \diagdown R^3}{\diagdown\diagup}} \qquad (21)$$

In the classification of the electro-organic reactions, the second category is the indirect reaction (Figure 2). In the direct reaction, only reactants possessing oxidation or reduction potentials in the range accessible to direct electron transfer can be activated at the electrode, though the oxidation and reduction potentials of many types of organic reactant that are familiar in organic synthesis are beyond this range. On the other hand, the indirect reaction using a mediator is an effective method for promoting the electro-organic reaction of such organic reactants. The comparison of the indirect reaction with the direct one is shown in Figure 6 (exemplified by the oxidative reaction).

As I have already mentioned above, the reactant is directly activated at the electrode in the direct reaction, provided that the oxidation or reduction potential of the reactant is in the range which allows direct transfer of an electron between the reactant and the electrode. Reactants with oxidation or reduction potentials outside the range

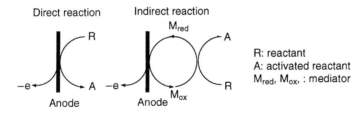

Direct reaction   Indirect reaction

R: reactant
A: activated reactant
M_{red}, M_{ox}, : mediator

Figure 6
A comparison of the indirect reaction with the direct one (exemplified by the oxidative reaction).

mentioned above are not activated by the direct reaction, but may be activated by the indirect reaction. In the indirect reaction, the reactant is not activated directly at the electrode; rather, a mediator (M_{red} in Figure 6) is activated at the electrode directly and the activated mediator (M_{ox} in Figure 6) activates (oxidation in Figure 6) the reactant in solution and M_{ox} is recovered as M_{red}. Therefore, an equimolar amount of the mediator is not necessary unless it is consumed by an unexpected side reaction.

The first indirect reaction found was the oxidation of an alcohol using the iodide ion as the mediator. Primary and secondary alcohols are oxidized to the corresponding carbonyl compounds and then the aldehyde, the product obtained from a primary alcohol, is further oxidized to an ester in the presence of alcohol.

$$R^1CHO + R^2OH \rightleftharpoons R^1\underset{OH}{\overset{|}{C}}HOR^2 \xrightarrow[-H^+]{I^+} R^1\underset{OI}{\overset{|}{C}}HOR^2 \longrightarrow R^1\underset{O}{\overset{||}{C}}HOR^2 + HI \qquad (22)$$

Although a variety of indirect reactions has been found, only the transformation of an oxime to a nitrile is shown in Figure 7. This is a unique example in which both oxidation at the anode and reduction at the cathode play important roles in one reaction. The oxime is oxidized to a nitrile oxide by the activated mediator (X^+, X is a halogen) and the nitrile oxide is reduced directly to a nitrile at the cathode.

When two types of appropriate mediator are used concertedly in one reaction system, their combined reactivity is greater than the reactivity of each mediator alone. In the indirect oxidation of an alcohol, for instance, the anodic potential required is higher than 1.6 V versus the standard calomel electrode (SCE) when thioanisol is used alone as the mediator. However, when the combination of sulfide and bromide is used as the mediator, the alcohol is oxidized at the oxidation potential of the bromide ion, that is, 1.1 V versus the SCE. Needless to say, an alcohol is not oxidized when bromide alone is used as the mediator. This reaction system is called a double-mediatory system and it is shown schematically in Figure 8.

The third category of electro-organic reactions is the electrochemical generation of active reagents, that is, the use of electrogenerated reagents. A typical example of an electrogenerated reagent is the electrogenerated base called EGB. The cathodic reduction of pyrrolidone in an aprotic solvent in the presence of a tetraalkylammonium salt forms the pyrrolidone anion having the corresponding tetraalkylammonium ion as the counter cation (equation (23)). This anion exhibits very unique reactivity as a base.

Figure 7

The transformation of an oxime into a nitrile is a unique example in which both oxidation at the anode and reduction at the cathode play important roles in one reaction.

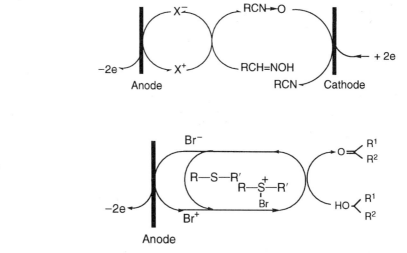

Figure 8

When the combination of sulfide and bromide is used as the mediator, alcohol is oxidized at the potential for oxidation of the bromide ion. This reaction system is called a double-mediatory system.

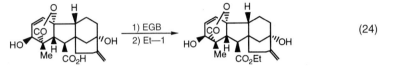

$$(23)$$

EGB

For instance, when a carboxylic acid is neutralized with the pyrrolidone anion in an aprotic solvent and then alkyl halide is added to the solution at room temperature, the corresponding ester is formed with a quantitative yield. Although the esterification of a carboxylic acid is generally an easy reaction, the esterification of a complex carboxylic acid such as gibberellic acid is not always easy. The method using the pyrrolidone anion is, however, always highly effective for the esterification of any type of carboxylic acid, including gibberellic acid.

$$(24)$$

When this method is applied to an ω-halocarboxylic acid, the corresponding lactone is selectively formed by intramolecular esterification.

$$Br-(CH_2)_{15}-CO_2H \xrightarrow{\text{EGB}} (25)$$

Besides the generation of active reagents, the electro-organic reaction promotes a unique chain reaction (Figure 9). In Figure 9, an A^- anion is formed when one electron is donated to AX at the cathode. Thereafter, an intermediate anion AB^- is formed by the reaction of A^- with an electrophile B. Finally, AB^- abstracts a proton from AH to yield the product ABH and regenerate the first anion A^- at the same time. Therefore, if one electron is donated to the system, this chain reaction must continue until all amounts of B and AH have been consumed. Typical examples of the combination of AH and AX are (i) $AX = CCl_4$ and $AH = CHCl_3$ and (ii) $AX = CCl_3CO_2CH_3$ and $AH = CHCl_2CO_2CH_3$ with $B = RCHO$. The chain length actually achieved is over 100. This chain reaction is remarkably effective for the stereoselective synthesis of carbohydrates.

The fourth category of electro-organic reactions is the reactions promoted by chemically reactive electrodes. As I have mentioned previously, the role of the electrode is generally just as a donor or acceptor of an electron and it is not involved in the reaction as something like a reagent. On the other hand, some electrodes made of chemically reactive materials promote unique reactions. One of the typical examples is shown below. When a chemically stable material such as platinum or carbon is used as the

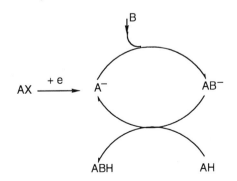

Figure 9
The electro-organic reaction promotes a unique chain reaction.

material of the electrode, the electroreduction is generally not an effective way to reduce an aliphatic carboxylic ester. The ester is, however, effectively reduced to the corresponding aldehyde or alcohol when the reduction is carried out with an electrode made of Mg in the presence of a proton donor such as *t*-butyl alcohol (equation (26)). It is clear that magnesium positively participates in the reduction in this reaction, though the mechanism is still unknown.

$$RCO_2Me \xrightarrow[\text{Mg anode and cathode}]{+\ e,\ t\text{-BuOH}} RCHO \text{ or } RCH_2OH \qquad (26)$$

The promotion of reactions with a chemically reactive electrode such as the Mg electrode will certainly be the most important aspect of electrosynthesis in the future.

Further Reading

1. T. Shono, *Electroorganic Chemistry as a New Tool in Organic Synthesis* (Springer-Verlag, Berlin, 1984).
2. T. Shono, *Electroorganic Synthesis* (Academic Press, New York, 1991).
3. H. Lund and M. M. Baizer, *Organic Electrochemistry*, Third edition (Marcel Dekker, New York, 1991).

Microwaves in Chemical Synthesis
D. Michael P. Mingos and David R. Baghurst

Introduction

As we approach the fiftieth birthday of the microwave oven, it is interesting to reflect that, although microwaves have been used in the preparation of food for around 40 years, it is only in the past decade that they have been exploited in the chemical laboratory. Legend has it that Percy Spencer discovered the influence of microwaves on foodstuffs in the 1940s when he accidentally leant against an open waveguide (a rectangular tube used to transmit microwaves) and the candy in his pocket melted! Early attempts to find industrial applications for the heating effect in areas as diverse as hardening of wood, sterilization of soil and blanching of mushrooms were largely unsuccessful and only the introduction of a restaurant microwave oven in 1955 generated any enthusiasm for the new technique. The home oven followed in the late 1960s and early 1970s when Japanese manufacturers Hitachi, Matushita, Toshiba and Sharp entered the market. By 1977, Sharp had produced some two million microwave ovens for domestic use.

The potential for microwaves to remove moisture had been recognized by this time and was used in a 150 kW (more than 150 times the power of a domestic oven) bacon dryer and pasta dryer, in which microwave drying was combined with conventional hot-air drying to reduce the processing time by 90%. Industrial applications of microwave heating included tempering meat (raising the temperature to -7 to -4 °C for slicing, dicing and re-packaging), curing and vulcanizing rubber, curing bacon, proofing doughnuts, curing wood and drying polymers.

Microwaves and Solutions

Application of microwave-heating techniques to the preparation of samples for analysis, the most developed chemical area to date, began in the mid-1980s. Microwaves have the ability to heat and dissolve small samples of biological and mineralogical specimens in strong acids such as nitric and hydrochloric acids quickly and cleanly. Special vessels made of Teflon, a microwave-transparent plastic, have been designed specifically for this work. They ensure that contamination from the laboratory environment is kept to a minimum, improving analytical results. The promising results from the analytical chemists prompted chemists in other areas to investigate other applications. Using the same Teflon vessels, organic and inorganic chemists began to uncover new microwave-enhanced chemistry.

That microwaves can be used to heat water is evident in the use of microwaves to heat food – all foodstuffs contain a greater or lesser fraction of water. It is less well known that microwaves can be used to heat other polar solvents such as methanol, ethanol and acetone. These solvents are commonly used in industry in the synthesis of organic and inorganic compounds.

The mechanism of microwave heating of solvents is related to the existence of an electrical dipole in the molecule of the solution. In water, for example, the dipole arises due to the differing affinities of oxygen and hydrogen for the available electron density and the angular shape of the water molecule. The electron density is most attracted to the more electronegative oxygen atom, which leads to there being a net dipole moment for the water molecule.

If we place the molecules in a strong electrical field, they will tend to align themselves parallel to this field. This new arrangement of molecules is rather higher in energy than the random arrangement found in the absence of the field. The molecules can be thought of as storing potential energy due to the application of the field. If the direction of the field is slowly changed, the molecules will rotate, keeping their alignment with the field.

The sum of the potential energy over all directions does not change. If we change the direction of the field more quickly, some of the molecules may be unable to remain aligned with the field's direction. The molecules may try to keep up with the field but they keep bumping into other molecules. The potential energy stored in the molecular arrangement no longer matches the applied field. In fact the excess energy is transformed into kinetic energy on collisions between the molecules and it is this effect which gives rise to microwave heating. It is noteworthy that, if the electrical field changes direction very rapidly, the molecules do not have time to react to the applied field and they remain randomly oriented. At these frequencies they are unable to interact with the applied field and no heating occurs. The microwave frequencies commonly used for heating match the time domain for which polar molecules can no longer follow the changing direction of the field. Note that the microwave-heating phenomenon is not limited to water molecules but is applicable to all polar molecules of a reasonable size.

One of the first chemical applications of microwave heating was in the field of chemical analysis. Whereas phenomenal advances have been made in analytical instrumentation, sample-preparation methods have not changed for many years. Sample preparation, in particular, digestion in acid, is a critical and rate-determining step in chemical analysis. The need to prepare large numbers of samples in less time and with greater efficiency is fostered by multi-element instruments that analyse samples in a fraction of the time needed to prepare them.

By microwave heating small samples in specially designed Teflon vessels, sample-preparation times have been reduced dramatically. In the determination of a zircon sample, for example, 100% recovery of the heavy element is achieved in only 2 h by a microwave autoclave digestion. During this time, the maximum pressure developed does not exceed 4 atm. Following 2 h heating on a conventional hotplate the sample does not digest. Traditionally recovery would be afforded by a high-pressure Krogh method, whereby pressures in excess of 40 atm are generated.

The success of the microwave sample preparation is mostly due to the high temperatures that are generated by processing the samples in sealed vessels. Under sealed conditions, the pressure inside the vessel increases as the solvent becomes more volatile. This pressure-cooker effect leads to an increase in temperature in the vessel, which enhances dissolution of the biological or mineralogical specimens. The microwave technique is clean and has been scaled up using pumped-flow systems with modified conventional microwave ovens.

The achievements of microwave sample preparation in sealed vessels led to studies first on organic and later on inorganic reactions in the same or similar vessels. Again the higher temperatures produced by carrying out the reactions under pressure almost always lead to reductions of reaction times. The only caveat to a more general statement is the possibility that the higher temperatures may encourage the decomposition of the desired products or lead to the formation of the thermodynamically more stable product in preference to the kinetic product.

In a typical example of an organic microwave reaction a mixture of 1.6 g of benzoic acid, 10 ml of 1-propanol and 0.1 ml of concentrated sulfuric acid was sealed in a Teflon bottle and heated in a microwave oven for 6 min. During this time the temperature inside the reaction vessel reached 135 °C. Following recovery, the reaction gave propyl benzoate in 79% yield. A conventional reflux reaction at 97 °C – the boiling point of 1-propanol – gave a similar 78% yield after 4 h, suggesting that the microwave procedure is some 40 times faster than the reflux procedure.

This basic methodology has now been applied to a very wide range of organic reactions. There are now more than 500 papers dealing with the acceleration of chemical reactions by microwave dielectric heating. Other classes of materials synthesized by pressurized microwave routes include the following.

(a) Organometallic compounds – a range of rhodium and iridium di-alkene dimers of this type:

have been synthesized in less than 1 min. Conventional reflux routes may take many hours. These materials are important starting materials in the synthesis of metal cluster compounds and precursors of catalysts since the alkene can readily be removed and replaced by other ligands.

(b) Zeolites – one of the challenges in zeolite chemistry is the synthesis of materials with a high silicon-to-aluminium ratio. These materials are hydrophobic and more suited to applications as catalysts in the petrochemical industry. The high Si/Al zeolite Y has been synthesized from silica, sodium aluminate and sodium hydroxide starting materials using microwave heating and has properties comparable to those of products synthesized conventionally.

(c) Minerals – small samples of minerals have been synthesized under microwave conditions that mimic the high pressures and temperatures achieved in the earth during natural synthetic processes. For example, samples of the iron and aluminium arsenates, scorodite and mansfieldite have been synthesized in sealed Teflon vessels at temperatures up to 200 °C and 15 atm pressure.

One specialized area of chemistry in which microwave heating has had a particular impact is in the synthesis of radio-pharmaceuticals. In these studies, the pharmaceuticals are selectively labelled with radioactive isotopes so that the mechanism of action and pathway of the drug through the body's metabolism can be traced by following the decay of radioactive nuclei. The nuclei used for these studies include ^{11}C, ^{13}N and ^{8}F. All these nuclei have very short half-lives (the half-life is the time taken for half the radioactive nuclei to decay) of the order of a few hours or minutes. For example, if we begin with 100% ^{11}C at time 0, by the time we have reached time 20 min, the strength of the decay signal will have fallen by 50%. Again, if we have to wait for several half-lives before the sample has been synthesized and purified so that it can be injected into a patient, we may no longer be able to detect the very weak radioactive decay. Given that the synthesis of most drugs consists of many steps (>10), we must be able to perform all transformations as quickly as possible and with as high a yield as possible.

One drug whose synthesis has been studied using microwave-heating techniques is temozolimide. This drug exhibits promising activity towards gliomas and malignant melanoma and has been labelled with the isotope ^{11}C either in the carbonyl or in the methyl position (shown with asterisks below). The final step in the synthesis of temozolomide is the reaction of radiolabelled methyl-isocyanide and a diazo precursor:

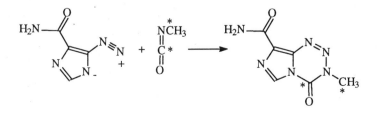

Figure 10
Cycloaddition of A gives a high yield of B.

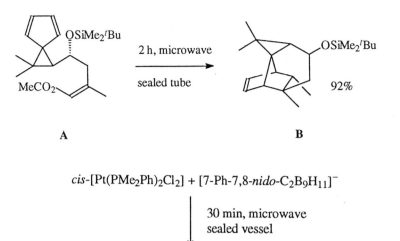

A   B

Figure 11
The synthesis of platina-carboranes by microwave sealed-vessel reactions.

$$cis\text{-}[Pt(PMe_2Ph)_2Cl_2] + [7\text{-}Ph\text{-}7,8\text{-}nido\text{-}C_2B_9H_{11}]^-$$

$$\downarrow \quad \text{30 min, microwave sealed vessel}$$

$$1\text{-}Ph\text{-}3,3\text{-}(PMe_2Ph)_2\text{-}3,1,11\text{-}PtC_2B_9H_{10} + 11\text{-}Ph\text{-}3,3\text{-}(PMe_2Ph)_2\text{-}3,1,11\text{-}PtC_2B_9H_{10}$$

This reaction is performed in dimethyl formamide at 60–80 °C. By comparing the yield and product distributions of microwave and conventionally induced reactions it has been found that the microwave technique leads to dramatic improvements. For example, whereas the conventional process gave a yield of only 30% in 60 min, 45% yield was achieved in 20 min (this corresponds to a six-fold increase in the radiochemical yield). Much smaller quantities of side products are produced in the microwave work, possibly due to the much faster initial heating rate achieved in microwave heating.

Although many earlier reports of microwave-accelerated reactions were concerned with microwave aspects of the work, more recently there have been several examples which mention the application of microwave-heating effects as part of a larger research strategy. The routine use of microwave-heating apparatus as part of everyday laboratory equipment is becoming more widespread and is set to increase as the quality and suitability of commercially available apparatus increase.

For example, in studies on cycloaddition routes to tricyclo[5.4.01,7.02,9] undecanes microwave heating has been shown to be applicable to two reactions that contribute to a larger total-synthesis project. In the first of these reactions, the cycloaddition of A had proven difficult and only a 10% yield of adduct B was obtained after 24 h of refluxing in toluene, whereas at higher temperatures decomposition predominated (Figure 10). By microwave heating a 0.05 M solution of A in toluene in the presence of 1 mol% equivalent of hydroquinone, excellent yields of B (92%) were recovered after only 2 h.

Similarly, performance of the reaction indicated in Figure 11 both by conventional reflux techniques and by microwave techniques has been studied. Whereas no insoluble quantities of material were recovered by using the former method, the latter produced excellent yields of the novel isomeric platina-carboranes. The reaction was completed in 30 min at a pressure of 10 atm with a final recorded temperature of around 135 °C.

Although microwave reactions in pressurised vessels provide the most impressive examples of accelerated chemistry, it has been found that even reactions at atmospheric pressure can be enhanced by using microwave heating. A simple experiment can be used to illustrate the general point. First we take a mug of water and place it in a microwave oven exposed to full power until it starts to boil. If we carefully remove the mug from the oven so that the boiling stops, we can then add a spoonful of instant coffee. The mixture suddenly froths up and the coffee appears to boil again. What we have observed is the phenomenon of microwave superheating. Polar solvents heated in a microwave oven almost always boil at temperatures above their accepted boiling points. Although

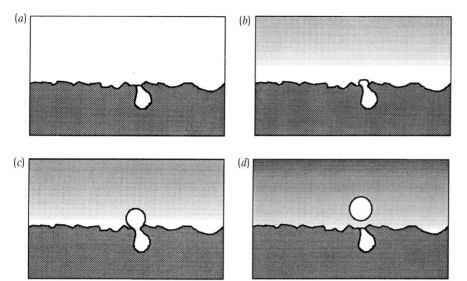

Figure 12
The mechanism of microwave superheating. (*a*) A gas bubble is trapped at the surface of the vessel. (*b*) As the solvent heats up (growth in the shaded region) the bubble expands. (*c*) The temperature continues to rise and the bubble grows to a critical size. (*d*) The temperature at the surface reaches the boiling point and the bubble detaches.

the effect on water and aqueous solutions is in general rather small, an increase by a maximum of 8 °C (superheating) in the expected boiling point, the effect on organic solvents can be more dramatic and increases in the boiling point of the order of 30–35 °C have been reported for the common solvents methanol, ethanol and acetonitrile.

In conventional heating, boiling begins at the surface of the reaction vessel where there are many tiny pits and scratches. The vessel walls are always significantly hotter than the refluxing solution itself and this encourages the formation of bubbles of solvent vapour in the pits or nucleation sites. When the bubble reaches some critical volume it detaches from the surface and boiling is observed. In microwave heating the vessel walls are generally cooler than the solution itself since they are made from poorly microwave-absorbing materials such as glass and Teflon. The areas of solvent close to the vessel walls are too cold to encourage the formation of bubbles and as a result the bulk of the solvent can be heated to temperatures much higher than the conventional boiling point before stable boiling is attained. This simple model (Figure 12) accounts for the rather poor superheating of water compared with organic solvents since the latter wet the surface of the vessel much more effectively, flooding the majority of the nucleation sites.

Microwave superheating has been used to synthesise the precursors of inorganic catalysts such as $RuCl_2(PPh_3)_3$. The conventional reaction between $RuCl_3$ and PPh_3 in refluxing methanol gave a yield of 74% of the complex after 3 h. Using open-vessel microwave-reflux techniques, an 85% yield of the purified product was recorded.

Microwaves and Solids

In the 1980s it was recognized that water could be removed from the zeolite molecular sieves that are often used as drying agents in chemical laboratories. In this manner the zeolites were regenerated quickly and for this reason alone many microwave ovens have found their way into chemical laboratories. For some zeolites and other solids though, more dramatic microwave heating effects were noted. Zeolite 13X, for example, melts when it is placed in a microwave oven and irradiated for short periods of time. Melting leads to destruction of the zeolite material and corresponds to temperatures in excess of 300 °C – much higher than the boiling point of water.

Although the aluminium–silicon–oxygen framework of zeolites is rigid, the structures also contain high percentages of relatively mobile cations, such as sodium, potassium, caesium and calcium, which sit in the channels created by the open framework.

Table 1	The effect of microwaves on a range of solids					
	A			B		
Chemical	Temperature (°C)	Time (min)		Chemical	Temperature (°C)	Time (min)
Al	577	6		CaO	83	30
C	1283	1		CeO_2	99	30
Co_2O_3	1290	3		CuO	701	0.5
CuCl	619	13		Fe_2O_3	88	30
$FeCl_3$	41	4		Fe_3O_4	510	2
$MnCl_2$	53	1.75		La_2O_3	107	30
NaCl	83	7		MnO_2	321	30
Ni	384	1		PbO_2	182	7
NiO	1305	6.25		Pb_3O_4	122	30
$SbCl_3$	224	1.75		SnO	102	30
$SnCl_2$	476	2		TiO_2	122	30
$SnCl_4$	49	8		V_2O_5	701	9
$ZnCl_2$	609	7		WO_3	532	0.5

Note:

Column A: 25 g samples in a 1 kW oven (2.45 GHz) with a 1000 cm³ vented water load.
Column B: 5–6 g samples in a 500 W oven.

When the sodium cations in zeolite 13X are changed to potassium or calcium, the rate of microwave heating first falls and then virtually disappears. This led chemists to examine which solids could be heated by microwaves and to the discovery that any solid with mobile charge carriers such as electrons or cations can absorb microwave energy. This has led to the development of a wide range of microwave-induced reactions between solids for which one or more components of the reaction mixture absorb microwaves strongly.

The effect of microwaves on a range of solids in microwave ovens is summarized in Table 1. Poor electrical conductors such as silica and alumina absorb very little microwave energy; room-temperature semiconductors such as graphite and nickel oxide are rapidly heated to very high temperatures. The origin of the microwave-heating effect is the interaction of the oscillating electrical field component of the electromagnetic field and the solid. Materials such as graphite that have mobile electron charge carriers interact strongly with the oscillating electrical field of the microwave radiation. The electrons are accelerated but there is a resistance to the flow of current in the material, which leads to heating. Solids such as zeolite 13X and the fast-ion conductor β-AgI that have loosely held and relatively mobile cations in their structures behave in a similar fashion. In insulating silica and alumina there are no free charge carriers. In order for them to absorb microwave energy they must be heated to temperatures in excess of 600 °C, by which time there are enough charge carriers available to cause some absorption. At higher temperatures the lattices begin to break down as areas of mobile cations and anions are formed.

Microwave heating attracted interest from ceramicists studying the transformations which materials undergo when they are heated to temperatures in excess of 1200 °C. At these temperatures, ceramic oxides tend to shrink and densify, producing harder and/or stronger materials (a process known as sintering). Microwave-heating experiments on ceramics have led to the discovery of differences between the technique and conventional furnace heating. In conventional furnaces heat is transported first of all to the surface of the material as the furnace walls are heated. As heating continues the temperature inside the sample gradually increases to match the surface temperature and equilibrium is achieved. In microwave heating an inversion of this temperature profile is possible, for many materials are effectively heated from the centre. Often, for the area of the sample's surface, the rate of absorption of microwaves leading to heating cannot compensate for the loss of heat from the sample. This effect is illustrated in Figure 13.

Inversion of the heating profile can lead to a reduction in the thermal stress in the sample. This can lead to a reduction of cracking and higher densities. Lithium hydride was sintered to more than 97% of the maximum theoretical density with less than 2% of open porosity. Similarly, $YBa_2Cu_3O_7$ has been sintered to more than 90% of the theoretical density, whereas conventional heating achieved at best only 85% of the theoretical density. Microwave-sintered samples of TiB_2 (with 3% CrB_2) have been shown to have significantly greater hardness and fracture toughness than do conventionally prepared samples.

Although Figure 13 shows a temperature inversion in a microwave-heated sample, not all samples will behave in this way. A temperature profile similar to that under conventional heating as well as an even temperature profile throughout the sample during heating are both possible. Among the factors influencing the type of heating profile are the size and geometry of the sample, the type of microwave apparatus, the complex effects of microwaves and the thermal properties of the sample. One interesting effect that has been reported is the resonant absorption of microwaves by cylindrical samples with specific dimensions.

For microwave radiation of frequency 2.45 GHz (the frequency of domestic microwave ovens world-wide) the free space of the radiation is approximately 12 cm. The wavelength of the radiation in a particular material depends on the material's microwave-dielectric properties; however, it can easily be appreciated that the wavelength of the radiation can approach the physical dimensions of the sample under investigation. When this happens the rate of heating of the sample can be dramatic because the microwaves effectively become trapped within the sample and resonant absorption occurs. If a slightly larger or smaller sample is heated then there will be no resonance and a modest rate of heating will be observed. This effect, which falls off as the size of the sample increases, is illustrated in Figure 14 for water samples. It was recorded by Ted Davis and his group at the University of Minnesota. The effect will also be observable and, indeed, more pronounced in solid samples, where the equilibration of temperature by random collisions that occurs in a mobile liquid is not possible.

Other chemists have used the microwave absorption of solids to accelerate chemical reactions in the solid state. Of particular interest is the synthesis of the mixed-copper-oxide high-temperature superconducting ceramics. These materials lose all electrical resistance below a certain temperature known as the critical temperature and have tremendous potential. The diverse applications proposed to date include low-loss power cabling and high-speed trains. Using microwave heating, an intimate mixture of copper oxide, yttrium and barium nitrate has been transformed into the superconducting phase of $YBa_2Cu_3O_7$ in 30 min in a microwave oven. Conventionally the synthesis involves lengthy calcination and annealing cycles which can take as long as 24 h.

Other important materials that have been synthesized using microwave techniques are the ternary chalcopyrites $CuInS_2$ and $CuInSe_2$ which are small-band-gap semiconductors with potential applications in solar cells. In a typical experiment stoichiometric quantities of the elements Cu, In and S are ground together using a pestle and mortar. The mixture is placed in a quartz tube and sealed under vacuum. The sample is exposed to microwave radiation for a total of 3 min in 1-min bursts with shaking in between to redistribute the sulfur which tends to sublime away from the reaction mixture. The product is obtained as a crystalline, blue–grey powder.

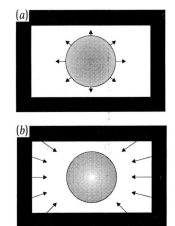

Figure 13

Microwave versus conventional heating. (a) In microwave heating the material starts to heat at the centre. The temperature of the oven increases as heat is lost from the sample's surface. (b) In conventional heating the material starts to heat from the surface as the furnace's temperature increases.

There are several noteworthy features of this reaction.

(a) Finely divided Cu and In metal particles can be safely placed in a microwave oven and heated. This is contrary to the popular view that metals should never be exposed to microwave energy and is a general feature of finely divided metal powders (larger metallic objects such as knives and forks do cause potentially destructive arcing).

(b) In a similar conventional process the quartz tube has to be heated slowly. The volatile sulfur leads to a build-up of pressure in the tube and may cause an explosion if the rate of heating is too fast. In the microwave reaction the sulfur reacts with the metals at the high temperatures generated almost instantaneously by the microwave-heating effect.

(c) In similar reactions stable plasmas with colours appropriate to the particular volatile non-metal in use have been obtained. The plasma provides a unique way of measuring the extent of the reaction since it disappears on completion of the reaction.

It is a general feature of the microwave heating of solids that some form of enhancement of the rate of processing is found. Studies on the sintering of Al_2O_3/TiC composites have shown that it is possible to achieve a density of 93% at 1750 °C, whereas conventionally a temperature of around 1880 °C would be required. Comparative data for microwave and conventional heating of alumina indicates that densification occurs more rapidly in a microwave field. The apparent activation energy for microwave sintering of alumina (160 kJ mol^{-1}) is less than a third that observed for conventional sintering (575 kJ mol^{-1}). Normal grain growth in dense fine-grained Al_2O_3–0.1% MgO has been studied both under conventional-furnace and under 28-GHz-microwave-furnace annealing conditions. The microstructural changes that occurred were the same for both sets of samples: bubble microstructures were observed and the aspect ratios and shape factors did not change during the annealing cycles. The kinetics of grain growth were greatly increased by the microwave annealing. For example, the rate at which grains grew at 1500 °C in the microwave furnace was the same as the rate at 1700 °C in the conventional furnace. The activation energy for growth of grains was reduced from 590 to 480 kJ mol^{-1}.

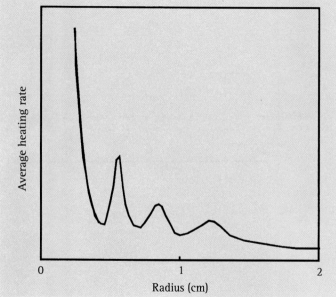

Figure 14
The measured rate of heating of cylinders of water as a function of their radius.

Some doubt has been cast on the interpretation of some of the results, mostly due to difficulties associated with measuring temperatures in the microwave field. Extreme care has to be taken to ensure that thermocouples do not interact directly with the electromagnetic field, leading to large and seemingly random fluctuations in the recorded temperature. Other remote-sensing techniques such as pyrometry can provide a poor indication of the reaction temperature in microwave experiments since they give only the temperature at the surface of the specimen. This remains an area of some controversy since the reason for those fluctuations is not understood but methods of temperature measurement have been improved by using fluoro-optical thermometers that contain no metallic parts and do not interact with the field.

The microwave technique is not without its difficulties. One of the pre-requisites for the successful extension of microwave-heating techniques to widespread chemical applications is the ability to control the reaction conditions and particularly the reaction temperature accurately and repeatably. One important effect of microwaves in this context is the thermal runaway which is illustrated in Figure 15. Whereas a sample of silica is heated relatively slowly by microwave heating, a similar sample of another inorganic oxide, chromium oxide, at first heats rather slowly and then suddenly the temperature begins to rise rapidly and uncontrollably. The temperature at which this effect takes place is called the critical temperature and it has been shown to depend not only on the material but also on the geometry of the sample. In some ways the effects of thermal runaway are beneficial in that samples can be heated to high temperatures using very small power levels. On the downside, overheating the reaction mixture can result in the formation of unwanted products, promote decomposition or even lead to melting of samples.

Another potential difficulty in microwave heating is the formation of hot spots in some samples. The electrical field distribution in a microwave oven is rather complex and rarely even. A combination of poor homogeneity of the field, poor packing of a sample and low thermal conductivity can lead to differences in temperature within the sample of several hundred degrees. Some areas may overheat while other areas will not react. There are several solutions to this problem, including the turntable seen in most domestic microwave ovens and the incorporation of mode stirrers or fan-shaped metallic paddles to stir up the magnetic field inside the oven. Moving to a higher frequency of microwave radiation (28 GHz rather than the normal 2.45 GHz) can also lead to improvements because the free-space length is reduced (from 12 cm to around 1 cm).

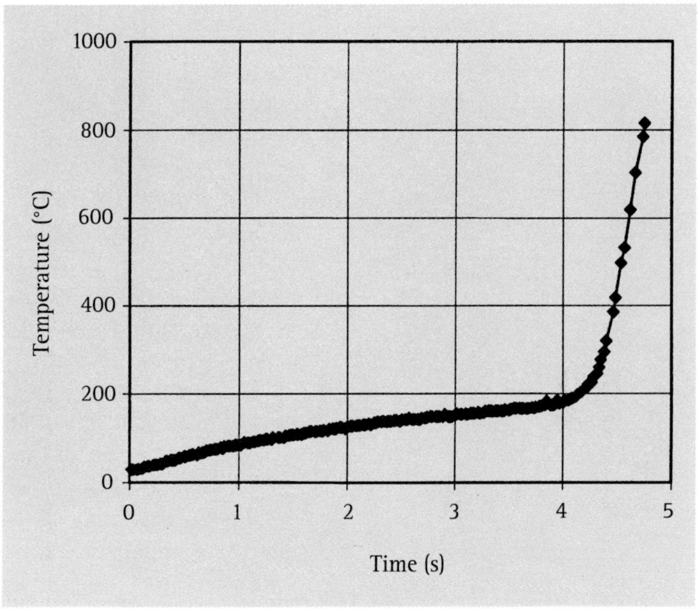

Figure 15
Thermal runaway in a sample of Cr_2O_3.

Summary

The microwave-heating technique provides the chemist with a new way of heating chemical reactions. Owing to differences between the mechanism of microwave heating and those of more conventional heating techniques, such as the use of oil baths and furnaces, improvements to synthetic procedures have been discovered. Applications for microwave techniques have been identified in many diverse areas of chemistry, including analytical chemistry, organic and inorganic chemistry, ceramics, pharmaceutical chemistry and catalysis.

Further Reading

1. R. Abramovitch, *Organic Preparations and Procedures Int.*, **23**, 685 (1991).
2. R. Gedye, F. Smith and K. Westaway, *J. Microwave Power Electromagnetic Energy*, **26**, 3 (1991).
3. D. M. P. Mingos and D. R. Baghurst, *Chem. Soc. Rev.*, **20**, 1 (1991).
4. D. Palaith and R. Silberglitt, *Am. Ceramic Soc. Bull.*, **68**, 1601 (1989).
5. W. H. Sutton, *Am. Ceramic Soc. Bull.*, **68**, 376 (1989).

Ultrasound in Chemical Synthesis
Paul D. Lickiss

Introduction

Ultrasound can be defined as sound with a frequency above that of the normal range heard by the human ear, i.e. above about 16 kHz. Its application to chemical synthesis (sonochemistry) has become widely popular only within about the last 15 years but the availability of such high-frequency sound has a much longer history. The first useful ultrasonic device was a whistle, developed in the early 1880s by Galton to investigate the human hearing frequency range. Such a whistle is a transducer, i.e. it converts one form of energy to another, in this case the motion of a gas into sound. A more useful transducer for the modern chemist relies on the piezoelectrical effect discovered by the Curies in 1880. This effect arises when some crystalline solids, for example quartz, are subjected to a sudden compression which causes a potential difference across opposite faces of the material. Modern ultrasonic generators rely on the opposite of this effect; a rapidly alternating potential difference is applied to the piezoelectrical material, which causes an oscillation in its dimensions, thus changing electrical into vibrational or sound energy.

If the average person is asked what they think ultrasound or ultrasonics might be used for the answers will probably not include chemistry but will cover such widely differing areas as foetal scanning in pregnancy, other medical diagnostic techniques, animal navigation and communication such as echo location by bats and dog whistles, the detection of cracks or flaws in solids and underwater echo location or SONAR. These applications tend to rely on the use of very high frequencies, typically in the 1–10 MHz range, and fall under the general umbrella of what is known as 'diagnostic' ultrasound. The important factors for diagnostic ultrasound are the high frequency used, which can give high-resolution data, and the low power which is unlikely to cause significant physical changes in the medium through which it passes, which is of particular importance in medical applications. In contrast, applications of ultrasound to chemically important systems in which physical and chemical changes are desired require relatively low-frequency ultrasound, usually in the range 20–100 kHz, and higher power. The range of frequencies used for sonochemistry is usually known as 'power' ultrasound and this is shown relative to other frequencies in Figure 16.

Equipment

There are two main types of ultrasonic equipment used by the synthetic chemist. One is commonly called an ultrasonic bath and comprises a water-filled stainless steel bath with ultrasonic transducers attached underneath such that when the electrical power is applied the high-frequency sound waves propagate through the water and any vessels that may be immersed in it, Figure 17. Ultrasonic baths are found in many laboratories and have traditionally been sold and used as cleaning baths, for the passage of ultrasound through a liquid leads to surface-cleaning effects which can be used for cleaning soiled glassware. This well-known cleaning effect can also be used to advantage in cleaning the surfaces of potentially reactive chemicals, see below.

The second common method by which ultrasound may be introduced into a system of interest to the chemist is via a probe or horn as shown in Figure 18. In this case the ultrasonic transducer is attached to a probe, usually made of titanium, which can be immersed directly into a reaction mixture. This type of equipment is also widely available in biochemical laboratories because it is commonly used to cause disruption of cell

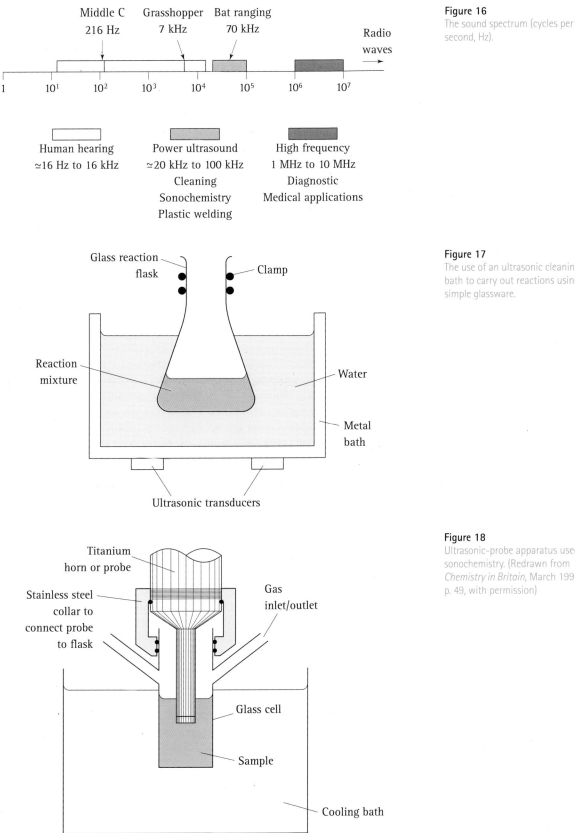

Figure 16
The sound spectrum (cycles per second, Hz).

Figure 17
The use of an ultrasonic cleaning bath to carry out reactions using simple glassware.

Figure 18
Ultrasonic-probe apparatus used for sonochemistry. (Redrawn from *Chemistry in Britain*, March 1996, p. 49, with permission)

walls. The large amounts of energy introduced into the reaction vessel using a probe mean that external cooling is required in order to stop boiling of the solvent used for the chemical reaction.

The ultrasonic bath has the advantages that it is relatively inexpensive and readily available, that it does not introduce anything new into a reaction mixture and that many types of ordinary reaction flask can be introduced into the bath without specialized equipment being required. Unfortunately, the amount of energy introduced into the reaction mixture is relatively low because the sound has to propagate through the base of the bath, the water in the bath and the wall of the reaction vessel before it can interact with the reaction mixture. The probe system can introduce a much greater intensity of ultrasound into the reaction vessel but it is relatively expensive, some specialized glassware is usually required, particles of the titanium horn may enter the reaction mixture and efficient cooling is needed. It is usually not possible to vary the frequency of the sound produced either by the bath or by the probe very much but this is not a problem for most chemical syntheses.

Physical Effects of Ultrasound

The question of what happens when power ultrasound in the frequency range 20–100 kHz passes through a reaction solution or any other liquid is one that is critical to sonochemistry. There is no direct interaction between the sound wave and any molecular vibrations because the wavelength of the sound is very much longer than any molecular dimension, for example, the wavelength of sound with a frequency of 20 kHz in water at room temperature is about 7.5 cm. The sound waves are longitudinal, that is particles vibrate in the same direction as that in which the wave is propagating and they comprise areas of alternating high and low pressure, compressions and rarefactions. If the intensity of the applied sound is high then the negative pressure exerted in the rarefaction part of the cycle may be sufficient to overcome the forces between molecules so that the bulk fabric of the liquid may be broken down at some weak point, such as an impurity, and microscopic bubbles or cavities are formed. Vapour or dissolved gas may then diffuse from the liquid into the cavity so that it grows. There is, of course, a compression part of the wave immediately following any rarefaction and this is likely to cause the collapse of any cavities that had been formed, more or less rapidly depending on the frequency and intensity of the sound. It is this formation and collapse of microscopic cavities, or cavitation, that gives rise to the chemical effects caused by ultrasound. The collapse of bubbles in the bulk liquid gives rise to very high temperatures and pressures, up to about 5000 °C and 10^8 Pa (approximately 1000 atm) within the bubble and molecules trapped either within the bubble or close to it will undergo reactions because of these physical extremes. If the cavity collapses at, or near to, a surface such as the wall of the vessel or some solid reagent in a reaction mixture, then it does so asymmetrically. This causes microstreaming of a jet of liquid onto the surface, which can cause the surface to be cleaned and fragments of material to be broken away from the surface and hence also causes rapid transport of material both to and from the surface. It is this surface abrasion that causes the well-known cleaning effect of ultrasonic baths and, together with the vigorous mixing caused by cavitation, it causes many of the effects of use to the synthetic chemist.

In common with more traditional synthetic methods, there are several experimental variables, apart from the intensity of the applied sound, to be considered when carrying out reactions under ultrasonic irradiation. It is not easy to vary the applied frequency of the ultrasound in a single reactor and, although frequency effects are observed, fortunately this does not usually seem to play a critical role. Very high frequencies are in fact not required, for the formation of bubbles requires a finite time and at high frequencies this time may be longer than that in the rarefaction part of the cycle, for example, at 20 kHz rarefaction lasts 25 μs (half the frequency) but at 20 MHz rarefaction last only 0.025 μs. The viscosity of the liquid through which the sound is passing is also not a very significant factor for common solvents, although cavitation in very viscous liquids is more difficult to achieve. The most important factor to consider is the bulk temperature of the reaction medium. As the temperature of the system is increased the viscosity and surface tension decrease, which allows cavitation bubbles to be formed more easily. Unfortunately, although it is easier to produce cavitation, its effects are greatly reduced, for the increase in vapour pressure on increasing the temperature has the effect of cushioning the collapse and thus the very high temperatures and pressures produced are reduced. Sonochemistry should, unlike many conventional syntheses, therefore be carried out at well below the boiling point of the solvent used, for example, the optimum temperatures for the solvents water, toluene and acetone are 35, 29 and −36 °C, respectively; their boiling points being 100, 111 and 56 °C, respectively. The content of dissolved gas in the solvent may also play a part insofar as an increase leads to there being more cavitation 'nuclei', leading to cavitation occurring more readily, but it will also lead to a cushioning effect on cavitational collapse.

There are three main classes of reaction mixture that may be significantly affected by the application of an ultrasonic field.

1. Heterogeneous, solid–liquid systems such as those found in the synthesis of many organometallic reagents containing magnesium, lithium, zinc etc. in which a solid metal reacts with organic reagents dissolved in a suitable solvent.
2. Heterogeneous, liquid–liquid systems comprising two immiscible liquids, for example water and styrene, in which the formation of an emulsion and the increase in interfacial interaction are important.
3. Homogeneous liquids such as water and alkanes or solutions.

In all these cases the effect of cavitational collapse can have several beneficial effects on reactions of synthetic importance. For example, reactions may occur more rapidly and at lower bulk temperatures, which can lead to simpler and safer synthetic routes, especially for syntheses of organometallic compounds. Shorter reaction times and lower temperatures may also lead to higher yields because the formation of by-products is often reduced. Ultrasonic activation may also cause a change in reaction pathway and hence cause different reaction products to be formed compared with normal synthetic methods by, for example, promoting a free-radical mechanism over an ionic one. Examples of all these effects are described below. The symbol))) is now commonly used to denote irradiation of a reaction by ultrasound and, although it does not give any indication of the frequency or source of the sound, it will be used in the equations below.

Heterogeneous Solid–Liquid Systems

The most widely used applications of ultrasound in synthetic chemistry so far have probably been in the preparation of organic derivatives of various metals such as lithium, magnesium, copper and zinc that have general utility. Although many compounds of this type can be prepared easily there are often problems with their synthesis due to the relatively inert layer of oxide or other unwanted material found on the surface of the metal. Such passivating layers can be broken down by chemical means or by preparing fresh metal surfaces by mechanically breaking down larger material into fine powder to increase the surface area. Unfortunately, there may still be an unpredictable time delay or induction time, between when the organic reagent is added to the metal and when a reaction takes place. There can also be problems, as the reaction progresses, with the build-up of reaction by-products such as metal halide salts on the metal surface, which again tend to protect the metal from reaction with the organic reagent. In this type of system the application of ultrasound can clean the surface of the metal before reaction starts, the particle size can be reduced such that the surface area for reaction is increased and the rate of transport of reagent to the surface and product away from the surface is increased. These beneficial effects have the effect of reducing both the time taken to initiate the reaction and the time for it to proceed to completion; extra chemical initiating reagents are not required and the temperature at which the reaction can be carried out can be reduced, which may also minimize the formation of unwanted by-products. For these reasons the application of ultrasound, often by immersing the usual reaction vessel in an ultrasonic bath, to the synthesis of common organometallic reagents is now common and has become one of the new range of methods that can be applied in synthetic chemistry. Many publications (see the list of further reading for comprehensive reviews) in this area have appeared during the last 5–10 years and only a few examples are shown below, equations (27)–(29), to give some idea of the improvements that ultrasound can give.

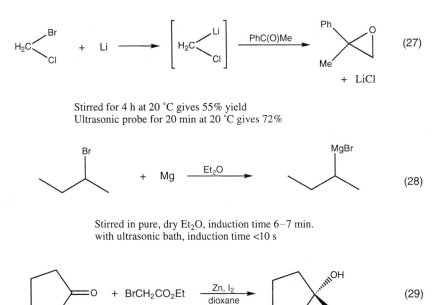

(27)

Stirred for 4 h at 20 °C gives 55% yield
Ultrasonic probe for 20 min at 20 °C gives 72%

(28)

Stirred in pure, dry Et_2O, induction time 6–7 min.
with ultrasonic bath, induction time <10 s

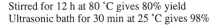

(29)

Stirred for 12 h at 80 °C gives 80% yield
Ultrasonic bath for 30 min at 25 °C gives 98%

A second type of heterogeneous solid–liquid system widely used in synthesis is one with a solid non-metal reagent, usually a metal salt or of some type, that reacts with a dissolved organic or inorganic reagent. Similar effects to those caused in the reactions with metals described above occur and these again derive from the removal of surface contaminants, the reduction in particle size and increase in mass transport both to and from the reactive surface. In addition, the breakdown of metal salts, e.g. $KMnO_4$, equation (30), into fine powders may enhance their solubility in solvents that are not traditionally used for such reagents. Another example, lithium aluminium hydride, is a commonly used reducing agent, which is normally used in ether solvents such as diethylether and tetrahydrofuran, but, if ultrasound is applied then it is possible to use non-polar solvents such as hexane, which can simplify the isolation of reaction products, equation (31). Equation (32) shows a rare example of the course of a reaction being changed completely (rather than the rate simply being increased) on applying ultrasound rather than mechanical agitation to a reaction system. Here the reaction mechanism changes from electrophilic aromatic substitution to nucleophilic aliphatic substitution on using ultrasound. This is thought to be due to the increase in contact between the Al_2O_3 and the KCN which decreases the catalytic activity of the Al_2O_3 for Friedel–Crafts-type reactions but enhances the nucleophilicity of the CN^- ion on the Al_2O_3 surface.

$$CH_3CH(CH_2)_5CH_3 \text{ (OH)} + KMnO_4 \xrightarrow[\text{5 h, 50 °C}]{\text{Hexane}} CH_3C(CH_2)_5CH_3 \text{ (O)} \qquad (30)$$

Stirred, 2.6%; ultrasonic bath, 92.8%

$$Me_3SiCl + LiAlH_4 \xrightarrow[\text{3 h, 40 °C}]{\text{Hexane,))) bath}} Me_3SiH \qquad (31)$$

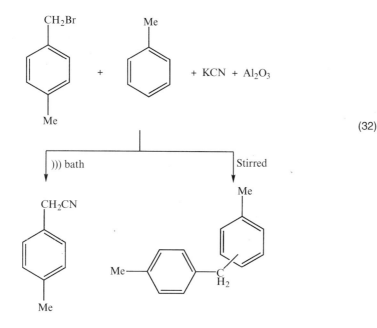

(32)

A third area of potential application to solid–liquid mixtures is the breakdown of polymeric solids so that they can dissolve in a solvent and then react with a dissolved reagent to give a product that is not otherwise readily available. For example, tin fluorides, which have a polymeric structure, are very insoluble in common solvents and are by-products from a range of organic synthetic processes. Their insolubility makes them difficult to recycle but if ultrasound is applied to them they can be converted into their corresponding bromide and chloride derivatives which are monomeric and hence soluble and easy to use, equation (33). Also silica, SiO_2, can be broken down to monomeric species much more readily under ultrasonic irradiation, equation (34).

$$[Me_3SnF]_n + n\,NaCl \xrightarrow{\text{THF}} n\,Me_3SnCl + n\,NaF \qquad (33)$$

Stirred for 3 days at 67 °C gives 50% yield
Ultrasonic bath for 10 h at 7 °C gives 74%
Ultrasonic probe for 2 h at 25 °C gives 81%

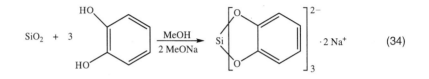

$$SiO_2 + 3 \quad \text{(catechol)} \xrightarrow[2\,MeONa]{MeOH} \quad \left[\text{Si(catecholate)}_3 \right]^{2-} \cdot 2\,Na^+ \tag{34}$$

A further application to solid–liquid systems that is increasing in popularity is the application of ultrasound to electrochemistry. Here the surfaces of electrodes are cleaned and there is rapid transport of ions both to and from the electrodes' surfaces, which can be used to good effect in electroplating and surface hardening.

Heterogeneous Liquid–Liquid Systems

The main application of ultrasound to liquid–liquid systems is to promote good mixing between the layers and the formation of emulsions with a small and even particle-size distribution. From a synthetic point of view such reaction systems are of use in emulsion polymerizations in which small droplets of a monomer are dispersed in water aided by the presence of emulsifiers and surfactants and a chemical initiator is added. The sonication of emulsion polymerizations such as styrene in water to give polystyrene has been found to prevent the particles adhering to the walls of the reaction vessel and to prevent the agglomeration of particles that can cause localized increases in temperature. Also, the sonochemically induced breakdown of an aqueous-phase initiator such as potassium persulfate may be increased, thus leading to an increase in the rate of polymerization.

The sonication of polymers has received considerable attention because not only may polymers be built up from monomers or oligomers but also they are degraded from high to lower relative molecular masses. If a macromolecule is trapped near a collapsing cavitation bubble then the rapid flow of molecules near the collapse causes very large shear forces and a long-chain molecule is likely to be broken close to the middle of the chain by such forces. For any given set of reaction conditions there is a tendency for a polymer with a wide range of relative molecular masses to be broken down into a narrower range of products of lower relative molecular masses. A narrower relative-molecular-mass distribution may well give rise to benefits regarding physical properties of the polymer.

Homogeneous Systems

The importance of surface effects on sonication of a liquid or a solution is much less than that for the heterogeneous systems described above; nonetheless, there are also significant effects to be observed. The effects on homogeneous systems arise from the very high temperatures and pressures generated on cavitational collapse. Any molecules trapped within the collapsing bubble in the vapour phase or in a thin shell of liquid immediately surrounding the cavity will be subjected to these physical extremes and will thus undergo reactions. Probably the most widely studied liquid is water, which on sonication undergoes homolysis to give $H^\bullet$ and $HO^\bullet$ radicals that combine to form hydrogen molecules and hydrogen peroxide, H_2O_2. If there are gases dissolved in the water then further reactions can occur, for example, sonication of N_2 in water affords HNO_2, HNO_3, NH_2OH, N_2O and NH_3; a mixture of H_2 and N_2 in water gives ammonia, NH_3; and a mixture of N_2 with CO, CH_4 or HCHO can give small amounts of amino acids. The formation of most of these molecules is probably due to a radical mechanism, started by the initial breakdown of the water molecules, which are, of course, present at a high concentration. Inorganic ions and molecules dissolved in water also undergo reactions, usually oxidations or reductions, on sonication, for example, Fe^{2+} ions are oxidized to give Fe^{3+} and OsO_4 is reduced to OsO_2. A wide variety of organic compounds dissolved in water has also been subjected to ultrasound and again many different products may be obtained, predominantly ones derived from free-radical mechanisms involving the initial formation of $H^\bullet$ and $OH^\bullet$. For example, sonication in water of carboxylic acids gives CO, CO_2 and hydrocarbons; sonication of aldehydes gives mainly carboxylic acids; and sonication of thioethers gives sulfoxides and sulfonic acids.

A further aspect of sonochemistry that has been studied most thoroughly for homogenous systems, particularly aqueous solutions, is that of sonoluminescence. This occurs as a weak emission of light when high-intensity ultrasound is passed through a liquid or solution containing dissolved gases. There are several theories that attempt to explain this phenomenon, the most popular of which attributes the production of light to the gaseous contents of collapsing cavitation bubbles being heated to incandescence or to emission from excited chemical species formed by cavitation.

Early workers in the field of sonochemistry came to the conclusion that sonochemistry could occur only with water as a solvent and that organic liquids could not be used as solvents for sonochemistry. This was an erroneous conclusion and it was drawn because the high vapour pressure of common organic solvents greatly reduces the intensity of cavitational collapse and hence the temperatures and pressures caused. It is now known, however, that, if the temperature of the solvent is kept well below its boiling point, so that its vapour pressure is relatively low, then all organic liquids will undergo sonochemistry, mainly via homolysis of bonds and the formation of free radicals. The sonication of alkanes leads to products similar to those found in high-temperature pyrolyses i.e. H_2, CH_4 and terminal alkenes; and carbon tetrachloride gives Cl_2 and C_2Cl_6, both derived from radical reactions.

One of the more recent and exciting areas of homogeneous sonochemistry has been developed by Ken Suslick and his group in the USA, who have sonicated transition metal complexes in organic solvents. A variety of reactions may occur in such systems. In the absence of added reagents that can also act as ligands to transition metals the build-up of clusters may be observed; if reagents are added then substitutions of ligands and sonocatalysis are seen and secondary reaction with solvent molecules may also sometimes occur, equations (35)–(37). These reactions are thought to proceed via an initial multiple dissociative loss of ligands such as CO (caused by the high temperature and pressure created by cavitational collapse) to give coordinatively unsaturated species that then react rapidly either with each other or with other potential ligands present. The metallic iron produced on sonication of $Fe(CO)_5$ is amorphous and is formed in this way rather than in the more usual crystalline forms due to the very rapid rates of cooling in the cavitational collapse, for the extremes of several thousand kelvins last only for a very short time. Amorphous iron formed in this way has unusual properties and can be used, for example, as an effective dehydrogenation catalyst.

$$Fe(CO)_5 \xrightarrow[\text{alkane}]{)))} Fe_3(CO)_{12} + Fe \qquad (35)$$

Cluster and amorphous
iron formation

$$Co(\eta^5\text{-}C_5H_5)_2 + CO \xrightarrow{)))} Co(\eta^5\text{-}C_5H_5)(CO)_2 \qquad (36)$$

Ligand substitution

$$Co_2(CO)_8 + \text{alkane} \xrightarrow{)))} Co_2(CO)_6(HC\equiv CH) \qquad (37)$$
$$+ Co_4(CO)_{10}(HC\equiv CH)$$

Secondary reaction
with solvent

Industrial Applications

There is a very wide range of applications of ultrasound in medicine and industry but many of them rely on the physical effects of the applied sound rather than chemical reactions caused by it. In medicine foetal imaging is now very common, as is, for example, the treatment of muscle strains and the cleaning and drilling of teeth. Biological applications include rapid homogenization and the disruption of cell walls so that the contents of cells can be extracted. In the food-processing industry there can be several beneficial effects such as more rapid nucleation leading to more uniform crystallization of, for example, chocolate and ice; and stable emulsions such as tomato sauce and mayonnaise can be formed. In the engineering industries, ultrasound can be directly applied to solid materials such that the rapid localized heating caused can be used for welding metals or plastics or for cutting and machining.

Conclusions

The application of ultrasound to chemical reactions gives the chemist a convenient means of applying physical extremes on a microscopic scale and being able to benefit from this on a bulk scale. However, although there are many industrial applications of ultrasound using the physical effects it causes and there is a wide range of laboratory applications for chemical synthesis, the transition from the laboratory to an industrial scale for chemical applications has been difficult. This was for a time due to the lack of suitable equipment but in recent years large baths, transducers attached to pipework and ultrasonic horns with flow attachments have all become available to help one to overcome these problems. A more difficult problem is that of measuring the sonochemical efficiency for a particular reactor. This in turn is allied to the more general problem of measuring the true sonochemical effect in a particular system. For such measurements it would be convenient to have some chemical 'dosimeter' such as those used in radiation chemistry. This is an area of considerable current activity and it is likely that reliable dosimetry will lead to an increase in the industrial use of sonochemistry.

Further Reading

1. J. P. Lorimer and T. J. Mason, *Chem. Soc. Rev.*, **16**, 239 (1987).
2. J. Lindley and T. J. Mason, *Chem. Soc. Rev.*, **16**, 275 (1987).
3. K. S. Suslick (editor), *Ultrasound. Its Chemical, Physical and Biological Effects* (VCH, Weinheim, 1988).
4. T. J. Mason and J. P. Lorimer (editors), *Sonochemistry. Theory, Applications and Uses of Ultrasound in Chemistry* (Ellis Horwood, Chichester, 1988).
5. S. V. Ley and C. M. R. Low, *Ultrasound in Synthesis* (Springer-Verlag, Berlin, 1989).
6. T. J. Mason (editor), *Chemistry with Ultrasound* (SCI, London, 1990).
7. T. J. Mason (editor), *Sonochemistry. The Uses of Ultrasound in Chemistry* (Royal Society of Chemistry, Cambridge, 1990).
8. G. J. Price (editor), *Current Trends in Sonochemistry* (Royal Society of Chemistry, London, 1992).
9. T. J. Mason (editor), *Advances in Sonochemistry*, Vols 1, 2, 3 and 4 (JAI Press, London and Greenwich CN, 1990, 1991, 1993 and 1996).
10. P. D. Lickiss and V. E. McGrath, *Chem. in Britain*, **32**, 47 (1996).

What, Why and When is a Metal?

Peter P. Edwards

5

Introduction

It is no exaggeration to say that, of all materials in common use, none has had a greater influence upon our development – *nay* upon our very existence – than metals. The combination of their excellent mechanical strength and ductility, their supreme electrical and thermal conductivities and their remarkable alloying properties guarantees metals a unique position in science and technology for which there is at present no substitute.

However, in 1967, Sir Ronald Nyholm commented that, with the notable exception of Linus Pauling, the subject of metals and the metallic state had been much neglected by chemists. My hope is that this present perspective for *The New Chemistry* will go some way towards catalysing the reversal of this state of affairs. This '*neglect of interest*' was certainly not the case in former times, for our chemist forefathers were always at the very heart of major developments and innovation in the science and application of metals. For example, at the dawn of the nineteenth century, Dimitri Mendeleev enthused that the discovery and isolation of the metals sodium and potassium in 1807 by the chemist Sir Humphry Davy was '*one of the greatest discoveries in science*' (Figure 1). What was it about this particular discovery of Davy that triggered Mendeleev's effusive compliments? (Figure 2). Think back to our predecessors' everyday knowledge and common experience of metals before 1800; of the forty or so known elements of the periodic classification, those established as metals (iron, gold, lead, silver, mercury, etc.) all had very high densities and modestly inert chemical characteristics. Davy's monumental discovery – so clearly favoured by Mendeleev – now posed a major threat to the very concept – and definition – of a metal. For '*the new metals – the alkali metals*' possessed most of the physical traits of the established metals (high conductivity, lustre, etc.), but they had exceptionally low densities and, of course, incredibly high chemical reactivities such as were unknown for the metals of the day! (Figure 1). Viewed from a modern-day perspective, it seems incredible to us that the alkali elements sodium and potassium were, in fact, designated '*metalloids*' (meaning '*metal-like elements*') by Paul Erman and Philip Simon in 1808, in an attempt to overcome this fundamental and perplexing dilemma. Davy's discovery, then, initiated a complete reappraisal of the most elementary – but engaging – philosophical question '*What is a metal?*'. It fell to Davy in his second Bakerian Lecture – delivered at the Royal Institution in London only two months after his remarkable discoveries – to answer the (then) controversial question of whether the newly discovered elements, sodium and potassium, could justifiably be called metals Davy, characteristically, identified the very essence of the problem. He answered that '*The great number of philosophical persons to whom this question has been put, have answered in the affirmative. They agree with metals in opacity, lustre, malleability, conducting powers as to heat and electricity, and in their qualities of chemical combination*'. The chemist Davy, then, was the instigator of this revolutionary development of '*the new metals . . . the alkali metals*'.

The discovery of the electron almost 90 years later by the mathematician turned experimental physicist Sir J. Joseph Thomson inspired the development of the highly successful '*free-electron theory*' of metals. It also provided a firm experimental basis for the natural understanding of the benchmark physical characteristics of metals. An

(a)

Figure 1

The new metals – the alkali metals. (a) A portrait of Sir Humphry Davy, the discoverer, in 1807, of 'the new metals', sodium and potassium. (b) A laboratory notebook entry for 19 October 1807. Courtesy of The Royal Institution of Great Britain. Having isolated potassium on 6 October of that year, Davy was anxious to demonstrate that potash was the oxide of this new metallic element. Electrolysing potash in a closed tube (see top right hand corner of notebook entry) yielded pure oxygen, establishing this to be so. In Davy's own words (at the bottom of this page of his notebook), 'Capital Experiment proving the decomposition of Potash'. (c) The reaction of potassium with water; this vigorous reaction produces so much heat that the hydrogen produced in the reaction is spontaneously ignited. Taken, with permission, from Chemistry, by Peter Atkins and Loretta Jones, W. H. Freeman and Co., New York. (d) The catastrophic reaction of caesium metal with water. The photograph is a still from the video The Elements produced by The Royal Society of Chemistry/Open University and used with permission.

(b)

inevitable consequence was thus an increase in interest in, and measurement of, the fundamental physical properties of metals. This burgeoning interest in what one would now term the physics of metals provided the experimental backdrop for the rigorous application, during the 1920s and 1930s, of the (then) new quantum mechanics to an understanding of the behaviour of electrons in solids. The result was the electronic band theory of metals and insulators, with its roots in the description of free electrons in momentum space (see p. 96). Curiously, however, until recently the precise physical mechanism(s) of just *how*, and indeed *why*, such valence electrons should become set free was largely neglected!

(d)

(c)

Figure 1 (cont.)

A distinctly more chemical-bonding (real-space) view of metals was championed by the great innovator Linus Pauling during the 1930s and 1940s (Figure 3). In 1938 Pauling outlined his chemical-bonding description of metals. At the heart of Pauling's perceptive ideas was the viewpoint that an atom in a metal exhibits great flexibility and versatility in the formation of its multitude of chemical bonds. He noted that there was very little difference among the stabilities of metal structures involving not only different types of bonds but also various types of coordination and, in consequence, an atom in a metal can readily form bonds after deformation of the original structure that are approximately as strong as the original bonds! This intrinsic atomic characteristic gives a metal its unique ability to heal itself after deformation, which underlies such characteristic metallic properties as malleability and ductility. In Pauling's '*metallic bond*' or '*metallic orbital*' viewpoint, the situation in metals is closely related to the covalent or electron-pair bond in a single molecule. Thus, some of the electrons of an atom in a metal are involved with those of neighbouring atoms in an interaction described as the formation of covalent bonds, with the bonds resonating among the available positions in the usual case that the number of positions exceeds the number of bonds. As we shall see, the chemical possibility recognized by Pauling in 1938, namely that it becomes possible to transfer valence electrons from one such orbital (atom) to the next in a metal, is at the heart of the physics of metals (see p. 107). Interestingly, many of these issues were also pursued with equal effectiveness by the physicist Sir Nevill Mott in a determined campaign, beginning around the same period.

Briefly, any substance will become a metallic conductor when the energy cost of transferring a bound, valence electron from a particular atom or chemical bond to another is more than offset by the gain in energy derived from liberating an electron from its atomically localized chemical bond into a '*free-electron state*'. From this very special vantage point, pioneered by Linus Pauling over half a century ago, the chemist should again be poised to make a substantive impact on the issue of localized (bond) versus itinerant (bond) electrons in solids and liquids. It is now recognized that a successful modern electronic theory of both metals *and* insulators will require the complementary development *both* of real-space *and* of reciprocal (momentum)-space viewpoints (see texts given on p. 114).

In essence, both the chemist and the physicist must continue to learn together about the remarkably deep and complex issues which conspire to produce either localization

Figure 2
Dmitri Ivanovitch Mendeleev, professor at the University of Leningrad, since 1865 of chemistry and since 1867 of inorganic chemistry *only*. The records show that Mendeleev discovered the periodic system on 17 February 1869. (Taken from R. D. van Spronsen, *The Periodic Table of the Elements*, and used with permission).

Figure 3
Linus Pauling, professor of chemistry at California Institute of Science and Technology. Pauling's book *The Nature of the Chemical Bond* is unquestionably one of the most influential scientific texts of this century. Pauling's *Resonating Valence Bond Theory* was an attempt at a chemical perspective on the metallic state. Photograph courtesy of California Institute of Technology.

or itinerancy of electrons in condensed phases. To achieve this, we outline just some of the essential science associated with the problem of what a metal is. In an attempt to set out these issues – I hope, transparently and unhindered by heavy numerical camouflage – we would wish to treble the scope of our enquiry to: '*What, why and when is a metal?*'.

Sir Isaac Newton is said to have acknowledged all of his discoveries by '*thinking at all times on these things*'. With respect to our approach outlined here, perhaps Newton's approach should now be abridged to '*questioning . . . and thinking at all times on these things*'! Our questions and enquiries relating to metals and the metallic state therefore include the following.

- What is a metal?
- Why – and how – do the splendid characteristics of a metal emerge?
- When can we transform an insulator into a metal (and vice versa)?
- What of '*the new chemistry of the new metals*' – new metals for old?

What is a Metal?

Solid substances can be sharply divided into metals, such as copper, silver, gold and aluminium, which are able to carry an electrical current efficiently, and non-metals such as sulfur, glass, common salt and wood, which are patently insulators. From our common experience, the combined properties which instantly epitomize a metal are those of high electrical and thermal conductivity, ductility, malleability and recognizable metallic lustre. The natural ductility of gold, for example, has permitted craftsmen, since time immemorial, to fashion exquisite ornamentation, such as that shown in Figure 4. Above all else, however, *the* characteristic hallmark of a metal or a metallic substance must surely be its supreme and unrivalled natural ability to transmit the flow of electricity. We all instinctively recognize that metals are '*magnificent conductors of electricity*' (Sir Alan Cottrell, Box 5). Ultimately it is this key, intrinsic property which sets metals apart from insulators or non-metals.

However, it is perhaps not always appreciated that the actual numerical range of electrical conductivities separating metals from non-metals or insulators is truly staggering. For example, pure metallic copper at room temperature has a measured dia-electrical conductivity of around 10^6 ohm^{-1} cm^{-1}. At the opposite end of the conductivity spectrum, a fluorinated ethene–propene copolymer (an excellent electrically insulating material) has a room-temperature conductivity of less than 10^{-17} ohm^{-1} cm^{-1}. At this low end of the conductivity spectrum, we also find another prototypical insulator – wood, with a conductivity of some 10^{-22} ohm^{-1} cm^{-1} (Figure 5). This range in room-temperature electrical conductivities – some 28 orders of magnitude separating a highly conducting substance such as copper from a stubbornly insulating material such as wood – is probably *the* widest variation of any (laboratory-based) physical property. Even larger ranges in electrical conductivities await us when we investigate the behaviour of metals and insulators at low temperature.

The class of substances or materials that we now routinely label '*metals*' is very much larger than that of the conventional metals found, for example, in the Periodic Table of the elements. There is an enormous range of inorganic and organic chemical compounds that exhibit all the traits associated with the prototypical elemental metals but, in many instances, they exhibit unconventional and surprising properties – sometimes properties not even found among the elemental metals themselves! For example, many transition metal oxides (e.g. $NaWO_3$, ReO_3 and CrO_2) and sulfides (e.g. TaS_2) have very high electrical conductivities, in certain cases approaching the values for copper metal. The oxides $NaWO_3$ and ReO_3 themselves exhibit a beautiful metallic lustre (Figure 6), entirely consistently with their classification as metallic '*bronzes*'.

Figure 4
In contrast to its alkali metal 'cousins' gold occurs in its elemental state. It was therefore one of the first metals used by ancient artisans. We show here a gold model of a chariot produced in the Bactrian area of western Asia around 500 BC. It was found in 1877, in the dry bed of the Oxus River, which now separates Afghanistan from Tajikistan. (Taken from *The Cambridge Guide to the Material World*, Rodney Cotterill, Cambridge University Press, and used with permission.)

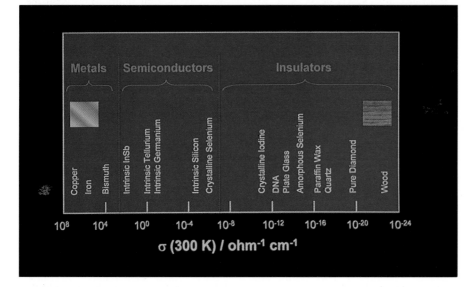

Figure 5
The electrical conductivities of various elements, compounds and materials. The figure illustrates the enormous range in electrical conductivities encountered for measurements at room temperature, with typical members of the classes conventionally described as metals, semiconductors and non-metals (insulators). At the absolute zero of temperature, the corresponding spread in conductivities would have only metals and superconductors, with finite and infinite conductivities, respectively, and non-metals or insulators with zero conductivity (see Box 3, p. 98). We highlight the enormous differences in electrical conductivty between copper and wood! This figure was kindly supplied by Dr Graeme B. Peacock of the University of Birmingham.

The introduction of sodium metal into anhydrous liquid ammonia produces an intensely coloured blue solution. At higher concentrations of metal, this solution transforms into a bronze-coloured metallic conductor with an electrical conductivity higher than that of liquid mercury! In the intermediate-concentration range, cooling of the homogenous solution gives rise to a fascinating liquid–liquid phase separation in which dilute (blue) and concentrated (bronze) phases co-exist (Figure 7). Remarkably, the more concentrated, bronze metallic phase floats on top of the dense – but less concentrated dark blue phase.

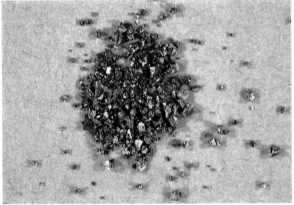

(a)   (b)

Figure 6

Photographs of the metallic oxides NaWO$_3$ (a) and ReO$_3$ (b), both having room-temperature conductivities comparable to that of elemental copper. Samples kindly provided by Dr Marten G. Barker, Nottingham University.

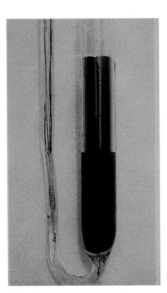

Figure 7

A solution of sodium metal in anhydrous liquid ammonia at a temperature of approximately −60 °C, showing the remarkable co-existence of metallic (bronze-coloured) and non-metallic (dark blue) solutions. Interestingly, the bronze, concentrated solution is less dense than the blue, dilute solution. Courtesy of Dr Pierre Chieux (Grenoble).

Complex multinary oxides of the transition elements are currently at the forefront of modern issues relating to metallicity. Let me illustrate. The very best conductor of electricity at the temperature of liquid nitrogen (77 K, −196 °C) is now no longer a traditional lustrous, ductile elemental metal, but rather a dull, black multinary ceramic oxide comprised of at least four metal constituents (Figure 8), namely the complex mercury–barium–calcium cuprate of chemical formula HgBa$_2$Ca$_2$Cu$_3$O$_{8+\delta}$. Below a characteristic temperature, T_c, these complex oxides not only become '*good*' metals but also achieve the status of '*the ultimate of metals – the superconductor*'. A superconductor is a remarkable material that has *no* electrical resistance, or, equivalently, *infinite* electrical conductivity below T_c.

Many highly conducting organic compounds are also known, prototype examples being iodine-doped polyethyne (CHI$_{0.25}$)$_n$ and the so-called '*charge-transfer*' organic molecular conductor TTF–TCNQ (Box 1). Some of these compounds also become superconducting, but at low temperatures.

Another defining characteristic of any metal or metallic substance is that it becomes a far superior conductor of electricity at low temperatures than it is at high ones. The complete opposite is true for insulators. I hope to illustrate this fundamental difference in conducting behaviour through a simple, but potent example. Consider the oxidation of elemental copper to form the non-metal (insulating) chemical compound copper(I) oxide, Cu$_2$O. This process can be followed experimentally by heating a thin film of metallic copper in air or in a stream of oxygen gas. John Adkins and his colleagues at Cambridge have perfected an ingenious way of measuring the electrical conductivity of thin films of metallic copper, whilst they undergo a continuous chemical oxidation to non-metallic Cu$_2$O.

By transferring one of the outer valence electrons of each copper atom to oxygen to form the binary oxide compound Cu$_2$O (most easily represented as (Cu$^+$)$_2$O^{2-}), a thin film of metallic copper will inevitably undergo a metal-to-non-metal or metal-to-insulator transition upon oxidation, driven by the overall change in chemical composition. If we now look at the details of the electrical-conductivity behaviour of thin films ranging from elemental copper to Cu$_2$O, one instantly recognizes the very high electrical conductivity characteristic of metallic copper and the fact that this conductivity increases substantially upon cooling to low temperatures (Figure 12). In stark contrast, the complete oxidation of the copper film to give Cu$_2$O presents us with a clear signature of insulating properties; *viz.*, a huge drop in the magnitude of the electrical conductivity at all temperatures relative to that of elemental copper and a conductivity that decreases significantly as the temperature decreases. These differences in conducting behaviour between elemental, metallic copper and non-metallic, insu-

Figure 8

A pellet of a high-temperature-superconducting cuprate ceramic. If such a ceramic is cooled below a certain characteristic temperature, usually approximately −190 °C, the compound exhibits the phenomenon of superconductivity. In this state, the material offers *no* resistance to the passage of electrical current. In the superconducting state the materials also exhibit a remarkable magnetic property by which all magnetic fields are repelled away from the superconductor. The cooled superconducting ceramic pellet thus levitates over a magnet, as shown here. This figure was kindly supplied by Hoechst Aktiengesellschaft Verkauf Chemikalien, and used with permission.

lating Cu_2O become even more exaggerated when the films are studied down to very low temperatures, a few degrees above the absolute zero of temperature ($T = 0$ K). If we now attempt to extrapolate the measured electrical-conductivity behaviour all the way down to $T = 0$ K, it now becomes abundantly clear that we arrive at an unambiguous working definition of a metal and an insulator (Figure 12). At $T = 0$ K, therefore, either we have an insulator, for which the electrical conductivity is zero (and I really do mean zero!), or one has a metal, for which the electrical conductivity is non-zero (i.e. finite). Obviously we also need to incorporate into our discussion the ultimate conductor of electricity – the superconductor – *viz.*, a substance having *infinite* electrical conductivity below a characteristic superconducting transition temperature. Of course, within this definition infinite is also non-zero and superconductors are metals (but see p. 113). It is amusing to note that our archetypal metal, copper, has never been made superconducting – even down to the lowest achievable temperatures. In contrast, its chemical cuprate 'cousin' (shown in Figure 8) is currently the very best high-temperature superconductor!

The situation at the absolute zero of temperature now reduces to a very simple issue. At $T = 0$ K, therefore, we have only *three* possible electronic states of matter, each characterized by the value of their '*zero-temperature*' conductivities (designated $\sigma(0)$); these are

- *metals* – having non-zero (but finite) conductivity, *viz.*, $\sigma(0) \neq 0$;
- *insulators* – having zero conductivity, *viz.*, $\sigma(0) = 0$; and
- *superconductors* – having infinite conductivity, *viz.*, $\sigma(0) = \infty$.

Can I also make the crucial point that one should therefore think not only of metals and insulators but also of the possible electronic transition between these two states of matter – i.e. *The metal-to-insulator or insulator-to-metal transition?* Now, one of the major themes of my contribution is to illustrate that the basic question of '*metal or not?*' becomes distinctly blurred when external conditions such as temperature, pressure and composition are varied. Thus, in principle, and, indeed, often in practice, it is possible to transform *any* stubbornly resistive insulator into a highly conducting metal and vice versa. We can turn wood into copper! (Figure 5).

Box 1 Organic metals

The discovery, in 1973, that a 1:1 complex (Figure 9) of tetrathiafulvalene (TTF) and tetracyanoquinodimethane (TCNQ) displayed metal-like electrical conductivity began a very fruitful period of research in organic metals. Other electron donors similar to TTF display interesting electrical properties, for example, superconductivity at low temperatures.

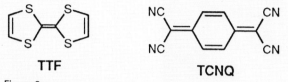

TTF   **TCNQ**

Figure 9
TTF–TCNQ.

Polyethyne, or polyacetylene, $(CH)_x$, is the simplest organic polymer. Figure 10 shows a metallic hue from a film of *trans*-polyethyne prepared in 1980 by Dr John Edwards and Professor W. (Jim) Feast of the University of Durham. Interestingly, in this condition polyethyne is not yet a '*metal*' in the conventional sense, since it has a conductivity in the range 10^{-8}–$10^{-5}\ \Omega^{-1}\ cm^{-1}$ (Figure 5). Upon exposure to oxidizing agents such as elemental iodine, however, the conductivity can be raised spectacularly into the 'metallic regime' with $\sigma > 10^3\ \Omega^{-1}\ cm^{-1}$.

In this metallic regime, I_2-doped polyethyne has a delocalized positive change that allows it to conduct like a metal. Conduction electrons extend along the polymer's backbone (shown here as silvery black atoms). The yellow triiodide dopants (I_3^-) are interspersed among four polymer chains (Figure 11).

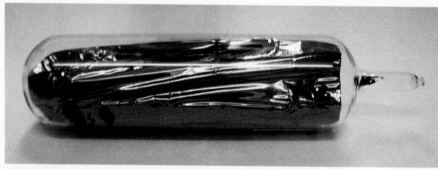

Figure 10
A film of polyethyne (Courtesy of Elizabeth Wood, John Edwards and Jim Feast, University of Durham).

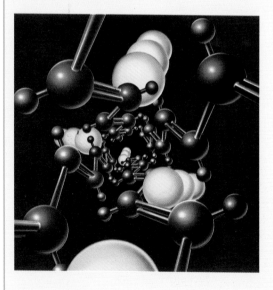

Figure 11
A colour picture of the crystal structure of metallic polyethyne $(CH)_x$. (Reproduced from *Scientific American*, February 1988, with permission.)

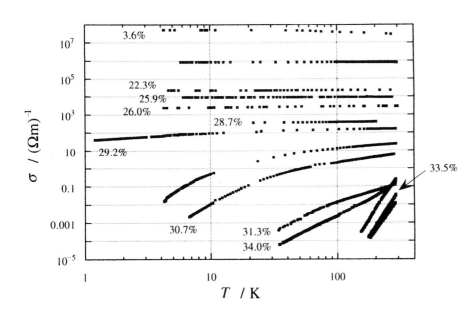

Figure 12
The electrical conductivites of a range of thin-film samples ranging from amorphous Cu metal to the semiconducting compound copper(I) oxide, Cu_2O (33.5%). The percentages on each curve represent the atomic compositions for each sample. If we extrapolate the data to $T = 0$ K, we clearly see the demarcation between metal and insulator. Courtesy of Dr John C. Adkins and colleagues, Cavendish Laboratory, Cambridge.

As one might expect, such electronic phase transitions have spectacular consequences not only for electrical conductivity but also for other physical properties. The phenomenon of the insulator–metal transition has been characterized by my colleague C. N. R. (Ram) Rao of Bangalore as '*turning wood into copper*'. The equivalent metal–insulator transition would then correspond to '*turning copper into wood*'! The metallization of fluid hydrogen is one such spectacular example; the lightest and most abundant element in the Universe transforms from a non-metallic, insulating fluid into a highly conducting liquid metal at ultra-high pressures (see p. 94 and Box 2). Another remarkable example is elemental sulfur. At room temperature and pressure, the electrical conductivity of yellow sulfur is at least 18 orders of magnitude lower than that of copper metal. However, at very low temperatures and high pressures, elemental sulfur becomes a far superior metal than even copper – it becomes a superconductor!

The intricate and pervading problem of the metal–insulator transition will underpin all of our forthcoming discussion of metals and the metallic state. I wish to convince you that the two questions: '*What is a metal*' and '*When does a metal transform into an insulator?*' are indeed completely miscible. The '*why and when*' of metals and the metallic state forms the major focus of this contribution. I now outline the microscopic origins of the electronic behaviour of metals and insulators and also the various mechanisms by which a transition from a metal to an insulator (and vice versa), induced by changes in temperature, pressure, density or chemical composition, can occur.

Why and When is a Metal?

The reason for the magnificent conductivity of metals has been known for most of the twentieth century and, indeed, part of the preceding century; *it is that metals contain free electrons*. I should stress that by '*free*' I do not mean '*inexpensive*'; rather that in a metal *all* of the constituent atoms spontaneously release one (or more) of their outer valence electrons to produce a highly concentrated gas of free or itinerant electrons. These electrons are called conduction electrons – the very essence of the characteristic high electrical conductivity of a metal. This startling innovation was first proposed by Paul Karl Drude soon after the discovery of the electron by Sir J. Joseph Thomson. It turns out that, in a good conductor such as copper or silver, each constituent atom has contributed just about one valence electron in the process of forming the metal.

Box 2 'Hydrogen shows its metal'

Hydrogen is the lightest and most abundant element in our universe and, until 1996, it had been seen only in its normal non-metallic form. The likelihood that hydrogen could be converted into a metallic form was first predicted by Eugene Wigner and Alan Huntington in 1935. They proposed that molecular diatomic, solid hydrogen – the 'conventional' form of the element under normal conditions – would undergo a transition to a metallic, conducting state at a pressure of about 250 000 atm; current predictions are in a range close to three million atmospheres. However, despite an unrelenting experimental assault at these ultra-high pressures, compressed *solid* hydrogen has so far defied all attempts at metallization. However, the situation in relation to compressed *fluid* hydrogen is remarkably different. In 1996, Samuel Weir, Arthur Mitchell and William Nellis utilized a two-stage (explosive-triggered!) gas gun at the Lawrence Livermore Laboratories, USA to create transient pressures in a thin film of fluid hydrogen of more than two million times the earth's atmospheric pressure.

Metallic hydrogen was formed in these experiments by the shock compression of a 0.5 mm layer of fluid hydrogen between two anvils of single-crystalline sapphire under a velocity impact of more than 25 000 km^{-1}, which generated pressures of up to 2 million atmospheres. The Livermore group measured the electrical conductivity of fluid hydrogen during its fleeting existence and found that it rose by more than five orders of magnitude as the pressure increased to 1.4 million atmospheres and then levelled off at a value of 2000 Ω^{-1} cm^{-1} up to 1.8 million atmospheres (Figure 13).

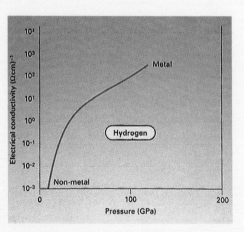

Figure 13
The conductivity of fluid hydrogen under external pressure.

These remarkable experiments highlight the tantalizing theoretical problem of metals and the metallic state – namely, how do we know that this measured conductivity of hot fluid hydrogen at high pressure is actually indicative of a *metallic* form of that element? We shall develop the theme that *the only rigorous criterion* for differentiating between a metal and a non-metal is the value of the electrical conductivity at the absolute zero of temperature, at which metals have a finite conductivity and non-metals are perfect insulators (they have zero conductivity). At temperatures close to 1700 °C – the conditions in the Livermore experiment – this clear and fundamental distinction becomes blurred and other

Consequently, one has an extremely dense assembly of conduction or itinerant electrons, more than 10^{22} of them in 1 cm^3 of a metal.

It is important to stress also that the loosely held outer (valence) electrons from each atom are completely detached from their parent atoms and shared amongst all the other atoms in the metal. Leaving the (parent) metal atoms in place as positive ions (cations) in a crystalline, periodic array, this high-density gas of itinerant conduction electrons then forms a pervasive 'glue' that permeates the network of ions and serves – so effectively – to bind them together. This simple idea is potent enough to explain many of the important physical characteristics of metals. For example, if the conduction electrons are free and mobile, the application of an external electrical field will draw the electron gas through a metallic wire. The presence of a high-density electron gas of free electrons is thus a pivotal feature of our picture of metals and the metallic state.

Now, one might then justifiably ask how it comes about that each atom in a metal spontaneously sees fit to donate its valence electron – in many cases its *only* valence electron – for the common good of all the remaining atoms. In other words we should ask the question '*Why is it that these conduction electrons are indeed free in a metal?*'. However, the question to be posed will have to be even more stringent than this; I have

Box 2 (cont.)

operational criteria are clearly needed. The recurring theme 'When is a metal?' will be developed in this chapter. Suffice it to say that the limiting electrical conductivity of compressed liquid hydrogen under these conditions is extremely close to that of fluid Rb and Cs undergoing a metal-to-non-metal transition at the same temperature. This clearly identifies hydrogen as the first – and the lightest – alkali metal!

The Livermore team's shock-compression experiments were designed to create unearthly pressures and temperatures. Remarkably, these conditions are representative of those on the planet Jupiter (Figures 14 and 15), which is more than 90% hydrogen. That planet, itself primarily a fluid, is composed of an outer layer of non-metallic, molecular hydrogen that continuously transforms to metallic fluid hydrogen as one approaches the inner core. The results give credence to the theory that the planet's magnetic field is produced not in the core, but much closer to the surface than had originally been thought.

Figure 14
Jupiter. (Taken from R. A. Kerr, *Science* **271**, 1667 (1996), with permission.)

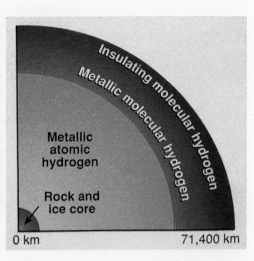

Figure 15
A schematic diagram of Jupiter. (Taken from R. A. Kerr, *Science* **271**, 1667 (1996), with permission.)

already shown you that a characteristic feature of a metal is that its electrical conductivity continuously improves on cooling to low temperatures, whereas that of an insulator continuously deteriorates as we approach a temperature of absolute zero (Figure 12). Thus, the very special and unique feature of a metal is not only that its conduction electrons are free but also that they continue to remain free, down even to the lowest possible temperatures!

Given that the valence electrons in a metal are free, it must therefore follow that those in an insulator are not! Why is this? This vexing and surprisingly complex question has a long and venerable history, dating back to the very dawn of the twentieth century. More recently, in 1940, John Bardeen (one of the famed inventors of the transistor) noted that '*there is no satisfactory explanation on any classical basis for the huge differences in electrical conductivity between metals and insulators*'. In 1964, Richard P. Feynman and colleagues also alluded to this problem in their celebrated *Lectures on Physics*, noting that

Some materials are electrical conductors – because their electrons are free to move about; others are insulators – because their electrons are held tightly to individual atoms. We shall consider later how some of these properties come about, but that is a very complicated matter.

Interestingly, it is not clear to me that Feynman did indeed fulfil his promise to revisit this *'very complicated matter'* later in this famous text! We must now take up this fascinating challenge noted by Feynman and ask *'Why is is that some substances and materials are metals, while others are insulators?'*. Our first such approach, generally designated the Bloch–Wilson model, begins with the remarkable assumption that *all* of the electrons – both valence and core – in a crystalline solid are in fact free! It then sets out to prove that, under certain circumstances, these free electrons can transport an electrical current (metals), whereas under others they cannot (insulators).

The Bloch–Wilson (Band) Model

The formulation of quantum mechanics by Erwin Schrödinger and its application to the problem of electrons in solids by Felix Bloch, Alan H. Wilson and others was a breathtaking attempt to provide a rigorous and mathematical basis for the fundamental differences between metals and insulators. Within this viewpoint, the underlying symmetry and periodicity of the crystalline solid is exploited mathematically. Bloch proposed that the electrons in any crystalline solid, – that is, a solid possessing long-range periodicity – are in a sense free, in that they can be described by the following wave function:

$$\Psi(x) = U_k(x)e^{ikx} \tag{1}$$

Here e^{ikx} represents a free-electron wave moving with the wave vector k in the x direction and $U_k(x)$ is a repetitive function possessing the characteristic periodicity of the underlying ordered crystalline lattice of ion cores (Figure 16).

By equation (1), Bloch had arrived at what is now recognized as perhaps *the* single most important result of the application of the (then) new quantum theory to solids – namely, that an underlying periodic field of the metal ion cores in an extended crystalline solid neither scatters nor destroys a free-electron wave (equation (1)). This so-called *'Bloch wave'* thus retains the key property of a free-electron wave extending continuously throughout the whole of the space available to it, i.e. in a crystalline and periodic lattice, we have electrons propagating throughout the entire crystal – be it a metal or an insulator! From this viewpoint, therefore, *all electrons in any perfect crystalline solid must be considered free.* It would then follow that, according to Bloch's theory, *all* solid, crystalline substances would be metals! In the words of Sir Alan Wilson, in 1980, recalling his own thoughts on the problem some 50 years earlier,

Bloch's theory had in fact proved too much. Before his paper appeared it was difficult to explain the existence of metals. Afterwards, it was the existence of insulators that required explanation.

Wilson took up the theoretical challenge. He proposed that Bloch's free electrons in crystalline substances, like valence electrons in isolated atoms, could form either open or closed shell configurations. In solids, therefore, open and closed atomic shells overlap to form partially filled and completely filled electronic bands, respectively. A clear distinction between metals and insulators could then be drawn, according to whether the resulting electronic bands in the solid were partially or completely filled with valence electrons. Only the first type of electronic band – a partially filled band – allows the free electrons in a crystalline solid to act as carriers of electrical current, making the substance a metal. The origin of the effect is that the periodic field of the metal ion cores (Figure 16) creates energy gaps or band gaps in the allowed quantum states – these are very specific ranges of energy in which there are *no* allowed Bloch-wave states (Figure 17). Thus, insulating behaviour arises because there occur gaps of forbidden energy in which no acceptable solution of the Schrödinger equation exists. Note that a completely filled electronic band will not exhibit a preponderance of elec-

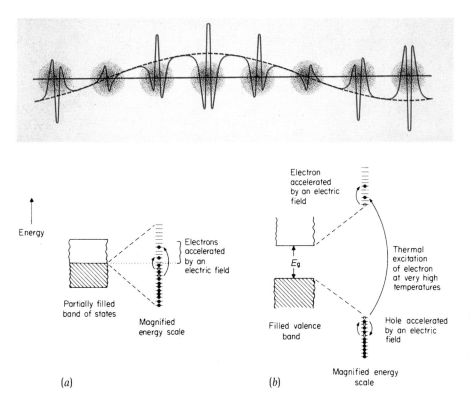

Figure 16
An electron wave moving through a periodic lattice of positive ions (grey spheres). In 1927, Felix Bloch showed that the characteristic periodic ionic lattice has a negligible influence on a travelling electron wave so long as the amplitude of the latter – shown here by a red line – is suitably modified when it is in the vicinity of each ion core. The electron wave thus traverses the ion cores so quickly that they have insufficient time to create a major hindrance to the motion of the electron. (Taken from *The Cambridge Guide to the Material World*, Rodney Cotterill, Cambridge University Press, and used with permission.)

Figure 17
Electronic energy bands in crystalline solids: (*a*) for metallic systems and (*b*) for semiconducting (small-gap) or insulating (large-gap) materials.

tron (Bloch) waves in any one direction; therefore, no electrical current can be induced by the application of an electrical field. An external electrical field (potential) can induce an electrical current only by exciting some electrons into accessible, vacant states of higher energy. In a completely filled band, by virtue of the Pauli exclusion principle, there are obviously no directly accessible vacant states into which electrons can be excited.

In contrast, an electrical current *can* flow in partially filled bands in the presence of an external electrical field. Here, the electron energy can be raised via the excitation of electrons by the electrical field into accessible and vacant energy states (Figure 17).

It therefore followed from the (combined) Bloch–Wilson model that any element within the Periodic Table with an odd-numbered chemical valency had to be a metal, whereas elements with even-numbered valencies would therefore be non-metals or insulators. Wilson's model seemed, at first sight, to predict that divalent elements such as calcium would be non-metals. This, of course, was clearly wrong and it was soon realized that a filled and an empty electronic band could overlap to produce a partially filled band – and hence metallic behaviour.

It was perhaps typical of the almost instantaneous acceptance of the new quantum mechanics in the 1930s that few doubted the Bloch–Wilson model of metals and insulators. Moreover, a basic approximation in the model was that each electron moved independently of all other electrons; that is, no account of the inescapable electron–electron Coulomb (repulsive) interactions in the dense gas of electrons was taken. Slowly, however, there emerged justifiable grounds for doubting the universal applicability of the Bloch–Wilson viewpoint. On the conceptual side, the theory led to an absurd conclusion in a hypothetical limit. To illustrate, if we take a crystal of, say, the highly conducting alkali metal caesium, it follows naturally from the Bloch–Wilson picture that we have a partially filled 6s electronic band and undoubted metallic status. However, imagine what would happen if we were able to expand our crystal of caesium

Box 3 Sir Nevill Mott: '*... a metal conducts and a non-metal doesn't*'

Over a period of more than 50 years, Professor Sir Nevill Mott pioneered the development of key concepts, models and theories for discussing the fundamental problem of metals versus insulators (non-metals). These issues occupied the thoughts of Sir Nevill until well into his nineties. We reproduce below his letter written to the present author on Thursday 6 May 1996 in which he notes '*Dear Peter, I've thought a lot about "What is a Metal" and I think one can only answer the question at T=0 (the absolute zero of temperature). There a metal conducts and a non-metal doesn't*'. Intuition at its most potent!

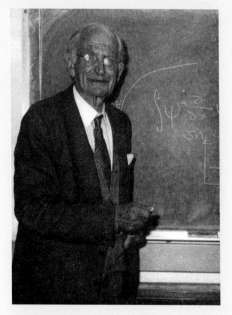

Figure 18
Sir Nevill Mott. Photograph courtesy of the Cavendish Laboratories, Cambridge.

Th

Dear Peter

I've thought a lot about "What is a metal" + I think one can only answer the question at $T=0$. Then a metal conducts, + a non-metal doesn't. This includes heavily doped semiconductors as metals, of course

I still don't know how to calculate the transition for $La_{2-x}Sr_xCuO_4$!

Yours
Nevill

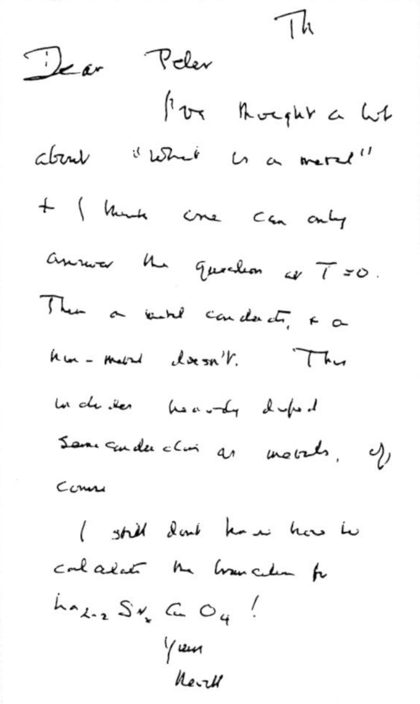

Figure 19
A letter from Sir Nevill Mott to P. P. Edwards. (Reproduced from *Nevill Mott; Reminiscences and Appreciations*, edited by E. A. Davis (Taylor & Francis, 1998).)

progressively, preserving its crystal structure intact, until the individual caesium atoms were, say, 1 m apart. According to the Bloch–Wilson theory, this massively dilated crystal of caesium would still be a metal! As Sir Nevill Mott (Box 3) so lucidly pointed out in 1961, '*This* [metallic status] *is against common experience and, one might say, common sense*'. Both from '*common sense*' and considering the experimental reality of the situation for expanded caesium (Box 4), once the average distance between the constituent caesium atoms exceeds a few tenths of a nanometre, the itinerant 6s electrons can significantly reduce their energies by simply condensing onto the constituent atoms – one electron per atom – so forming an array of neutral caesium atoms, with *no* free electrons. In this situation elemental caesium becomes a non-metal or insulator. The beautiful experiments of Friedrich Hensel and colleagues at Marburg provide a particularly clear-cut example of the inevitable transition from metallic to non-metallic behaviour as elemental caesium is continuously expanded (Box 4). Remember, of course, that this is the same element, caesium, behaving either as a metal or as a non-metal! This illustrates the key issue of density and the metallic state.

It was also realized in the late 1930s that an entire class of crystalline chemical compounds, typified by nickel oxide, NiO, which must be characterized by partially filled

Box 4 Expanded fluid metals

Most elements of the Periodic Table are metals. These elements remain metallic when they are heated through their melting point at normal pressure. Now, from common intuition one knows that a liquid metal such as mercury will conduct electricity, but its vapour patently will not. This simple – but potent – observation points to the all-important role that the density of an element plays in dictating whether it behaves as a metal or a non-metal. However, what if the non-metallic vapour of, for example, caesium were compressed, or liquid caesium were expanded? At what stage would we see a transition between the metallic and non-metallic states of, for example, mercury or caesium? In such a study one needs to probe what happens beyond the metallic element's critical point – the point at which the temperature is too high to sustain a liquid and the pressure is too high to permit a vapour. Above the critical temperature, T_c, and pressure, P_c, the liquid and vapour no longer co-exist and the fluid is said to be super-critical (Figures 20(a) and (b)). In this supercritical state the elemental density can be continuously varied at will, allowing one to determine the electrical (and other) physical properties of elements over a very wide range of densities.

The DC electrical conductivity of expanded caesium at supercritical temperatures is plotted as a function of the density of atoms in Figure 21; these data are taken from work in the laboratory of Professor Friedrich Hensel and colleagues, in Marburg University, Germany, and are used with permission.

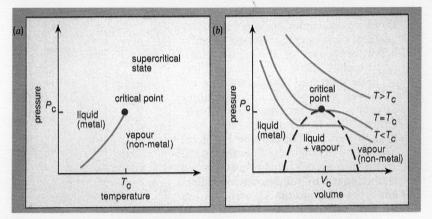

Figure 20
Phase diagrams for a typical liquid metal. Taken from
Fittensel and P. P. Edwards, *Physics World*, April 1996, p. 43.

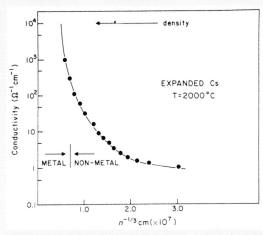

Figure 21
The electrical conductivity of expanded Cs; note that this is
the same element, caesium, being continuously transformed
from a metal to an non-metal via expansion in the
supercritical state. (F. Hensel, used with permission.)

energy bands, should therefore be metals according to the Bloch–Wilson model. In fact the overwhelming majority of these binary transition metal oxides are insulators! This observation was seen as even more perplexing in view of the relatively small differences between metal–metal separations in the insulating transition metal monoxides and those in the corresponding (highly conducting) parental transition metals.

In summary, therefore, the Bloch–Wilson theory is characterized by undoubted mathematical rigour, but, as we have seen, it predicts to be metals substances that are known in fact to be insulators! As Sir Alan Cottrell (Box 5) has succinctly pointed out,

> It [the Bloch–Wilson band theory] *would seem to have chosen not quite the right criterion for deciding between metals and insulators.*

In 1937 Rudolph Peierls, Nevill Mott and others realized that a drastic reappraisal of the Bloch–Wilson approach was necessary in order to account for these fundamental failings. Most notably, it was demonstrated that the Coulomb repulsion between electrons – which had completely been neglected in the original Bloch–Wilson formulation – could lead, under certain circumstances, to the complete localization of valence electrons at individual atomic sites in an insulator.

In the following sections we outline various other approaches that attempt to set out the physical basis of the distinction between metals and insulators and the associated metal–insulator or metal–non-metal transition.

The Polarization Catastrophe

Probably the first attempt to explain the microscopic electronic origins of the differences between metals and insulators – and with it the first discussion of the associated metal–insulator transition – was made by Dimitry A. Goldhammer in 1913 and Karl F. Herzfeld in 1927. The Goldhammer–Herzfeld classical view thus predates the Bloch–Wilson quantum-mechanical description of metals and describes the idea of – and the specific conditions for – a so-called polarization, or dielectric catastrophe, which leads to the release of *all* bound valence electrons from constituent atoms to yield the metallic state.

The Goldhammer–Herzfeld approach sets out a detailed scenario by which localized valence electrons 'tied' to their parent atoms can become set free as a result of mutual attractive interactions among *all* of the atoms in the material. It is salutory to note that the importance of this polarization approach – recall that it is based on classical, rather than quantum-mechanical, arguments – was recognized early on in the development of the quantum theory of solids, by no less a visionary than Arnold H. Sommerfeld, a pioneer of the quantum-mechanical description of metals. In a paper presented at the 1928 meeting of the German Chemical Society, Sommerfeld noted Herzfeld's (then) recent contribution and further suggested that one could easily understand how the valence electrons in atoms will experience an instability (in German, '*ein instabiler Zustand*') at sufficiently high atomic densities and thus become detached (ionized) from their parent atoms.

Sommerfeld therefore concluded that it was now (then!) possible, following the approach of Herzfeld, to predict which elements of the Periodic Table will assume a metallic character in their solid or liquid states. With this fundamental issue apparently resolved, Sommerfield obviously then felt able to concentrate his talent and efforts on the many and varied physical consequences of the resulting gas of free electrons – with quite remarkable success! (See, for example, the texts by Ashcroft and Mermin, and Kittel, p. 114 for detailed references.)

Let us now examine the physical basis of the Goldhammer–Herzfeld model. Think again of our assembly of caesium atoms with an average atomic spacing so large that

Box 5 Sir Alan Cottrell: *'Metals contain free electrons'*

Sir Alan Cottrell is internationally renowned as one of the founding fathers of modern metallurgy. His 1955 book *Theoretical Structural Metallurgy* marked a milestone in the quantitative explanation of metallurgical phenomena. Sir Alan has been responsible for transforming metallurgy and materials science into rigorous disciplines through his pioneering contributions to research and teaching. The quotation *'Metals contain free electrons'* is the first sentence of his remarkably perceptive article of 1960 entitled 'The metallic state'. Interestingly, in 1946 (aged 27), Dr Alan Cottrell, as he then was, presented to the Birmingham Metallurgical Society a lecture entitled 'What is a Metal?'!

Figure 22
Sir Alan Cotterell. Photograph courtesy of the Institute of Metals, London.

each 6s valence electron unquestionably remains firmly bound to its parent atom; the entire system under these conditions is unquestionably non-conducting (Box 4). The Herzfeld approach is a classical description in which each caesium atom is modelled in terms of an elastic spring connecting the valence 6s electron to its parent ion's core (Cs+). The strength of the valence-electron binding can be gauged by the 'stiffness' of this atomic spring and hence by the characteristic oscillations of the electron about its equilibrium position. In the low-density regime, the 6s electron executes only small spatial extensions about its equilibrium position. Hence its electronic polarizability – its ability to respond to an external electrical field – is relatively small.

Now, at the finite densities encountered in solid or liquid caesium a progressive weakening of the spring stiffness is brought about via the multitude of caesium–caesium atom interactions initiated by the screening of any individual Cs+-e− interaction. This process leads to a reduction in the characteristic frequency of the 6s valence electron, producing larger and larger displacements away from the parent ion's core. With this increase in spatial extent, the system becomes progressively more polarizable and the dielectric constant consequently increases. With further increases in density, we rapidly encounter a runaway situation by which a so-called 'polarization catastrophe' occurs. This ensures that each valence electron then becomes detached from its parent atom and is now set free to take its place in the conduction electron gas.

Herzfeld argued in 1927 that the characteristic vibrational frequency, ϑ_0, of the valence electron in a free atom will now be diminished in the condensed phase (be it liquid or solid) to the value ϑ, whereby

$$\vartheta = \vartheta_0 (1 - R/V)^{1/2} \qquad (2)$$

where V is the molar volume (at the particular density in question) and R is the molar refractivity of the isolated atomic state. Now R is related to the atomic (electronic) polarizability, α_0, via

$$R = \tfrac{4}{3} \pi N_A \alpha_0 \qquad (3)$$

where N_A is the Avogadro number. In the simple Herzfeld approach, R is assumed to be density-independent, but clearly V, the molar volume, is inversely proportional to the density of the condensed phase.

Now what happens to our isolated atom when we place it in a system containing, say, up to 10^{22} other identical atoms? As the density continually increases (viz., V decreases), the valence electrons become progressively less tightly bound to their parent ions' cores. This process continues until a critical value of the density (given by $R/V = 1$) is reached, whereupon the characteristic vibrational frequency (2) then drops catastrophically to zero, the valence electron is no longer subjected to any restoring force from the parent ion, and the electron becomes set free. The critical condition

$$R/V = 1 \qquad (4)$$

thus signals the transition from an insulator to a metal. In the Goldhammer–Herzfeld model this critical condition marks a polarization or dielectric catastrophe at the transition point for the transformation from localized to itinerant electrons. Depending upon the relative values of R for the isolated atom and V (at any given density) we can now rationalize the existence of metallic ($R/V \geq 1$) and insulating ($R/V < 1$) behaviours.

Following Herzfeld's arguments of 1927, the key feature which critically determines metallic versus insulating behaviour for any element relates to the ratio of the force holding the valence electron in place (which, approximately, could be taken as the first ionization energy of an atom) to the density of the valence electrons in the solid or liquid. For example, the element mercury has two 6s electrons in its valence shell and an extremely large ionization energy (≈ 10 eV), approximately three times that of atomic caesium. A single mercury atom therefore holds its valence electrons much more strongly than does a caesium atom. Nevertheless, elemental mercury at room pressure and room temperature is a metal because of its extremely high density, giving a low molar volume and consequently $(R/V) \simeq 1$. As Herzfeld pointed out in 1927, if mercury had, in the liquid or solid phase, the extremely large molar volume (low elemental density) of caesium it would be a non-metal! The transition from the metallic to the non-metallic state of elemental mercury has also been observed by Friedrich Hensel and colleagues. The variation of the electrical conductivity of '*expanded mercury*' is also shown in Box 4; from these remarkable experiments one can readily see the very drastic effect of varying the elemental density on the electrical conductivities both of mercury and of caesium. Note that, even at high temperatures, the transition from metallic to non-metallic behaviour occurs over a relatively narrow range of elemental density.

In contrast, the alkali metal caesium is metallic under ambient conditions of room temperature and pressure in spite of its low elemental density (large molar volume), in relation to other metals in the Periodic Table. The reader will recall the controversy following Davy's discovery of sodium and potassium (see p. 85); the approach pioneered by Goldhammer and Herzfeld now provides a natural explanation of the importance of key atomic properties (e.g. polarizability) that make an element a metal in the condensed phase.

This simple criterion (equation (4)) requiring only knowledge of the free atom's electronic polarizability and the elemental density (molar volume) was used successfully by Herzfeld *in 1927* to explain why elements in the Periodic Table are metallic, non-metallic or borderline when they are in the condensed phase. A clear and continuing demonstration of the great utility of this approach can be seen by referring to Figure 23, which shows (as a backdrop) the values of R/V for a range of elements at their normal solid or liquid densities. For example, the group VIIA elements (the halogens) and the noble gases all have values of R/V substantially less than unity and non-metallic behaviour is predicted and, of course, observed for these elements under normal conditions. The Herzfeld criterion can further allow one to predict for a (normally) non-metallic element, i.e. one having $R/V < 1$, the critical value of V (*viz.*, the density and pressure) which would produce a polarization/dielectric catastrophe at $R/V = 1$. This simple but incredibly powerful criterion has long been used to guide experimentalists towards estimates of the high pressures necessary for the metallization of nominally 'non-metallic' elements. Perhaps the most spectacular example is that of the pressure-induced metallization of fluid hydrogen reported recently by Bill Nellis and colleagues at Lawrence Livermore Laboratories in Berkeley, California (Box 2).

It has long been presumed that the element hydrogen at sufficiently high density (pressure) would eventually succumb to metallization. In Figure 24 we show the recently measured electrical conductivities for shock-compressed fluid hydrogen compared with the corresponding conductivities of the expanded fluid metals rubidium and

Figure 23

A plot of the ratio R/V for elements from various groups of the Periodic Table under ambient conditions on this planet. Here R is the molar refractivity and V is the molar volume at the density/pressure in question. The insert shows Professor Mike (Michell) J. Sienko, of Cornell University who was given the American Chemical Society Award for Chemical Education, in 1983. (Taken from P. P. Edwards and M. J. Sienko, *J. Chem. Education*, **60**, 691 (1983).)

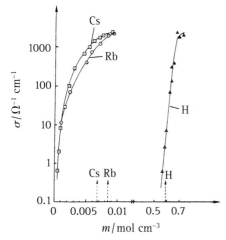

Figure 24
The measured electrical conductivities of fluid caesium, rubidium and hydrogen as functions of the molar atom density (m). The arrows indicate the predicted metallization densities for each element, based on the Goldhammer–Herzfeld model for a dielectric catastrophe at the onset of metallization. (Taken from F. Hensel and P. P. Edwards, *Chemistry: A European Journal*, **2**, 1201 (1996).)

caesium, measured for all of these elements at comparable temperatures (2000–3000 K) over a wide range of molar (atomic) densities. It is manifestly obvious that these three elements, hydrogen, rubidium and caesium, all undergo a continuous density-induced transition from a non-metallic to a highly conducting metallic state above some critical threshold density. In the case of fluid hydrogen at these high temperatures, pressures close to 2×10^6 atm are necessary in order to effect the transition to a highly conducting metallic state. The '*conventional*' alkali metals rubidium and caesium, which are unquestionably metallic at room pressure and temperature (Figure 1), now continuously transform into a state having exceptionally low electrical conductivity upon expansion to low density.

The metallization densities predicted from the Herzfeld criterion (equation (4)) for the three elements in question are shown as arrows along the molecular density axis for each element shown in Figure 24. These predictions are in excellent agreement with the experimental densities at which the three elements attain a limiting electrical conductivity close to 2000 Ω^{-1} cm^{-1}. The modern experimental data clearly reveal the onset of metallization in these elements at densities close to Herzfeld's 1927 estimates! The relatively small value of the electronic polarizability for the hydrogen atom necessitates the unusually high elemental densities required for the metallization of fluid hydrogen. In contrast, the large polarizabilities for atomic rubidium and caesium – and indeed all the alkali elements – ensure that the group I elements acquire metallic status at (elemental) densities readily commensurate with room pressures and temperatures on our planet.

The Mott Transition

A fundamentally different approach to the problem of metals versus insulators was pioneered by Sir Nevill Mott almost 50 years ago (Box 3). Mott's viewpoint relates to the dielectric screening properties of a gas of itinerant electrons and how such properties change as a metallic system transforms into an insulator and vice versa. To illustrate, I will consider the technologically important example of the doped group IVA semiconductors, e.g. phosphorus or arsenic donors or dopants in a pure silicon or germanium host crystal.

Phosphorus and arsenic atoms each have five valence electrons compared with the four associated with silicon and germanium. Thus, each phosphorus or arsenic donor or impurity atom doped into the group IVA semiconductors will possess one additional valence electron – it is the fate of this electron that will concern us here. For such an isolated donor atom in silicon or germanium, the additional valence electron moves in the Coulomb attractive field of the parent donor ion (e.g. P$^+$ or As$^+$) in a manner reminiscent of a 1s electron orbiting around the proton in an isolated hydrogen atom. However, since the parent phosphorus or arsenic cation is itself completely embedded within the silicon dielectric host, this potential field, $V(r)$, between the additional electron and the parent ion core (P$^+$ or As$^+$) is significantly reduced from $-e^2/r$ (the case for the hydrogen atom) to $-e^2/(\epsilon r)$, due to the presence of the background (dielectric constant ϵ) of the silicon or germanium lattice. For silicon, ϵ is approximately 13. As a result, the extra valence electron moves in a highly expanded Bohr orbit whose Bohr radius, a_H^*, is now considerably larger than that of the corresponding 1s electron orbit in the hydrogen atom (Figure 25).

Under such conditions, for example, silicon lightly doped with phosphorus (hereafter Si:P) will be insulating, having a DC electrical conductivity that goes to zero as the temperature approaches absolute zero. In contrast, at a sufficiently high density of dopant atoms (equivalently, at a small average interparticle or interdonor separation), we will have a substantial overlap of the constituent electronic wave functions. The

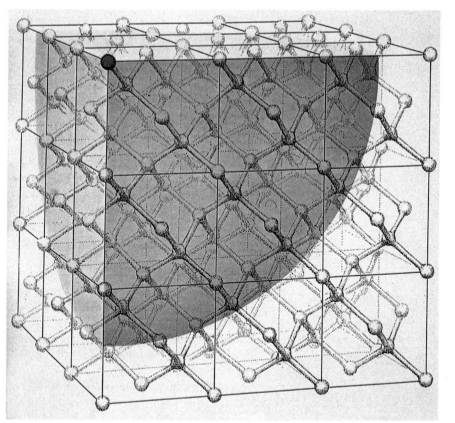

Figure 25
Charge carriers (electrons) can be supplied to the semiconductors silicon or germanium by doping the crystal with small amounts of, say, phosphorus or arsenic. The dopant atoms (P or As) have five valence electrons to the four of silicon and germanium. The fifth electron is only weakly bound to its parent (P or As) atom and extends over very considerable distances from it, indicated by the large, light-blue-coloured volume. We have, then, something akin to a 'swollen' hydrogen atom in the silicon lattice, and localized around the dopant ion (P$^+$ or As$^+$) (Taken from H. Ehrenreich, *Sci. Am.*, **217**, 195 (1967) and used with permission.)

valence electrons can therefore tunnel from one expanded orbital to another and screen the Coulomb interaction between the orbiting electron and the P$^+$ core. This screening then ionizes further dopant electrons – which themselves enhance the screening – and a runaway situation results in the ionization of *all* valence electrons at the transition from an insulator to a metal. Mott presented a simple but elegant analysis of the problem based on the screening properties of a conduction electron gas. According to the so-called Thomas–Fermi description, the ion-core–dopant-electron interaction can be represented by a screened Coulomb potential of the form

$$V(r) = \left(\frac{-e^2}{r} \right) \exp(-qr) \tag{5}$$

where q^{-1} is defined as the characteristic Thomas–Fermi screening length which is critically dependent upon the electron density, n, of the uniform electron gas, *viz.*

$$q^{-1} = \frac{4me^2(3n\pi)^{1/3}}{\hbar^2} \tag{6}$$

The condition for metallic behaviour is that the screening length be less than the characteristic Bohr radius; *viz.*

$$q^{-1} \lesssim a_H^* \tag{7}$$

giving

$$n_c^{1/3} a_H^* \gtrsim 0.25 \tag{8}$$

where n_c is now the critical density of dopant phosphorus atoms which marks the location of the electronic transition from a metal to an insulator and vice versa.

Mott further proposed that, at this critical carrier density, a first-order (discontinuous) transition from a metal (having $\sigma(0) \neq 0$) to an insulator (characterized by $\sigma(0) \neq 0$) would occur at $T = 0$ K. This conclusion, namely that a doped semiconductor such as Si:P should inevitably undergo a metal–insulator transition as a function of doping, was perhaps not surprising; what was remarkable, however, was Mott's conjecture that, at the critical density, n_c, *all* valence electrons would become free at once, not just a few of them. In essence, *either none of the electrons is free to move, or all are!* The words of the poet John Dryden[†] penned over 300 years ago, '*Either be wholly slaves or wholly free*', beautifully describe Mott's vision of the metal–insulator transition. This kind of discontinuous electronic phase transition from a metal to an insulator (and vice versa) at $T = 0$ K has now widely become known as '*The Mott Transition*'. The ramifications for a '*Gedankenexperiment*' for a doped semiconductor at $T = 0$ K are sketched in Figure 26, which reveals two possible scenarios for the DC electrical conductivity ($\sigma(0)$) of a system such as Si:P undergoing a composition-induced metal-to-insulator transition. Here the relevant experimental control parameter is the average distance, d, between donor atoms. Obviously, this distance d is intimately related to n, the donor density, via $d \simeq n^{-1/3}$, and d_c is now the (average) critical distance between dopant centres *at* the metal–insulator transition.

For very large interparticle distances ($d > d_c$) we have an insulating sample (with $\sigma(0) = 0$) and complete localization of the dopant-electron wave function at individual atom sites. According to Mott, we would then have a discontinuation transition at $d = d_c$ from a metal to an insulator as we exceed the (average) critical distance between phosphorus donor states (Figure 26).

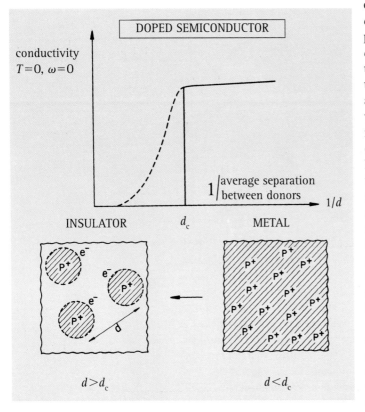

In contrast, Eliu Abraham and colleagues have predicted *a continuous* metal-insulator transition. The two extreme scenarios of continuous and discontinuous metal–insulator transitions at $T = 0$ K are compared in Figure 26.

Mott further argued that, *at the very transition from a metal to an insulator*, there exists a so-called '*minimum metallic conductivity*', σ_{min}, which signifies the *minimum* value of the DC electrical conductivity associated with genuine metallic behaviour. Thus, at $T = 0$ K, the conductivity would jump (discontinuously) from a value of zero to a value of σ_{min} as the system suddenly transforms from an insulator to a metal. A discontinuous metal–insulator transition is a remarkable tantalizing theoretical prediction – and has long been sought by experimentalists and theoreticians alike!

In fact, the experimental situation is curiously equivocal – even for measurements close to $T = 0$ K. In a remarkable series of experiments, Tom Rosenbaum, Gordon Thomas and colleagues at Bell Telephone Laboratories (now Lucent Technologies) carefully measured the DC electrical conductivities of large, single-crystal ingot samples of Si:P of various dopant concentrations, down to temperatures of just 30 mK – i.e. only

Figure 26
A schematic representation of the situation at $T = 0$ K for a single crystal of Si:P at both high and low donor-atom densities. We illustrate two possible scenarios for the metal–insulator transition, namely a discontinuous (solid line) and a continuous (dotted line) transition.

[†] *The Hind and the Panther*, John Dryden, 1687.

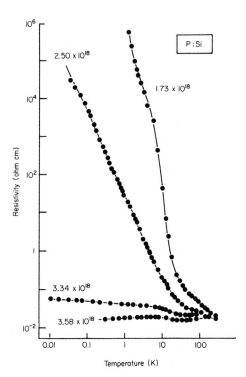

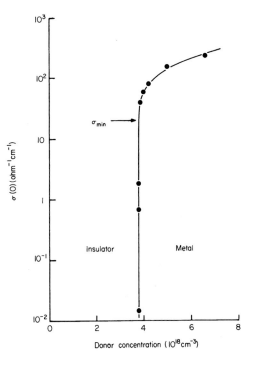

Figure 27 (left)

The measured electrical resistivities of a range of single crystals of phosphorus-doped silicon (Si:P) as functions of temperature. The electron (or donor-atom) content supplied by phosphorus donor atoms is indicated on each curve in units of electrons cm^{-3}.

Figure 28 (right)

A semi-logarithmic plot of the zero-temperature conductivity, $\sigma(0)$, versus the donor atom content for a range of single-crystal samples of silicon doped with phosphorus. The transition from insulator to metal is extremely sharp, but probably continuous. Also indicated is Sir Nevill Mott's estimate for the so-called *minimum metallic conductivity*, σ_{min}, for the Si:P system. (Taken from T. F. Rosenbaum, K. Andres, G. A. Thomas and R. N. Bhatt, *Phys. Rev. Lett.*, **45**, 1723 (1980).)

0.03 K above absolute zero! In Figure 27 we reproduce their experimental data, which show the measured electrical resistivities as a function of temperature for a series of single crystals of differing phosphorus content. For samples with phosphorus-atom concentrations (n) of 1.73×10^{18} and 2.50×10^{18} cm^{-3}, the system is undoubtedly insulating, since the electrical resistivity (ρ, the inverse of the conductivity) goes towards infinity (*viz.* $\sigma \rightarrow 0$) as the temperature tends towards absolute zero. For samples having $n = 3.34 \times 10^{18}$ and 3.58×10^{18} cm^{-3}, the system appears metallic, since $\sigma \neq 0$ in the low-temperature limit. We therefore witness a metal–insulator transition emerging at very low temperatures as the dopant concentration is changed. But, is this unquestionable transition from a metal to an insulator *discontinuous* or *continuous* at $T = 0$ K?

Now, these conductivity measurements extend 'only' down to 0.03 K. These scientists have also attempted to extrapolate their experimental data all the way to the temperature of absolute zero in order to obtain a true value for the '*zero-temperature conductivity*', $\sigma(0)$, for each of these samples (See Box 3).

The results of such '*zero-temperature*' experiments are shown in Figure 28, which shows values of $\sigma(0)$ for these Si:P samples of different donor concentrations traversing the metal–insulator transition. It is immediately apparent that the transition from insulator to metal at $T = 0$ K in Si:P is unquestionably sharp – but possibly continuous – since we see several samples having $\sigma(0)$ values intermediate between zero and σ_{min}. Mott's estimate for σ_{min} and also the critical density, n_c (from $n_c^{1/3} a_H^* = 0.25$, equation (8)), are also indicated in Figure 28. It is remarkable that Mott's 'fingerprint' parameters for the location of the metal–insulator transition, *viz.*, n_c and σ_{min} are indeed excellent indicators for the experimental situation close to $T = 0$ K.

However, even these apparently '*simple*' experimental extrapolations – from 0.03 to 0 K – are still highly controversial; it has even been stated that a temperature of 0.03 K might be too '*high*' for one to carry out a meaningful extrapolation of the data to a temperature of absolute zero! The search for a genuine *Mott Transition* continues, even today.

What is abundantly clear is that the Mott criterion (equation (8)) and Mott's minimum metallic conductivity, σ_{min}, appear to survive as very reasonable indicators

Figure 29

A three-dimensional representation of the Mott–Hubbard U (in electronvolts) for atoms of the Periodic Table of the elements. It can be seen that high values of the atomic U correspond to insulating elements under standard conditions, whereas low values of U typify the metallic elements. (Taken from P. P. Edwards, R. L. Johnston, F. Hensel, C. N. R. Rao and D. P. Tunstall, *Solid State Phys.*, **52**, 229 (1999).)

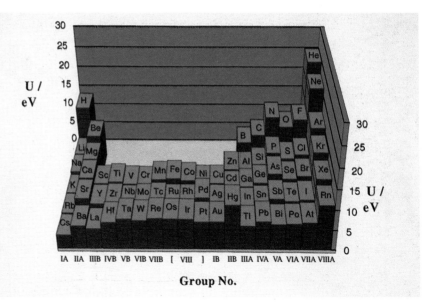

for the demarcation '*metal or insulator*' in this doped semiconductor, even though the conductivity itself may go continuously to zero as the transition is crossed, in disagreement with Mott's original idea of a discontinuous (first-order) transition (Figure 26).

The Mott–Hubbard Transition

Mott's critical insights opened the way to several important formalizations of the problem of the metal–insulator transition; a landmark development was made by John Hubbard in the mid-1960s. Let me return to the issue of our '*expanded*' alkali metal such as caesium. Imagine once again that the caesium atoms are sufficiently far apart – previously we have said 1 m! – that the 6s valence electrons are unquestionably bound to their parent (neutral) atoms. For any electrical conduction to occur, a 6s electron must be taken away from one caesium atom and (instantaneously) attached to another, distant caesium atom, leading to the production of the two charged species Cs^+ and Cs^-. Removing the 6s electron from one caesium atom inevitably costs energy – the first ionization energy (I), some 3 eV – and giving it to another caesium atom at least returns an electron affinity energy, E_A (for Cs, this is 0.5 eV). We assume that the 6s electron, with opposite spin, joins the 6s electron already present in the outermost orbital of the other caesium atom. Hence, there is a Coulomb repulsion, e^2/r, between the two electrons at any average electron–electron separation, r.

Now, since $I > E_A$, the energy of the highly expanded array of Cs atoms would have to be raised should this transfer of electrons (charge) take place. Thus, there is an energy barrier, or energy gap, $(I - E_A)$ inhibiting any electron transfer; this is customarily denoted by U, denoting the so-called Mott–Hubbard gap. This Mott–Hubbard energy gap naturally preserves the atomically neutral, non-conducting state of our caesium at large interatom separation. Under these conditions the system would patently be an insulator (see Figure 21 in Box 4).

Now, if the distance between caesium atoms is gradually reduced (which would be equivalent to an increase in the elemental density), the inevitable overlap of the constituent 6s atomic orbitals will allow a degree of (spontaneous) electron transfer from one caesium atom to another. In quantum mechanics, this electron transfer is specified by a so-called *overlap or hopping integral*. Of course, it is this interaction which leads to molecular binding in diatomic states. In our extended assembly of many caesium atoms, this interaction leads to a characteristic electronic bandwidth, Δ. This orbital overlap (embodied in Δ) serves to oppose the localizing influence of the Mott–Hubbard gap, U, and encourages the delocalization of the 6s valence electron between the constituent atoms.

Eventually, at some critical density of caesium, the electronic bandwidth becomes so large as to overcome U and spontaneous ionization of each caesium atom takes place – at no energy cost; caesium now becomes a metallic conductor. This is known as the Mott–Hubbard Transition from an insulator to a metal, and vice versa.

The Mott–Hubbard correlation energy has been used to attempt a natural demarcation of the elements of the Periodic Table into metals and insulators (Figure 29). On this basis, metallic elements have values of the Mott–Hubbard U of about 8–10 eV or below, whereas non-metallic elements have values of U in excess of 9 eV. This is once again a good coarse-grained indicator of the importance of atomic properties (in this case, I and E_A) in dictating the metallic versus insulating status of elements of the periodic classification. One can easily discern a link with Pauling's view of metals.

The detailed theory of the Hubbard model also leads to the simple criterion $n_c^{1/3} a_H^* \simeq 0.2$ for the metal–insulator transition, which is obviously very close to the value derived from Mott's dielectric-screening approach (equation (8)), considering that the underlying physical arguments are completely different. In the Mott–Hubbard viewpoint, therefore, a system is a metal or an insulator depending upon whether the electronic bandwidth, Δ, is large or small, respectively, relative to the repulsive Coulomb energy, U, for the interaction among the outermost valence electrons. With $\Delta > U$ we have a metal, whereas a Mott–Hubbard insulator has $\Delta < U$.

Disorder: The Anderson Transition

No material will ever exhibit the perfectly regular structure of an ideal solid epitomized earlier in Figure 16. However, our discussion of metals and insulators has so far centred on such an assumption of a perfect crystalline structure throughout, even though all of the experimental examples (expanded Cs and doped semiconductors) are unquestionably disordered. We must now enquire what the consequences to the electronic structure of introducing *disorder* into the system are. Once we no longer depend on the implicit periodicity of the system to assist us in deciding whether it is a metal or an insulator (see p. 96), we will find that the distinction between the crystalline and non-crystalline arrangements of states becomes blurred. Indeed, the venerable example of the metal–insulator transition represented by Donald Holcomb of Cornell University in Figure 30 is of a phosphorus-doped silicon crystal in which the dopant phosphorus atoms are randomly arranged throughout the host crystal lattice. This cartoon is a two-dimensional representation that is thought to be closely representative of the situation in Si:P at a donor concentration just below the critical density, n_c, for the metal–insulator transition. The large physical dimensions of the isolated dopant atoms (a_H^*), in comparison with the (average) nearest-neighbour separation means that the system is far removed from the (idealized) regular and ordered lattice of hydrogenic dopant states pictured in our earlier treatments (e.g. Figure 16). We stress that this electronic disorder will inevitably be present – even though the material itself is perfectly crystalline! In this case disorder arises because of the natural random substitutional doping of the P atoms into the host (Si) single crystal (Figures 25, and 26).

The presence of substantial disorder in a nominally crystalline system can have the dramatic effect of localizing all the electronic wave functions, leading to insulating behaviour. This remarkable innovation was first suggested by Philip W. Anderson in 1958 in a highly influential paper, which, the author has mischievously noted, '*is often cited, but seldom read*'! (Box 6). One person who clearly *did* read this paper – and recognize its widespread importance – was Sir Nevill Mott, who was instrumental in developing and applying Anderson's seminal ideas to the issue of metals versus insulators in disordered

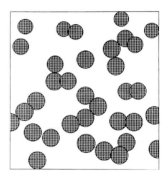

Figure 30

A two-dimensional representation of the doped semiconductor Si:P at a dopant concentration just below n_c, the critical density for the metal–insulator transition. The radius of each shaded circle is approximately the Bohr radius (a_H^*) of the dopant or impurity phosphorus atom. The system obviously forms a highly disordered, overlapping state at this composition close to n_c. The corresponding electrical conductivity behaviour for a system close to the transition has been shown earlier, in Figures 27 and 28. (Taken from D. F. Holcomb in *Metal–Insulator Transitions Revisited*, edited by P. P. Edwards and C. N. R. Rao, Taylor and Francis Ltd, London (1995), p. 65.)

Box 6 Professor P. W. Anderson *'Anderson localization ... and a theory of white paint!'*

In 1970, P. W. Anderson (Figure 31) noted *'About a dozen years ago I wrote an obscure paper, now widely quoted but perhaps not as widely read, on "Absence of diffusion in certain random lattices".'* The central idea of this ground-breaking paper was to suggest that, under certain circumstances, an electron in a disordered system may become spatially localized, in that its wave function decays exponentially from some point in space (this phenomenon is illustrated in Figure 34 overpage). This is known as Anderson localization. It is now anticipated that a similar phenomenon will occur for electromagnetic waves.

Diederick Wiersma, Paolo Bartolini, Ad Lagendijk and Roberto Righini have recently reported direct experimental evidence for Anderson localization of light in optical experiments on very strongly scattering semiconductor powders (*Nature*, **390**, 671 (1997)). Wiersma and colleagues have demonstrated that light can be forced to stand still through the process of multiple scattering and wave interference. In Figure 32 we can see a phase transition in light waves from classical diffusion to localization. In Figure 32(*a*) we have classical diffusion through a scattering medium (such as TiO_2 particles in white paint), for which the intensity distribution exhibits a multitude of fluctuations due to interference. Diffusion is possible since the peaks overlap. In Figure 32(*b*) we see that, when the scattering becomes strong enough, the light is localized in discrete separate peaks; the lack of significant spatial overlap between the peaks prohibits the transport of light throughout the system. One wonders what colour the paint would be in such a situation!

Figure 31
P. W. Anderson. (Photograph courtesy of the Cavendish Laboratory, Cambridge.)

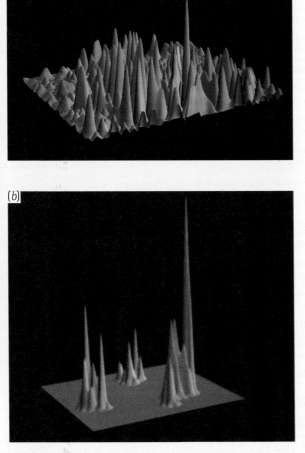

Figure 32
The localization of light. (Taken from J. Sajeev, *Nature*, **390**, 661 (1997), with permission.)

systems. A fascinating and important consequence of this disorder is a remarkable transformation from metal to insulator; the term *Anderson localization* thus refers to the localization of electronic wave functions brought about by the presence of disorder in the system. This is a wave-mechanical phenomenon that arises in other areas of wave-propagation science as well, such as acoustics and scattering of light (Box 6). How does Anderson localization come about?

Consider our simple model system in which each electron moves in a perfectly periodic potential, as shown in Figure 23. As we have noted earlier (p. 96), in these situations, each electron wave will itself be subjected to a periodic potential – and this Bloch wave will extend throughout the entire system as a travelling wave. Anderson then posed the key question: '*What happens if the potential energy wells of such a perfectly crystalline system are then made random?*', as shown in Figure 33. Here we have allowed the depths of the potential energy wells to assume completely random values, with a corresponding spread of energy values of magnitude V_0. Under these circumstances, what happens to the Bloch electronic wave function?

The mathematical analysis of this problem is particularly complex, but Anderson was able to show that, as the variation in the depths of the wells becomes larger, there comes a point where a Bloch electron wave could no longer extend throughout the system, Figure 16, but would be localized in a certain region of the system. In his masterly 1958 article entitled 'Absence of diffusion in certain random lattices', Anderson proved this conclusion. The results can be qualitatively seen from the following discussion. For a nominally disordered system in which the random potentials are weak, *viz.* $V_0 \ll \Delta$, the conduction (valence) electrons are still essentially free to propagate throughout the system (Figure 34(a)). If we now gradually increase the degree of disorder such that $V_0 < \Delta$, then the scattering of conduction electrons becomes stronger. The amplitude of the wavefunction begins to fluctuate through the system and this will lose phase coherence over a distance usually termed the electronic mean free path (Figure 34(b)).

Eventually, as the disorder increases sufficiently, the mean free path of the electron is shortened to such a degree that it now becomes comparable to the average interatomic distance, d, between the potential wells. If the disorder (scattering) becomes even more dominant, the electron wave function will become *completely* localized in a discrete region of space; *viz.* its amplitude will fall off exponentially with distance, so that a typical envelope wave function would appear like Figure 34(c). These exponentially localized states ultimately signify the Anderson (disorder-induced) transition from a metal to an insulator. According to Anderson's work, there exists a critical value of (V_0/Δ) for which, at zero temperature, diffusion of electrons throughout the material is impossible; this is the situation on account of which a nominally metallic system now becomes insulating because of the pressure of disorder!

New Metals for Old

In the preceding sections I have briefly reviewed just some of the natural physical processes and interactions which determine whether a chemical element or chemical compound behaves as a metal or as an insulator. As I hope you can now see, even the simplest of enquiries . . . '*metal or not?*' . . . belies deep and complex chemical and physical issues relating to localization versus itinerancy of electrons in crystalline and disordered systems.

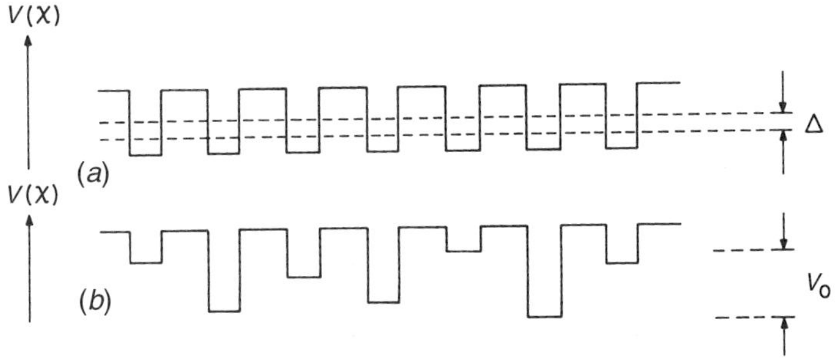

Figure 33

The potential energy of an electron within (a) a periodic field and (b) a random potential field. Here Δ is the one-electron bandwidth in the absence of the random potential V_0.

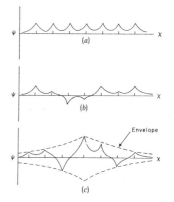

Figure 34

A typical electron wave function Ψ in an Anderson lattice. (a) An extended wave function, having Bloch character. Here the mean free path of electrons far exceeds the separation between potential wells. (b) A wave function, still extended, but subject to the effects of disorder (in the potential) and $\Delta > V_0$. (c) A weakly localized function for which $\Delta \approx V_0$, with the overall form of the (localized) electron wave function. (Adapted from N. F. Mott, *Metal–Insulator Transitions*, Taylor and Francis Ltd, London, 1990.)

Part of the problem – and, indeed, part of the enduring fascination of the subject – is that a variety of competing, yet complementary, effects must always be taken into account. The effects of electron–electron interactions, screening of electrons and disorder are just three key facets identified here. In the majority of cases, the accompanying physical models are basically descriptions of the situation at $T = 0$ K, at which, at least in our *Gedankenexperiment*, the fundamental difference between metals and insulators is crystal clear. However, at finite temperatures – temperatures for the experimentalist (!) – the theoretical issues become even more daunting.

The scale of the theoretical problem at hand is hinted at in Figure 35, which is a representation of three important facets of the problem, namely temperature, electron-electron interactions and disorder. David Logan, Yolande Szczech and Michael Tusch remind us that most (indeed probably all) of conventional band theory is unfortunately confined to a single point at the origin of Figure 35 – this represents the situation in which one has *no* electron-electron interactions, *no* disorder *and* a temperature of absolute zero. *All* real experimental systems lie away from the

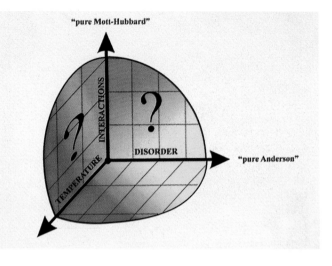

Figure 35

A representation of the various competing – but complementary – effects of interelectron interactions, disorder and temperature in relation to the metal–insulator transition. *All* experimental systems fall somewhere in the three-dimensional space away from the origin, which represents the (idealized) limit of a zero-temperature, non-disordered and non-interacting situation. This figure was kindly redrawn by Dr Martin O. Jones, the University of Birmingham. (Taken from D. E. Logan, Y. H. Szczech and M. A. Tusch in *Metal–Insulator Transitions Revisited*, edited by P. P. Edwards and C. N. R. Rao, Taylor and Francis Ltd, London (1995), p. 343.)

origin of Figure 35. In these regions (shaded in Figure 35) the answer to the question: '*Metal or not?*' will depend critically upon the subtle interplay of *all* of these key, underlying physical interactions.

It will now be obvious to the reader that the closer one gets to the actual region of the transition from a metal to an insulator (and vice versa) the greater the scale and complexity of the theoretical problem. This is why Mott's famous *Gedankenexperiment* at $T = 0$K (Box 3) is so beautifully tantalizing . . . '*Either be wholly slaves (insulators) or wholly free (metals)!*'. The scale and complexity arises since it is inevitable that *all* such competing effects and interactions will become of comparable magnitude as we move closer and closer to the transition from a metal to an insulator.

Paradoxically, I have also illustrated how simple, but highly effective criteria, such as the Goldhammer–Herzfeld polarization catastrophe, the Mott metallization condition, the Mott minimum metallic conductivity etc., continue to serve as very effective guides for rationalizing the issue: '*Metal or insulator?*'.

What of the future? My personal view is that the new frontier – '*the new metals*' – will surely be . . . '*the new, unnatural metals*'. These will be complex chemical compounds formed by deliberate chemical synthesis – 'self-assembly' in the current vogue (!) – with highly specific and spectacular target electronic properties in mind. These solid-state architectures almost certainly will not have been found previously in the natural or man-made worlds. Roald Hoffmann's visionary comment from just over a decade ago is then particularly apt in our look forwards: '*Chemists must be able to reason intelligently about the electronic structure of compounds they make in order to understand how these properties and structures may be tuned*'. (Ref. 3 in Further Reading, p. 114.)

It will be interesting to see to what extent the theories and physical models of electronic structure outlined here are directly applicable to the rapidly evolving new breed of metals . . . increasingly complex multinary oxides, nitrides, oxide halides and the like. In this millennium we now look set to enter the era of *the chemical electronic engineering of the new metals*. Let me finish with just a few comments on just one of the current issues, namely, chemical compounds lying close to the metal–insulator transition.

A fundamental property of a conventional metal such as copper or gold is its superb ability to conduct electricity. However, as Sir Alan Cottrell has recently noted, '*much more exotic happenings appear in materials that are on the borderline between metals and insulators*'.

In particular, I wish to highlight the spectacular case of just one clan of exotic chemical compounds that can be '*engineered*' to sit at the very borderline between metals and insulators. These are the high-temperature (high-T_c) oxide superconductors; remarkably these compounds – the ultimate metals – have even been termed '*marginally metallic*' by C. N. R. (Ram) Rao or equivalently '*barely metallic*' by Yozuru Tokura. We now even have a new class of '*half-metallic*' oxides!

The discovery of high-temperature superconductivity in 1986 by J. George Bednorz and K. Alex Müller of IBM Laboratory, Switzerland now stands as one of the greatest experimental discoveries in science. It also ranks alongside the discovery and isolation of sodium and potassium by Sir Humphry Davy as a development that forces us to examine our very concept of metals and the metallic state. For, in this dull-black ceramic oxide of the complex chemical composition $HgBa_2Ca_2Cu_3O_{8+\delta}$, we have unquestionably the ultimate metal – the very best conductor of electricity at temperatures below 135 K (Figure 36). Paradoxically, the same chemical compound at room temperature could not justifiably be called a metal! This must surely be self-evident to the reader even from a cursory glance at the levitating black pellet which is currently the very best high-temperature superconductor (Figure 8)!

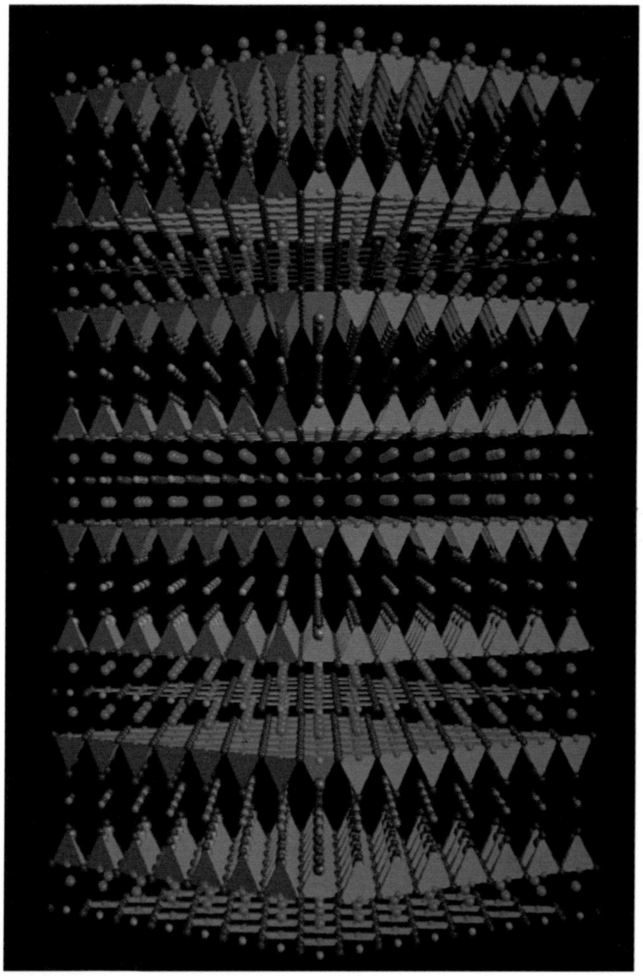

Sir Nevill Mott, Sasha Alexandrov of Loughborough University and I have noted that, in their so-called '*normal*' state (i.e. at temperatures above T_c), these multinary oxide superconductors are characterized by electrical conductivities typical of Mott's *minimum metallic conductivity*, σ_{min}. Recall that σ_{min} represents the absolute minimum value of the electrical conductivity of a compound associated with what we would conventionally regard as '*metallic*' behaviour (see p. 106). It is immediately transparent, therefore, that these high-T_c oxide superconductors thus lie on the tantilizing borderline between metals and insulators, i.e. close to – but not at – the point of complete electron localization. Yet, remarkably, they transform from this barely metallic state to the ultimate of metals – the high-temperature superconductor – when cooled to a temperature below 135 K!

The microscopic origins of high-temperature superconductivity are still highly controversial, but Sir Alan Cottrell's perceptive assessment firmly places these remarkable compounds '*in the schizophrenic state between metals and insulators*'. This statement underpins my earlier comment that, to understand metals and insulators, one must also understand the transition between them. I would now argue that, to understand high-temperature superconductivity, one must simultaneously understand the occurrence of, and the transitions between, the metallic, insulating *and* superconducting states of matter in such complex chemical systems.

Figure 36

A high-temperature superconductor composed of mercury, barium, calcium, copper and oxygen ($HgBa_2Ca_2Cu_3O_{8+\delta}$) exhibits the highest transition temperature discovered thus far. Below a temperature of some 133 K, this compound offers no resistance whatsoever to the flow of electricity. In keeping with all other '*high-T_c*' cuprate ceramic superconductors, the mercury compound contains extended planes of copper and oxygen atoms that allow conduction – and superconduction (!) – of electrons through the material. (Kindly supplied by Dr Graeme B. Peacock.).

Our story, therefore, has now come full circle. The discovery of sodium and potassium by Sir Humphry Davy almost two centuries ago necessitated a complete reappraisal of the enquiry *'metal or not?'*. The oxide high-temperature superconductors and other marginally metallic chemical compounds now prompt us to initiate another enquiry, namely one into *the new chemistry of the new metals . . . 'metal or not, . . . or what?'*.

Further Reading

1. Sir Nevill Mott, *Metal–Insulator Transitions* (Taylor and Francis, London, 1990). An extensive account of metal-insulator transitions across a wide range of systems – it sets the scene for the present chapter – Mott's book assumes a working knowledge of modern solid state physics (see below).

2. N. W. Ashcroft and N. D. Mermin, *Solid State Physics* (Saunders College, Philadelphia, 1976). A classic text in solid state physics – extremely useful for background issues relating to the nature of electrons in solids.

3. R. Hoffmann, *Solids and Surfaces: A Chemist's View of Bonding in Extended Structures* (VCH Publishers, Weinheim, 1988). This beautifully crafted little book is centred on the idea that the chemist's intuitive approach to bonding in molecules could, and indeed should, be useful to the solid state physicist. The idea is proven and the book, a joy to read!

4. C. Kittel, *Introduction to Solid State Physics* (John Wiley & Sons, New York, 1976). This pioneering text gives a comprehensive coverage; a classic text for modern solid state physics.

5. A. H. Cottrell, *Introduction to the Modern Theory of Metals* (The Institute of Metals, London, 1988). Another classic, this text attempts (and succeeds!) to incorporate and apply the modern theory of metals to metallurgy. Highly readable, complete but concise.

6. C. N. R. Rao and J. Gopalakrishnan, *New Directions in Solid State Chemistry*, second edition (Cambridge University Press, Cambridge, 1997). A visionary text which highlights modern solid state chemistry and indicates new directions for the subject in a very concise manner. The book is most useful for all practitioners of modern solid state science – whether chemists, materials scientists or physicists!

7. A. P. Sutton, *Electronic Structure of Materials* (Clarendon Press, Oxford, 1993). An excellent book centred around the modern approach that both the reciprocal-space *and* real-space viewpoints should be used for a realistic description of the electronic structure of materials.

8. R. Cotterill, *The Cambridge Guide to the Material World* (Cambridge University Press, Cambridge, 1985). This beautiful book – lavishly illustrated – is, to quote, '. . . *intended for those who seek a concise but comprehensive picture of the world at the microscopic and atomic levels.*' It succeeds at presenting a non-mathematical description of the physics, chemistry and biology of nature's materials. Ours is a material(s) world, and this book conveys key issues of that world!

Two books that attempt to present reviews of the vast array of experimental systems traversing the transition between the metallic and non-metallic states of matter, are:-

9. P. P. Edwards and C. N. R. Rao (editors), *Metal–Insulator Transitions Revisited* (Taylor and Francis, London, 1995).

10. P. P. Edwards and C. N. R. Rao (editors), *The Metallic and Nonmetallic States of Matter* (Taylor & Francis, London, 1985).

11. P. P. Edwards and M. J. Sienko, 'What is a Metal?' *Int. Rev. Phys. Chem.*, **3**, 83 (1983). An attempt to survey simple predictive models and operational definitions of the metallic state. Obviously an 'early' attempt at answering one-third of the trilogy of questions posed by the present chapter!

12. Paul Davies (editor), *The New Physics* (Cambridge University Press, Cambridge, 1989). See, in particular, the articles by Anthony Leggett and David Thouless which present an excellent background to many of the issues raised here.

The Clothing of Metal Ions: Coordination Chemistry at the Turn of the Millennium

6

Malcolm Chisholm

Introduction

Metals comprise well over half of all known elements and their function is fundamental to our existence. Metals play a pivotal role in biological and environmental chemistry, ranging from harvesting light and fixation of nitrogen in plants to controlling respiratory and muscular systems in mammals. Metal ions also play a dominant role in the economic development of our society. They are involved in the production of common chemicals such as methanol and acetic acid, in the formation of polymers and plastics, in the cracking, reforming and cleaning of fossil fuels, in solar-energy conversion, in pigments and paints and advanced optoelectronic and electromagnetic devices. An understanding of how metal ions influence their environment and how their environment influences their own properties requires an understanding of modern coordination chemistry, a field of study which finds its roots in the work of chemists such as Werner, who, at the turn of the nineteenth century, was attracted to the study of the colourful complexes of cobalt. It is from these humble beginnings that we have now come to an age in which we can unravel the mysteries of the role of metal ions in biology and we can design new materials and devices that rival and even surpass those evolved in nature.

In the Beginning

Metal ions exhibit a marked preference for their environment in nature (Figure 1). It is no accident that minerals and gems have the form that they do. The co-existence of the aluminium(3+) and chromium(3+) ions within a hexagonal oxide lattice gives rise to rubies. Similarly, the ability of iron to combine with oxygen and sulfur yields magnetite and iron pyrites (fool's gold), respectively. That magnesium(2+) and iron(2+) can be accommodated within a porphyrin ligand allows the harvesting of light in plants and the uptake and release of oxygen in warm-blooded mammals. It is, however, only through an understanding of the behaviour of metal ions in solution that we advanced to the position we are in today.

The most common and abundant solvent is of course water, and much of our current knowledge came about through the studies of the reactions and the properties of metal ions in aqueous media. When, for example, a crystal of salt is dissolved in water the now well-known sodium chloride lattice structure is broken up as the Na^+ and Cl^- ions are solvated by water molecules. The interactions between the Na^+ and Cl^- ions and water are dipolar in nature (electrostatic) and these ions are often represented as $Na^+(aq)$ and $Cl^-(aq)$ because the water molecules bonded to each Na^+ ion and Cl^- ion are in rapid exchange with the water molecules present in solution. The actual rate of exchange of bound and free water molecules is more than a million times per second, and for large cations such as caesium, Cs^+, and barium, Ba^{2+}, the rate of exchange may approach the limiting rate of diffusion, about 10^9 s^{-1} at room temperature. This situation may be contrasted to that of, say, chromium sulfate, $Cr_2(SO_4)_3$, which, when it is dissolved in water, contains the well-defined $Cr(H_2O)_6^{3+}$ ion wherein each Cr^{3+} ion is surrounded by six water molecules in an octahedral environment. The water molecules bound to the Cr^{3+} ion do not undergo rapid exchange with the water molecules of the solution. Indeed, the rate of exchange of H_2O is about 10^{-6} s^{-1} at room temperature,

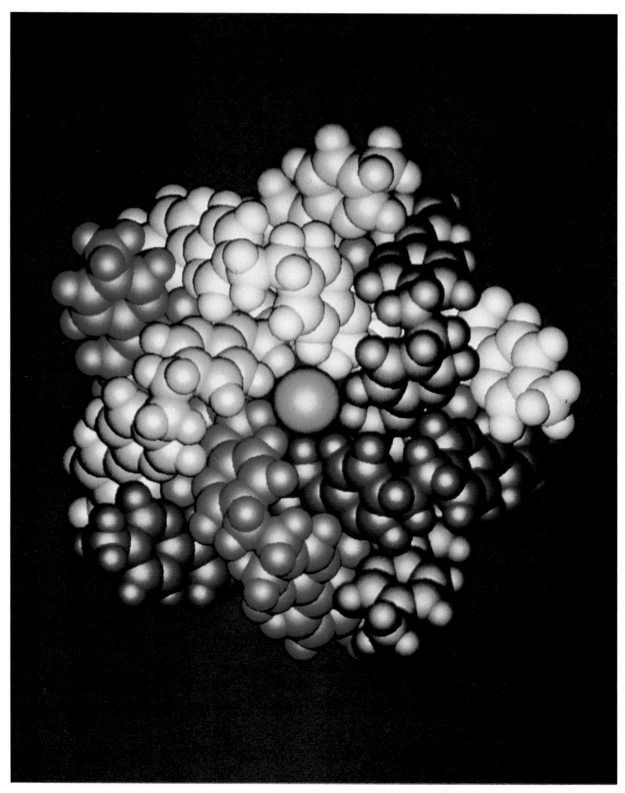

Figure 1
A chloride anion surrounded by a circular helicate is the complex cation $[(FeL)_5Cl]^{9+}$ which is formed by self-assembly from the reaction between $FeCl_3$ and five *tris*-bpy ligand strands in ethylene glycol. Each *tris*-bpy strand is shown in a different colour and the iron atoms which are octahedrally coordinated are shown in purple.

which means that, if $Cr(H_2{}^{16}O)_6^{3+}$ were dissolved in ^{18}O-labelled water, much of the $Cr(H_2{}^{16}O)_6^{3+}$ would still be present in solution at the turn of the next century. If the salt K_2PtCl_4 is dissolved in water the potassium ions K^+ are solvated in a manner similar to that of the sodium ions described above for NaCl, i.e. $K^+(aq)$, but the platinum prefers to retain its bonds to chlorine, giving the $PtCl_4^{2-}$ anion in solution. If $Co(NH_3)_6^{3+}PO_4^{3-}$ is dissolved in an acidic aqueous solution then equation (1) is thermodynamically in favour of the formation of $Co(H_2O)_6^{3+}$ and NH_4^+. However, the $Co(NH_3)_6^{3+}$ persists in solution virtually indefinitely at room temperature.

$$Co(NH_3)_6^{3+} + 6H^+ + H_2O(aq) = Co(H_2O)_6^{3+} + 6NH_4^+ \qquad (1)$$

From the above we can learn to make two important observations. First, the form a species presents in the solid state may, but need not, persist in solution. Secondly, in solution the rates at which bonds are formed and broken vary very dramatically with the nature of the metal ion and its ligands, the groups to which the metal is bound. This leads to an important distinction, namely the classification of metal ions in certain environments as being labile or inert. For a metal ion to be classified as inert does not mean that it is stable in the thermodynamic sense; it merely means that it is kinetically persistent. To be kinetically persistent yet thermodynamically unstable means that there is a significant energy of activation that must be overcome before bonds are broken and new bonds are formed. Within this context it is important to recognize that the species present in the solid state, which can these days often be determined very rapidly by a variety of physical techniques (X-ray diffraction, infrared and Raman spectroscopy, solid-state magic-angle spinning NMR spectroscopy), may, but need not, be present in solution. Furthermore, whereas water was originally the most common solvent for the study of metal ions and metal complexes, today numerous researchers employ a wide variety of organic solvents (acetone, methanol, ether, hexane, benzene, chloroform, methylene chloride, etc.). Also, whereas much of the original work was done in an environment of air, now controlled inert atmospheres, e.g. argon and dinitrogen, are more commonly employed since many of the species in the air may be reactive towards the complex being studied.

Fundamental Concepts of Ligation

Metal complexes can generally be viewed as a positively charged metal ion, which acts as a Lewis acid, bonded to ligands that are Lewis bases. The ligand may be negatively charged such as a halide ion or may be neutral such as H_2O, NH_3, a tertiary phosphine PR_3, where R is alkyl or aryl, and carbon monoxide. The bonding may be viewed as ionic such as that in salts such as NaCl, SrO and TiF_2 or covalent such as that in $TiCl_4$, UF_6 and $W(CH_3)_6$. The coordination of neutral ligands (H_2O, NH_3, R_3P, CO, etc.) may be understood in terms of dipolar or dative bonds and these may be kinetically labile or persistent, as noted above. In the simple ionic or electrostatic model a central positive metal ion will attract negative or neutral ligands in keeping with requirements on size and charge neutrality. The coordination number of a metal ion is defined as the number of ligand-to-metal bonds supported at the metal centre. For example, for NaCl and CsCl the coordination numbers of Na^+ and Cs^+ are six and eight, respectively. In solution the large Cs^+ ion can attract 10–12 oxygen atoms from the like number of water molecules. These concepts of coordination number, size and dipolar or dative bonding have given rise to the development of metal-ion-specific ligands. For example, crown ethers have been designed to range in ring size and number of oxygen atoms in such a way as to be specific for particular metal ions (see Figure 2). Thus potassium permanganate, an ionic salt, may be dissolved in benzene by the addition of 18-crown-6, yielding 'purple benzene' solutions. Paramagnetic lanthanide ions may be so tightly sequestered by certain macrocyclic ligands that they may be injected into the body for the purpose

Figure 2

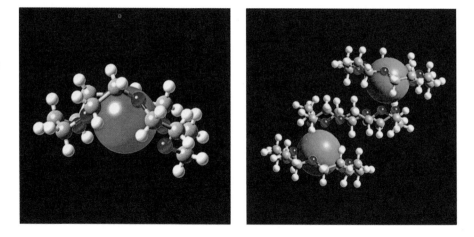

of medical diagnosis by magnetic resonance imaging (MRI) and subsequently excreted from the body through the urinary system. This complexation of the lanthanide cation by the macrocyclic ligand masks its otherwise high toxicity.

If the bonding is covalent in nature then the coordination numbers are restricted by the use of available atomic orbitals of the metal. For transition metals with valence s, p and d orbitals this leads to a maximum coordination number of nine if each orbital is used in an electron-precise manner. The ReH_9^{2-} anion, in which rhenium adopts a tricapped trigonal prismatic geometry, is a prime example. Many covalent transition metal complexes conform to the so-called 18-electron rule, according to which that the coordination of a metal ion is saturated when each of its metal valence, s, p and d orbitals is occupied by an electron pair. The analogy with the octet rule that often is obeyed by the lighter first-row elements such as carbon and oxygen can be easily seen here. For steric reasons bulky ligands L, where L is PR_3, NR_2, SiR_3, OR or SR, are often employed to generate so-called coordinatively unsaturated complexes, ML_n, which do not employ 18 electrons and thus are considered coordinatively unsaturated. These complexes may exhibit reactivity towards small molecules or even nearby carbon–hydrogen bonds in order to increase both the coordination number and the electron count at the metal centre. It should be noted that, although 18 electrons may be viewed as the magic number and many trends in coordination chemistry can be explained by considering that systems are conforming to it, e.g. early transition metal ions in high oxidation states, such as Zr(4+) and Nb(5+), strive for high coordination numbers such as eight or nine whereas later transition metals such as Pt(2+) and Ag(+) prefer low coordination numbers, four or even three or two, it is not the rule. Many complexes that can be viewed as quite stable have fewer or more than 18 electrons, e.g. $Ti(CH_2{}^tBu)_4$ and $Ni(H_2O)_6^{2+}$, in which the metal ions formally have 12 and 20 electrons, respectively. However, covalent molecules with small ligands often attain an 18-electron valence shell. Prime examples are seen in isoelectronic series of mononuclear metal carbonyls ($Ni(CO)_4$ and $Co(CO)_4^-$, in which the metal ions have four ligands in a tetrahedral geometry; and $Ti(CO)_6^{2-}$, $V(CO)_6^-$, $Cr(CO)_6$ and $Mn(CO)_6^+$, which contain six-coordinate octahedral metal ions) and in the trigonal bipyramidal $M(CO)_5$ compounds, where M is Fe, Ru and Os. In each of these carbonyl complexes the number of valence electrons of the metal can be calculated by taking the sum of the electrons on the metal centre in its specific oxidation state, dn, and two from each carbonyl ligand; e.g. for $Mn(CO)_6^+$, $n = 6$ and 6CO provide 12, giving a total of 18 valence electrons. As will be described later, generating coordinative unsaturation is key to many catalytic reactions and to the generation of fascinating, unusual molecules that feature novel modes of bonding.

Metal ions and ligands are sometimes classified as hard or soft. Hard ligands bond through oxygen or nitrogen atoms and bind to 'hard' metals, in preference to 'soft' ligands, which coordinate by sulfur or phosphorus and bind to 'soft' metals. The classification can be extended to the halides, of which fluoride and chloride are hard but bromide and iodide are soft. In the hard–hard match there is a strong ionic or electrostatic component to the bonding. The soft–soft bonding mode generally involves more covalent interactions when the ligand orbitals are more polarizable. It is not a simple classification insofar as many transition metal ions may be viewed as hard or soft depending upon their oxidation state and the other attendant ligands. However, it is true that the alkali and alkaline earth metals and the lanthanides and actinides may be classified as hard since they exhibit marked preferences for oxygen over sulfur and for fluoride over iodide. Conversely, cadmium(2+) and platinum(2+) are soft and prefer sulfur donors to oxygen donors and iodide in preference to fluoride.

A very common geometry for six-coordinate complexes is based on the octahedron. For complexes with a mixed set of ligands this leads to isomers, e.g. for a compound of formula MA_2B_4, where A and B represent different ligands, *cis* and *trans* isomers may exist. For a complex of formula MA_3B_3 mer (meridial) and fac (facial) isomers are possible. For compounds with chelating ligands such as $Cr(oxalate)_3^{3-}$ D and L optical isomers are possible. These are non-superimposible molecules and are related to each other by being mirror images.

Four-coordinate metal complexes are typically tetrahedral but, for platinum(2+), palladium(2+), gold(3+) and other transition metal ions with a d^8 metal configuration, a square-planar geometry is preferred. Therefore, MA_2B_2 complex may exist in *cis* and *trans* isomeric forms for square complexes such as $PtCl_2(NH_3)_2$, or be tetrahedral such as $(Ph_3P)_2NiBr_2$. Five-coordinate complexes are most often trigonal bipyramidal or square-based pyramidal.

Examples of these common geometries are shown in Figure 3. In solution the conversion of one isomeric form into another may be slow, requiring days or months, or rapid, occurring within a fraction of a second, leading again to the recognition that a complex may be viewed as inert or labile.

So far we have considered only that ligands can act as electron-pair donors by virtue of their having lone pairs or negative charge, e.g. F^-, CH_3^- and NH_3. Some ligands contain either more than one lone pair or have energetically accessible vacant orbitals that may become involved in bonding. There are in these cases opportunities for multiple bonding between a metal and the ligand. Prime examples of this are seen for the oxo and carbonyl ligands which are complementary an cylindrical π-donors and π-acceptors, respectively. Their interactions with metal d orbitals are shown in Figure 4. The classification π is used to denote that there is a nodal plane in the resultant molecular orbital that contains the internuclear axis. Whereas oxo and carbonyl ligands are cylindrical π-donors and π-acceptors, other ligands may be single-faced π-donors or π-acceptors, such as amides, NR_2^- and alkenes, $R_2C{=}CR_2$, respectively. π-Donor ligands are capable of stabilizing metal ions in high oxidation states, e.g. Os(8+) in OsO_4, whereas π-acceptors stabilize metal ions in low oxidation states, e.g. $V(CO)_6^-$, which contains vanadium in the formal oxidation state of -1.

The oxidation state of a metal is a formalism that arises from the view that all ligands are Lewis bases. In many instances classification of the oxidation state can become a moot point and it must be remembered that the detailed partitioning of charge can really be determined only by detailed spectroscopic or molecular-orbital calculational procedures. A good example of this type of problem is seen in the coordination of orthoquinone to a metal atom, since this may bind as a neutral ligand A, a semiquione, B, or as a catecholate, C.

Figure 3
Common geometries found for mononuclear (one-metal-atom-containing) complexes. (*a*) Tetrahedral and square-planar complexes of the type ML$_4$ for which, if two different types of ligand are present, denoted by A and B, there are two isomers, *cis* and *trans*, as shown. (*b*) Trigonal bipyramidal and square-pyramidal geometries that are seen for five-coordinated ML$_5$ complexes. (*c*) Octahedral complexes of the form MA$_2$B$_4$ give rise to *trans* (Left-hand side) and *cis* (right-hand side) isomers. (*d*) Octahedral complexes of the type MA$_3$B$_3$ give rise to meridial (left-hand side) and facial (right-hand side) isomers. (*e*) Chelating ligands such as oxalate C$_2$O$_4^{2-}$ and ethylenediamine form enantiomeric pairs of isomers in pseudo-octahedral M(L–L)$_3$ complexes.

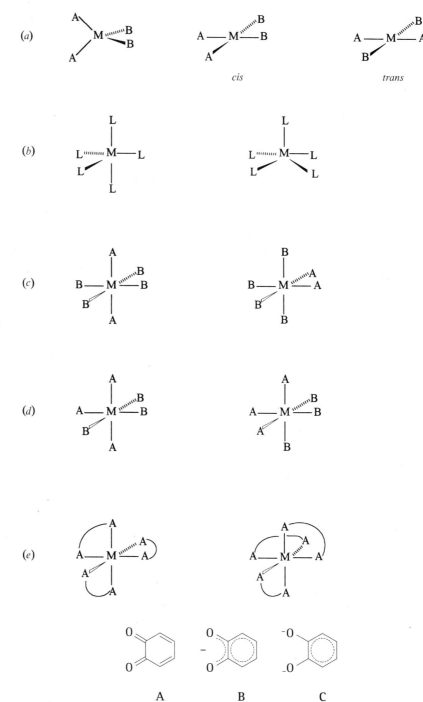

So, for a complex of formula M(C$_6$H$_4$O$_2$)$_3$, the question of the metal's oxidation state and the ligand's charge can be a matter of conjecture. This situation is quite common for coordination complexes in which there are ligands that are termed 'non-innocent'. This term merely implies that the oxidation state is not a given. For example, in metal–alkene and metal–acetylene complexes the ligand may be viewed as a neutral two-electron donor or, if the π-acceptor properties are viewed to be more important, then the metallacyclopropane or metallacyclopropene form, **D** or **E**, respectively, may be used.

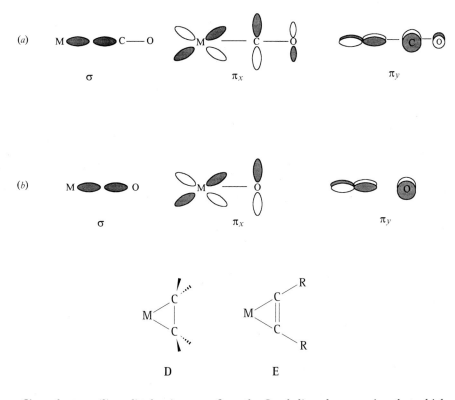

Figure 4
Metal–ligand σ and π interactions for a CO ligand (top) and an oxo ligand (bottom). CO is a π-acceptor interacting with filled metal d_π obitals whereas oxo, O^{2-}, is a π-donor to vacant metal d_π orbitals.

Since the term 'ligand' takes its name from the Greek *ligandos*, meaning that which binds, one might ask what the limits are when it comes to metal coordination. The answer is that almost anything can bind and therefore be a ligand to a metal under the appropriate circumstances. We shall return to this matter shortly.

Sandwich Compounds

Carbocyclic ligands with unsaturation form an important group in organometallic chemistry, that is the field of chemistry which deals with complexes having metal–carbon bonds. So called 'sandwich compounds' such as ferrocene, $(\eta^5\text{-}C_5H_5)_2Fe$, and dibenzenechromium, $(\eta^6\text{-}C_6H_6)_2Cr$, were revolutionary at the time of their discovery. In general, these compounds are covalent and the $\eta^n\text{-}C^nH^n$ ligand is not labile in solution. (The term η^n refers to the number, n, of carbon atoms directly bonded to the metal.) So $(\eta^6\text{-}C_6H_6)_2Cr$ may be dissolved in benzene-d_6, C_6D_6, and no exchange of free and coordinated C_6H_6 (benzene) can be detected by ^1H NMR spectroscopy. Several sandwich compounds are shown in Figure 5. From this it should be noted that the metal ions that form sandwich compounds are by no means limited to the transition metal elements but include also main group and lanthanide and actinide elements. Also the η^n-carbocyclic systems are readily extendable to heterocyclic ligands such as pyridine, thiophene and even planar P_4 and cage systems such as the carbolide $\eta^5\text{-}C_2B_9H_{11}^{2-}$. Probably the best way to think of the bonding in these compounds is in terms of the frontier orbitals of the metal and the sandwich ligand. In this manner we can see the formation of σ, π and even δ interactions, as shown in Figure 4. Most, but not all, sandwich compounds of the transition metal elements attain 18 valence electrons. When this is not so the complex is likely to be either less thermodynamically stable or more kinetically labile. Thus, whereas $(\eta^5\text{-}C_5H_5)_2Fe$ is inert both to air and to water, $(\eta^2\text{-}C_5H_5)_2M$, where M is Cr and Co, are extremely air-sensitive. $(\eta^5\text{-}C_5H_5)_2Co$, for example, is readily oxidized and the redox potential of reaction (2) is close to that of sodium metal.

Figure 5

Selected sandwich compounds of the form $(\eta^n\text{-}C_nH_m)M(\eta^m\text{-}C_mH_m)$. In the electron counting shown the $\eta^n\text{-}C_nH_n$ ligand donates n electrons to a metal centre. The combinations of these n C 2p orbitals can be classified according to symmetry giving rise to σ, π and δ interactions with metal s, p and d orbitals, as shown below for ferrocene, the prototypical sandwich compound $(\eta^5\text{-}C_5H_5)_2Fe$. In many sandwich compounds the metal conforms to the 18-electron rule but this is not necessarily true. When the 18-electron rule is not satisfied the compounds are generally more reactive towards O_2 and H_2O.

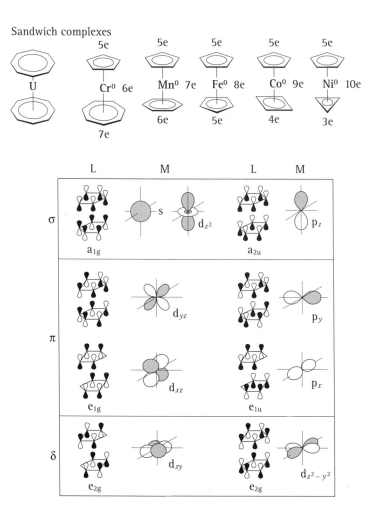

Sandwich complexes

Ligands such as benzene may also coordinate to ions such as potassium and lanthanum in the solid state but in these cases the benzene is labile in solution. In this regard it should be noted that almost any molecule, no matter how chemically inert/stable, can be found to coordinate to a metal ion under the appropriate circumstances. An electron-deficient metal may therefore show so-called agostic M --- H–C interactions with ligands that are within close proximity. These, for a metal–alkyl group, may involve α, β or γ interactions, as shown in F.

$$(\eta^5\text{-}C_5H_5)_2Co = (\eta^5\text{-}C_5H_5)_2Co^+ + e^- \qquad (2)$$

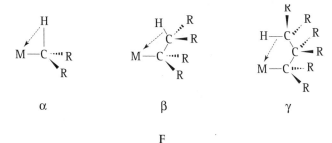

F

One ligand that is η^n-bonded to one metal may also be η^n-bonded to another. This is commonly seen in the crystal structures of $(\eta^5\text{-}C_5Me_5)_2M$ complexes of the heavier group IIA metals, e.g. Ba, giving rise to infinite polymeric structures in the solid state. Even N_2 and H_2 are known as ligands and each was considered quite remarkable at the time of its discovery. The bonding of N_2 in $Ru(NH_3)_5(N_2)^{2+}$ may be considered to be related to that of a η^1-CO ligand, albeit that N_2 is both a poorer σ-donor and a poorer π-acceptor ligand. Similarly, the bonding of η^2-H_2 in *trans*-$(P(^cC_6H_{11})_3)_2W(CO)_2(\eta^2$-$H_2)$ may be compared to that of an η^2-alkene complex with both bonding forms shown in G being important.

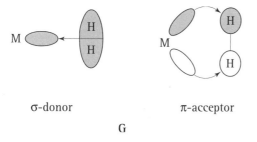

σ-donor π-acceptor

G

The formation of an η^2-H_2 complex can therefore be seen as an intermediate along the reaction pathway leading to the reduction of H_2 by a reducing metal centre to give two metal–hydride ligands: $M(\leftarrow H_2) \rightarrow M(H)_2$ and the two forms shown in H represent limiting redox descriptions of the ligand.

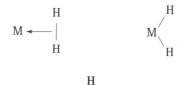

H

This having been stated, it should come as little surprise that photochemically generated unsaturated molecules such as $M(CO)_5$, where M is Cr, Mo or W, will, in a matrix or solvent at low temperatures, bind weakly to ligands (molecules) such as methane, ethane and cyclohexane and even to large polarizable atoms such as xenon. Therefore, under the appropriate circumstances almost anything can be found to bond (to ligate) to a metal atom but one has to recognize once again the all-important matter of lability and stability of the bond. Metal bonds to ligands such as xenon and methane are not only thermodynamically weak but also kinetically labile, but an η^6-benzene ligand may be thermodynamically unstable with respect to solvolysis but kinetically inert or labile depending on the metal, the attendant ligands and the electronic configuration of the metal. Similarly, a ligand such as $C_5H_5^-$ may bond in an η^5, and η^3 or an η^1 (σ) mode, as shown in I.

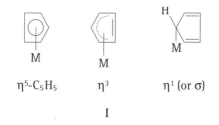

η^5-C_5H_5 η^3 η^1 (or σ)

I

The flexible bonding mode of the cyclopentadienyl ligand shown in I is shared by numerous other variable electron donors such as $(\eta^n\text{-}C_nR_n)$, OR, SR, NR, NR_2 and NO and makes these ligands particularly attractive since they may easily respond to the electronic requirements of a metal centre upon the uptake or release of a substrate.

Multiple Bonds to Metals

The ability of ligands with filled p_π orbitals to interact with a metal with vacant d_π orbitals has already been noted for the case of oxo. In a related manner nitrido, N^{3-}, and alkylidyne ligands, CR^{3-}, may form triple bonds to metal atoms. Alkylidene, CR_2^{2-} and imido, NR^{2-}, ligands can also form double bonds and, in the case of the imido ligand, the presence of the additional N p_π lone pair can allow the formation of an effective triple bond just like in the case of oxo. The bond's polarity will depend on the relative orbital energetics of the metal's d_π orbitals and the ligand's p_π orbitals. If the heteroatom is much more electronegative then the M—X multiple bond will be polarized in the manner $M^{\delta+}$ and $X^{\delta-}$. This will favour ligand–bridge formation and also influence the reactivity of the M—X multiple bond. The M—X functionality may thus have a variety of bonding possibilities, as shown in Figure 6 for metal nitrides. Structural examples of each are known and a particularly fascinating situation emerges for metal nitrides in $[L_nMN]_x$ compounds, which are seen in the solid state to form infinite polymers with alternating triple and dative bonds, cyclic trimers either with equivalent M—N distances implying a delocalization of π-orbital electron density such as in borazines or with alternating double and single bonds, or dimers with planar bridged rhomboidal M_2N_2 units. Which structure is preferred is influenced in very subtle ways by the nature of the specific L_nM fragment.

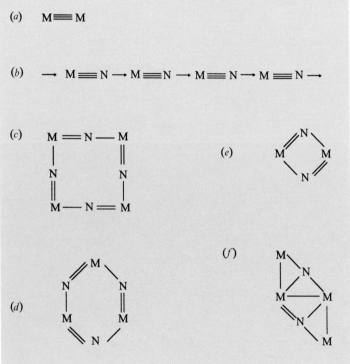

Figure 6
Some bonding modes of the nitride ligand in molybdenum and tungsten nitrides of the compounds of formula L_3MN, where L is alkoxide, siloxide, aryloxide or halide.

One metal may also bind with another to form a multiple bond. With the use of valence s, p and d orbitals, this can lead to a sextuple bond in the naked diatomic Mo_2 molecules with the calculated valence configuration $\sigma_1^2\pi^4\delta^4\sigma_2^2$. The Mo_2 molecule, though it can be synthesized by condensation of metal atoms in an argon matrix at low temperatures and by sputtering in the gas phase, is a kinetically labile molecule that readily oligomerizes to molybdenum metal. However, there are many complexes of molybdenum and neighbouring metal atoms in the Periodic Table (V, Cr, W, Re, Te, Ru, Nb and Ta) that form discrete dinuclear complexes with M—M multiple bonds. A well-known family of d^4-d^4 dinuclear complexes for Cr_2, Mo_2, W_2, Tc_2 and Re_2 having an eclipsed M_2X_8 core is shown in Figure 7. These include carboxylates $M_2(O_2CR)_4$, where M is Cr, Mo and W, halides $M_2X_8^{4-}$ for M being Cr, Mo and W and $M_2X_8^{2-}$ for M being Re and Te and mixed-ligand systems such as $M_2X_4L_4$, where M is Mo or W and L is a tertiary phosphine. These have M—M quadruple bonds of configuration $\sigma^2\pi^4\delta^2$ as a result of the d–d overlap shown in Figure 8. Other examples of multiply bonded complexes include the so-called ethane-like dimers M_2X_6 of molybdenum and tungsten, where X is a bulky alkyl, amido, alkoxide, thiolate or selenate, and other geometries with bridging ligands such as those shown in Figure 9.

Many of these dinuclear complexes exhibit reversible redox chemistry (electron transfer) leading to stepwise changes in the M–M bond order. For example, $M_2(O_2CR)_4$ can be oxidized to $M_2(O_2CR)_4^+$ when M is Mo or W. This leads to a change of the bond order from 4 to 3.5 when the M–M bonding configuration changes from $\sigma^2\pi^4\delta^2$ to $\sigma^2\pi^4\delta^1$. $Ru_2(O_2CR)_4$ complexes have the electronic configuration $\sigma^2\pi^4\delta^2\delta^{*2}\pi^{*2}$ with a formal double bond that, upon oxidation to Ru_2^{5+}, gives $\sigma^2\pi^4\delta^2\delta^{*1}\pi^{*2}$ with a M–M bond order of 2.5.

Multiple bonds within the main-group elements, the so-called p block, were once known only for the lighter elements of the first row, carbon, nitrogen and oxygen, but are now known for the heavier elements Si, Ge, Sn, P and As. The key to the synthesis of these compounds has been the use of bulky ancillary ligands, L, such as $CH(SiMe_3)_2$, $N(SiMe_3)_2$ and $O(2,6\text{-}^iPr_2C_6H_4)$. Rather interestingly the Si–Si, Ge–Ge and Sn–Sn double bonds do not lead to planar structures like those seen for ethene and alkenes but rather the L_2E groups bend back as shown in J. From spectroscopy it can be shown that the π-to-π^* energy separation decreases quite dramatically on going down the series $E = C > Si > Ge > Sn$. This is because the np–np π overlap becomes poorer as n gets larger and the angular structure J reflects the fact that, with very bulky

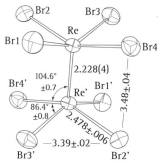

Figure 7
A view of the $Re_2Cl_8^{2-}$ anion showing the eclipsed geometry of the Cl_8 unit and the short Re–Re distance unbridged by any other atoms. It was this structural determination that brought about the recognition of the δ bond and that M–M bonds could be quadruple, higher in order than those known between main-group elements.

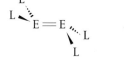

groups L, monomeric carbene-like L_2E fragments can be isolated and are found to exist in the singlet form, unlike R_2C, for which the triplet lies lowest. Therefore the bonding in these L_2E 'dimers' may be viewed as the sum of two singlet carbene couplings as shown pictorially by reaction (3)

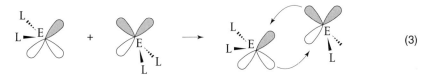

$$(3)$$

For E being Sn and Pb the E–E bond strength is so weak that the bonds may break reversibly in solution and in one case a crystal structure of a monomeric E_2Sn molecule grown from vacuum sublimation has been obtained, whereas from solution the dimer $L_2E–EL_2$ was characterized.

Multiple bonding between hetero-atoms of the main-group elements has also been seen; examples include P–B double bonds supported by bulky ancillary ligands and so-called hypervalent molecules such as $N\equiv SF_3$ and XeO_3, in which sulfur and xenon exceed the octet in their valence shell. In these and related oxo anions such as SO_4^{2-} and ClO_4^- a molecular-orbital approach to an understanding of the bonding is more fruitful than a simple valence-bond description involving the formation of localized double bonds or dative bonds.

Multiple bonds between the heavier main-group elements and transition elements are also now known. Again the use of bulky spectator ligands, such as $L = N(SiMe_3)(2,6\text{-}^iPr_2C_6H_3)$, is important in the isolation of species such as $L_3M\equiv E$, where M is Mo or W and E is P or As. The steric pressure at the metal centre suppresses the otherwise probably favourable reaction of bridge formation by the E group.

Aside from eliciting interest in the bonding and spectroscopies of M–L and M–M multiple bonds, these species provide reactive inorganic or organometallic groups. We shall consider that aspect of their chemistry shortly.

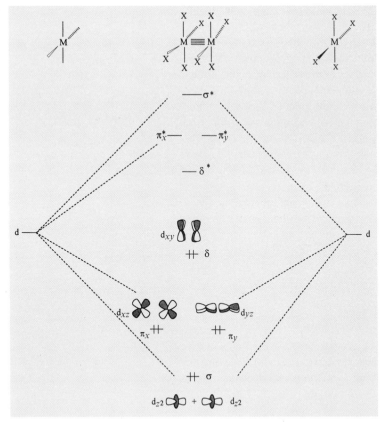

Figure 8

A schematic molecular orbital diagram showing the formation of a M–M quadruple bond for a d^4–d^4 interaction between two square ML_4 fragments. The order σ below π below δ is to be expected from the overlap of the d–d orbital interactions. The ordering of the antibonding orbitals follows similar arguments with δ^* below π^* which in turn is below σ^*.

Metal Cluster Compounds

Although metal–metal bonding is weaker than metal–ligand bonding (if it were not, there would be no coordination chemistry) there is now a large class of compounds that contain both metal–metal bonds and metal–ligand bonds. A metal cluster compound is defined as having two or more metal atoms that are bonded to one another in addition to a set of attendant ligands. Thus a general formula of M_xL_y could apply, although the cluster might contain more than one type of metal atom, in which case it would be termed a heterometallic cluster. Also the ligand set L_y may be comprised of several different ligands, ranging from halides, alkoxides and thiolates to neutral species such as CO and tertiary phosphines. Commonly metal cluster compounds may have hydrocarbyl (organic) fragments that are bonded to them in a manner not possible in mononuclear chemistry.

A now well-established class of clusters is that of the carbonyl metal clusters. These are often formed by the condensation of reactive carbonyl metal fragments during a thermal or photochemical reaction wherein CO is evolved. A simple example is shown in equation (4), where M is Fe, Ru or Os.

$$M(CO)_5 \xrightarrow[\text{or heat}]{\text{light}} M(CO)_4 + CO \qquad (4i)$$

$$M(CO)_5 + M(CO)_4 \longrightarrow M_2(CO)_9 \qquad (4ii)$$

$$M_2(CO)_9 \xrightarrow[\text{or heat}]{\text{light}} M_2(CO)_8 + CO \qquad (4iii)$$

$$M_2(CO)_8 + M(CO)_4 \longrightarrow M_3(CO)_{12} \qquad (4iv)$$

Here the $M(CO)_4$ fragment is reactive, having only 16 valence electrons. Its reactivity has been likened to that of a carbene and the valence frontier orbitals and electron occupation support this analogy. Two fragments are said to be isolobal if their frontier orbitals are of like symmetry, have the same electron occupation and are of similar energy. This principle has been used to elucidate the bonding in many organometallic systems. Thus the formation of $M_3(CO)_{12}$ in (4) can be compared to the hypothetical formation of cyclopropane from the combining of three methylene units.

The $M_3(CO)_{12}$ compounds are themselves labile to further loss of CO to give reactive fragments $M_3(CO)_{11}$ and $M_3(CO)_{10}$ that may be isolated with ligation of solvent molecules such as CH_3CN. This has proven one fascinating entry point to the activation of substrates at a three-metal centre.

Under pyrolysis $Ru_3(CO)_{12}$ and $Os_3(CO)_{12}$ form a remarkable series of clusters of higher nuclearity. Under some conditions the carbonyl ligands undergo a disproportionation reaction at the cluster $(2CO \rightarrow C + CO_2)$ and the carbide atom becomes an integral part of the cluster.

Representative carbonyl cluster compounds and carbonyl cluster anions are shown in Figure 10. One of the intriguing features of these clusters of higher nuclearity is that they represent a bridge linking the metallic state, the chemistry of metallic surfaces and classical coordination complexes. In many instances the ligands on the surface of a metal cluster undergo a dynamic site exchange whereby they roam from one metal centre to another. This process is termed a fluxional one. Moreover, the cluster itself may undergo rearrangements of the metal-atom skeleton. The presence of several metal atoms affords the possibility of multi-site activation of substrates and many of these structurally characterized compounds provide models for how substrates bind to metal surfaces and subsequently break apart and form bonds.

If the definition of a cluster is extended somewhat to include multimetallic complexes that are strongly coupled magnetically and electronically, but do not have formal metal–metal bonds, then a number of biologically relevant systems are of note. So, for example, there are iron–sulfur proteins that contain Fe_2 and Fe_4 cores, haemocyanin with a Cu_2 centre and the nitrogenase enzyme that fixes nitrogen ($N_2 + 6H^+ + 6e^- \rightarrow 2NH_3$), which contains a core with six Fe and one Mo. The key feature that the cluster unit offers is the potential for multi-electron redox reactions and multi-site activation of substrates.

Reactivity

A metal ion serves in many ways to bring about useful transformations of organic molecules within its coordination sphere. These transformations may be stoichiometric or catalytic and the function of the metal ion may be summarized as follows.

1. It serves as a collecting point for the binding or assembly of substrates.
2. It may polarize a substrate in a manner that brings about an unusual type of reaction for that substrate. Alternatively, but within the same context, it may mask the usual reactivity of the organic ligand.
3. The usual conformation of a substrate may be changed upon complexation with a metal.
4. The metal centre may be redox active and thus capable of oxidizing or reducing a ligand.
5. The ligand set will influence the orbital energy of the frontier orbitals of the metal and may even change its spin states (magnetic properties).

Consequently, the influence of the ligand on the metal and of the metal on the ligands is symbiotic. Some examples of metal-mediated/modified organic reactions are given below.

The hydrolysis of esters and amides may be promoted in a catalytic manner by their coordination to a cationic metal centre. The role of the metal is to function as an electrophile and facilitate the nucleophilic attack of H_2O/OH^- on a coordinated ketonic functionality. The attack of the hydroxyl group may involve a water molecule bound to a metal as a *cis*-migration. Such is believed to be the role of the Zn^{2+} ion in carbonic anhydrase, because the equilibrium involving $CO_2 + H_2O$ and H_2CO_3 is very slow in the absence of the metal-promoted reaction catalysed by the enzyme.

Nucleophilic attack on arene rings also normally requires extreme conditions, such as those in the hydrolysis of chlorobenzene by water at about 200 °C. However, the arene ring in (η^6-arene)Cr(CO)$_3$ is readily susceptible to attack by nucleophiles. Similarly, alkenes and alkynes, upon complexation to a cation, can readily be attacked by nucleophiles, as is seen in the Hg^{2+}-catalysed conversion of terminal alkynes, $RC\equiv CH$, to ketones, RCH_2COCH_3, in aqueous media.

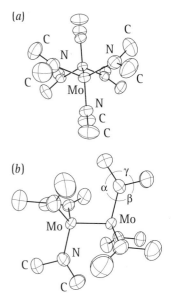

(a)

(b)

Figure 9
Two views of the $Mo_2(NMe_2)_6$ molecule. The top view is looking down the M–M axis and reveals the ethane-like geometry. The bottom view is looking perpendicular to the M–M triple bond of distance 0.222 nm. The Mo–Mo–M angles are 103°.

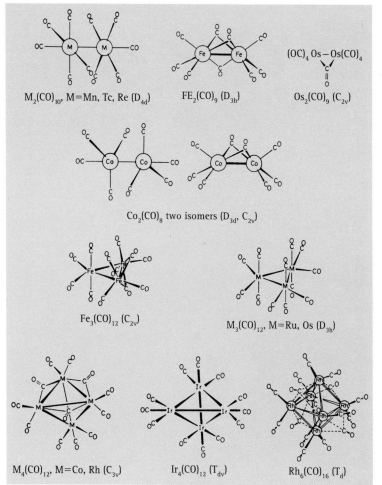

$M_2(CO)_{10}$, M=Mn, Tc, Re (D_{4d})

$FE_2(CO)_9$ (D_{3h})

$(OC)_4Os-Os(CO)_4$

$Os_2(CO)_9$ (C_{2v})

$Co_2(CO)_8$ two isomers (D_{3d}, C_{2v})

$Fe_3(CO)_{12}$ (C_{2v})

$M_3(CO)_{12}$, M=Ru, Os (D_{3h})

$M_4(CO)_{12}$, M=Co, Rh (C_{3v})

$Ir_4(CO)_{12}$ (T_{dv})

$Rh_6(CO)_{16}$ (T_d)

Figure 10

Representative structures of some simple metal carbonyl cluster compounds. The di-, tri-, tetra- and hexa-nuclear clusters shown are all neutral. Families of homometallic and heterometallic carbonyl clusters and cluster anions, some incorporating interstitial ligands such as hydride and carbide, in which the number of metal atoms exceeds 40 are known.

The above examples take advantage of the ability of a metal ion to polarize the electron density of the ligand in the normal sense that might be expected when the ligand acts as a Lewis base upon complexation to a Lewis-acidic metal cation. There are, however, numerous examples in which the metal is relatively electron-rich and can facilitate electrophilic attack on a ligand. A classic example of the latter is seen in the stabilization of cyclobutadiene, C_4H_4, by complexation with a Fe(CO)$_3$ fragment in the complex $(\eta^4\text{-}C_4H_4)Fe(CO)_3$. In solution with D_2SO_4 the C_4H_4 protons are readily exchanged to give $(\eta^4\text{-}C_4D_4)Fe(CO)_3$. Similarly, with ferrocene, $(\eta^5\text{-}C_5H_5)_2Fe$, the rings are susceptible to Friedel–Crafts electrophilic substitution, which is common for aromatic rings. So, for example, with AlCl$_3$/acetyl chloride one may replace all ten of the cyclopentadienyl protons to give $(\eta^5\text{-}(CH_3CO)_5C_5)_2Fe$.

The template effect of a metal ion is seen in the synthesis of macrocyclic ligands such as corrin and porphyrin rings at a metal centre. Similarly, the condensation of salicylaldehyde with an amine to give a chelating imine, a so-called Schiff-base ligand, at a metal centre takes advantage of the template effect and electronic polarization by the metal cation. Metal ions may be incarcerated within a ligand assembly, a cryptand, so that the metal cannot escape. Also, in the design of a supramolecular coordination chemistry, metal ions may be assembled in order and linked within a molecular framework like within a daisy chain. An example is shown in Figure 11.

The ability of transition metal ions to exhibit variable oxidation states allows so-called oxidative-addition and reductive-elimination reactions. Schematically these are presented by the addition of an X−Y substrate to a metal centre in equation (5i) and by its microscopic reverse, elimination, in equation (5ii). In equation (5i) the metal centre is formally oxidized by giving two electrons to the substrate as the X−Y bond is reductively cleaved. In the reverse reaction the X−Y bond is formed and the metal is reduced by two electrons.

$$L_nM + X-Y \longrightarrow L_nM(X)(Y) \tag{5i}$$

$$L_nM(X)(Y) \longrightarrow L_nM + X-Y \tag{5ii}$$

The substrate X−Y may be a carbon–halogen or carbon–hydrogen bond, or a simple diatomic molecule such as Cl_2 or H_2. Note that uptake of hydrogen to give $L_nM(H)_2$ is classified as an oxidative process with respect to the metal centre since the hydride ligand is counted as a two-electron donor, i.e. H−.

The reaction (5) when coupled with a reaction commonly called migratory insertion, wherein an unsaturated group, un, reacts with a coordinated ligand R (often an alkyl or hydrocarbyl group), equation (6), form the basis of numerous catalytic cycles.

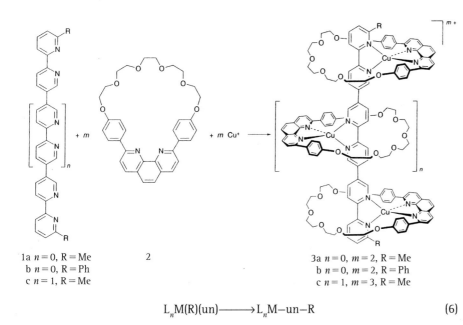

1a $n = 0$, R = Me
  b $n = 0$, R = Ph
  c $n = 1$, R = Me

2

3a $n = 0$, $m = 2$, R = Me
  b $n = 0$, $m = 2$, R = Ph
  c $n = 1$, $m = 3$, R = Me

$$L_nM(R)(un) \longrightarrow L_nM-un-R \qquad (6)$$

If R is CH_3 and un is CO then the new ligand formed by migratory insertion is an acyl. If R is H and un is an alkene then reaction (6) generates a metal–alkyl bond.

The coupling of reactions (5) and (6) forms the basis for the Monsanto process for the production of ethanoic acid employing one of the world's most expensive and precious metals, rhodium. The reaction sequence involves the use of an anionic rhodium(I) iodide complex abbreviated as [Rh]:

$$MeOH + HI \rightleftharpoons MeI + H_2O \qquad (7i)$$

$$[Rh] + MeI \rightleftharpoons [Rh](Me)(I) \qquad (7ii)$$

$$Rh(Me)(I) + CO \longrightarrow [Rh](CO)(Me)(I) \qquad (7iii)$$

$$[Rh](CO)(Me)(I) \longrightarrow [Rh](COMe)I \qquad (7iv)$$

$$[Rh](COMe)I \longrightarrow [Rh] + MeCOI \qquad (7v)$$

$$MeCOI + H_2O \longrightarrow MeCOOH + HI \qquad (7vi)$$

The net transformation is $MeOH + CO \rightarrow MeCOOH$.

In a similar manner phosphine complexes of Rh and Ru can hydrogenate alkenes to alkanes in a sequence described by equations (8), where [M] is a complex of Rh or Ru.

$$[M] + H_2 \longrightarrow [M](H)_2 \qquad (8i)$$

$$[M](H)_2 + alkene \longrightarrow [M](H)_2(\eta^2\text{-alkene}) \qquad (8ii)$$

$$[M](H)_2(\eta^2\text{-alkene}) \longrightarrow [M](\eta^1\text{-alkyl})(H) \qquad (8iii)$$

$$[M](\eta^1\text{-alkyl})(H) \longrightarrow [M] + alkane \qquad (8iv)$$

If the [M] centre is chiral by virtue of its ligand set (chiral chelating diphosphine ligands are often employed) then the hydrogenation of the prochiral substrate, the alkene, will lead to enantioselectivity. Asymmetric hydrogenation of functionalized alkenes has proved of particular importance in the pharmaceutical industry; for example, in the catalytic formation of L-Dopa, which is used in the treatment of Parkinson's disease. The use of a chiral chelating phosphine such as that named chiraphos shown in **K** leads to enantiofacial selectivity in the binding of an alkene. It is this that leads to asymmetric hydrogenation of the alkene but it need not be the thermodynamically favoured form of the alkene adduct that is preferentially hydrogenated since the overall selectivity reflects not only the binding of the alkene but also the subsequent relative rates of insertion and reductive elimination.

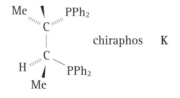

chiraphos   **K**

Other industrially important catalytic processes involving the redox properties of transition metal ions include du Pont process for the hydrocyanation of 1,3-butadiene to adiponitrile which is catalysed by nickel phosphite complexes:

$$CH_2{=}CH{-}CH{=}CH_2 + 2HCN \xrightarrow{\text{[Ni]}} NCCH_2CH_2CH_2CH_2CN \tag{9}$$

The redox reactions in (9) involve Ni(0) and Ni(2+) and the key feature of the reaction is that one obtains control of the stereochemistry of the additions of HCN to the carbon–carbon double bonds. Adiponitrile is one of the molecules employed in the synthesis of nylon.

The hydroformylation of alkenes is also of considerable commercial interest and is typically catalysed by carbonyl/tertiary phosphine complexes of cobalt or rhodium, equation (10). It is, of course, much cheaper to employ a cobalt-based catalyst but this has to be compared with the selectivity of the reaction (the minimization of side reactions) and also the relative rates of reaction at specific temperatures and pressures.

$$RCH{=}CH_2 + CO + H_2 \xrightarrow[\text{[Rh]}]{\text{[Co] or}} RCH_2CH_2CHO \tag{10}$$

The desired product in (10) is the terminal aldehyde, not the methyl ketone or the product of the direct hydrogenation of the alkene, the alkane. Thus, the relative concentration (partial pressure) of H_2 versus CO and the rates of the migratory insertion, equation (6), and of the oxidative-addition and reductive-elimination steps are absolutely crucial to the optimization of the desired product.

The production of polymers with a high degree of stereoregularity and a narrow polydispersity (distribution of relative molecular mass) is currently an area of great commercial and academic interest and there are numerous examples of metal complexes that are proving effective in this area. The tailoring of the ligand set is a key component in the production of such stereoregular polymers. For example, high-density linear polyethylene with a narrow distribution of molecular mass is produced when the rate of insertion of ethene into the growing polymer k_p is comparable to the rate of the initiation process and the rates of chain termination by hydrogenation and chain transfer by β-hydrogen elimination are very slow. The stereochemistry of the polymer is thus simply $(CH_2CH_2)_n$ and the end groups are methyl. The relative molecular mass is controlled by the ratio of partial pressures of C_2H_4 and H_2 since H_2 is normally used for chain termination and the formation of a metal hydride that then starts a new growing chain by insertion of a molecule of ethene. The overall process is represented by the reaction sequence shown in (11), where k_i is the rate of initiation, k_p is the rate of propagation and k_t is the rate of termination.

$$[M]\text{-}H + CH_2=CH_2 \xrightarrow{k_i} [M]-CH_2CH_3 \qquad (11i)$$

$$[M]-CH_2CH_3 + CH_2=CH_2 \xrightarrow{k_p} [M]-CH_2CH_2-(P) \qquad (11ii)$$

$$M-CH_2CH_2-(P) + H_2 \xrightarrow{k_t} [M]\text{-}H + CH_3CH(P) \qquad (11iii)$$

Typically early transition elements such as titanium, zirconium, vanadium and chromium are used in synthesis of high-density polyethylene. However, some cationic rhodium complexes have recently been found to polymerize ethene to give a highly branched form of the polymer. These forms of polyethylene have very different properties and the branching arises because β-hydrogen elimination is reversible, leading to the formation of *n*- and *sec*-alkyl chains, both of which then grow by further insertion of ethene.

Workers at Shell have also found an interesting catalytic cationic palladium(2+) system, $[[(\text{diphos})Pd-(P)]]^+$, which is active in the copolymerization of ethene and carbon monoxide; diphos is $Ph_2PCH_2CH_2PPh_2$ and $-(P)$ is the growing polymer chain. The polymer has the form of a polyketone, $(-CH_2CH_2C(O)-)_n$, and the end groups are alkyl groups formed by chain termination with hydrogen and initiation by insertion of ethene into a Pd–H (or Pd–Me) bond.

In the case of α-alkenes, $CH_2=CHR$, the key to successful formation of a polymer lies in the control not only of the distribution of relative molecular mass but also of the tacticity, since the isotactic, syndiotactic and atactic forms of the polymer (see L) have vastly different properties.

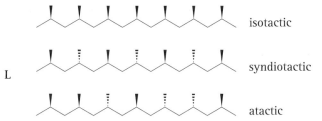

L

isotactic

syndiotactic

atactic

A family of C_2 symmetric cationic zirconium alkyl complexes based on bent metallocenes has been prepared and these have been used successfully in the production of stereoregular polymers derived from propylene and styrene. Indenyl cationic polymerization of propylene for example proceeds by enantiofatial selectivity in the binding of the prochiral propene as shown in M below. The preference for the binding of the alkene either in the *re* or in the *se* form followed by rapid insertion into the growing polymer chain yields control of the stereoregularity of the polymer.

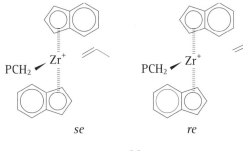

M

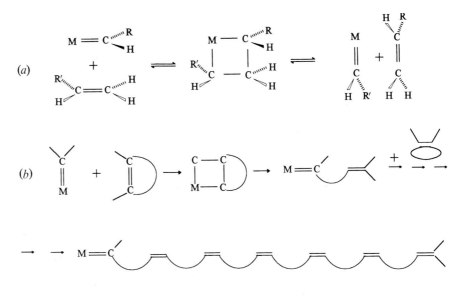

(a)

(b)

Metal–ligand and metal–metal multiple bonds are labile with respect to addition reactions and particularly important are the reactions of certain metal carbenes (metal alkylidenes) with alkenes. A fascinating reaction is that of alkene metathesis wherein an equilibrium of the type shown in equation (12) is established.

$$RCH=CH_2 \underset{}{\overset{[catalyst]}{\rightleftharpoons}} RCH=CHR + CH_2=CH_2 \qquad (12)$$

Metal–carbon triple bonds in alkylidyne complexes are also known to be able to catalyse metathesis of alkynes:

$$RC\equiv CR + R'C\equiv CR' \rightleftharpoons 2RC\equiv CR' \qquad (13)$$

Again the selection of the attendant ligands is crucial to the achievement of effective catalysts for (12) and (13). $((CF_3)Me_2CO)_2(Ar'N)W=CH^tBu$, where Ar is $2,6\text{-}^iPr_2C_6H_3$, is used as an effective catalyst initiator for metathesis of alkenes and $(^tBuO)_3W\equiv C^tBu$ as a catalyst for metathesis of alkynes.

Metal carbenes also find use in the ring-opening metathesis polymerization (ROMP) of strained cyclic alkenes such as cyclopentene, norbornene, norbornadiene and dicyclopentadiene. The polymerization of dicyclopentadiene, DCP, to polydicyclopentadiene by tungsten phenoxide alkylidyne catalysts is extensively used in the production of 'plastic' car bumpers and other body parts by ring-injection moulding.

For some time the detailed mechanism of metathesis of alkenes and acetylene was the subject of intense debate but it is now recognized that they proceed by 2 + 2 addition reactions. Metallacycles are key intermediates. The same pathway is responsible for the ring-opening polymerizations noted above. These are depicted in Scheme 1. A well-defined alkylidyne catalyst such as $(RO)_2(Ar'N)W=CH^tBu$ exists in two forms, *anti* and *syn*, as shown in N below. These react with C–C double bonds at different rates and the equilibrium between the *anti* and *syn* isomers depends on R. Again this

N

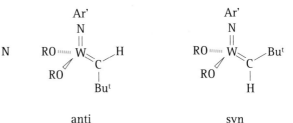

anti   syn

emphasizes the importance of the selection of a given ligand system to achieve a desired catalytic reaction.

Multiple bonds between metal atoms are capable of effecting the reductive cleavage of alkynes, equation (14), nitriles and even carbon monoxide. Again the importance of the selection of the attendant ligands is seen. For example, for equation (14) the reactions involving R′ being Me, Et and ^{n}Pr proceed rapidly in hydrocarbon solvents when R is ^{t}Bu but for R being $SiMe_2Bu^t$ no reaction is observed.

$$W_2(OR)_6 + R'C{\equiv}CR' \longrightarrow 2[R'C{\equiv}W(OR)_3] \tag{14}$$

The mechanism of the reductive cleavage of alkynes is believed to involve a $2 + 2$ addition akin to that shown in Scheme 1 for reactions involving alkenes and metal–carbon double bonds. In this regard we recognize that, although such $2 + 2$ cycloaddition reactions are forbidden in organic chemistry, the presence of metal ions with d orbitals relaxes this constraint. Indeed, the reversible coupling of two M–M triple bonds has been observed for the case of $W_2(OR)_6$, where R is ^{i}Pr, giving $W_4(O^iPr)_{12}$, which has a diamond-shaped (rhomboidal) W_4 core and is an inorganic analogue of cyclobutadiene, equation (15) (here O denotes O^iPr).

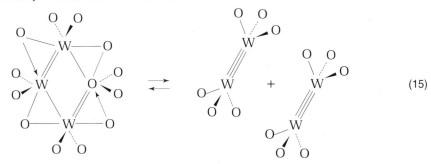

$$\tag{15}$$

The entropy of activation $\Delta S^{\ddagger}$ for the coupling of the two W≡W bonds is about −168 eu, as might be expected for bringing two molecules together in a highly ordered fashion, but the enthalpy of activation is only 42 J mol^{-1} K, clearly indicating that there is no electronic barrier to the coupling.

Macromolecular Coordination Chemistry

A popular theme in current synthetic molecular chemistry is the design of larger assemblies of coordination complexes and clusters in an orderly manner. One may blend the architecture of a metal having a specific electronic signature with that of an organic system. For example, strained linked sandwich compounds may be polymerized upon thermolysis according to

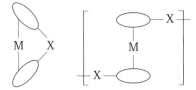

$$\tag{16}$$

Quadruply bonded metal–metal complexes may be coupled through the use of rigid rod spacers such as oxalate and 1,4-$(CO_2)_2C_6H_4$ to give perpendicular one-dimensional polymers, whereas with 1,8-anthracenedicarboxylate or 2,7-dioxynaphthiridine the linking group generates parallel one-dimensional polymers. These are schematically represented by **O** and **P**, respectively, where n is the bond order, $n = 2$ being a double bond.

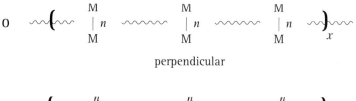

perpendicular

$$P \quad \left(\sim M \overset{n}{\longrightarrow} M \sim\sim M \overset{n}{\longrightarrow} M \sim\sim M \overset{n}{\longrightarrow} M \sim \right)_x \sim$$

parallel

Pyrolysis of an $(\eta^5\text{-benzene})(\eta^5\text{-}C_3R_3B_2R_2)Rh$ complex at 200 °C leads to the smooth loss of benzene and the polymeric compound $[(\mu\text{-}\eta^5,\eta^5\text{-}C_3B_2R_5)Rh]_\infty$:

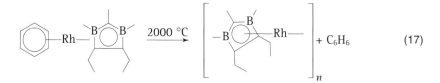

$$\text{(17)}$$

Organoplatinum complexes have been used in the synthesis of dendrimer polymers as shown in Scheme 2 and M_3, M_4 and M_6 cluster compounds can be linked together to give nano-size clusters, $[M_6]_n$, which in turn may be condensed further within an organic or organosilicon matrix to make composite matrices.

Photochemically active complexes such as $Ru(bpy)_3^{2+}$, where bpy is 2,2′-bipyridine, have been attached to electrode surfaces and polymers and incorporated into metalloproteins and DNA. This has allowed the study of long-range electron transfer within a framework relevant to the site-specific cleavage of DNA.

Our knowledge of the coordination chemistry of metal ions has facilitated the synthesis of solid-state materials ranging from synthetic bones to organically modified concrete. As our understanding of the synthesis and properties of volatile metallo-organic compounds increases we are becoming able to design molecules with a built-in Achilles' heel so that, under appropriate conditions, the molecule may self-destruct in an orderly fashion. In this way, for example, certain $Cu(+)$ complexes have been used for the deposition of metallic copper films. By surface modification the deposition may be made substrate selective and may occur at temperatures in the range 150–200 °C. This and related advances in the applications of metallo-organic vapour decomposition offer advances towards the low-temperature, substrate-selective deposition of dielectric, semiconducting and conducting thin films for microelectronic circuitry.

The field of liquid-crystalline technology has yet to employ the use of any metal ion. Consequently, the field of metallomesogens (metal-containing liquid crystals) is one that is in its infancy but offers great potential for synthetic and materials chemists insofar as the incorporation of metal ions may provide new opportunities in the design of optoelectronic materials with specific functional properties. The mesophase of metallomesogens may be determined either by the architecture of the metal centre or by that of the organic ligand so that control of the thermotropic properties of the mesophases may be almost infinitely tunable.

Concluding Remarks

Modern coordination chemistry is only a century old, having its roots in Alfred Werner's classic studies of cobalt($3+$) complexes, from which he was able to define the terms primary valence, the oxidation state of the metal, and secondary valence, the coordination number of the metal. From the number of isomers of a compound of a

Scheme 2

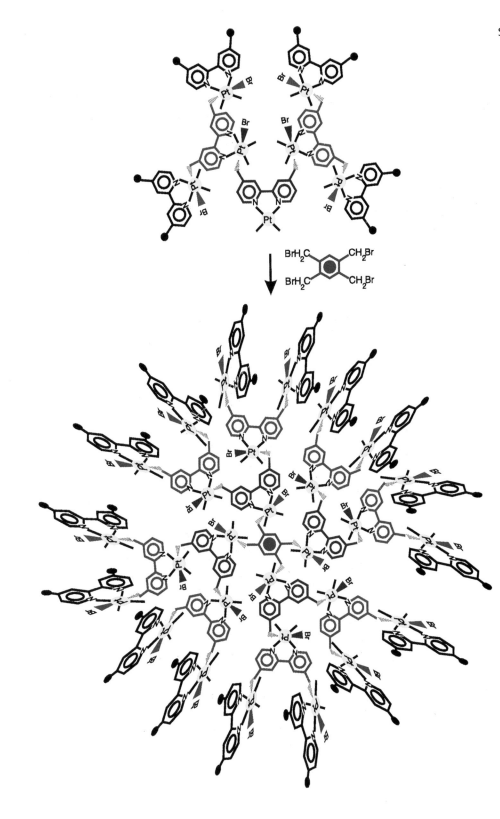

given formula he determined that the coordination geometry for the six-coordinate complex must be octahedral. It could not be planar, benzene-like, or trigonal prismatic. Similarly, the determination that certain compounds of platinum(2+) of formula type MA_2B_2, where A and B are ligands, e.g. A is NH_3 and B is Cl^-, exist in the then so-called α and β forms ruled out the possibility that Pt(2+) complexes of this type were tetrahedral. The determination of dipole moments revealed that one isomer had $\mu = 0$. This led to the recognition that Pt(2+) complexes are nearly always square planar and that the isomers of PtA_2B_2 are *cis* and *trans* forms.

From these now seemingly modest origins we have come to be able to synthesize and characterize new complexes with remarkable speed. Having learned the basic principles, we can logically proceed in synthesis and we can reasonably expect to design new or better catalytic procedures. The world of macromolecular or supramolecular 'coordination chemistry' is opening up before our eyes, as is the use of coordination chemistry for the construction of new materials, polymers and composites. In medicine, radioactive metal coordination complexes find uses in organ-specific imaging. The future promise for the 'clothing of metal ions' is limited only by the imagination, time and effort that we devote to this field of scientific endeavour.

Further Reading

1. F. A. Cotton and G. Wilkinson, *Advanced Inorganic Chemistry*, sixth edition (Wiley Interscience, New York, 1999).
2. Ch. Elschenbroich and A. Salzer, *Organometallics: A Concise Introduction*, second edition (VCH, Weinheim, 1988).

Surface Chemistry

Gabor A. Somorjai

7

Introduction

Surfaces have two major uses in chemistry.

(1) They protect the condensed phase against chemical and physical changes. Coatings, like a coat of paint on a house or hand cream to protect dry and rough skin, come to mind for this application.

(2) Surfaces can also accelerate chemical and physical changes. Catalysts that speed up a chemical reaction, brake pads that stop a moving car when they are pushed against the wheel and lubricated, highly slippery surfaces that reduce friction between the piston and engine wall all possess this surface property.

Without surfaces, either as chemical protectors or as agents for accelerating chemical change, our life would be very different.

Surfaces may be studied through investigation of their fundamental characteristics; their atomic and electronic structures, composition, oxidation state, the nature of their chemical bonds and the motion of atoms and molecules residing on surfaces. They may also be studied by investigating the applications of one or more of their unique properties. These may be classified as chemical, mechanical, electrical, optical and magnetic properties. Some of these depend mostly on making or breaking chemical bonds at a local surface site (chemical and mechanical properties) whereas the others involve the behaviour of many surface atoms in concert.

Both nature and technology utilize internal and external surfaces. Microporous solids such as molecular sieves (mostly alumina silicates, although aluminium phosphates and borates are also prepared in this manner) and high-surface-area carbon have most of their atoms internally. Such materials can be prepared with hundreds of square metres of surface area per gram of solid. They are outstanding absorbents and often are very good catalysts. Alternatively, external surfaces have most (or all) of their atoms at the boundary of a gas, dense solid or liquid. A single crystal with a surface area of 1 cm^2 is bounded by such an external surface. Although the total surface area is low in this circumstance, there are many important applications of surfaces in this more readily accessible geometry. For example, all microelectronic circuits are constructed on such surfaces, usually starting with a single crystal of silicon with (100) orientation.

The development of techniques that can monitor various surface properties (composition, oxidation state, atomic and electronic structure, transport) on the atomic scale is crucial to the rapid advancement of surface chemistry. Without proper instrumentation, none of the properties of surfaces could have been studied adequately. During the last 30 years, over 50 surface-science techniques have been developed; which, in part, account for the large gains made in the understanding of interfaces on the molecular level and the proliferation of new surface technologies (xerography, magnetic hard discs, sensors and new catalysts just to mention a few). A list of these, together with their acronyms, brief descriptions of the principles of their operation and the important surface information they produce is displayed in Table 1. The references are for recent reviews that may be consulted by those wanting to learn about these techniques in more detail.

An example of the level of sophistication of modern surface chemistry is shown in Figure 1, where the structure of a monolayer of benzene co-adsorbed with carbon monoxide onto a single-crystalline rhodium surface is shown. The top part of Figure 1 displays the surface structure of this co-adsorbed monolayer system determined by low-energy electron diffraction (LEED) surface crystallography. The bottom part of Figure 1 shows the same co-adsorbed system imaged by the scanning tunnelling microscope (STM) at room temperature.

Table 1	Commonly used surface science techniques	
Name	Description	Primary surface information
Adsorption or selective chemisorption	Atoms or molecules are physisorbed and their concentration is used to measure the total surface area. Chemisorption of atoms or molecules onto sites yields surface concentrations of selected elements and atomic sites.	Concentration and composition of surface-area sites
Auger-electron spectroscopy (AES)	Core-hole excitations are created, usually by 1–10 keV incident electrons, and Auger electrons of characteristic energies are emitted through a two-electron process as excited atoms decay to their ground state. AES gives information on the near-surface chemical composition.	Chemical composition
Atomic-force microscopy (AFM)	Similar to STM. An extremely delicate mechanical probe is used to scan the topography of a surface by measuring forces exerted by surface atoms. This is designed to provide STM-type images of insulating surfaces or to detect mechanical properties at the molecular level.	Atomic structure
Angle-resolved photoemission spectroscopy (ARPES)	A general term for structure-sensitive photoemission techniques	Electronic structure, surface structure
Electron-energy-loss spectroscopy (EELS)	Mono-energetic electrons (about 5–50 eV) are scattered off a surface and the energy losses are measured. This gives information on the electronic excitations of the surface and adsorbed molecules (see HREELS).	Electronic structure, atomic structure
Ellipsometry	Used to determine the thickness of an adsorbed film. A circularly polarized beam of light is reflected from a surface and the change in the polarization characteristics of the light gives information about the surface film.	Layer thickness
Extended X-ray absorption fine structure (EXAFS)	Mono-energetic photons excite a core hole. The modulation of the adsorption cross section with energy 100–500 eV above the excitation threshold yields information on the radial distances to neighbouring atoms. The cross section can be monitored by fluorescence as core holes decay or by the attenuation of the transmitted photon beam. EXAFS is one of many 'fine-structure' techniques.	Local surface structure and coordination numbers
Field-emission microscopy (FEM)	A strong electrical field (of the order of 1–10 V Å^{-1}) is applied to the tip of a sharp, single-crystalline wire. The electrons tunnel into the vacuum and are accelerated along radial trajectories by Coulomb repulsion. When the electrons impinge on a fluorescent screen, variations of the electrical field strength across the surface of the tip are displayed.	Atomic structure

Table 1 (*cont.*)

Name	Description	Primary surface information
Field-ionization microscopy (FIM)	A strong electrical field (of the order of 1–10 V Å^{-1}) is created at the tip of a sharp, single-crystalline wire. Gas atoms, usually He, are polarized and attracted to the tip by the strong electrostatic field and then are ionized by electrons tunnelling from the gas atoms into the tip. These ions, accelerated along radial trajectories by Coulomb repulsion, map out the variations in the electrical field strength across the surface with atomic resolution, showing the surface topography.	Atomic structure and surface diffusion
Fourier-transform infrared spectroscopy (FTIR)	Broad-band IRAS experiments are performed and the IR adsorption spectrum is deconvoluted by using a Doppler-shifted source and Fourier analysis of the data. This technique is not restricted to surfaces.	Geometry and strength of bonding
High-energy ion-scattering spectroscopy (HEIS)	High-energy ions, above about 500 keV, are scattered off a single-crystalline surface. The 'channelling' and 'blocking' of scattered ions within the crystal can be used to triangulate deviations from the bulk structure. HEIS has been used especially to study surface reconstructions and the thermal vibrations of surface atoms. (See also MEIS and ISS.)	Atomic structure
High-resolution electron-energy-loss spectroscopy (HREELS)	A mono-energetic electron beam, usually about 2–10 eV, is scattered off a surface and energy losses below about 0.5 eV to bulk and surface phonons and vibrational excitations of adsorbates are measured as functions of angle and energy (this is also called EELS).	Bonding geometry, vibrations of surface atoms
Infrared reflection adsorption spectroscopy (IRAS)	Mono-energetic IR photons are reflected off a surface and the attenuation of the IR intensity is measured as a function of the frequency. This yields a spectrum of the vibrational excitations of adsorbed molecules.	Molecular structure
Infrared emission spectroscopy (IRES)	The vibrational modes of adsorbed molecules on a surface are studied by detecting the spontaneous emission of infrared radiation from thermally excited vibrational modes as a function of energy.	Molecular structure
Ion-scattering spectroscopy (ISS)	Ions are inelastically scattered from a surface and the chemical composition of the surface is determined from the transfer of momentum to surface atoms. At higher energies this technique is also known as Rutherford back-scattering (RBS).	Atomic structure, composition
Low-energy electron diffraction (LEED)	Mono-energetic electrons with energies below about 500 eV are elastically back-scattered from a surface and detected as a function of energy and angle. This gives information on the structure of the near-surface region.	Atomic structure and molecular structure
Low-energy positron diffraction (LEPD)	Similar to LEED with positrons as the incident particles. The interaction potential for positrons is somewhat different from that for electrons, so the form of the structural information is different.	Atomic structure
Medium-energy ion scattering (MEIS)	Similar to HEIS, except that energies of incident ions are about 50–500 keV.	Atomic structure
Neutron diffraction	Neutron diffraction is not an explicitly surface-sensitive technique, but neutron-diffraction experiments on large-surface-area samples have provided important structural information on adsorbed molecules and also on surface phase transitions.	Molecular structure

Table 1 (cont.)

Name	Description	Primary surface information
Near-edge X-ray absorption fine structure (NEXAFS) or X-ray absorption near-edge structure (XANES)	A core hole is excited just like in fine-structure techniques (see EXAFS), except that the fine structure within about 30 eV of the excitation threshold is measured. Multiple scattering is much stronger at low electron energies, so this technique is sensitive to the local three-dimensional geometry, not just the radial separation between the source atom and its neighbours.	Atomic structure
Nuclear magnetic resonance (NMR)	NMR is not an explicitly surface-sensitive technique, but NMR work on large-surface-area (≥ 1 m^2) samples has provided useful data on molecular adsorption geometries. The magnetic moment of a nucleus interacts with an externally applied magnetic field and provides spectra highly dependent on the nuclear environment of the sample. This method is limited to the analysis of magnetically active nuclei.	Chemical state
Rutherford back-scattering (RBS)	Similar to ISS, except that the main focus is on depth-profiling and composition. The momentum transfer in back-scattering collisions between nuclei is used to identify the nuclear masses in the sample and the smaller, gradual momentum loss of the incident nucleus through electron–nucleus interactions provides depth-profile information.	Atomic structure, composition
Reflection high-energy electron diffraction (RHEED)	Mono-energetic (about 1–20 keV) electrons are elastically scattered from a surface at glancing incidence and detected as a function of angle and energy for small forward-scattering angles. Back-scattering is less important at high energies and glancing incidence is used to enhance surface sensitivity.	Composition
Surface-enhanced Raman spectroscopy (SERS)	Some surface geometries (rough surfaces) concentrate the electrical field of the Raman scattering cross section so that it is surface sensitive. This gives information on surface vibrational modes and some information on geometry via selection rules.	Atomic structure
Surface extended X-ray absorption fine structure (SEXAFS)	A more surface-sensitive version of EXAFS in which the excitation cross section's fine structure is monitored by detecting the photoemitted electrons (PE-SEXAFS), Auger electrons emitted during core-hole decay (Auger SEXAFS) or ions excited by photoelectrons and desorbed from the surface (PSD-SEXAFS).	Molecular structure
Surface-force apparatus (SFA)	Two bent mica sheets with atomically smooth surfaces are brought together to within a separation in the nanometre range. The forces acting on molecular layers between the mica plates perpendicular and parallel to the plates' surfaces can be measured.	Forces acting on molecules squeezed between mica plates are measured
Sum-frequency generation (SFG)	Similar to SHG. One of the lasers has a tunable frequency that permits variation of the second-harmonic signal. In this way the vibrational excitation of adsorbed molecules is achieved.	Atomic structure
Second-harmonic generation (SHG)	A surface is illuminated with a high-intensity laser beam and photons are generated at the second-harmonic frequency through non-linear optical processes. For many materials only the surface region has the appropriate symmetry to produce a SHG signal. The non-linear polarizability tensor depends on the nature and geometry of adsorbed atoms and molecules.	Molecular structure

Table 1 (cont.)

Name	Description	Primary surface information
Secondary-ion mass spectroscopy (SIMS)	Ions and ionized clusters ejected from a surface during ion bombardment are detected with a mass spectrometer. The surface's chemical composition and some information on bonding can be extracted from distributions of SIMS-ion fragments.	Electronic structure, molecular orientation
Scanning tunnelling microscopy (STM)	The topography of a surface is measured by mechanically scanning a probe over a surface with Ångström-unit resolution. The distance from the probe to the surface is measured from the probe's surface tunnelling current.	Magnetic structure
Transmission electron microscopy (TEM)	TEM can provide surface information about carefully prepared and oriented bulk samples. Real images of the edges of crystals for which surface planes and surface diffusions have been observed have been formed.	Atomic structure
Thermal desorption spectroscopy (TDS)	An adsorbate-covered surface is heated, usually at a linear rate, and the desorbing atoms or molecules are detected with a mass spectrometer. This gives information on the nature of adsorbate species and some information on adsorption energies and the surface's structure.	Atomic structure, composition, heat of adsorption
Temperature-programmed desorption (TPD)	Similar to TDS, except that the surface may be heated at a non-uniform rate to obtain more selective information on adsorption energies.	Composition, heat of adsorption, surface structure
Ultraviolet photoemission spectroscopy (UPS)	Electrons photoemitted from the valence and conduction bands are detected as a function of energy in order to measure the electronic density of states near the surface. This gives information on the bonding of adsorbates to the surface.	Valence-band structure
X-ray photoemission spectroscopy (XPS) or electron spectroscopy of chemical analysis (ESCA)	Electrons photoemitted from atomic core levels are detected as a function of energy. The shifts of core-level energies give information on the chemical environments of the atoms.	Composition, oxidation state
X-ray diffraction (XRD)	X-ray diffraction has been carried out at extreme glancing angles of incidence at which total reflection ensures that one has surface sensitivity. This provides structural information that can be interpreted by well-known methods.	Atomic structure

Surface Chemistry and Catalysis

The use of surface chemical processes on a large industrial scale began in the early part of the nineteenth century. The application of catalysis started with the discovery of the platinum-surface-catalysed reaction of H_2 and O_2 in 1823 by Döbereiner. He used this reaction in his portable flame source, of which he sold a large number. By 1835 the discovery of heterogeneous catalysis was complete and the phenomenon had been described by Berzelius. After the discovery of the battery by Volta in 1796, studies by

Gabor A. Somorjai

Figure 1

The surface structures of benzene and carbon monoxide co-adsorbed onto the Rh(111) single-crystalline surface. At the top is the structure obtained by low-energy-electron-diffraction surface crystallography. At the bottom is the structure obtained by scanning tunnelling microscopy.

$$Rh(111)\text{-}(3\times3)\text{-}C_6H_6 + 2\ CO$$

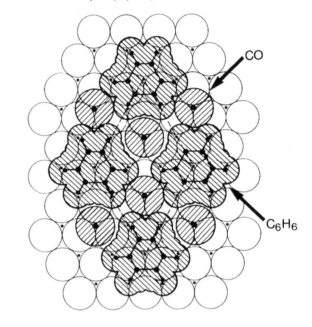

CO

C_6H_6

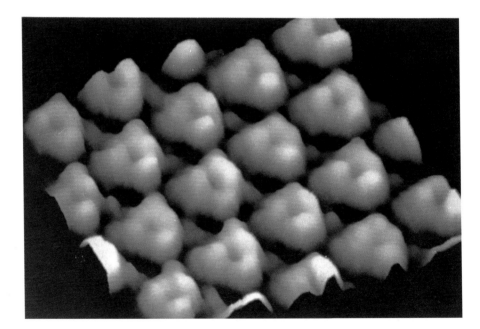

Davy and Faraday, as well as others, had defined the field and applications of electrochemistry by the end of the first quarter of the nineteenth century. It was about this time that the Daguerre process was introduced for photography. The study of tribology, which includes friction, lubrication and adhesion, also started around this time, coinciding with the industrial revolution as machinery with moving parts became prevalent. Scientists were fascinated by the mechanical properties of ice and snow, why glaciers migrate by plastic sliding and deformation and the reasons for skating on ice and skiing on snow being possible. Measurements of surface tension and the recognition of the concept of equilibrium led to the development of surface thermodynamics by Gibbs (1877). The existence of amorphous aggregates that diffuse slowly (such as gelatin) was recognized by Graham (1861), who called these systems 'colloids'. The colloid sub-field of surface chemistry gained prominence at the beginning of the twentieth century with the rise of the paint industry and the production of synthetic rubber. Catalysis became a major industrial enterprise through its application in the Deacon process ($2HCl + \frac{1}{2}O_2 \rightarrow H_2O + Cl_2$), the oxidation of SO_2 to SO_3, steam 'reforming' of methane ($H_2O + CH_4 \rightarrow CO + 3H_2$), oxidation of ammonia, hydrogenation of ethene and the synthesis of ammonia during the period 1860–1912. The studies of high-surface-area gas-absorber materials for the gas mask and other gas-separation technologies and investigations of the lifetime of the light-bulb filament led to the determination of the probabilities of dissociation and adsorption of many diatomic molecules on surfaces as functions of the gas pressure (adsorption isotherms) and temperature (Langmuir, 1915). The properties of chemisorbed and physisorbed monolayers were studied. Studies of electrode surfaces in electrochemistry led to the detection of the surface space charge. Surface diffraction of low-energy electrons was discovered by Davisson and Gernier in 1927 and the diffraction of atoms (helium) from surfaces somewhat later. Preparation for World War II brought major advances in catalyst-based technologies for producing fuels using acids or metal surfaces.

By the 1950s, frontier research in chemistry had shifted to studies of gas-phase molecular processes that could be studied by the newly developed spectroscopic techniques and the X-ray crystallography of solids. During that period it was not possible to study the chemistry of surfaces on the molecular level due to the lack of suitable techniques. The development of surface-based energy conversion and chemical technologies continued at a high rate as crude oil became the dominant feedstock. Then, by the end of the 1950s, the rise of the solid-state-device-based electronics industry and of space technology provided surface chemistry with new tools, challenges and opportunities, resulting in an explosive growth of molecular surface science that has continued unabated until the present. Since the speed of the electronic devices depended on their size, miniaturization became a way of making them faster. As the surface-to-volume ratio increased, surface science became dominant in controlling solid-state device technology. Because of the development of space sciences, ultra-high vacuum became available at a reasonable cost; and, for the first time, clean surfaces of single crystals could be studied with relative ease. The energy crisis in the early 1970s again focused attention on the need to develop new methods of catalyst-based chemical energy conversion using new feedstocks (natural gas and coal). New surface instrumentation and techniques have been developed that permit the study of surface properties on the atomic scale. Most of these techniques are listed in Table 1.

As a result of this sudden availability of surface-characterization techniques, macroscopic surface phenomena (adsorption, bonding, catalysis, oxidation and other surface reactions, diffusion, desorption, melting and other phase transformations, growth, nucleation, charge transport, atom, ion and electron scattering, friction, hardness and lubrication) are being re-examined on the molecular scale. This has led to a remarkable growth of surface chemistry that has continued uninterrupted until the present. The discipline has again become one of the frontier areas of chemistry. The newly gained knowledge of the molecular ingredients of surface phenomena has given birth to a steady stream of high-technology products, including new hard coatings that passivate surfaces; chemically treated glass, semiconductor, metal and polymer surfaces to which treatment imparts unique surface properties; newly designed catalysts, chemical sensors and carbon fibre composites; surface-space-charge-based copying; and new methods of electrical, magnetic and optic signal processing and storage. Molecular surface chemistry is being utilized increasingly in the biological sciences.

Segregation at Surfaces

For a bulk that contains more than one type of atom or molecule, the composition of the surface differs from the bulk's composition. There are sound thermodynamic reasons that give rise to segregation at surfaces. In a multicomponent system, the constituent with the lowest surface energy will segregate to the surface. Metals usually have the highest surface energy ($\simeq 1$ J cm^{-2}, proportional to the heat of sublimation) followed by oxides, sulfides and halides ($\simeq 0.1$ J cm^{-2}). Organic solids and liquids have relatively low surface energies (0.01–0.08 J cm^{-2}), with fluorinated compounds at the bottom of the list. The atom or molecule with the lowest combined surface energy and interfacial energy relative to the atoms underneath will always accumulate at the surface in large excess (Figure 2). This phenomenon is responsible for the detergent action of certain organic molecules that displace soil at the surface of cotton, wool and synthetic fibres. Steel is rendered stainless by the segregation at its surface of chromium oxide, which displaces iron oxide when material is heated under appropriate conditions. Other steels can be chemically protected by the surface segregation of aluminium oxide that forms a non-porous coating.

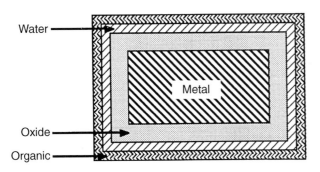

Figure 2

A representation of materials of lower surface energy with coating materials of higher surface energy, leading to a net reduction of the total surface energy ($\gamma_{\text{new surface}} + \gamma_{\text{interface}} < \gamma_{\text{old surface}}$).

Phase Diagrams in Two Dimensions

Clusters of nanometre size have many interesting properties that bulk systems of the same composition do not have (Figure 3). Since every atom in the cluster is a surface atom, there is no bulk phase. Because of the low coordination, their electronic structures, optical properties and melting points are very different from those of bulk systems made up of the same atoms. Here we focus on clusters composed of two different metal atoms (which are often called bimetallic clusters). These often exhibit complete miscibility even when these atoms do not mix in the bulk phase. For example, copper and osmium or gold and iridium are immiscible in three dimensions, whereas they are completely miscible in two dimensions. Changes in phase diagrams, including the temperature at which the miscibility gap appears, with changes from three to two dimensions have been predicted by theory and observed by experiments. Thus, mixing metals in the form of nano-cluster-sized particles could lead to novel chemical, physical and mechanical properties that these materials would not exhibit when they are mixed in the bulk phase.

Surface Versus Bulk Structures

Two dominant phenomena occur at clean surfaces of materials that distinguish their atomic structures from that in the bulk; namely relaxation and reconstruction. Figure 4 shows the nature of the relaxation of surface atoms at metal surfaces. The first layer of atoms moves inwards and this contraction leads to a great shortening of the interlayer spacing between the first and second layers of the surface. The more open the surface the larger the relaxation. Often the contraction in the first layer (indicated by a negative sign) is followed by a small expansion in the second layer (but not always). At rough edges, such as stepped surfaces, the atoms at the step relax by a large amount in order to smooth the surface irregularity. This is shown schematically in Figure 5.

At ionic surfaces, the nature of surface relaxation is very different. Figure 6 shows what happens at iron oxide surfaces. Iron oxide, in its bulk structure, has alternating layers of oxygen ions and iron ions, such that the iron ions are in tetrahedral or octahedral positions. At the surface, the two metal ions move in such a way that the positive and negative ions become almost co-planar. Probably because of the necessary

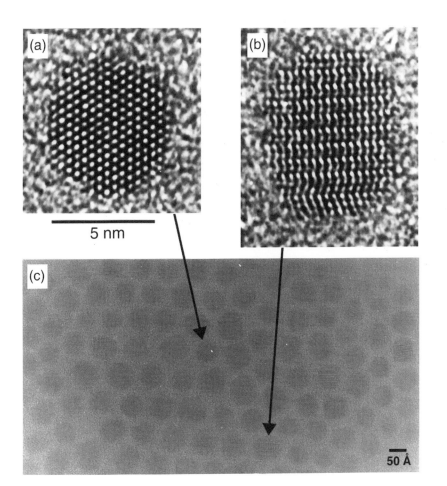

Figure 3
Nanocrystals of CdSe imaged at atomic resolution. The unit cell is a hexagonal prism: (*a*) and (*b*) are enlarged views of the crystals indicated in (*c*). In (*a*), the hexagonal base of the unit cell seen in the interior has imposed a hexagonal shape on the crystal as a whole. In (*b*), the long axis of the prism results in an elongation of the crystal along that direction. (TEM micrographs taken at the National Center for Electron Microscopy, Berkeley, CA. Courtesy of Andreaas Kadavanich and A. Paul Alivisatos.)

condition of charge neutrality, this type of surface structure is thermodynamically more stable than having alternating oxygen-ion and iron-ion layers. Such an expansion at the surface is clearly a property of ionic solids and future studies will prove how general this type of relaxation is.

Because of the directionality of bonding in most solids, such contraction or relaxation at the surface moves atoms away from their positions of optimum bonding. As a result, the atoms move not only in a direction perpendicular to the surface but also parallel to the surface. This leads to the formation of new surface unit cells. This phenomenon is called surface reconstruction. Perhaps the most celebrated example is the (7×7) surface structure that forms on the Si(111) crystal face. Figure 7(*a*) shows the LEED patterns from this structure and Figure 7(*b*) shows this complex surface structure. The unit cell has 49 different locations of surface atoms that are distinguishable, as shown in Figure 7(*b*). The Si(100) surface also undergoes reconstruction. It exhibits the formation of staggered dimers that are different from the arrangement of Si atoms in the bulk near the surface (Figure 8).

Pt, Au and Ir (100) surfaces that should have square unit cells reconstruct to form hexagonal surface unit cells. This is shown in Figure 9. STM can image these reconstructed surfaces when they co-exist with domains that are unreconstructed because of contamination by adsorbates of various types.

Molecules at the liquid–vapour interface also undergo restructuring, as shown in Figure 10 for the arrangement of molecules at liquid–alcohol surfaces. The alcohol molecules are oriented with their O—H bonds pointing inwards, probably in order to

Figure 4
A schematic diagram showing the relaxation inwards of the topmost layer of atoms on metal surfaces when the surface is clean.

Figure 5
Relaxation at the Cu(410) stepped surface.

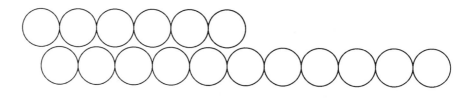

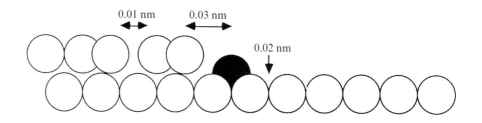

Figure 6
Side and top views of the $Fe_3O_4(111)$ surface structure with the spacing relaxations shown. The corresponding bulk values are $\Delta_1 = 0.063$ nm and $\Delta_2 = 0.0044$ nm and $d_{12} = d_{23} = 0.119$ nm. The A and B layers are strongly expanded by about 0.046 nm.

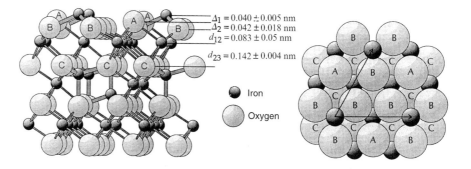

provide optimum hydrogen bonding. The alkane chain sticks out from the surface. This orientation is readily detectable by nonlinear laser-optics sum-frequency generation.

Adsorbate-induced Restructuring

When clean surfaces are covered with nearly a monolayer of chemisorbed molecules, the structure of the surface undergoes profound alterations. This was perhaps best shown in the field-ion-microscopy (FIM) studies carried out by Kruse *et al.* with rhodium field-emission tips (Figure 11). When carbon monoxide is chemisorbed on these tips, every crystal face restructures as shown in Figure 11. This massive restructuring is reversible if CO is removed when the surface is heated in vacuum.

A new STM system that is placed in an environmental cell that can be pressurized and heated to elevated temperatures reveals surface restructuring of the Pt(110) surface when this surface is exposed to atmospheric pressures of hydrogen, then oxygen and then carbon monoxide. When these surfaces are heated, the surface restructures from a reconstructed ordered structure (exhibiting the so-called 'missing row' reconstruction) in the presence of hydrogen, to large (111) orientation facets in the presence of oxygen and then again to smooth (110) unreconstructed surfaces in the presence of carbon monoxide (Figure 12).

Low-energy electron-diffraction surface crystallography indicates the detailed atomic-level nature of such reconstructions. When carbon is adsorbed onto the Ni(100) surface, it occupies a four-fold hollow site (Figure 13). As a result of the formation of

(a)

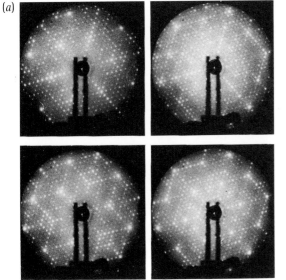

(b)

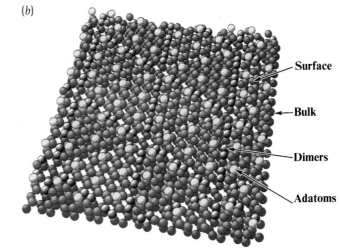

Surface

Bulk

Dimers

Adatoms

Figure 7
(a) Low-energy-electron-diffraction patterns of the reconstructed Si(111) crystal face, exhibiting a (7 × 7) surface structure, taken at four different electron energies. (b) The Si(111)-(7 × 7) reconstruction that occurs after the desorption of H which caps the Si(111)-(1 × 1) surface.

Top view: Si(100)

Ideal

p(2×1) reconstructed

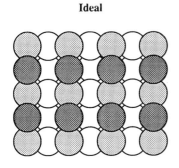

Side view: Si(100)

Ideal

p(2×1) reconstructed

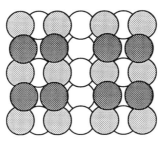

0.031 nm

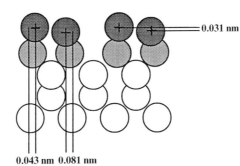

0.043 nm 0.081 nm

Figure 8
Top and side views of the Si(100)-(1 × 1) and (2 × 1) reconstructions, showing the dimer pairing of the top-layer Si-surface atoms, which results in the removal of half of the clean-surface dangling bonds.

Figure 9
Top and side views of the Ir(100)-(1 × 5) surface reconstruction. The more open square (100) lattice is reconstructed into a close-packed hexagonal overlayer, with a slight buckling, as shown from the side.

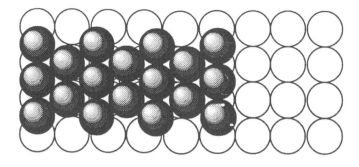

Ir(100)–(1×5) reconstruction   **Ir(100)–(1×1)**

Top view

Side view

Figure 10
The normal alcohols exhibit substantial ordering at the liquid–vapour interface. The OH groups of the alcohols extend into the liquid, forming a hydrogen-bonded network, while the alkyl chains are oriented away from the liquid. No evidence of an odd–even effect is seen for this system.

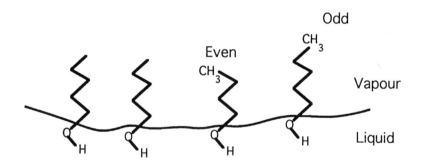

the carbon–metal chemisorption bonds, the surface metal atoms move away from the adsorption site, probably to give more space to the carbon atom so that it can sink deeper into the surface, thereby forming bonds with second-layer nickel atoms underneath (Figure 13). This expansion around the chemisorption site induces strain that is relieved by rotation of the surface unit cell as shown in Figure 13.

When sulfur is chemisorbed onto the Fe(110) crystal face, the S atom pulls the neighbouring Fe atoms to positions equidistant from the chemisorption site to form four equal Fe–S bonds. The strength of these bonds pays for the weakening of the metal–metal bonds as a result of the restructuring. When NO is adsorbed onto the Ni(111) surface, the molecule occupies a threefold hollow site; a so-called hexagonal close-packed (HCP) hollow site. This means that there is a metal atom directly under-

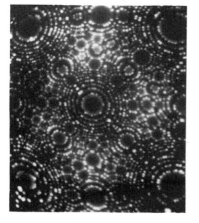
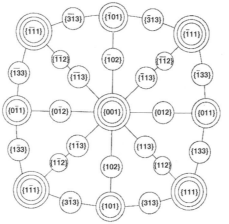

Figure 11
Field-ion micrographs (imaging gas Ne; $T = 85$ K) of a (001)-oriented Rh tip before (top left-hand side) and after (bottom left-hand side) reaction with 10^{-4} Pa CO for 30 min at 420 K. The stereographic projections on the right-hand side demonstrate the change in the morphology from nearly hemispherical to polygonal (the scheme on the bottom right-hand side indicates the coarsening of the crystal and the dissolution of a number of crystallographic planes due to the reaction with CO).

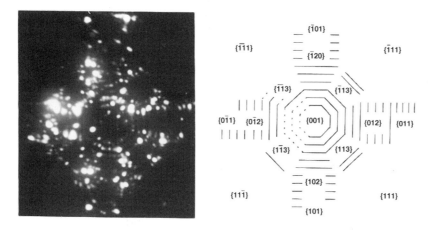

neath the chemisorption site in the second layer of metal atoms. Chemisorption induces an upwards movement of this metal atom in the second layer and rumpling of the metal surface. When ethene is adsorbed onto the Rh(111) surface, it occupies a hollow site (in this case, again, a HCP hollow site). This is shown in Figure 14. The metal atoms move away from the carbon atom bound to the hollow site to allow the carbon to bond to the Rh atom directly underneath the carbon in the second layer. On the Pt(111) surface, ethene forms an ethylidyne molecule; again in a threefold hollow site, but in this case it is a face-centred cubic (FCC) hollow site. That is, there is no metal atom directly underneath the carbon in the second layer of metal atoms. Under this circumstance, the surface metal atoms move inwards, probably to provide as strong a bond to the carbon as possible, and metal–metal distances on the surface are altered as well. The second metal atom next to the chemisorption bond moves downwards to produce a corrugated surface (Figure 15). It appears that surface bonding is cluster-like, in that nearest-neighbour metal atoms that surround the adsorbate move in such a way as to optimize the surface chemical bond. The heat of adsorption, which is always exothermic, pays for the weakening of the next nearest-neighbour metal–metal bonds which are altered as a result of the movement of the metal atoms nearest to the chemisorption site. Figure 16 shows several multi-nuclear organometallic clusters that have bonding similar or identical to that found after chemisorption of the same molecules or molecular fragments onto metal surfaces.

Hydrogen: 1.7 atm, 73 nm × 70 nm

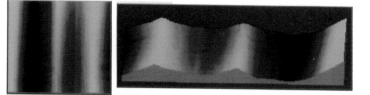

Oxygen: 1 atm, 90 nm × 78 nm

Carbon monoxide: 1 atm, 77 nm × 74 nm

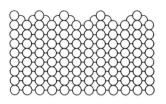

'Nested' missing-row reconstruction

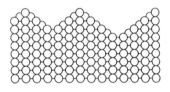

(111) microfacets

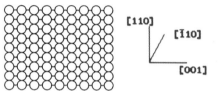

Unreconstructed (1 × 1) terraces
separated by multiple height steps

Figure 12
An *in situ* high-pressure STM picture showing adsorbate-induced surface reconstructions of Pt(110) under atmospheric pressure.

The Flexible Surface

As a result of all these studies that indicate that adsorbate-induced restructuring and cluster-like bonding occur, the new model of the surface which has been adopted is that of the so-called 'flexible surface'. In the past it was assumed that the metal atoms at the surface are rigid and occupy equilibrium sites dictated by the bulk unit cell. On adsorption, their location would not be altered. Instead, the flexible surface is one on which the metal atoms move into new sites, dictated by the chemisorption bond, in such a way as to optimize that bond. The lower the coordination of metal atoms (the fewer nearest neighbours) the more easily they restructure to optimize the surface-adsorption bond. Thus, atoms on rough surfaces or at steps move more readily and of course small clusters of atoms where the coordination is much reduced are the most flexible. It is not surprising therefore that we use nano-clusters in the field of catalysis (and in many instances chemisorption) to optimize chemical effects, such as chemical reactions and adsorption.

The flexible-surface model explains why rough surfaces and defect sites at surfaces are so active in surface chemistry. Bond breaking and catalysis most frequently occur at low coordination sites such as steps and kinks or at defect sites such as oxygen vacancies on oxide surfaces. Table 2 shows that the H_2–D_2 exchange at surface steps of a platinum crystal has unity reaction probability, whereas on defect-free flat Pt(111) surfaces the rate of breaking of the H–H bonds that this reaction probes is below the detection limit.

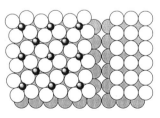

Ni(100)–p4g(2×2)–2C Ni(100)–(1×1)

Figure 13
The restructuring of the Ni(100) surface induced by chemisorption of carbon.

Thermal Activation of Bond Breaking

It is often found that, when a molecule is adsorbed at a surface at low enough temperature, its structure is unaltered compared with its gas-phase molecular structure. As

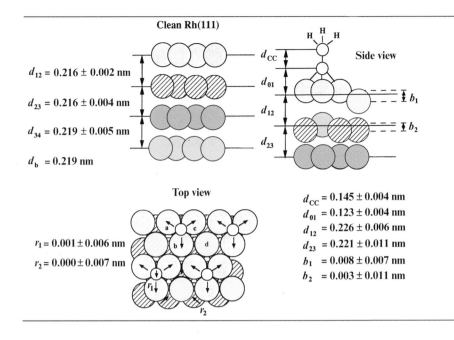

Figure 14
The structure of ethylidyne on Rh(111). Ethylidyne is bonded on the HCP three-fold hollow site. This site has a metal atom right underneath the carbon-bonding site in the second layer. The molecule-induced distortion in the top metal layers pulls the nearest-neighbour metal atoms up out of the surface plane.

Clean Rh(111)

$d_{12} = 0.216 \pm 0.002$ nm

$d_{23} = 0.216 \pm 0.004$ nm

$d_{34} = 0.219 \pm 0.005$ nm

$d_b = 0.219$ nm

Side view

Top view

$r_1 = 0.001 \pm 0.006$ nm

$r_2 = 0.000 \pm 0.007$ nm

$d_{cc} = 0.145 \pm 0.004$ nm
$d_{01} = 0.123 \pm 0.004$ nm
$d_{12} = 0.226 \pm 0.006$ nm
$d_{23} = 0.221 \pm 0.011$ nm
$b_1 = 0.008 \pm 0.007$ nm
$b_2 = 0.003 \pm 0.011$ nm

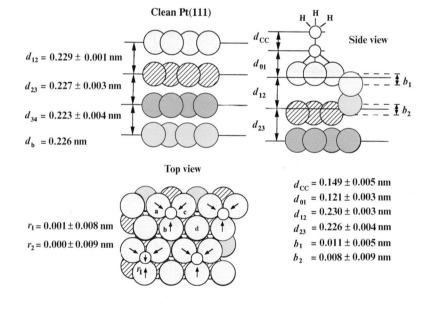

Figure 15
The structure of ethylidyne on Pt(111). On the platinum surface, ethylidyne bonds in the FCC three-fold hollow. The molecular adsorption-induced distortions in the top metal layer are clearly shown.

Clean Pt(111)

$d_{12} = 0.229 \pm 0.001$ nm

$d_{23} = 0.227 \pm 0.003$ nm

$d_{34} = 0.223 \pm 0.004$ nm

$d_b = 0.226$ nm

Side view

Top view

$r_1 = 0.001 \pm 0.008$ nm

$r_2 = 0.000 \pm 0.009$ nm

$d_{cc} = 0.149 \pm 0.005$ nm
$d_{01} = 0.121 \pm 0.003$ nm
$d_{12} = 0.230 \pm 0.003$ nm
$d_{23} = 0.226 \pm 0.004$ nm
$b_1 = 0.011 \pm 0.005$ nm
$b_2 = 0.008 \pm 0.009$ nm

Table 2	The sensitivity to structure of $H_2 \Leftrightarrow D_2$ exchange at low pressures ($\simeq 10^{-6}$ torr)

Surface	Probability of reaction
Stepped Pt(332)	0.9
Flat Pt(111)	$\simeq 10^{-1}$
Defect-free Pt(111)	$\leq 10^{-3}$

Figure 16
The structures of small organic molecules on surfaces and in organometallic clusters. Note the similarity of bonding.

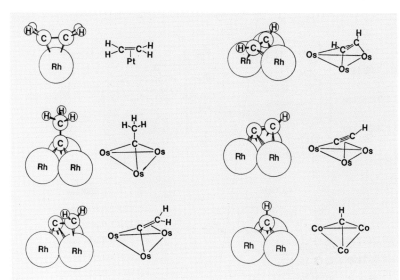

Figure 17
Thermal fragmentation pathways for *p*-xylene and *o*-xylene.

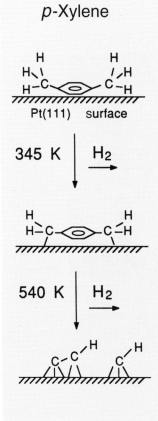

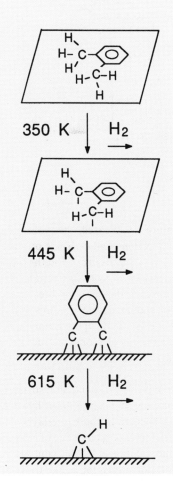

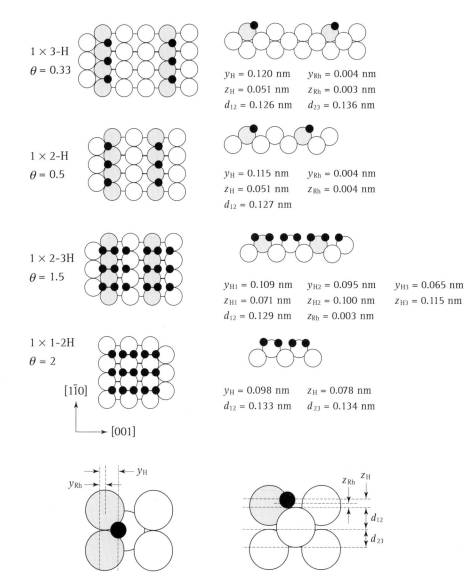

Figure 18
Real-space configurations of the
Rh(110) surface for different
coverages of hydrogen. Rh atoms
are shaded to represent buckling in
the Rh layer.

1×3-H
$\theta = 0.33$

$y_H = 0.120$ nm $y_{Rh} = 0.004$ nm
$z_H = 0.051$ nm $z_{Rh} = 0.003$ nm
$d_{12} = 0.126$ nm $d_{23} = 0.136$ nm

1×2-H
$\theta = 0.5$

$y_H = 0.115$ nm $y_{Rh} = 0.004$ nm
$z_H = 0.051$ nm $z_{Rh} = 0.004$ nm
$d_{12} = 0.127$ nm

1×2-3H
$\theta = 1.5$

$y_{H1} = 0.109$ nm $y_{H2} = 0.095$ nm $y_{H3} = 0.065$ nm
$z_{H1} = 0.071$ nm $z_{H2} = 0.100$ nm $z_{H3} = 0.115$ nm
$d_{12} = 0.129$ nm $z_{Rh} = 0.003$ nm

1×1-2H
$\theta = 2$

$[1\bar{1}0]$

$[001]$

$y_H = 0.098$ nm $z_H = 0.078$ nm
$d_{12} = 0.133$ nm $d_{23} = 0.134$ nm

the temperature is increased, the molecule undergoes sequential bond breaking. This is shown for *p*-xylene and *o*-xylene on the Pt(111) surface (Figure 17). At well-defined temperatures H—C bonds break and the molecule assumes a different surface structure. This structure is stable within a certain temperature range. When the temperature is increased, however, further bond breaking takes place. This sequential bond breaking occurs at well-defined temperatures for a given substrate–adsorbate system. The sequential bond breaking, upon minor changes in temperature, can be explained if one assumes that the metal atoms restructure at that temperature and the relocation of surface atoms breaks molecular orbital symmetries, causing bond breaking to commence.

Adsorbate-coverage-dependent restructuring of the surface is often observed. Figure 18 shows that, on chemisorption of hydrogen, the surface metal atoms that are contracted with respect to the bulk interlayer spacing when the metal surface is clean move outwards with increasing coverage of hydrogen. This observation indicates that surface atoms assume a more bulk-like configuration due to chemisorption and that they will move into locations not unlike those of atoms in the bulk. However, this simple

Figure 19
Structures of sulfur on Re(0001). The formation of clusters as the sulfur coverage is increased is observed.

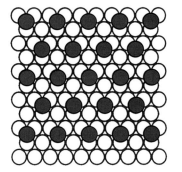

p(2×2)
$\theta = 0.25$

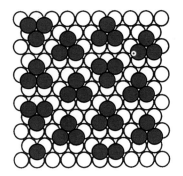

$(2\sqrt{3}\times2\sqrt{3})R30$
$\theta = 0.70$

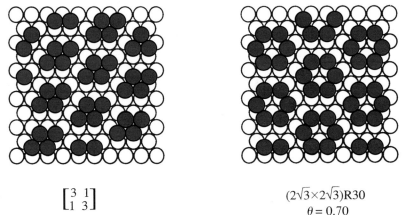

$\begin{bmatrix} 3 & 1 \\ 1 & 3 \end{bmatrix}$
$\theta = 0.50$

$(2\sqrt{3}\times2\sqrt{3})R30$
$\theta = 0.70$

physical picture does not apply for every adsorbate–substrate chemisorption system. For example, when sulfur is chemisorbed onto the Re(0001) crystal surface, with increasing coverage one finds the formation of sulfur clusters (trimers, tetramers and hexamers) as shown in Figure 19. The reason for this is that the rhenium metal atoms reconstruct. By moving outwards near the chemisorption site, they neutralize the repulsive interaction between adjacent sulfur atoms which would otherwise prevent clustering.

Co-adsorption

When two molecules are adsorbed on the surface, adsorbate-induced reconstruction may be very different from that when only one or the other molecule is chemisorbed. This is shown for co-adsorption of benzene and CO onto the Rh(111) surface (Figure 20). When these molecules form a mixed unit cell, the metal atoms under the benzene are closer to their bulk-like configuration than are those under the CO molecule. There is a rumpling of the surface that occurs because of the differences in chemical bonding of the co-adsorbed species to the substrate metal atoms. When benzene is adsorbed

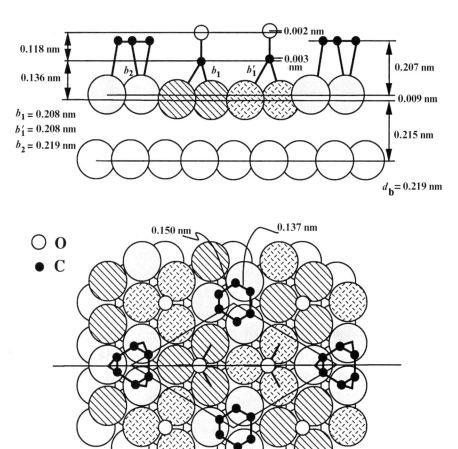

Figure 20
The surface structure with benzene
and carbon monoxide co-adsorbed
onto the Rh(111) crystal face
obtained by low-energy-electron-
diffraction surface crystallography.

alone it is bent. Four of the C atoms are in one type of surface site while two of the others are in different types of surface sites. When co-adsorption occurs, the benzene molecule is flattened out.

Co-adsorption of two molecules, one an electron donor to the metal (like most organic molecules), the other an electron acceptor (like CO or NO are to many transition metal substrates), leads to the formation of ordered surface structures with both molecules in the same surface unit cell. There appears to be an attractive donor–acceptor interaction among the co-adsorbed molecules that stabilizes these ordered surface structures. When the co-adsorbed species are both electron donors or electron acceptors to the substrate, the repulsive adsorbate-adsorbate interaction usually causes separation of the two adsorbates into islands that destroy the long-range order in the adsorbate layer.

Gas–Surface Energy Transfer

The dynamics of adsorption and exchange of energy upon impact of an incident molecule on the surface can often be studied by a variety of spectroscopic techniques. The changes of translational energy, rotational energy and vibrational energy of the incident molecules upon scattering from the surface have been measured. These studies are an important part of developing the understanding of elementary surface processes (adsorption, surface diffusion and desorption) that occur during gas–surface interactions.

These investigations usually employ molecular beams and ordered single-crystalline surfaces. The beam of molecules is prepared in such a way that the kinetic energy and internal energy of the molecules is uniform and the angle of incidence on the surface is well defined. The angular distribution and the energy distribution of the scattered molecules are measured using time-of-flight mass spectroscopy and optical spectroscopy techniques.

Most often gas molecules undergo only partial energy transfer upon incidence on surfaces. If they lose a large fraction of their kinetic energy, they can be trapped on the surface in so-called 'precursor states' that permit diffusion along the surface, followed either by desorption or by strong chemisorption at certain surface sites. Their vibrational and rotation energies are also altered, depending on their adsorption geometries and thermal energies.

Heterogeneous Catalysis

A catalyst that is used in chemical technologies is usually a microporous solid of high surface area (often in the range $1–300 \text{ m}^2 \text{ g}^{-1}$). On this material, metal atoms that carry out catalytic chemical reactions are adsorbed in small clusters. Irrespective of whether the active catalyst material is deposited onto a high-surface-area support, its high surface area is necessary because the rate of formation of desired product molecules per second is proportional to the surface area. Thus, the higher the area the higher the number of molecules produced per unit time.

The catalyst surface should be able to break or weaken the chemical bonds of reactant molecules that adsorb. After appropriate chemical rearrangements, however, the product molecule that formed should be able to desorb from the surface so that the same surface chemical process can be repeated. Thus, a good catalyst will form chemical bonds with the adsorbed reactants, reaction intermediates and product molecules that are of intermediate strength. If the surface chemical bonds are too weak, there is no surface chemistry; if they are too strong, the adsorbed species cannot be removed and the catalyst surface is deactivated or poisoned. A good catalyst surface should be able to perform the same reaction over and over again to achieve millions of turnovers without deactivation.

Catalysts are usually transition metals that form covalent bonds with reactant molecules, or acids and bases that are capable of electron/proton transfer to/from the adsorbed reacting species. Often, both are utilized in an active catalyst system that may have a metal and an acid 'reaction channel'. There are solid oxides that have acid–base properties superior to those of concentrated sulfuric acid and HF.

Model catalyst systems are usually employed in surface-science studies of heterogeneous catalysis. These are often single crystals or evaporated thin-film deposits that order themselves and become crystalline when treated properly. Their surface areas are roughly 1 cm^2 and their structures and compositions can be controlled and studied more easily than can those of dispersed, microporous catalyst systems. As a result, the correlations between structure (or other atomic properties of surfaces) and catalytic activity can be more readily established. Such model systems may be studied in the apparatus shown in Figure 21. In this system, the sample is placed in an ultra-high-vacuum chamber where it is properly cleaned by ion bombardment, heat and chemical treatments and then characterized by a variety of surface techniques that can determine the structure and composition of the surface. The sample is then enclosed in a stainless steel tube that separates it from the ultra-high vacuum. The tube can be pressurized and catalytic reactions carried out at well-defined temperatures. The product distribution and kinetics of the process can be monitored using a gas chromatograph or a mass spectrometer. After the reaction the gases are pumped out and the sample is again characterized in ultra-high vacuum.

Most surface-science studies of heterogeneous catalysis have focused on transition

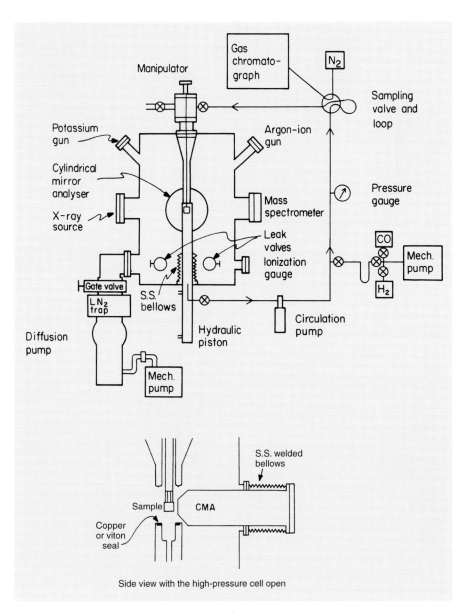

Figure 21
A high-pressure/low-pressure chamber for studying surfaces of catalysts.

metals that play significant roles in carrying out reactions at optimum rates. The rate may change by orders of magnitude when one changes the transition metal. This is shown in Figure 22 for reactions that involve the breaking of C–C, C–N and C–Cl bonds.

Catalytic reactions that occur on transition metals may be classified into structure-sensitive and structure-insensitive categories. The synthesis of ammonia for example, is a structure-sensitive reaction. As shown in Figure 23, various crystal faces of iron (one of the best catalysts for this reaction) carry out the conversion of N_2 and H_2 to ammonia at rates that may vary by two orders of magnitude. Conversely, the hydrogenation of ethene is a structure-insensitive reaction. The rates are identical irrespective of whether it is carried out on single crystals, foils or dispersed Pt clusters. Figure 24 shows a hydrodesulfurization reaction by which thiophene is converted into butane and butenes. One metal exhibits structure insensitivity for this reaction (molybdenum): the reaction rates on each single-crystalline surface are identical. On rhenium,

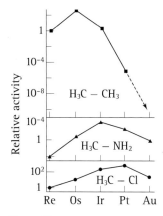

Figure 22
Catalytic activities of transition metals across the Periodic Table for the hydrogenolysis of the C—C bond in ethane, the C—N bond in methylamine and the C—Cl bond in methyl chloride.

however, the reaction is definitely structure-sensitive. Thus, the structure sensitivity or insensitivity of a reaction depends not only on the reaction but also on the catalytic material.

The catalysts are almost always modified by the addition of structure and bonding modifiers. For example, in the synthesis of ammonia, aluminium oxide and potassium oxide are added to Fe to optimize the reaction rate. The role of aluminium oxide is one of a structural promoter or structure modifier. Alumina forms an iron aluminate compound with Fe and, during the reaction, Fe restructures and grows epitaxially, forming crystallites of (111) orientation on this iron aluminate surface. The (111) orientation of iron is the best for producing ammonia at the maximum possible rate. Because all other crystal faces of Fe are restructured and converted to the most active (111) crystal face, a most active catalyst is produced in the presence of alumina.

Potassium is an electron donor to the iron surface. When it is co-adsorbed with ammonia, the product of the hydrogenation of nitrogen, it weakens the bonding of ammonia to the Fe surface since both K and ammonia are electron donors. This decreases the heat of adsorption of ammonia by about 10 kJ mol^{-1}. Thus, ammonia, which is by far the stickiest molecule in terms of adhesion to the catalyst's surface during the synthesis of ammonia has its residence time on the iron surface shortened. Therefore it does not block active metal sites and does not impede the dissociative adsorption of dinitrogen as atomic nitrogen to the same extent as it does in the absence of co-adsorbed potassium.

One product only is formed during the synthesis of ammonia. However, in most other catalysed reactions there are many products that may form, all of which are thermodynamically feasible. We often desire to optimize the formation of one of these products while inhibiting the formation of others. For example, n-hexane is converted to other species that may be aromatic (like benzene or methylcyclopentane) or to various types of isomers (Figure 25). We often desire to obtain only one of these products. When producing high-octane petrol we may want to produce a branched isomer or an aromatic molecule. Both of these molecules would provide a high octane number because of their favourable rates of combustion. Platinum is an excellent selective catalyst for carrying out these processes if the catalyst is properly structured. Figure 26 shows the most active surface for the conversion of n-hexane to benzene, the nature of which was uncovered by a combination of surface chemistry and studies of catalytic reaction rates using single-crystalline platinum surfaces. Both the presence of steps of atomic height and terraces where the Pt atoms have hexagonal arrangements are necessary if one is to obtain benzene selectively.

During reactions involving hydrocarbons the metal surface is often covered with a stagnant, tenaciously adsorbed carbonaceous deposit that does not turn over during the reaction. The question of how the reaction occurs in the presence of such a tenaciously adsorbed layer arises. We can image these hydrocarbon fragments on the surface by using STM. It has been found that adsorption occurs as long as one can compress the stagnant chemisorbed molecular species, thereby creating a hole where adsorption of reactants and subsequent reaction occurs. Upon desorption of the product, the hydrocarbon fragments diffuse back to their original location. Thus, as long as the activation energy for surface diffusion is small, mobility along the surface provides the opportunity for compression of this stagnant layer upon adsorption of other species, allowing one to carry out chemical reactions in the holes that are created.

Table 3 lists the catalysts and the catalysed reactions that are carried out in very large volumes in the chemical and petroleum industries. Most chemicals, irrespective of whether they are produced in large or small quantities, encounter one or more catalyst systems during their preparation. Heterogeneous catalysis is one of the most important applications of surfaces and surface chemistry.

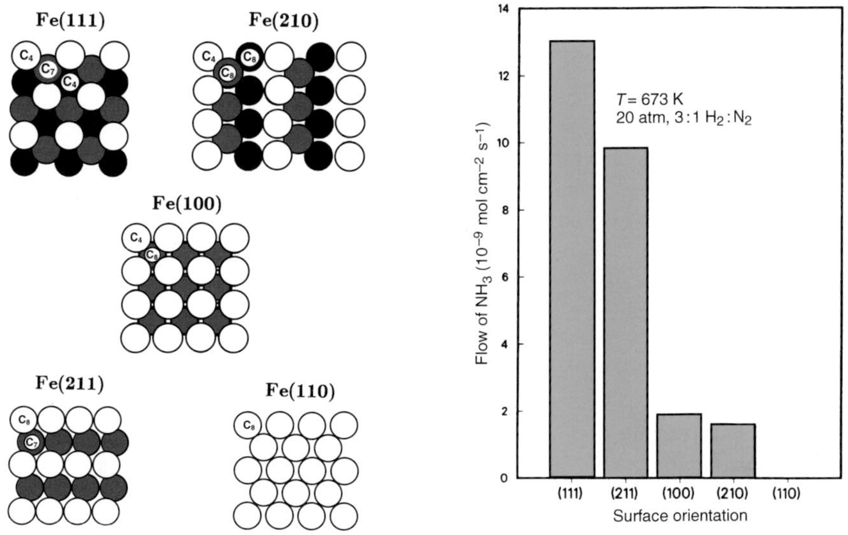

Fe(111)

Fe(210)

Fe(100)

Fe(211)

Fe(110)

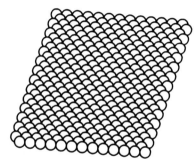

$T = 673$ K
20 atm, 3:1 $H_2 : N_2$

Flow of NH_3 (10^{-9} mol cm^{-2} s^{-1})

Surface orientation

(111) (211) (100) (210) (110)

Figure 23
Schematic representations of the idealized surface structures of the (111), (211), (100), (210) and (110) orientations of single crystals of iron. The coordination of each surface atom is indicated.

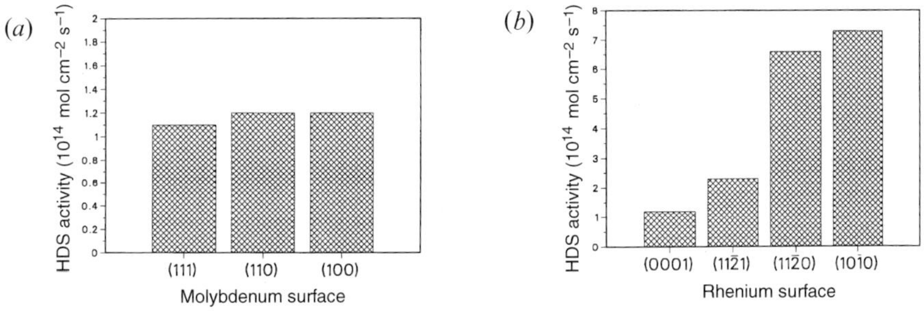

(a)

HDS activity (10^{14} mol cm^{-2} s^{-1})

(111) (110) (100)
Molybdenum surface

(b)

HDS activity (10^{14} mol cm^{-2} s^{-1})

(0001) (11$\bar{2}$1) (11$\bar{2}$0) (10$\bar{1}$0)
Rhenium surface

Figure 24
The sensitivities to structure of the catalytic hydrodesulfurization of thiophene on single-crystalline surfaces of molybdenum and rhenium. Molybdenum (a) exhibits insensitivity to structure, whereas rhenium (b) exhibits sensitivity to structure for this important reaction.

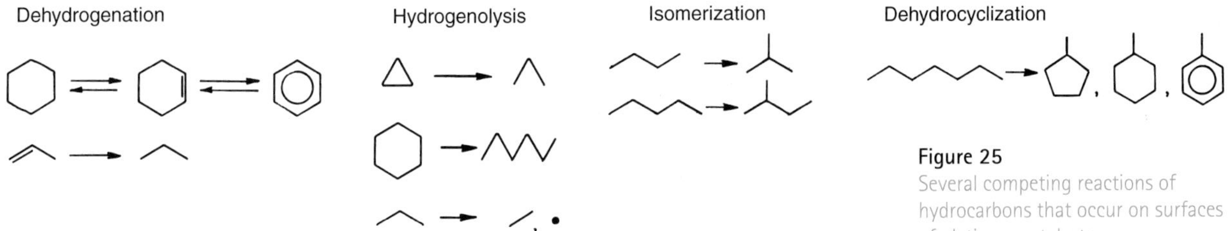

Dehydrogenation

Hydrogenolysis

Isomerization

Dehydrocyclization

Figure 25
Several competing reactions of hydrocarbons that occur on surfaces of platinum catalysts.

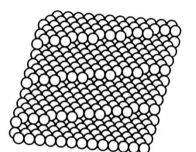

FCC (111)

FCC (10, 8, 7)

Figure 26
Idealized atomic surface structures for the flat Pt(111) and the stepped Pt(10, 8, 7) surfaces.

Reactions	Catalysts
Oxidation of CO and hydrocarbons in car exhaust gases	Pt, Pd on alumina
Reduction of NO_x in car exhaust gases	Rh on alumina
Cracking of crude oil	Zeolites
Hydrotreating of crude oil	Co–Mo, Ni–Mo, W–Mo
Re-forming of crude oil	Pt, Pt–Re and other bimetallics on alumina
Hydrocracking	Metals on zeolites or alumina
Alkylation	Sulfuric acid, hydrofluoric acid
Steam re-forming	Ni on support
Water-gas-shift reaction	Fe–Cr, CuO, ZnO, alumina
Methanation	Ni on support
Synthesis of ammonia	Fe
Oxidation of ethyne	Ag on support
Nitric acid from ammonia	Pt, Rh, Pd
Sulfuric acid	V oxide
Acrylonitrile from propene	Bi, Mo oxides
Vinyl chloride from ethene	Cu chloride
Hydrogenation of oils	Ni
Polyethene	Cr, Cr oxide on silica

Table 3 Chemical processes that are the largest users of heterogeneous catalysts at present and the catalyst systems they utilize most commonly

Tribology

In Greek, the word *tribein* means rubbing and tribology has come to mean the science of interfaces in relative motion. The mechanical properties of interfaces that are scrutinized in tribology include friction, sliding, lubrication and wear.

There is evidence that in Egypt a lubricant was poured in front of the sledges used during the transport of statues and stones. The Romans used iron nails in the soles of their *caligae* (shoes) to reduce wear. They used metal sleeves for crushing olives and in corn-grinding mills, indicating that they appreciated that rolling was easier than sliding.

The use of wheels and axles prompted extensive application of animal fat and vegetable oil for lubrication. In the Middle Ages, stone studs were employed to reduce the wear of wooden ploughs. During the Renaissance there was a burst of development of machinery with moving parts. Leonardo da Vinci gained an understanding of the laws of friction and designed rolling-element bearings. Amontons reported the first comprehensive study of friction in 1699. In the same period, the role of the surface texture in determining frictional resistance was realized and utilized; and the concept of deformation during sliding and adhesion was suggested. During the industrial revolution, mineral oil was discovered to be an excellent lubricant. Since that time the formulation of lubricating oils has made the development of fast, steam-turbine-driven ship, automobile and aircraft engines possible. Because of the importance of mechanical properties of surfaces, centres of tribology research have been established in many countries. From the development of synthetic lubricants used between moving parts at high speeds and at high temperatures to the design of steel-belted radial tyres and modern skis, from solid lubricants for applications in ultra-high vacuum and space research to lubricants for computer disc drives, the control of friction between moving surfaces is one of the most challenging problems in the design of modern devices and machinery.

While the macroscopic concepts of hardness, adhesion, friction and sliding evolved over the last two centuries, atomic-level understanding of the mechanical properties of surfaces eluded researchers. The discovery of the atomic-force microscope in recent years promises to change this state of affairs. Being able to measure forces as small as 10^{-9} N and as large as 10^{-2} N over a very small surface area (a few atoms) and by virtue of simultaneously providing atomic spatial resolution, this technique permits the study of deformation (elastic and plastic), hardness and friction on the atomic scale. The buried interfaces between moving solid surfaces can be studied with spectroscopic techniques on the molecular level. The study of the mechanical properties of interfaces is, again, a frontier research area of surface chemistry.

Let us consider the surface contact on an atomic scale. A surface atom is bound to its neighbours by strong chemical bonds that add up to about 600 kJ mol^{-1}. We would like to estimate the mechanical force necessary to break these bonds. If the atom behaves as a harmonic oscillator, then stretching its bonds by 10^{-2} nm about its equilibrium position – fives times its root-mean-square displacement of about 2×10^{-3} nm – would certainly break the bonds. Thus, the force needed to stretch the bond by 10^{-2} nm is given by

$$F = E/\Delta x = 10^{-18} \text{ J}/10^{-11} \text{ m} = 10^{-7} \text{ N}$$

Therefore, we need about 10^{-7} N of force per atom to break the roughly 600 kJ mol^{-1} (10^{-18} J per atom) surface-atom bond.

The atomic-force microscope (AFM) can be used to investigate contact and hardness on the atomic scale. Analogously to the STM, the AFM uses a feedback loop to control the distance between the sample and a probe tip at the end of a cantilever arm. Rather than a tunnelling current, however, the AFM monitors an optical signal as feedback to measure the level of deflection. Thus, both attractive and repulsive interactions of the tip and sample can be monitored. As the microscope tip approaches the surface, attractive forces are first exerted on the tip by the surface and can be measured to as small a value as 10^{-9} N. Upon contact with the surface, further motion of the tip results in repulsive forces between the tip and the sample. This procedure is capable of producing loads that overlap the forces encountered in macroscopic mechanical measurements. After a force of known magnitude has been applied, that area of contact with the tip is scanned for signs of permanent damage. For a smooth gold surface, permanent damage has been detected only for forces as large as about 5×10^{-5} N, as shown in Figure 27. This result suggests that the metal surface responds to the approaching metal tip by elastically 'bending' in the range of 10^{-9} to 5×10^{-5} N. This range of forces is the regime of elastic deformation. Only on application of a force greater than 5×10^{-5} N did plastic deformation (the irreversible breaking of metal–metal bonds) begin. Measurements of this type permit one to determine the forces needed for elastic and plastic deformation on the atomic scale and to correlate the results to those obtained by macroscopic studies.

The coefficient of friction μ is defined as the force F required to initiate and maintain sliding divided by the load W applied perpendicular to the surface: $\mu = F/W$. The coefficient of friction is found to be independent of the area of contact over a wide range of loads that cause plastic deformation. The reason for this is that both the force and the load are proportional to the same area of interfacial contact. For plastic deformation, we have already shown that $W = Ap$, where p is the average pressure over the contacts and A is the area of contact. At the contacts, chemical bonds are made and must be sheared or broken as one solid surface slides over the other. If the shear strength s is constant, the force S required to shear these contacts equals the area of contact times the shear strength ($S = As$). If materials of comparable hardnesses are brought into contact, the sliding force is equal to the shear force and proportional to

Figure 27

The microhardness of gold determined by using the atomic-force microscope. At 3.4×10^{-5} N load there is no sign of plastic deformation (i.e. permanent damage). At 6.7×10^{-5} N plastic deformation occurs. The measurements were carried out on a fold of film deposited onto a mica substrate to produce atomically smooth metal surfaces. (Courtesy of C. M. Mate and G. S. Blackman.)

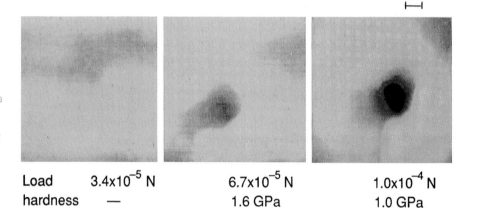

100 nm

Load	3.4×10^{-5} N	6.7×10^{-5} N	1.0×10^{-4} N
hardness	—	1.6 GPa	1.0 GPa
		(1.6×10^{4} atm)	(1.0×10^{4} atm)

the area of contact. Therefore we can write $\mu = F/W = s/p$; this observation that the frictional force is independent of the contact area is known as *Amontons' law*.

An additional phenomenon can occur at the sliding interface between two materials of different hardnesses. In this case, the chemical bonds that are continuously formed and broken at the interface compete with chemical bonds within the softer material. As a result, parts of the softer of the two sliding partners are transferred to the side of the harder surface. This phenomenon is known as *ploughing* and is usually associated with the presence of wear tracks, the transport of material and the accumulation of debris at the interface. Here, the total frictional force is the sum of the shear force and the force required to displace the softer material.

The frictional forces that operate when a wheel or a ball bearing rolls over a surface are different from the friction between two flat sliding surfaces. When it is in contact with a surface, the wheel is temporarily deformed elastically and also causes deformation of the other side of the interface. The energy stored in producing the deformation is partially recovered as the wheel moves. However, the deformation is not completely elastic and friction due to adhesion also occurs, which tends to increase with increasing velocity. Rolling friction plays an all-important role in permitting us to walk and run. It has been found that the sliding force needed to initiate motion is usually greater than the sliding force needed to maintain a given speed of sliding. Thus, there is both a static and a kinetic friction coefficient, μ_s and μ_k, respectively, with $\mu_s > \mu_k$. The coefficients of friction (μ_s) of several surfaces in contact are listed in Table 4. The μ values are usually in the range 0.4–1.0.

Lighting a match by rapidly moving the match head against a rough surface is a good example of how heat is generated by friction. The rise in temperature in this circumstance is large enough to ignite the coating material, which has a low flash-point. Considering the same geometry, if all the frictional work is converted to heat, the amount of heat q is $q = \mu W v$, where μ is the kinetic coefficient of friction, W is the load and v is the velocity of sliding. High temperatures, in the range 200–600 °C, can be reached before the heat is dissipated by thermal conduction.

Melting at a sliding interface is a mechanism that can be invoked in some instances to rationalize measured coefficients of friction. In the case of a ski sliding on snow, for example, the formation of a film of water by local heating at the interface is thought to be responsible for the low coefficients of friction on snow and ice. Thus it is the interaction with water, rather than that with ice, that determines the degree of friction. On looking at the range of μ for various materials on ice, those that cannot be wetted

| Table 4 | Coefficients of static friction for various contacts between materials |

Materials	μ_s
Glass on glass, clean	0.9–1.0
Diamond on diamond, clean	0.1
Hard carbon on steel	0.14
Tungsten carbide on steel	0.4–0.6
Polystyrene on steel	0.3–0.35
Brick on wood	0.6
Brake material on cast iron, clean	0.4
Brake material on cast iron, wet	0.2
Copper on copper, in air	2.8
Iron on iron, in air	1.2
Molybdenum on molybdenum, in air	0.8
Ice on ice, 0°C	0.05–0.15
Aluminium on snow, 0°C	0.35
Ski on snow, 0°C	0.04

Source:
F. P. Bowden and D. Tabor, *The Friction and Lubrication of Solids* (Oxford University Press, New York, 1971).

by water, such as fluorocarbons, are found to have the lowest coefficients of friction. Snow skis are accordingly coated with such materials in order to enhance their performance.

The rapid passage of a weighted wire through ice was first reported in 1872. This interesting phenomenon has been used to cut large blocks of ice to smaller pieces. Both Faraday and Thomson suggested the presence of a liquid-like layer on the surface of ice below the freezing point to account for the rapid motion of the wire through the ice. Recent studies indicate that this is indeed the correct explanation, rather than melting being caused by pressure or by the heating due to friction. The liquid-like surface layer on ice appears to exist down to -30 °C, its thickness decreasing with decreasing temperature. Thus the cutting of ice with a wire would not be possible below this temperature.

Materials that lower the coefficient of friction are commonly used in tribology under all conditions of relative motion. These materials are called *lubricants*. Long-chain organic molecules with polar groups at one end of the chain are commonly employed to lubricate moving parts, with excellent results – often a decrease by an order of magnitude in μ. Table 5 lists well-known and commonly used lubricants. It is thought that the polar groups improve binding and adhesion to the solid while the long hydrocarbon chain provides solubility in the lubricant liquid. In addition, the coefficient of friction decreases with increasing relative molecular mass of the hydrocarbon chain. One of the most successful lubricant systems uses the dialkyl-dithio zinc phosphates, where R may be either alkyl or aryl groups, as additives. The reasons for its lubricating ability are being studied on the molecular level in several laboratories.

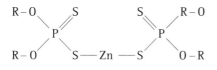

Table 5 Commonly used lubricants

Name	Structure
Diethyl phosphonate	C_2H_5-O $\diagdown$ P $\diagup$ $\diagdown$ O; C_2H_5-O $\diagdown$ $O-H$
Cyclohexyl amine salt of dibutyl phosphate	C_4H_9-O $\diagdown$ P $\diagup$ $\diagdown$ O; C_4H_9-O $\diagdown$ $O-{}^+H_3N\,C_6H_{11}$
Diphenyl disulfide	$C_6H_5-S-S-C_6H_5$
Di-(methylethanoate)disulfide	$C_2H_5\overset{\displaystyle O}{\overset{\|}{C}}-OCH_2-S-S-CH_2O-\overset{\displaystyle O}{\overset{\|}{C}}C_2H_5$
Zinc O,O-diisopropylphosphorodithioate	$Zn\,[-\overset{\displaystyle S}{\overset{\|}{P}}-(O\,CH\,(CH_3)_2)_2]_2$

The process of lubrication varies with the degree of interaction of the two surfaces. At low load levels (fluid lubrication), there are many layers of organic molecules sandwiched between the solid surfaces. In this circumstance, the friction depends on these molecules sliding by each other – that is, on the viscosity of the liquid. When the load is high and/or the speed is low, this pressure will not be sufficient and the surfaces come into close contact, resulting in the elastic deformation of the asperities. We are then in the 'elastohydrodynamic' lubrication regime. If the load and/or speed are further increased, the asperities deform plastically and the thickness of fluid is reduced to molecular dimensions. This regime is called 'boundary lubrication'. In this regime the molecular layer of lubricant is often not continuous, because high pressures at the asperities or narrow contact points squeeze out the lubricant from that area. A recent model of a lubricated interface is shown in Figure 28.

Liquids are the most common lubricants used to increase the performance of moving mechanical components over long periods. However, in certain environments, in vacuum or at high pressures, liquids cannot survive. In these cases, we must rely on solid lubricants.

Molybdenum disulfide, MoS_2 is an excellent solid lubricant. MoS_2 is a lamellar compound in which each layer is composed of two planes of sulfur atoms and an intermediate plane of molybdenum atoms tightly bound by covalent bonds. These layers are held together by van der Waals bonds. The weakness of these bonds allows easy shearing at this interlayer. Sputtered MoS_2 films are adherent, with a long endurance life and low coefficients of friction (<0.1). They are used for precision tribo-elements and exhibit extremely high endurance and low friction in high vacuum (spacecraft). The durability and adherence of MoS_2 films in air can be increased by modifying the process of their synthesis.

PTFE (e.g. Teflon) is widely used in the form of composites as self-lubricating cages for ball bearings. A thin film of polymer is transferred from the cage to the rolling/rubbing surface, providing continuous lubrication for the system.

The surface of a solid can be changed to modify the friction and wear properties.

Hard coatings are used to control friction and wear in such a way as to extend the life of a part. Ideally, a coated part should fail by a mechanism other than wear. If two metallic surfaces are sliding on each other, the dissipation of frictional energy can lead to frictional welding at points of real metal-to-metal contact and wear will follow by detachment of material from welded junctions. If one of the surfaces is coated with a hard coating, exhibiting high levels of covalent bonding, the probability of metal-to-metal welding is lowered, as is the subsequent wear. Cubic carbides and nitrides of transition metals are commonly used to lower this type of wear. We can estimate the level of hardness necessary to obtain wear resistance. To prevent important wear of a part sliding on a hard steel piece (maximum Vickers hardness 900), it is sufficient to coat this part with a material having a Vickers hardness in the range 1500–2000. Many oxides, nitrides, carbides, silicides and borides possess Vickers hardnesses greater than 1500. Because the rate of wear in the presence of the coating is remarkably low, thick hard layers are not necessary. Thin films will be sufficient for most applications as long as they adhere strongly enough to the substrate.

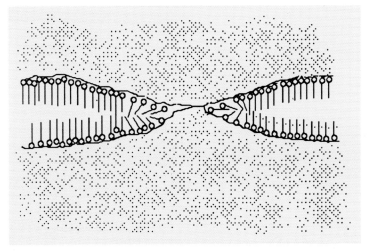

Figure 28
A model of a lubricated interface under high loads. The lubricant may be squeezed out at the narrow contact points.

Thin coatings are not only used for mechanical protection and lubrication but also widely applied to protect materials against corrosion and chemical reaction. The steel body of a car is first covered by a layer of zinc, which is further treated in a phosphate bath, to promote the adhesion of the paint. Damaging this sandwich structure, by breaking one of these layers, results in quick corrosion of the car's body; namely rusting. The improvement in the quality of this protective coating during the last 20 years is shown by the increase in the duration of corrosion warranties proposed by car manufacturers (from 2 years less than 15 years ago to 8 years nowadays).

Coatings can form naturally by reaction with the surrounding atmosphere: aluminium is quickly covered by a layer of aluminium oxide and hydroxide in the presence of oxygen and water. This inert and protective layer can be formed artificially, as is done for window and door frames. A porous layer of alumina can be formed by reaction with oxygen. Pigments can be included in the pores for decorative purposes. After inclusion of the pigments, a suitable treatment of the layer transforms its porous structure into a continuous non-porous one (sealing). These coated frames can withstand relatively harsh conditions for years.

Coatings are also very common in food packaging. An example is tomato products that are corrosive to canning materials (such as iron), which hence must be protected by polymer coatings. Otherwise the food will be contaminated by metallic ions and rupture of the can may occur.

Frontiers of Surface Chemistry

There is a great deal of information about the chemical behaviour of molecules at the solid–gas interface on the molecular level. In electrochemistry one deals with the properties of the solid–liquid interface, where, in addition to the presence of reactants, there is an electrolyte and also an external potential. In the future, the bridge between our knowledge of the solid–gas and electrochemical interfaces will be made by studying solid–liquid interfaces and correlating this information to that for solid–gas interfaces. The next step will then be to study the interaction of the reactant–electrolyte solution with solid surfaces, followed by applying an additional external potential. These sequential studies are necessary if one is to bring electrochemical interfaces to the same level of understanding as that which we have of many phenomena at the solid–gas interface.

The study of solid–solid interfaces is another frontier of modern surface science. Quantum-well devices have unique optical and electronic properties that are being increasingly utilized. They are composed of single-crystalline thin films of semiconductor materials no more than 10 nm thick. Because the mean free path of electrons is greater than the film's thickness, these materials have physical and chemical properties not unlike those of the nano-clusters that were mentioned above. Many of their unique properties arise from the structure at the solid–solid interface. These must be studied by modern surface-science techniques. Biomaterials or composites will be another challenging area of solid–solid interfaces to be studied by molecular surface science. The spine of the sea urchin is one of the toughest materials we know. It is composed of single crystals of calcium carbonate that are quite brittle dispersed in protein biopolymers. From molecular-level studies of such systems (with atomic-force microscopy, for example) we will be able to learn why such materials, which are composed of brittle and soft constituents, become uniquely tough when they form a composite. The brain and other biological systems are composed of solid–solid and solid–liquid interfaces. The molecular structure and chemistry of their interfaces will be studied increasingly in the future using modern surface-science techniques.

Environmental issues will have a major impact on developments in chemical technology during the 1990s and beyond. Catalytic combustion is a most important technology that would provide us with combustion rates at 1200–1500 °C equal to those used at present for energy conversion at 1800 °C without a catalyst. At the higher temperature, however, large concentrations of NO_x form per unit of energy produced. At the lower temperature one can minimize the formation of NO_x as part of the process. The production of chiral molecules in the pharmaceutical industry is most important because the body knows the difference between left-handed and right-handed optically active molecules. On producing the racemic mixture, undesirable (in fact dangerous) compounds are formed, which can be physiologically harmful. The pharmaceutical industry will greatly benefit from learning how to produce just one of the two stereoisomers catalytically. Finally, natural gas will take over from oil as a feedstock for fuels and chemicals in the future. The development of new technologies to make liquid fuels and alkenes (ethene and propene) from natural gas (methane, ethane, propane and butane) is a major challenge.

There are many new applications of surfaces that utilize their unique physical–chemical properties. The development of hard discs for magnetic storage of information and microelectronic devices that pack more components into ever decreasing volumes of space requires the ability to control atomic composition and structure on the nanometre scale. Techniques for atom-by-atom characterization of surfaces of area 1 nm^2 must be developed. This is one of the challenges of surface analysis.

We should be able to monitor the motion of surface atoms and adsorbed molecules with better time resolution *under reaction conditions*. Using optical techniques (nonlinear laser optics, nuclear magnetic resonance and synchrotron-radiation-based spectroscopies) the motion of atoms and molecules could be monitored in the 10^{-12}–10^{-6} s time range. In this way we could learn about the elementary molecular rearrangements on surfaces that accompany the breaking and making of chemical bonds and charge transfer between the adsorbate and the substrate.

By controlling the preparation of surfaces on the nanometre scale we could produce catalysts that exhibit 100% selectivity and optimum rates of reaction with negligible deactivation. These will be clusters of atoms, all of the same size and structure, dispersed in well-defined spatial arrangements. Techniques of electron-beam and X-ray lithographies, atomic- and molecular-beam epitaxies and chemical vapour deposition could help us to achieve this goal.

The preparation of polymers with tailored mechanical properties using surface treatments is one of the exciting new directions of materials science. Polymer surfaces that are as hard as diamond or as soft and lubricating as graphite can be produced in this way. Nano-clusters of silicon and carbon, not to mention other insulator and semiconductor materials, have novel optical, electrical and chemical properties that will be exploited by surface technologies.

Further Reading

1. D. P. Woodruff and T. A. Delchar. *Modern Techniques of Surface Science* (Cambridge University Press, New York, 1986).
2. G. A. Somorjai and M. A. V. Hove. In *Structure and Bonding*, edited by J. D. Dunitz, J. B. Goodenough, P. Hemmerich, J. A. Albers, C. K. Jorgensen, J. B. Neilands, D. Reinen and R. J. P. Williams, (Springer-Verlag, Berlin, 1979).
3. M. Prutton. *Surface Physics* (Oxford University Press, New York, 1975).
4. G. Ertl and J. Kupper. *Low Energy Electrons and Surface Chemistry* (VCR Verlagsgesellschaft, Weinheim, 1985).
5. G. A. Somorjai. *Introduction to Surface Chemistry and Catalysis* (John Wiley, New York, 1994).
6. H. Ibach and D. L. Mills. *Electron Energy Loss Spectroscopy and Surface Vibration* (Academic Press, New York, 1982).
7. J. B. Pendry. *Low Energy Electron Diffraction* (Academic Press, New York, 1974).
8. D. Dowson. *History of Tribology* (Longman Group, London, 1979).
9. J. B. Hudson. *Surface Science: An Introduction* (Butterworth-Heinemann, Boston, 1992).
10. *Clean Solid Surfaces. The Chemical Physics of Solid Surfaces and Heterogeneous Catalysis.* D. A. King and D. P. Woodruff (editors). (Elsevier, New York, 1981).

8 New Roads to Molecular Complexity

K. C. Nicolaou, E. W. Yue and T. Oshima

Introduction

Few other endeavours can claim both art and science and none commands the centrality and scope of organic synthesis. Imagine the power of being able to produce continuously new materials like and unlike those we encounter every day: foods, clothes, medicines, polymers, plastics, fuels, dyes, perfumes and other high-value and high-tech substances. This virtually unlimited kingdom of chemicals remains relatively untapped and its size depends only on time and the imagination and creativity of synthetic chemists. Today this awesome power of being able to create new substances is rapidly changing the world around us and constantly alters the way we live, the way we communicate and the way we die. How did we acquire this powerful capability and how can we use it to our benefit while at the same time avoiding catastrophes and the destruction of our environment? In this chapter we will discuss these questions and provide a description of the state of the art of organic synthesis, particularly insofar as it pertains to the construction of complex organic molecules such as natural products and designed compounds.

Organic Synthesis

Organic synthesis (or synthetic organic chemistry) is the science (some call it an art) of constructing molecules from atoms and from other (usually simpler) molecules. Organic synthesis is, therefore, not only a creative science but also a most useful one. The beneficial impact of organic synthesis on society becomes obvious when we stop to recount the many applications of this science to human health care, advanced technology and everyday life: pharmaceuticals, plastics, foods, clothing, perfumes and high-technology materials for construction, engineering and electronics, not to mention sources of energy. Most of these are artificially created from other materials through organic synthesis. The centrality of chemical synthesis relative to the other sciences and its crucial impact on biology and medicine, in particular, merit special emphasis and will become even more enabling and evident as we move into the next century.

The goal of organic synthesis is to assemble a defined organic compound (the target molecule, usually composed of atoms of the elements C, H, O, N, S, P, Cl, Br, I and B) from readily available starting materials in the most efficient way. A high degree of control of the stereochemistry and low waste are two important criteria in judging the efficiency and practicality of a synthesis. Unusual reactions and novel strategies add to the elegance of a synthesis. The science of organic synthesis is continually being enriched by new chemical inventions and discoveries. However, despite the impressive strides that have been taken, organic synthesis is still a youthful science. Nevertheless, this science is a powerful tool for several other disciplines, including biology, medicine, physics and materials science. The advancement of the science of organic synthesis for its own sake is, however, its most exciting and exhilarating aspect, as the masters of the art amply demonstrated during the past few decades. As a field, organic synthesis can be subdivided into several subdisciplines, as illustrated in Figure 1. Thus, the invention, discovery and development of new synthetic reactions, reagents and catalysts is collectively referred to as synthetic technology or methodology or methods-oriented synthesis, whereas the pursuit of a given molecule (natural or

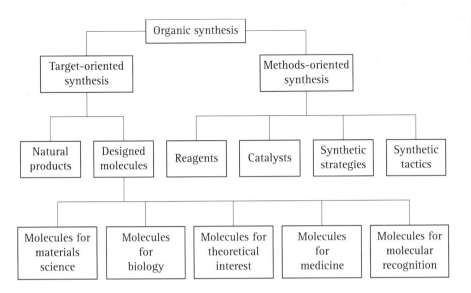

Figure 1
Organic synthesis in perspective
(modified from *Classics in Total
Synthesis*, VCH, Weinheim, 1996).

designed) is classified as target-oriented synthesis (total synthesis is included in this category).

From a historical perspective, the science of organic synthesis saw its dawn in the nineteenth century. By most accounts, its beginning was marked by the synthesis of urea, $(NH_2)_2CO$, by the German chemist Wöhler in 1828. This synthesis was followed by several other landmark achievements, shown in Figure 2, until the Woodward era (from the 1940s to the 1960s), which constituted a major leap forward. The science was elevated to even higher levels of sophistication and effectiveness by E. J. Corey and his school in the decades to follow. This was brought about by the introduction of the concept of retrosynthetic analysis and the emphasis on new synthetic technology as part of the synthetic programme. R. B. Woodward received the Nobel Prize for chemistry in 1965 for 'achievements in the art of organic synthesis'. Corey received the Nobel Prize for chemistry in 1990 for his 'development of the theory and methodology of organic synthesis'.

Today, the advancement of organic synthesis continues unabated. The science is driven by the continuous stream of fascinating and challenging structures provided by nature, by the endless imagination of chemists designing new molecules and by the need to provide organic compounds by efficient, cost-effective and environmentally benign methods.

Practising Total Synthesis

With its goal-oriented nature and requirement for deep and visionary thinking, total synthesis can be compared to the game of chess. They have in common glorious moments, setbacks and frustrations. The objective in the game of chess is to checkmate the opponent's king, which usually is achieved by a series of allowed moves played out in such a combination and sequence as to outmanoeuvre the opponent. In a similar manner, total synthesis, whose objective is to produce the target molecule, achieves its goal by a series of reactions (allowed by nature) carried out in the proper combination and sequence in such a way as to overcome natural barriers. Considering and applying the moves (reactions) to capture the king (make the molecule) is then the object of the game (total synthesis). Inventing new moves (reactions) adds extra excitement to the venture of total synthesis and expands the power and scope of the endeavour. The practice of the art and science of total synthesis usually involves, and heavily depends on, the following stages: (a) selecting the target, be it a natural product or a

Figure 2

Selected landmark total syntheses of natural products from 1828 to 1944 (reproduced from *Classics in Total Synthesis*, VCH, Weinheim, 1996).

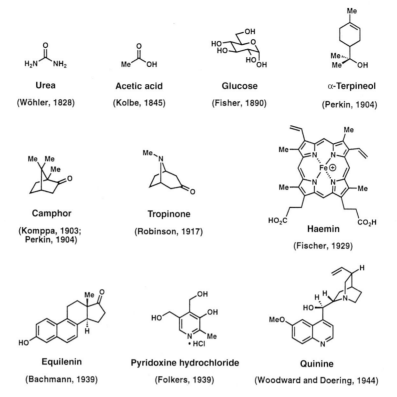

Urea

(Wöhler, 1828)

Acetic acid

(Kolbe, 1845)

Glucose

(Fisher, 1890)

α-Terpineol

(Perkin, 1904)

Camphor

(Komppa, 1903; Perkin, 1904)

Tropinone

(Robinson, 1917)

Haemin

(Fischer, 1929)

Equilenin

(Bachmann, 1939)

Pyridoxine hydrochloride

(Folkers, 1939)

Quinine

(Woodward and Doering, 1944)

designed molecule; (b) designing the synthetic strategy by applying retrosynthetic analysis; (c) selecting proper reagents and conditions for each step of the projected synthesis, thus defining the tactics; and (d) executing the total synthesis in the laboratory. Smooth as they may appear, these stages often pose serious challenges and present unexpected hurdles. Only constant intellectual input, perseverance and skill can solve these problems and lead to eventual success. Often, the target may still be the same, but the final strategy and tactics of its total synthesis may look much different from the original plan.

Selecting Target Molecules from Nature

Target molecules are classified into two categories: naturally occurring substances (natural products) and designed molecules. Nature has been amply generous to the synthetic chemist by providing a wealth of structural diversity in natural products. These substances come in all sorts of sizes, shapes and forms and are often biologically relevant and useful. Those with novel molecular architectures, important properties, interesting mechanism of action and medical potential are the most opportune targets. Their great complexity adds to the synthetic challenge and elevates the exercise of the total synthesis to a test of the state of the art. Indeed, it is the molecular complexity and diversity of natural products that make their syntheses both optimum opportunities for inventing new methods and strategies (and thus pushing the boundaries of chemical synthesis forwards) and the true litmus tests of the power of the science. It is for these reasons that accomplishments in the field of total synthesis are often accompanied by exhilaration, excitement and satisfaction, for they speak directly of the creative and powerful nature of the science. From a practical point of view the total synthesis of a natural product may: (a) prove or confirm its chemical structure; (b) provide it in large quantities for biological or other investigative purposes; and (c) pave the way for the synthesis of designed analogues.

Since the synthesis of urea in 1828, synthetic organic chemists have made gigantic strides in terms of coping with molecular complexity. Progress was initially steady, but the second half of the twentieth century saw dramatic leaps forward. Strychnine, vitamin B_{12}, gingkolide B, Taxol™, palytoxin and brevetoxin B (Figure 3) are but some examples of the most complex molecules synthesized to date. Such landmark accomplishments in total synthesis are often accompanied by discoveries and developments in the sister field of synthetic technology and add considerably to the advancement and sharpening of organic synthesis as a tool. However, all its impressive achievements notwithstanding, organic synthesis may still be viewed as just emerging from the stone age in comparison with nature's routes to complex molecules. The latter statement, extreme though it may seem, can easily be appreciated by comparing human achievements with the biosyntheses of natural products by plants, bacteria and animals, often at ambient temperatures and under environmentally benign conditions. These processes often proceed with 100% yield and with little, if any, waste. Enzymes are, of course, the magic catalysts capable of carrying out these marvelously efficient syntheses. It is, therefore, no wonder that synthetic chemists have adopted several of nature's enzymes for laboratory transformations. Chemists have also been engaged, with varying degrees of success, in mimicking nature through the design and synthesis of artificial enzymes such as enzyme models, enzyme mimics and catalytic antibodies. Although such designs have proven quite challenging and still require refinement, the field of enzyme engineering is exciting and highly promising.

Designed Molecules as Synthetic Targets

Although natural products may present the ultimate and most attractive synthetic challenges, designed molecules provide exciting opportunities beyond those associated with their syntheses. Generating new substances and investigating their properties gives the synthetic chemist an awesome power and a sense of freedom of expression that is normally reserved for artistic endeavours. Even though they were formulated centuries ago, the following quotations from Democritus and Leonardo da Vinci convey the essence of modern molecular design and chemical synthesis:

The endless changes of reality are due to the continual aggregation and disaggregation of atoms.

Democritus

Where nature finishes producing its own species, man begins, using natural things and in harmony with this very nature, to create an infinity of species.

Leonardo da Vinci

Indeed, the essence of chemistry lies in its ability to manipulate matter and create new substances. An almost infinite number of chemical structures can be imagined and millions of them have already been synthesized in the laboratory. It is from these new chemicals that useful inventions eventually emerge.

Molecules are designed for all sorts of purposes depending on the interest of the synthetic chemist. Molecular design is usually based on principles of organic chemistry and is often aided by computer or manual modelling and computational chemistry. The purpose of the design may vary, ranging from purely theoretical or aesthetic reasons to the need to produce materials for practical and useful applications. Tetrahedranes, prismanes and cubanes have, for example, been targetted for their geometrical significance and to investigate their properties. Myriads of designed molecules are synthesized and tested for their physical and biological properties. Compounds with beautiful colours and odours are used in clothing and cosmetics, whereas polymers and other

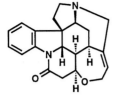

Strychnine (1954)

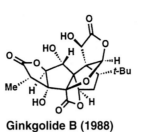

Ginkgolide B (1988)

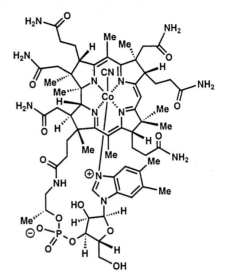

Vitamin B$_{12}$ (1973)

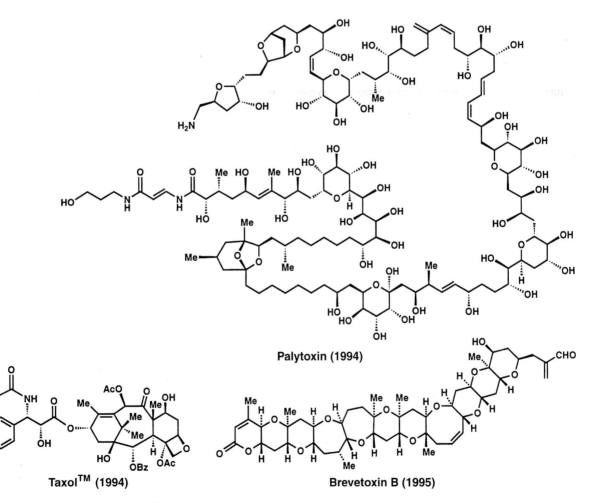

Palytoxin (1994)

Taxol™ (1994)

Brevetoxin B (1995)

Figure 3
Some of the most complex natural products synthesized to date (reproduced from *Classics in Total Synthesis*, VCH, Weinheim, 1996).

substances with special properties become construction materials for engineering, housing and high-tech industry. Designed molecules find their way into food, fuels and clothing. Finally, the magic of molecular design and chemical synthesis serves as the central component of the process of discovering drugs. Most of today's drugs are designed molecules whose origins can be traced back to natural products or simply to human imagination. The field of drug design is constantly enriched and fuelled by new discoveries of molecular receptors and of biologically active natural products. More recently, combinatorial chemistry and solid-phase synthesis have augmented considerably the scope and power of synthetic chemistry in terms of the speed of production and diversity of structures of designed, or even randomly assembled, molecules. These powerful techniques together with the rapidly emerging genomic information and high-throughput screening capabilities bode well for an acceleration of breakthroughs in medicine in terms of the scope of treatments of diseases and the pace of the discovery and development of drugs.

A special chapter of organic chemistry, that of supramolecular chemistry, emerged primarily due to the concepts and advances in molecular design and chemical synthesis. The ingenious and sometimes complex molecular designs in this fascinating field include cryptands, spherands, self-assembling and self-replicating systems, model enzymes, artificial receptors, peptide constructs, DNA helices and nanotubes.

Synthetic Strategy

Many outside the field of organic synthesis admire the achievements of the practitioners of this art and often wonder how it is done. The answer is not simple, of course, but central and crucial to a successful synthesis is an appropriate synthetic plan or synthetic strategy. Faced with a complex molecular target, the synthetic chemist often has to consider carefully the construction and maintenance of ring frameworks, sensitive functionalities, stereochemical groupings and unusual bond connectivities and to identify appropriate starting materials. Depending on the complexity, the conception and development of a strategy for the total synthesis may require intellectual and visionary input of the highest order. In the early days of the science when the target molecules were relatively simple, the synthetic chemist would try to connect the target with known starting materials mentally, via known chemical reactions. Weighing the various options, he or she would then choose one or more strategies that looked reasonable and try them in the laboratory until success could be achieved. The starting materials were often close relatives of the target molecules and the known reactions were rather limited, allowing only modest complexity of the targets. Progress was relatively slow, but, as the science of chemistry progressed and the target molecules became more complex, synthetic chemists rose to the challenge by devising new reactions and ingenious strategies for moving towards the new molecules. As the field of total synthesis took shape and received recognition as highly demanding, the endeavour was met with excitement, admiration and euphoria. After World War II, the unprecedented and brilliant contributions of R. B. Woodward sharply defined the field and set the pace for what was to come during the second half of the twentieth century: a cataclysm of developments in organic synthesis that would make the total synthesis of many natural products achievable. Major developments in electronic theory, understanding of mechanisms of reactions and conformational analysis together with advances in analytical, chromatographic, crystallographic and spectroscopic techniques and instrumentation greatly facilitated developments in the field of organic synthesis and allowed it to flourish and become more productive in scope and yield.

As a strategist, the synthetic chemist became a master of the art of constructing nature's most complex structures as well as a designer of molecules for chemical, physical and biological investigations. Formidable and long-sought targets fell one after the other; in this sense the 1950s and 1960s will forever remain a glorious era and a turning point for the art and science of chemical synthesis. Milestones such as the total syntheses of strychnine (Woodward, 1954), reserpine (Woodward, 1956), penicillin V (Sheehan, 1957), colchicine (Eschenmoser, 1959), chlorophyll (Woodward, 1960) and cephalosporine (Woodward, 1965) stand as glistening monuments to the brilliance and character of the synthetic chemists of that era. Ingenious as the thinking of this generation was, however, synthetic planning was not formulated on a systematic footing until the concept and philosophy of retrosynthetic analysis was introduced by E. J. Corey during the 1960s.

Retrosynthetic Analysis

The advent of retrosynthetic analysis, which was introduced by E. J. Corey, constituted a major advance in the strategic planning of total synthesis. Ever since its introduction, this concept has been adopted by almost all synthetic chemists, who use it consciously or unconsciously. This concept facilitated tremendously the advancement of the field. With the following summary, E. J. Corey defines the concept of retrosynthetic analysis for which he received the Nobel Prize for chemistry in 1990:

Retrosynthetic (or antithetic) analysis is a problem-solving technique for transforming the structure of a synthetic target (TGT) molecule to a sequence of progressively simpler structures along a pathway which ultimately leads to simple or commercially available starting materials for a chemical synthesis. The transformation of a molecule to a synthetic precursor is accomplished by the application of a transform, the exact reverse of a synthetic reaction, to a target structure. Each structure derived antithetically from a TGT then itself becomes a TGT for further analysis. Repetition of this process eventually produces a tree of intermediates having chemical structures as nodes and pathways from bottom to top corresponding to possible synthetic routes to the TGT.

E. J. Corey

Today, complex molecular structures are routinely disassembled by retrosynthetic analysis and synthetic strategies are planned using this process. Indeed, it now seems almost unimaginable that, prior to the 1960s, chemists would devise complex synthetic strategies without using these techniques, at least unconsciously.

When the target molecule is not overwhelmingly complex, it is relatively easy to ponder its structure and figure out suitable starting materials for its construction. When the complexity of the synthetic target places its structure far away from those of readily available starting materials, however, the task of defining the initial building blocks becomes increasingly difficult. Instead of attempting to bridge such major structural gaps at once, retrosynthetic analysis allows the synthetic chemist to do this in stages. Thus, the practitioner focuses on the target molecule, identifying strategic bonds for imaginary disconnections and generating progressively simpler intermediates that become the new targets. Reiteration of the process generates a retrosynthetic tree from which several synthetic routes can be designed and evaluated. At each step of retrosynthetic analysis the chemist evaluates both the viability of each intermediate generated and the chances of being able to carry out the transformation involved in the direction of the synthesis. This can be an exciting and highly inspirational experience, for it is during this process that the chemist is faced with the opportunity to design new strategies and invent new reactions. Persistence and thorough consideration of as many disconnections as possible and elucidating subtle structural features of the target molecule are often the keys to the development of elegant and efficient synthetic strategies. Rushed decisions and compromises have no place at this stage of the planning and should be avoided. What is usually rewarding is the continuous upgrading of the strategies by searching for novel and unprecedented manoeuvres and ideas and by insisting on the desirability of shorter routes and on the invention and application of new reactions.

The following passage from *Classics in Total Synthesis* describes vividly the process that follows the retrosynthetic analysis, leading to the execution of the total synthesis and beyond.

Having exhausted all retrosynthetic possibilities, the strategist is then in a position to evaluate the possible paths uncovered and devise the most attractive synthetic strategy for the construction of the target molecule. The strategy may dictate the invention of new reactions, reagents, and tactics, and may require model studies before synthesis on the real target can start. This is usually a good practice, for it is destined, more often than not, to result in new synthetic technology, a vital feature of a novel total synthesis, and to pave the way for a projected total synthesis. Other attractive features of a planned synthetic strategy are: (a) efficient synthetic reactions; (b) brevity; (c) readily available and inexpensive starting materials; (d) practical and convenient conditions; (e) flexibility of modification in case of pitfalls; (f) adaptability to the synthesis of other members of the structural family, be they naturally occurring or designed molecules; and (g) novelty, elegance, and artistry!

It is of paramount importance to recognize that in total synthesis the achievement itself is not always the prize or the most significant advance. Rather, it is the journey towards the target molecule that becomes the essence and significance of the exercise. The invention and development of new synthetic technology and strategies, and the molecular design, chemical synthesis, and biological investigation of bioactive compounds structurally related to the target are two emerging and important aspects of modern total synthesis. Chemists and biologists will no doubt be busy for a long time harvesting the benefits of this newly emerging field of investigation that combines the best of chemistry and biology.

K. C. Nicolaou and E. J. Sorensen

Having discussed how synthetic strategies for moving towards target molecules are evolved, we will now focus on some newer strategies and tactics that helped shape the science of chemical synthesis as we know it today.

New Synthetic Technology

As a rule, nature produces molecules of one handedness even though often molecules both with 'right' and with 'left' handedness could exist. Such molecules are called chiral and their two forms, termed enantiomers, are related in the same way as our two hands are, that is to say they are not identical: one is the mirror image of the other. Although nature can produce enantiomerically pure compounds with great ease using its chiral catalysts, synthetic chemists are still addressing this issue with unprecedented vigour. The synthesis of enantiomerically pure compounds is of paramount importance, especially for the pharmaceutical industry, since biological receptors respond differently to different enantiomers. Initially, chemists realized that they could separate enantiomers by reacting racemic mixtures with enantiomerically pure compounds to form diastereoisomers, which upon separation can be used to liberate pure enantiomers. This process, known as resolution, was the only widespread method for achieving this goal until recently. However, its wasteful nature (at least half of the material is wasted in the process) prompted great efforts at developing methods of so-called *asymmetric induction* by which potentially chiral compounds could be synthesized in enantiomerically enriched or pure form. Discoveries came slowly at first, but then the pace accelerated quickly, as will be discussed below.

Asymmetric Synthesis

From among the many asymmetric reactions available today, those involving catalytic amounts of chiral ligands (catalytic asymmetric reactions) are generally preferred.

Scheme 1

Asymmetric aldol and alkylation reactions with Evans' auxiliaries.

Scheme 2

Enzymatic resolution of meso compounds.

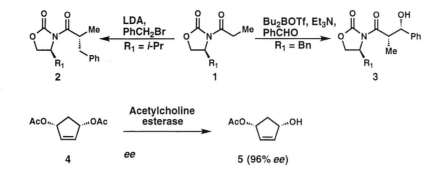

Reactions utilizing stoichiometric chiral auxiliary groups are, however, also quite useful. A prominent example of a chiral auxiliary system is the one developed by Evans (Scheme 1). However, it is the catalytic asymmetric processes that attract the most attention and often offer the most practical solutions for the production of enantiomerically pure compounds. The use of enzymes in this respect has been quite fruitful and is widely practised today. Scheme 2 shows a recent example of an enzymatic reaction leading to a desymmetrization of a *meso* substrate to afford an enantiomerically pure substance. More recently, the advent of catalytic antibodies has expanded the scope and promise of biological catalysis.

Within the field of catalysis, however, the metal- or heteroatom-based catalysts have elicited the most activity and dramatically enhanced our power to construct molecules of great complexity in pure form. Such organometallic or organo-element-based catalysts can direct the stereochemical course of several synthetically important organic reactions.

One of the first examples of a practical catalytic asymmetric reaction was the asymmetric catalytic hydrogenation of unsaturated amidocarboxylic acids which led to the Monsanto process for the manufacture of L-DOPA (**8**) (Scheme 3), an effective agent against Parkinson's disease. This development was inspired by Wilkinson's milestone discovery of the homogeneous hydrogenation of alkenes using the soluble catalyst $(Ph_3P)_3RhCl$. Replacement of the achiral triphenylphosphine units of the catalyst by chiral diphosphine ligands led to the development of the cationic catalyst $[Rh((R,R)\text{-}DiPAMP)COD]^+BF_4^-$ which catalyses the asymmetric hydrogenation of enamides to form highly enriched enantiomers of amino acids, such as L-DOPA (**8**) (Scheme 3). Similar principles were used in devising a commercial process for the production of aspartame (**11**) (Scheme 4) commonly known as Nutrasweet™. The crucial asymmetric hydrogenation of the enamide **9** proceeds in ethanol with a substrate:catalyst ratio of 15000:1 and in 83% enantiomeric excess (*ee*). Subsequent recrystallization raises the *ee* of the product to 97%.

The discovery of the powerful Sharpless asymmetric epoxidation reaction in the early 1980s was another milestone in asymmetric catalysis and stimulated the ensuing rapid developments in the field. This exceptionally useful reaction entails the conversion of allylic alcohols to enantiomerically enriched epoxides via the action of *tert*-butyl hydroperoxide and catalytic amounts of a titanium/tartrate complex as depicted in Scheme 5. The practicality and generality of this reaction coupled with the versatility of the resulting epoxides in organic synthesis make the process one of the classics in the field and one that made a difference to our ability to produce molecules of great complexity in enantiomerically pure form.

The cyclopropanation of alkenes is another important reaction whose catalytic asymmetric version has enjoyed both practical and commercial success. A notable example is Aratani's Sumitomo process for the production of the enantiomerically

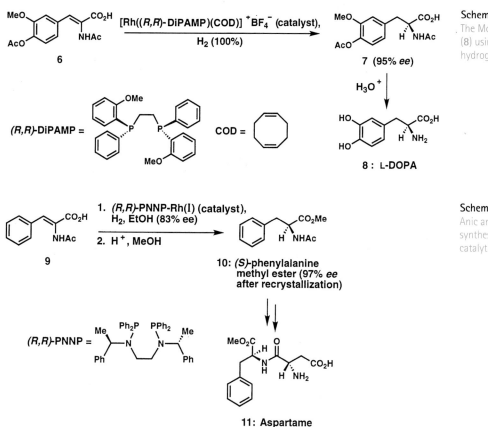

Scheme 3
The Monsanto synthesis of L-DOPA (8) using catalytic asymmetric hydrogenation.

6

7 (95% *ee*)

(*R,R*)-DiPAMP =

COD =

8 : L-DOPA

9

10: *(S)*-phenylalanine methyl ester (97% *ee* after recrystallization)

(*R,R*)-PNNP =

11: Aspartame

Scheme 4
Anic and Enichem's commercial synthesis of aspartame (11) using catalytic asymmetric hydrogenation.

enriched cyclopropane derivative **15**, which is used for the manufacture of cilastatin (**16**) (MK-0791) by Merck (Scheme 6). The latter compound is an *in vivo* stabilizer of the labile β-lactam antibiotic imipenem and is used clinically to combat infectious disease in combination with the antibiotic.

A spectacular example of an application of asymmetric homogeneous catalysis is found in the Takasago process for the production of (−)-menthol (**26**) (Scheme 7). The key step in this process is the establishment of an asymmetric centre via the isomerization of diethylgeranylamine (**21**) to the isomeric enamine (**22**) using a rhodium (I) catalyst containing the *S* form of the BINAP ligand. The enamine intermediate is obtained in essentially quantitative yield and >98% *ee*.

The Sharpless asymmetric dihydroxylation of alkenes was another major advance in the area of asymmetric synthesis. This osmium-catalysed process (Scheme 8) utilizes the alkaloid-derived chiral ligands (DHQ)$_2$PHAL (**27**) and (DHQD)$_2$PHAL (**28**) to direct the delivery of two oxygen atoms from one face of the alkene preferentially, furnishing enantiomerically enriched 1,2-diols that serve as useful synthetic intermediates. Extensions of this process to 1,2-hydroxyamination reactions have also been developed.

The reduction of ketones is an important reaction in synthetic organic chemistry, so the ability to perform its asymmetric version is highly desirable. A most practical and useful catalyst, (*S*)- or (*R*)-oxazaborolidine (**29**), was developed by Corey for effecting this transformation with boranes. This useful process is exemplified by the asymmetric reduction of the prostaglandin intermediate **31** (Scheme 9). The powerful Diels–Alder reaction can also be made to proceed asymmetrically by the use of certain catalysts, as shown in Scheme 10.

Scheme 5
The Sharpless catalytic asymmetric epoxidation.

Scheme 6
The Sumitomo Chemical Company's catalytic asymmetric synthesis of ethyl (S)-(+)-2, 2-dimethylcyclopropanecarboxylate (15), an intermediate in Merck's commercial synthesis of cilastatin (16).

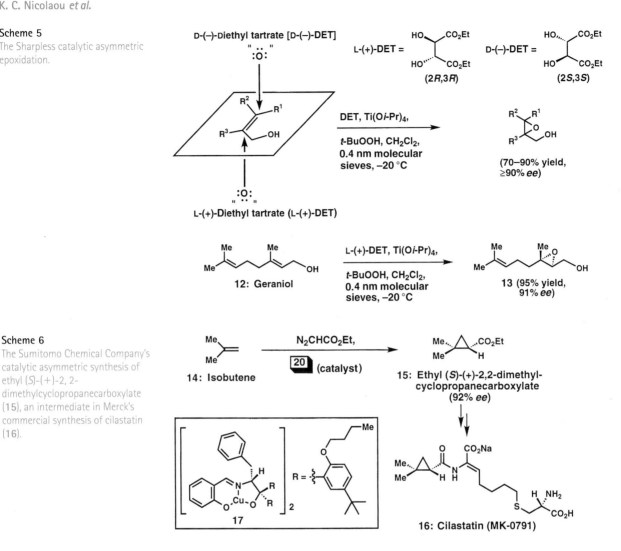

The profound achievements in asymmetric synthesis, particularly those involving catalysis, have considerably extended and influenced the scope of chemical synthesis. The rich chemistry of transition metals combined with ingenious molecular designs promises even more exciting opportunities for asymmetric synthesis in the future. In terms of molecular complexity, these developments will, no doubt, allow us to push the boundaries further and further.

Tandem Reactions and Synthetic Strategies

The synthesis of complex molecules requires not only efficient chemical reactions but also powerful synthetic strategies. In this section we will discuss a group of strategies that are particularly powerful for building complex molecular structures from rather simple ones and in a minimum number of operations.

Reaction sequences that proceed in a single synthetic step or in 'one pot' are called tandem, cascade or domino reactions. Since they combine several transformations of the same molecule and often incorporate added components they constitute powerful avenues to complex molecules. In addition to being highly efficient and economical, such cascade reactions are often viewed as spectacular and aesthetically pleasing events. These cascades are usually designed by the synthetic strategist on the basis of

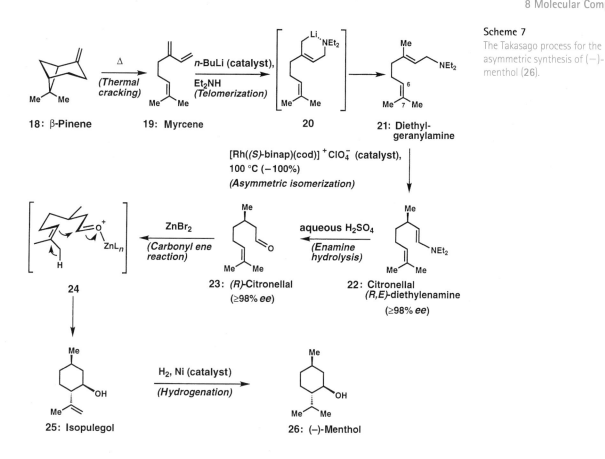

Scheme 7
The Takasago process for the asymmetric synthesis of (−)-menthol (26).

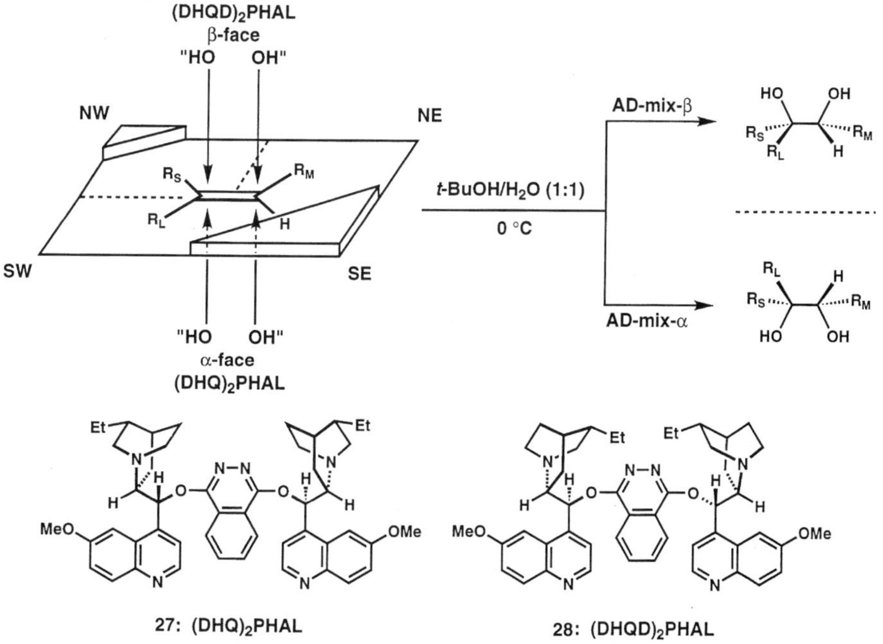

18: β-Pinene

19: Myrcene

20

21: Diethyl-geranylamine

[Rh((S)-binap)(cod)]⁺ClO₄⁻ (catalyst), 100 °C (−100%)
(Asymmetric isomerization)

Δ
(Thermal cracking)

n-BuLi (catalyst), Et₂NH
(Telomerization)

ZnBr₂
(Carbonyl ene reaction)

aqueous H₂SO₄
(Enamine hydrolysis)

24

23: (R)-Citronellal (≥98% ee)

22: Citronellal (R,E)-diethylenamine (≥98% ee)

H₂, Ni (catalyst)
(Hydrogenation)

25: Isopulegol

26: (−)-Menthol

Scheme 8
The Sharpless catalytic asymmetric dihydroxylation.

(DHQD)₂PHAL
β-face
"HO   OH"

NW

NE

"HO   OH"
α-face
(DHQ)₂PHAL

SW

SE

t-BuOH/H₂O (1:1)
0 °C

AD-mix-β

AD-mix-α

27: (DHQ)₂PHAL

28: (DHQD)₂PHAL

Scheme 9
Corey's catalytic asymmetric
carbonyl reductions.

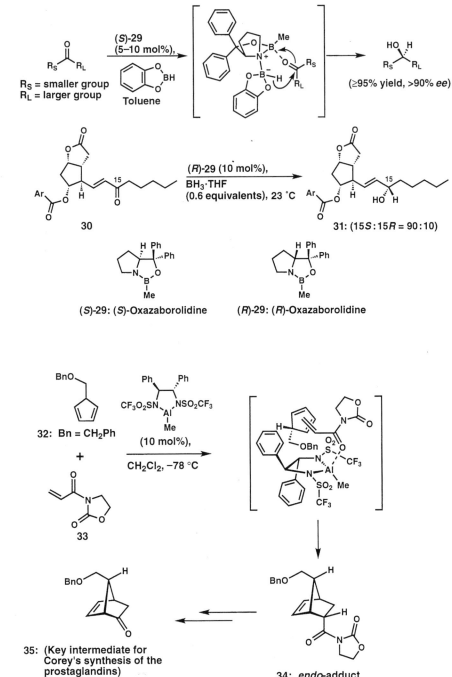

Scheme 10
Corey's catalytic asymmetric
Diels–Alder reaction.

rational mechanistic considerations and they proceed rapidly upon initiation, leading
to the final product through a number of intermediates. Thus, upon triggering the reaction sequence, a reactive intermediate may be formed, which proceeds to the next stage
via an intramolecular or intermolecular reaction leading to a new compound which,
in turn, may find itself in a favourable situation for further reaction, thus generating
a product of greater complexity. Depending on the steps involved, such tandem reactions can form several bonds, rings and stereocentres. The selectivity, in terms of

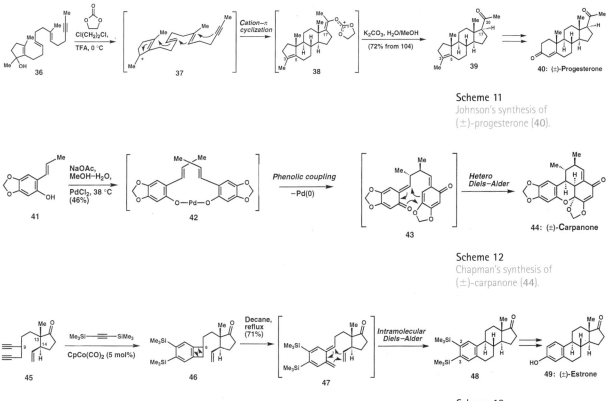

Scheme 11
Johnson's synthesis of
(±)-progesterone (40).

Scheme 12
Chapman's synthesis of
(±)-carpanone (44).

Scheme 13
Vollhardt's synthesis of (±)-estrone
(49).

molecular structure and stereochemistry, is often excellent because of the intramolecular nature of at least some of the bond-forming reactions.

Nowadays, tandem reaction strategies are finding increasing use both in total synthesis and in the construction of complex intermediates. Such strategies incorporate an ever-increasing variety of reactions, including cationic and radical cyclizations, aldol reactions, pericyclic reactions, palladium-catalysed reactions, carbene reactions and photochemical reactions. Below we illustrate this road to molecular complexity with a number of notable examples.

Many tandem strategies are inspired by biosynthetic schemes through which nature makes its complex molecules. An early example of such a case is the spectacularly successful synthesis of progesterone in its racemic form by W. S. Johnson. This synthesis, shown in Scheme 11, involves a cationic π-electron polycyclization leading from the monocyclic system 36 to the tetracyclic intermediate 39 in a single step via intermediates 37 and 38. Intermediate 39 was then converted into the target molecule (40) via a two-step process involving ozonolysis and Robinson annulation. Although the planning of this strategy was based on biosynthetic considerations and on the Stork–Eschenmoser hypothesis, its experimental execution was both daring and highly rewarding.

Another example of biomimetic total synthesis of a complex natural product is that of racemic carpanone (44) reported by Chapman *et al.* (Scheme 12). Again, starting from a hypothesis regarding how nature may be constructing this structure, the Chapman strategy entailed a remarkable cascade mediated by a palladium (II) reagent. Phenol 41 was oxidatively coupled with palladium dichloride to form *bis*(quinodimethide) 43 via the intermediate 42. A spontaneous intramolecular Diels–Alder reaction then furnished the natural product (44) in good yield.

Scheme 14

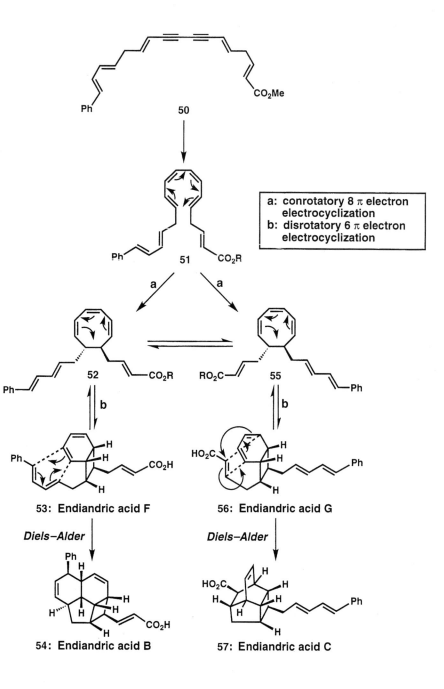

a: conrotatory 8 π electron
electrocyclization
b: disrotatory 6 π electron
electrocyclization

53: Endiandric acid F

56: Endiandric acid G

Diels–Alder

Diels–Alder

54: Endiandric acid B

57: Endiandric acid C

The versatile cobalt-catalysed cyclotrimerization of alkynes developed by Vollhardt made a spectacular and novel approach to synthesis of the steroid nucleus possible. Racemic estrone **49** was synthesized in very short order by the route depicted in Scheme 13. This efficient process involves a cobalt-induced 2 + 2 + 2 cycloaddition of substrate **45** and bis(trimethylsilyl)ethyne to afford, initially, benzocyclobutene **46**, which, under the reaction conditions, is transformed to the tetracycle **48** via ring opening and intramolecular Diels–Alder reaction. In this cascade, three new rings and two chiral centres are constructed, furnishing an advanced intermediate compound, **48**, which is converted to estrone **49** in two steps involving protonolysis and oxidation.

The architecturally novel natural products endiandric acids B (**54**) and C (**57**), found

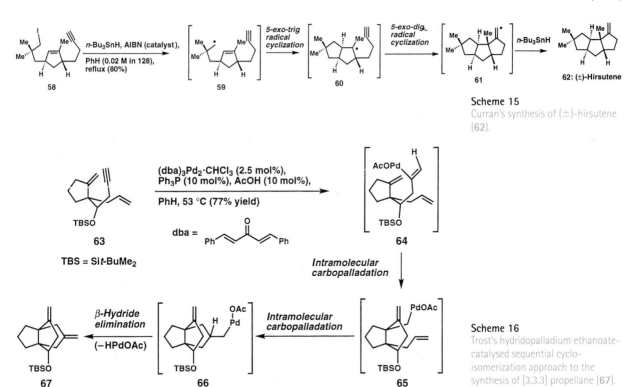

Scheme 15
Curran's synthesis of (±)-hirsutene (62).

TBS = Sit-BuMe₂

dba =

Scheme 16
Trost's hydridopalladium ethanoate-catalysed sequential cyclo-isomerization approach to the synthesis of [3.3.3] propellane (67).

in the endemic Australian trees *Endiandra introsa*, provided an intriguing challenge to synthetic chemistry in the 1980s. In an attempt to explain their occurrence in nature in racemic form, D. St. C. Black put forward a brilliant hypothesis for their 'biosynthesis' postulating acyclic non-chiral precursors and a series of electrocyclizations for their formation without the aid of any enzymes (Scheme 14). Inspired by this interesting hypothesis and in order to test its validity, we designed a one-step strategy for moving towards these compounds involving the cascade shown in Scheme 14. The stable diacetylenic precursor 50 was selectively hydrogenated to afford the corresponding conjugated tetraene system (51) which underwent a smooth 8π-electron cyclization to the cyclooctatrienes 52 and 55, which, in turn, entered into 6π-electron electrocyclizations to furnish the bicyclic systems 53 and 56, respectively. The latter compounds underwent facile intramolecular Diels–Alder reactions leading to endiandric acids B (54) and C (57). It is noteworthy that this process results in the construction of two rather complex polycyclic natural products each containing eight chiral centres from an acyclic, non-chiral precursor and in one operation.

The emergence of radical-based reactions as synthetically useful processes allowed the design of several ingenious strategies for the synthesis of complex molecules starting from relatively simple compounds. An elegant example of such chemistry is the total synthesis of (±)-hirsutene (62) from a monocyclic precursor (58) (Scheme 15). In addition to building two new rings, this tandem process introduces, stereoselectively, two new chiral centres in a single step.

The power and scope of organopalladium chemistry for the formation of carbon–carbon bonds has enhanced our ability to build complex frameworks considerably. Palladium-catalysed molecular reactions have recently been applied with increasing regularity and with great success to the construction of complex molecules. Particularly effective are tandem palladium-induced reactions such as that illustrated in Scheme 16, in which a tricyclic, propellane-type structure (67) is formed from a

Scheme 17
An application of the Pd(0)-mediated coupling reaction.

Suzuki coupling (Suginome–Suzuki)

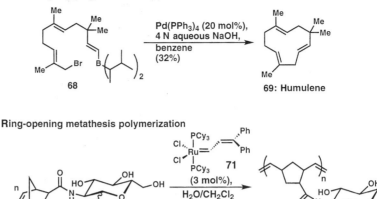

Scheme 18
An application of the alkene metathesis reaction.

Ring-opening metathesis polymerization

monocyclic system. It is clear that palladium-catalysed cascade reactions offer tremendous opportunities to chemists wanting to synthesize complex molecules and that imaginative new strategies based on tandem sequences will lead us to even more complex systems in the future.

Transition Metal Catalysts in Organic Synthesis

The use of transition metal catalysts to construct carbon–carbon bonds in recent years has brought about a remarkable revolution in organic synthesis. In addition to the cascade reactions involving palladium discussed above, there are examples using chromium/nickel, palladium, ruthenium and molybdenum. See Schemes 17 and 18. Furthermore, the potential of such catalysts appears to be enormous and their future applications should be even more impressive.

Total Syntheses of Natural Products

The confidence gained by synthetic chemists from their successes after World War II and the ensuing developments in synthetic methodology and analytical, chromatographic and spectroscopic techniques offered unique opportunities for total synthesis. Indeed, many unusual and complex target molecules have been synthesized during the last few decades, the list becoming longer and longer and the boundaries of attainable complexity continuously being pushed forward. In this section we will highlight the total syntheses of a selected number of complex natural products and describe one of them in detail to illustrate the process of constructing such molecules.

The total synthesis of vitamin B_{12} (**73**), completed in 1973, stands as a landmark achievement in organic chemistry. This elegant synthesis was accompanied by a rich harvest of new synthetic methods, novel strategies, intriguing biosynthetic hypotheses and principles of conservation of orbital symmetry. This accomplishment was the result of a collaborative effort involving the groups of Woodward and Eschenmoser working at Harvard and the ETH Zürich, respectively, and defined at the time the cutting edge of molecular complexity that could be reached by total synthesis. The total synthesis of vitamin B_{12} can be better understood if we trace the logic of its conception, which was based on the principles of retrosynthetic analysis (Scheme 19). Since cobyric acid (**74**) had already been converted to vitamin B_{12} (**73**), the former compound served as a convenient and logical precursor. Rupture of the corin ring in **74** allowed the generation of the secocorin structure **75**. The observation that compound **75** was able to encapsulate the cobalt atom and thus organize itself in advance ready for ring closure to **74**, was crucial to the success of this reaction. Further dis-

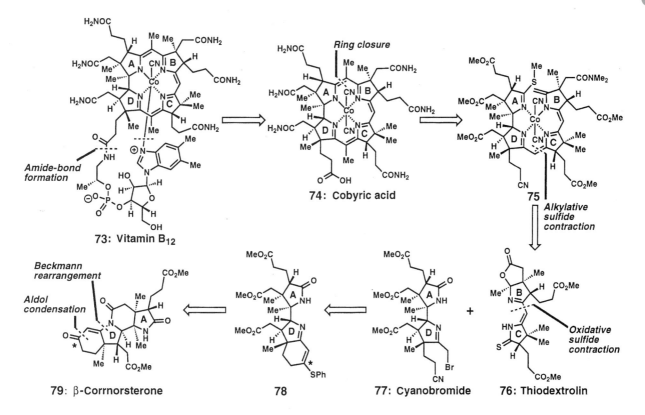

Scheme 19

A retrosynthetic analysis of the Woodward–Eschenmoser synthesis of vitamin B₁₂ (**73**).

connection of **75** furnished intermediates **76** and **77**, termed thiodextrolin and cyanobromide, respectively, as potential precursors to **75**. The latter intermediates were then simplified further until simple starting materials for the total synthesis had been traced. A complete account of this complex synthesis is beyond the scope of this chapter, but can be found elsewhere in all its details.

Corey's total synthesis of ginkgolide B (**80**), a constituent of the ginkgo tree, *Ginkgo biloba*, was another milestone achievement in total synthesis (Scheme 20). Despite the relatively small size of ginkgolide B, its structure is highly complex and challenging to the synthetic chemist. Its unique cage features make up a central spiro[4.4]nonane carbon framework, three γ-lactones, one tetrahydrofuran ring, a *tert*-butyl group and 11 stereogenic centres. This molecular complexity was achieved by the power of modern synthetic organic chemistry and the brilliant strategies and tactics of the members of Corey's group who completed the total synthesis of ginkgolide B in 1979. The key steps of this synthesis are outlined retrosynthetically in Scheme 20. Central to the success of this strategy were an intramolecular ketene–alkene[2 + 2] cycloaddition reaction and a number of selective oxidations.

Molecular complexity of a different kind is found in calicheamicin γ_1^I (**89**), a prominent member of the enediyne class of anti-tumour antibiotics produced by the bacterium *Micromonospora echinospora calichensis*. The unusual framework of **89**, with its conjugated enediyne system within a ten-membered ring, the novel oligosaccharide and the trisulfide moiety, constituted an unprecedented molecular architecture at the time of its disclosure. Combined with its phenomenal anti-tumour activity and fascinating mechanism of action, this new structure presented itself not only as a formidable challenge to the synthetic chemist but also as an opportunity to create new science in chemistry and biology. The first total synthesis of calicheamicin γ_1^I was accomplished in 1992 in our laboratories. The initial stage of the retrosynthetic analysis of this molecule is shown in Scheme 21. The synthetic plan called for the construction of

Scheme 20

A retrosynthetic analysis of ginkgolide B (80).

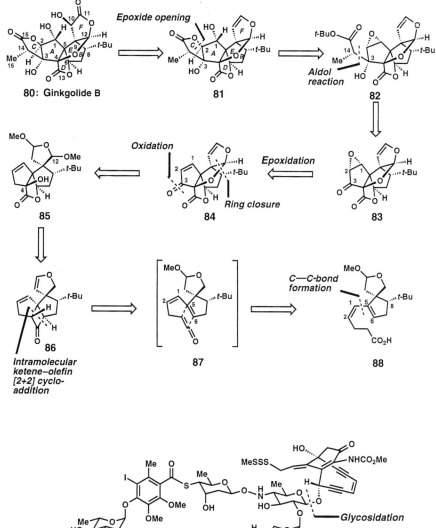

Scheme 21

A retrosynthetic analysis of calicheamicin γ_1^I (89).

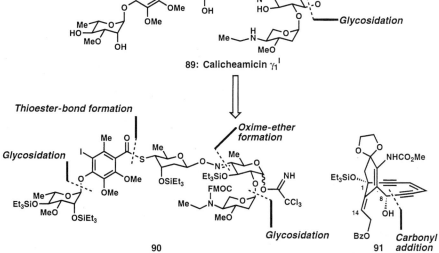

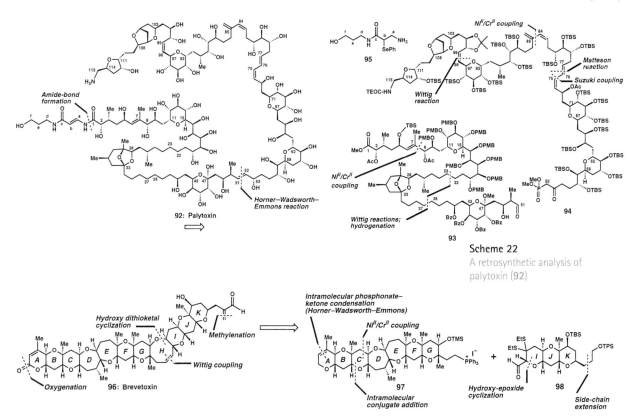

Scheme 22
A retrosynthetic analysis of palytoxin (**92**)

Scheme 23
A retrosynthetic analysis of brevetoxin B (**96**).

the oligosaccharide fragment (**90**) and the enediyne core (**91**) in protected forms, their coupling via a stereoselective glycosidation reaction and final elaborations. This convergent synthesis allowed the preparation not only of the natural compound but also of a highly potent designed analogue termed calicheamicin γ_1^I. A second total synthesis of calicheamicin γ_1^I was accomplished in 1994 by Danishefsky's group.

Owing to its size and plethora of asymmetric centres, palytoxin **92** (Scheme 22), a marine natural product isolated from sponges of the genus *Palythoa*, presented a formidable challenge both to structural and to synthetic chemistry. In 1995, Kishi announced the total synthesis of this target, the largest organic molecule to have been synthesized in the laboratory to date, excluding nucleic acids and proteins. Although the details of this monumental synthesis are beyond the scope of this chapter, its highlights are indicated in retrosynthetic fashion in Scheme 22. The synthesis utilized a convergent strategy in which several smaller fragments were constructed in their enantiomerically pure forms and joined together using powerful coupling reactions such as the amide-bond-forming reaction, the Horner–Wadsworth–Emmons reaction, the Matteson reaction, the Suzuki coupling reaction and the $NiCl_2/CrCl_2$ coupling reaction.

Brevetoxin B (**96**, Scheme 23) is a molecule of unparalleled architectural complexity and beauty. This neurotoxin found in 'red tides' is secreted by the unicellular alga *Ptycodiscus brevis* (*Gymnodunium breve* Davis) and causes massive numbers of fish to die and food poisoning among humans. Its total synthesis by our group took 12 years to complete and was accompanied by many useful discoveries in synthetic technology and synthetic strategy. The successful synthetic plan is highlighted retrosynthetically in Scheme 23. The target molecule was reached from 2-deoxy-D-ribose in 83 steps (the longest linear sequence applied to date) with an overall yield of 0.0043 (the average yield per step was 91%). With these developments the road to what some may call the most complex natural product, maitotoxin (Figure 4), may now at least be charted on paper.

Scheme 24
Retrosynthetic analysis of Taxol™ (99).

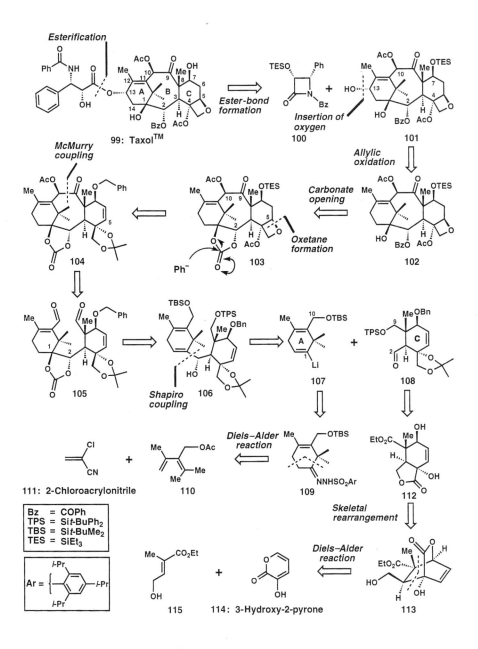

The Total Synthesis of Taxol

Taxol (**99**, Scheme 24), one of the newest weapons against cancer, stood as a symbol of defiant molecular complexity from the time of its discovery in 1971. Despite the many and often ingenious strategies designed for its synthesis, the goal of synthesizing taxol remained unapproachable until 1994 when two total syntheses were completed, with several more to follow. The first to be reported was the one performed in our laboratories. Below we will discuss this synthesis in some detail in order to illustrate the modern approach to synthesizing complex molecular structures.

The taxol molecule (**99**, Scheme 24) is characterized by a 6–8–6 tricyclic carbon skeleton onto which there is attached, through an ester bond, a phenylserine side chain; it has also an oxetane ring and a dense array of oxygen functionalities. The challenges facing the strategist interested in the total synthesis of taxol include not only the stereoselective construction of its 11 stereogenic centres but also the formation of

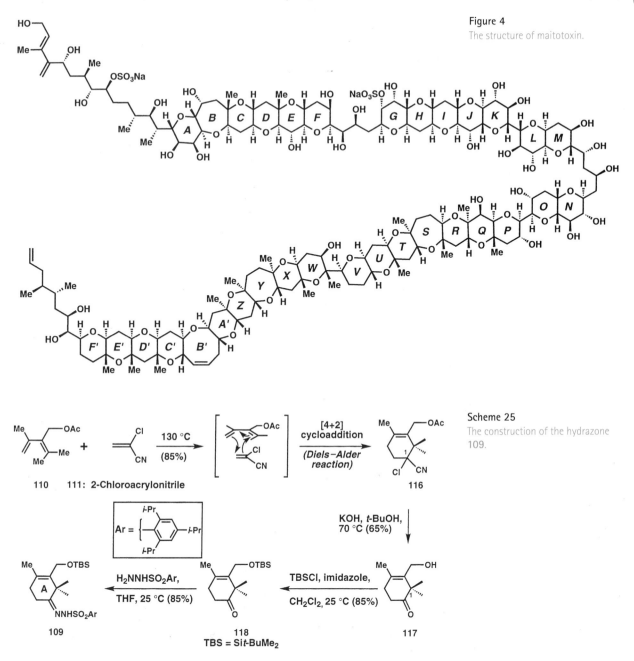

Figure 4
The structure of maitotoxin.

Scheme 25
The construction of the hydrazone 109.

its polycyclic framework, particularly the strained eight-membered ring. By using retrosynthetic analysis (Scheme 24) we will see how sequential disconnections of strategic bonds, manipulations of functional groups and structural simplifications led to the synthetic plan adopted.

The first disconnection (see Scheme 24) removed the ester side chain from the main framework and relied on procedures with good precedents for the attachment of taxol's side chain in the synthetic direction. This disconnection, followed by manipulations involving protecting groups, led to the β-lactam **100** and the key intermediate **101** as potential precursors. To achieve further simplification, the oxygen atom at C-13 was excised, furnishing the compound **102** as an important staging intermediate. At this juncture the cyclic carbonate group was chosen as a protecting device for the 1,2-diol system and as a potential means of achieving the pre-organization deemed important for the projected ring closure to form the eight-membered ring. Furthermore, it was expected that the carbonate functionality could be used to introduce the 1-hydroxy-2-benzoate system of taxol at the appropriate stage. Attack of the carbonyl group by phenyllithium, followed by regioselective ring opening, was expected to give the desired product as shown in Scheme 24 (**103** → **102**). Such a chemoselective transformation may seem unlikely, but, as will be seen later, steric shielding of the other three carbonyl groups in **103** made this reaction not only feasible but also practical.

Proceeding along the retrosynthetic path of simplification, it was then decided to disconnect the oxetane system and introduce a double bond within ring C in order to allow a Diels–Alder reaction to be used for the construction of this region of the molecule. Whereas the change in protecting-group chemistry on going from **103** to **104** can easily be accommodated in the synthetic direction, the double bond in **104** would have to be regioselectively and stereoselectively hydroborated to afford the desired hydroxy group at C-5. The eight-membered ring in **104** was then disconnected through a McMurry-type coupling reaction, allowing the generation of dialdehyde **105** as a potential precursor. In the synthetic direction, a pinacol-type ring closure using titanium (0) was expected to lead to a diol whose conversion to **104** was considered reasonable. Appropriate manipulations of functional groups and disassembly of the carbonate ring led to the compound **106** which can be traced back to the Shapiro coupling reaction of fragments **107** and **108**. Further analysis led to **109** and **110** by standard chemistry. Compounds **109** and **112**, being cyclohexenes, were envisioned to be assembled via Diels–Alder reactions (**112** was first rearranged to **113**), furnishing four simple and prochiral compounds as starting materials (**110**, **111**, **114** and **115**).

This retrosynthetic analysis indicated a plausible strategy for the total synthesis of the target molecule. The execution of the synthesis would require the selection of appropriate tactics and considerable experimentation, which we will discuss next.

With the overall strategy for the total synthesis defined, the construction of the two requisite cyclohexene rings A and C could begin. The synthesis of the hydrazone **109** is shown in Scheme 25. It involved the powerful Diels–Alder reaction which, in this case, converts the alkeneic systems **110** and **111** into the six-membered ring **116**. The latter molecule contains all the necessary functionality for its transformation to compound **109** via intermediates **117** and **118**. The details for all these conversions are indicated in Scheme 25. The hydrazone **109** was obtained in multi-gramme quantities as a colourless crystalline solid and could be used to generate a vinyllithium species suitable for coupling with an appropriate electrophile carrying the C-ring fragment of taxol.

Of the two cyclohexanoid fragments of taxol, the C-ring structure is clearly the most complex, containing as it does four contiguous asymmetric centres. The potential of using the Diels–Alder reaction to create not only cyclohexene rings but also up to four contiguous stereogenic centres in a single stereospecific step weighed heavily in favour of its use as a means of constructing this system. Thus a careful analysis of the structure of compound **108** led to the adoption of a highly efficient strategy for its synthesis featuring the Diels–Alder reaction (Scheme 26). Success came when 3-hydroxy-2-pyrone (**114**) and the α,β-unsaturated ester **115** were combined in the presence of phenylboronic acid at 90 °C to give the fused bicyclic lactone **112** in 61% yield. In this most impressive and efficient process, which is based on the pioneering work of Narasaka, two rather simple achiral materials attach themselves, through their hydroxyl groups, to the boron atom of phenylboronic acid to give a mixed boronate ester (**119**). Thus, the boron atom serves as a template bringing together the two reactants for interaction in an intramolecular version of the Diels–Alder reaction and controlling by this mechanism the regiochemical outcome of the process. Incidentally, this

Scheme 26
The construction of the aldehyde
122.

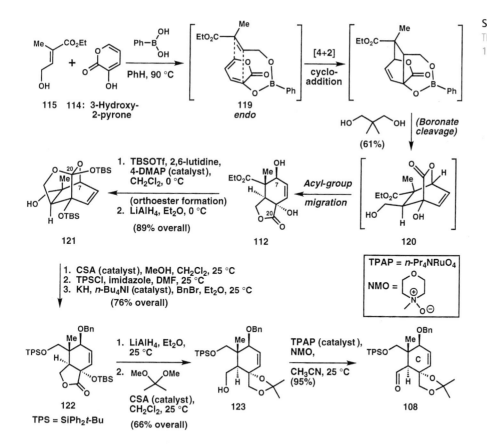

reaction is one among a class of diverse transformations that are facilitated and controlled through the use of disposable tethers. The initially formed intermediate **120** apparently undergoes a rapid rearrangement (migration of the lactone) to the most stable fused system **112**. It is also noteworthy that the cycloaddition reaction exhibits complete *endo* stereoselectivity. Although compound **112** is obtained as a racemate, it possesses all the required functionality and stereocentres for its conversion into taxol.

Proceeding with the synthesis, it was observed, rather unexpectedly, that exposure of the lactone **112** (Scheme 26) to excess *tert*-butyldimethylsilyl triflate followed by reduction with lithium aluminium hydride (LAH) leads to the formation of the orthoester alcohol **121** with a high overall yield. Treatment of the latter compound with catalytic amounts of camphorsulfonic acid (CSA) cleaves one of the *tert*-butyldimethylsilyl ethers selectively, leading to a γ-lactone 1,3-diol that was then protected sequentially, as shown, to afford the fully protected γ-lactone **122**. Reduction of the γ-lactone functionality in **122** was accompanied by cleavage of the *tert*-butyl dimethylsilyl ether, furnishing the corresponding triol which was then converted to the acetonide **123** by treatment with 2,2-dimethoxypropane under anhydrous acidic conditions. Finally, the desired C-ring aldehyde **108** was obtained by oxidation with tetrapropylammonium perruthenate (TPAP)/4-methylmorpholine *N*-oxide (NMO).

The union of intermediates **108** and **109** was brought about in a stereoselective manner as outlined in Scheme 27. The hydrazone **109** was converted to vinyllithium **107** by the action of *n*-butyllithium and then reacted with **108** to afford the secondary alcohol **106** with yield 82%. This remarkable stereoselectivity has been attributed to the combination of complexation and steric effects shown in structure **124**. Compound **106** is now poised for a regioselective and stereoselective epoxidation at the allylic hydroxyl site, a process that proceeds efficiently in the presence of

Scheme 27
The synthesis of the intermediate
(±)-**128**.

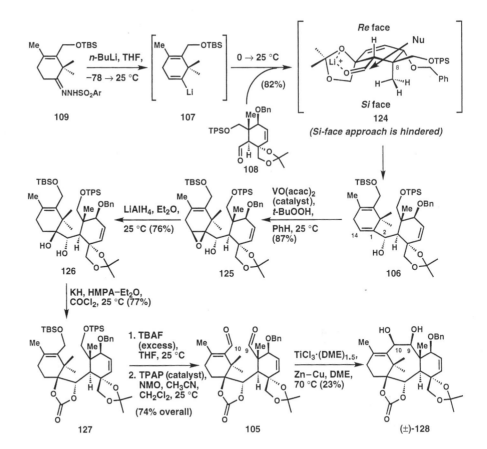

Scheme 28
The resolution of the intermediate
(±)-**128** and synthesis of the
intermediate **132**.

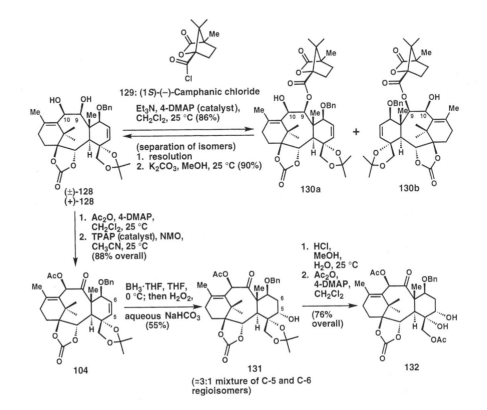

Scheme 29
The synthesis of Taxol™ (99).

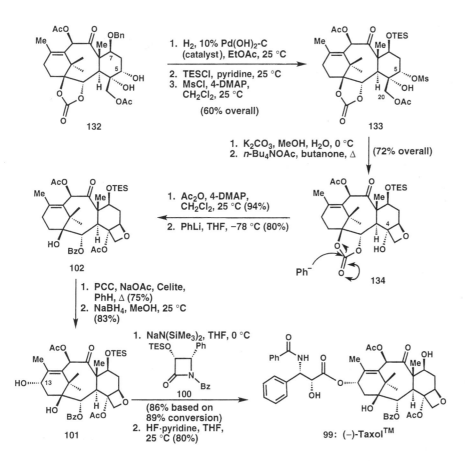

tert-butylhydroperoxide and a vanadium catalyst. The resulting epoxide **125** is then opened regioselectively with LAH, furnishing diol **126** in good yield. The next step was to engage the 1,2-diol in a cyclic carbonate system, an operation that would, at the same time, pre-organize the acyclic precursor for the crucial ring closure ahead. This was accomplished by treatment of compound **126** with phosgene under basic conditions to afford the carbonate **127** with yield 77%. The silicon-protected primary alcohols in the latter compound were then liberated by exposure to fluoride ions and oxidized to the dialdehyde **105** with TPAP/NMO, with an overall yield of 74%.

The stage was now set for the crucial cyclization reaction leading towards the eight-membered ring of the taxoid skeleton. Reductive coupling of the dicarbonyl system in **105** using Ti(0) according to the McMurry protocol afforded, albeit in modest yield, the cyclized product **128** as a single stereoisomer. Despite this relatively poorly yielding step, this sequence of reactions was short enough to allow the synthesis of the key taxoid intermediate **128** in sufficient quantities to allow us to proceed. Before that, however, a resolution of the two enantiomers of **128** was undertaken. To this end, (±)-**128** was reacted with (1S)-(−)-camphanic chloride to afford two diastereoisomers of the 9-camphanoate ester that we separated chromatographically (Scheme 28). After identification, by X-ray crystallographic techniques, the diastereoisomer with the correct absolute stereochemistry (**130a**) was converted into (+)-**128** by basic hydrolysis and the synthesis was resumed as shown in Scheme 28. Regioselective acetylation of (+)-**128** at the C-10 position followed by oxidation with TPAP/NMO afforded the ketoethanoate **104** with an overall yield of 88%. Regioselective and stereoselective hydroboration of the 5,6-double bond in **104**, followed by the usual work-up with basic hydrogen peroxide, resulted in the formation of the 5-hydroxy compound **131**

Scheme 30

The multi-enzyme synthesis of a sialyl Lewis^x derivative from an *N*-acetyl glucosamine derivative.

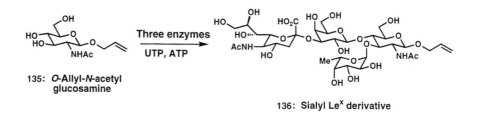

135: *O*-Allyl-*N*-acetyl glucosamine

136: Sialyl Le^x derivative

as the major isomer. In preparation for the construction of the oxetane, it was then decided to remove the acetonide protecting group from **131**, which task was accomplished by the action of methanolic hydrochloric acid, and to protect the resulting primary alcohol as an ethanoate, leading to formation of the intermediate **132** with an overall yield of 76%. At this stage, the protecting group at C-7 was switched to a triethylsilyl group because it became known that the anticipated allylic oxidation at C-13 would be accompanied by an undesirable oxidation of the benzylic position of the benzyl ether at C-7 (see **102**→**101**, Scheme 29). After this sequence, the secondary alcohol (C-5) was mesylated selectively to afford compound **133** (overall yield 60%). The ethanoate group was then removed from **133** by treatment with potassium carbonate in methanol and the resulting dihydroxy mesylate was exposed to tetra-n-butylammonium ethanoate in refluxing butanone to give the desired oxetane compound **134** (overall yield 72%). The remaining tertiary hydroxyl group was then acetylated and the carbonate ring was opened with phenyllithium regioselectively to afford benzoate **102** with an excellent overall yield. The final stages of the synthesis involved conversion of **102** to the baccatin III derivative **101** (by allylic oxidation with pyridinium chlorochromate, followed by stereoselective reduction with sodium borohydride) and attachment of the side chain by following a known sequence (generation of the alkoxide and quenching with the β-lactam **100** followed by desilylation with fluoride ions). The total synthesis of taxol (**99**) had been accomplished.

With the total synthesis of taxol (**99**) by three groups, synthetic chemists extended the boundaries of the molecular complexity they could reach. In the process, much was learned about new synthetic chemistry and new strategies for chemical synthesis and one more class of complex molecules was conquered synthetically.

The Semisynthesis of Pharmaceuticals

The complexity of natural products often makes their large-scale production by total synthesis impractical even if a laboratory synthesis has been demonstrated. On the other hand, complex target molecules, be they naturally occurring or designed ones, can often be reached from other, more readily available natural products. The sequence leading from one natural product to another or to a related target is often referred to as semisynthesis or partial synthesis. Such processes are commonly short, start from relatively cheap and plentiful compounds and are amenable to large-scale production, in contrast to some total syntheses, which may be characterized by elegant and pioneering strategies but, nevertheless, suffer from involving rather long sequences and expensive operations that make them impractical. Several semisynthetic processes for manufacturing pharmaceuticals and other useful products have been developed. Amongst them are steroidal contraceptives, β-lactam and macrolide antibiotics, Zocor™ and Taxol™.

Modern advances in organic synthesis also made it possible to manufacture by total synthesis relatively complex pharmaceuticals that are either totally unrelated to natural products or not readily available from natural sources. For example, beyond aspirin, zantac and biotin came the antibiotic thienamycin and its derivatives and the new anti-AIDS drug arixovir (Figure 5). It is expected that more and more complex

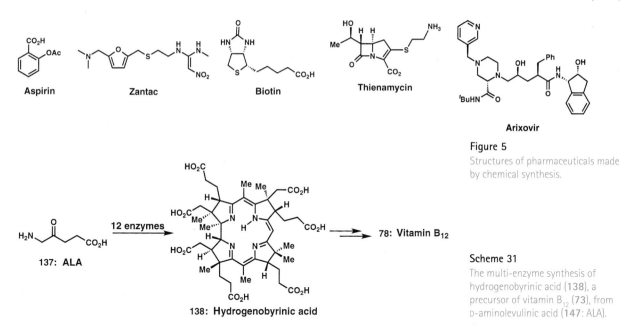

Aspirin

Zantac

Biotin

Thienamycin

Arixovir

Figure 5
Structures of pharmaceuticals made by chemical synthesis.

137: ALA   12 enzymes →   **138: Hydrogenobyrinic acid**   →   **78: Vitamin B$_{12}$**

Scheme 31
The multi-enzyme synthesis of hydrogenobyrinic acid (**138**), a precursor of vitamin B$_{12}$ (**73**), from D-aminolevulinic acid (**147**: ALA).

structures will come within reach of total synthesis as the science and technology of chemical synthesis moves forwards.

Biosynthesis

Nature provides a wealth of molecular architecture with such efficiency and calm that it leaves the synthetic chemist admiring in awe. Indeed, on comparing all our advances in the sciences of chemical synthesis with the power of nature in these matters, it becomes clear that we are undoubtedly in the early stages. Elucidating nature's pathways to complex molecules, which chemists and biologists have done in the past in numerous cases, gave us a glimpse of how enzymatic mechanisms work *in vivo*. Reproducing these pathways via catalytic networking *ex vivo* has been the dream of researchers in the field and it is only recently that this dream has started to be translated into reality. Such a strategy for the production of complex natural products would have tremendous applications not only in the manufacture of natural substances but also in the engineering of new ones. This, of course, requires basic knowledge pertaining to the individual genes involved in each step of the biosynthetic pathway and the ability to clone and express these genes as well as handle the enzymes for the, preferably, one-pot production of the target molecule from simple and readily available starting materials. Some recent successes in this field bode well for the future. The early work of Whitesides and Wong in the multi-enzyme synthesis of carbohydrates found its most sophisticated application in the recent one-pot assembly of a sialyl Lewis^X derivative by Wong's group from simple building blocks (Scheme 30).

Perhaps the most spectacular accomplishment in this area is the one-pot formation of hydrogenobyrinic acid, a corrin related to vitamin B$_{12}$, from β-aminolevulinic acid by a cocktail of 12 enzymes (Scheme 31). This success is a testimony to the power of this method for attaining high molecular complexity in such a direct way and it also stands as an example of the fruits of research that combines chemistry and biology. It became possible, after a long campaign by three groups working in the UK, France and the USA, to elucidate the biosynthetic pathway by which nature makes vitamin B$_{12}$. Given the momentum in this field and the advances in chemical biology and analytical techniques, more impressive accomplishments such as the ones described above are to be expected.

Scheme 32
The solid-phase synthesis of
epothilone A (144).

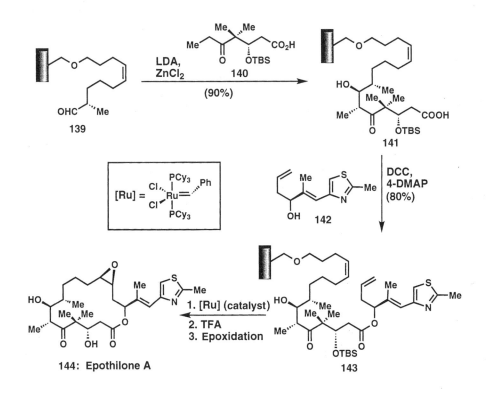

Combinatorial Synthesis

The traditional goal of the synthetic chemist has been the stereospecific construction of a given target molecule. In the pursuit of such goals the chemist was driven to achieving high levels of efficiency and molecular complexity and, as we have seen so far in this chapter, succeeded admirably in doing both. Recent advances in biology presented the synthetic chemist with a new challenge, namely the generation of large numbers of compounds to satisfy the demands of high-throughput screening against an ever-increasing number of newly isolated biological receptors. Thus, a new concept, that of combinatorial chemistry emerged, whose goal is the simultaneous construction of as many compounds as possible. Various strategies for achieving this goal, ranging from the parallel synthesis of pure compounds to the generation of multicomponent mixtures in the same pot, have been utilized. Perhaps the most powerful technique in combinatorial chemistry is the so-called split-and-pool method originally proposed by Furka (Figure 6). Like most strategies, the split-and-pool method involves solid-phase chemistry, the advantages of which include the ability to perform multi-step sequences without isolation or purification and the ease of removal of excess reagents and by-products from the polymer-bound compounds. This technique is demonstrated in Figure 6 with a pool of polymer beads split into three subpools, pooled again and re-split after common operations such as deprotection and/or addition of a common building block. Although this technique has the potential of delivering huge numbers of individual compounds, it is associated with a major challenge, that of structural identification of each compound. To address this problem several encoding methods allowing one to follow the history of the construction of each member of the chemical library have been proposed. Notable among them are those involving oligonucleotide tags, halocarbon tags and radiofrequency tags. The first two, despite their being brilliant in conception, have inherently the possibility of causing certain chemical complications, whereas the radiofrequency-based technology enjoys the benefits associated with the simplicity of storing and retrieving radiofrequency signals within

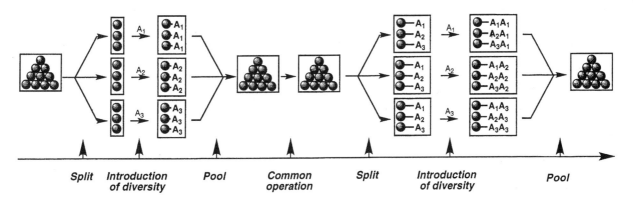

Figure 6

Spilt-and-pool combinatorial synthesis.

special microreactors equipped with memory chips. The latter strategy also has the advantage of allowing full exploitation of the split-and-pool combinatorial strategy in that microreactors can be sorted out precisely prior to splitting, leading to there being one compound per microreactor without the chance of statistical misses, such as is the case for the chemical tagging techniques. Among the many applications of this already widely used technology is the synthesis of a combinatorial library of epothilones (e.g. **144**, Scheme 32), a new class of promising anti-tumour agents.

Combinatorial synthesis, a phenomenon of the 1990s, has been hailed as a revolutionary development in medicinal chemistry promising to shorten considerably the process of discovering drugs. A great many chemical libraries have already been generated and screened in a variety of biological assays. Among them are peptide and oligonucleotide libraries, both of which are useful and relatively easy to construct, and libraries of small organic molecules that are useful and more challenging to synthesize in terms of chemical efficiency and numbers of compounds.

Combinatorial chemistry has established itself as a routine and powerful technique for the generation of diverse and complex molecules. Taken together with the wealth of nature in terms of biologically relevant molecular diversity and complexity, this technique is sure to have a major impact both on chemistry and on the discovery of drugs.

Conclusion

The constant challenge of producing molecular structures, particularly those derived from nature, helped to drive chemical synthesis to its present state of sophistication. Although some may proclaim that synthetic chemists can make anything given enough man-power and time, the need for further advances in the field remains acute. Reaching for more complex molecules is a great goal of our science, but it is not enough. We need to reach such complexity with higher efficiency, economy and purity. The twenty-first century will see even greater strides in these directions than the twentieth century witnessed.

Further Reading

1. K. C. Nicolaou and E. J. Sorensen, *Classics in Total Synthesis* (VCH, Weinheim, 1996).
2. E. J. Corey and X.-M. Cheng, *The Logic of Chemical Synthesis* (Wiley, New York, 1989).
3. C.-H. Wong and G. M. Whitesides, *Enzymes in Synthetic Organic Chemistry* (Pergamon, Oxford, 1994).
4. J.-M. Lehn, *Supramolecular Chemistry* (VCH, Weinheim, 1995).
5. S. R. Wilson and A. W. Czarnik, *Combinatorial Chemistry* (Wiley, New York, 1997).

6. I. Ojima (editor), *Catalytic Asymmetric Synthesis* (VCH, Weinheim, 1993).

7. R. Noyori, *Asymmetric Catalysis in Organic Synthesis* (Wiley, New York, 1994).

8. R. Noyori (editor), *Organic Synthesis in Japan, Past, Present, and Future* (Tokyo Kagaku Dozin, Tokyo, 1992).

9. H. B. Kagan and O. Riant, *Chem. Rev.* 1992, **92**, 1007.

10. K. Mikami and M. Shimizu, *Chem. Rev.* 1992, **92**, 1021.

11. T.-L. Ho, *Tandem Organic Reactions* (Wiley, New York, 1992).

12. R. Grigg (editor), *Cascade Reactions* (Pergamon, Oxford, 1996).

13. T. Lindberg (editor), *Strategies and Tactics in Organic Synthesis*, Vol. 1–3 (Academic, San Diego, 1984, 1989 and 1991).

14. J. Tsuji, *Palladium Reagents and Catalysts* (Wiley, New York, 1995).

Medicines from Nature

Jim Staunton and Kira Weissman

9

Introduction

Primitive medicine men and women brewed strange potions from plants and small animals, because ancient wisdom had it that drinking these crude liquids would alleviate pain and cure disease. Many of the compounds they unwittingly employed, for example caffeine, cocaine and morphine (Figure 1), are still in use today both in their native state and in the form of simple chemical derivatives. In this chapter we will focus on the beneficial aspects of these so-called 'natural products' as pharmaceuticals.

Until the beginning of the nineteenth century, man continued to use a very wide range of distillates as medicines and as mind-altering drugs without understanding their 'magic' properties. As science advanced, however, it became possible to determine rigorously the active components of these extracts through painstaking and laborious chemical methods: the metabolites, termed natural products, are organic compounds that are unique to one organism or a few closely related organisms. This rational approach to the discovery of drugs inaugurated an era of bio-prospecting, that is, raiding nature's storehouses of plant and microbiological life. Bio-prospecting literally involves exploring the forests, diving in the oceans and digging in the dirt to obtain

Figure 1
A selection of important natural products and their biological activities.

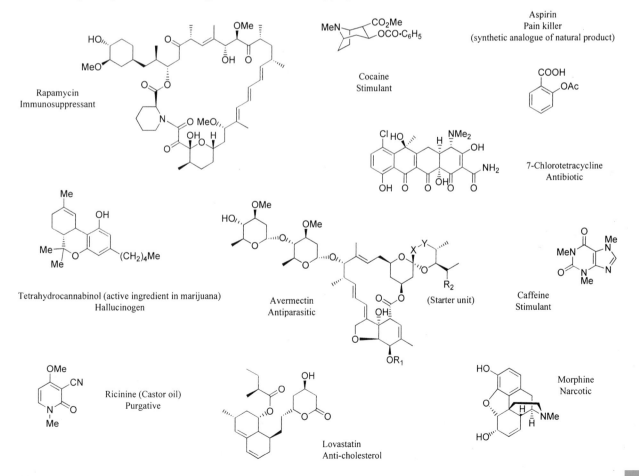

Rapamycin
Immunosuppressant

Cocaine
Stimulant

Aspirin
Pain killer
(synthetic analogue of natural product)

7-Chlorotetracycline
Antibiotic

Tetrahydrocannabinol (active ingredient in marijuana)
Hallucinogen

Avermectin
Antiparasitic

(Starter unit)

Caffeine
Stimulant

Ricinine (Castor oil)
Purgative

Lovastatin
Anti-cholesterol

Morphine
Narcotic

environmental samples. The study of the compounds discovered by these methods has become a major area of research in organic chemistry, and has led to the isolation and identification of thousands of different structures. Most have been extracted from plants and, more recently, micro-organisms, with the animal kingdom contributing rather sparsely to the total.

Most of the known natural products exert powerful physiological effects in one form or another on the human body. To be acceptable for medical use, however, a natural product, like a candidate synthetic drug, has to be shown to exert an acceptably high level of beneficial action at the same time as having a reasonably low level of toxic side effects. Unfortunately, only a small proportion of these compounds satisfies the rigorous criteria set by the regulators of the modern pharmaceutical industry. Even so, natural products and chemical derivatives prepared from them account for around half of all medicines. These drugs are used to combat a wide range of conditions, especially infectious diseases and cancer. A selection of important natural product structures and their biological effects is presented in Figure 1. Most readers of this chapter will have benefited from the use of one or more of these compounds at some time in their lives.

The Search for New Medicines

Given the enormous medical and economic impact of natural-product pharmaceuticals, we have obvious incentives both to improve the activities of existing compounds and to discover new metabolites. Again, until recently, most drug candidates came directly from nature. An important recent success from the plant kingdom is the discovery of taxol, a constituent of the pacific yew tree, and its subsequent development as a treatment for cancers of the ovary, breast, head and neck (Figure 2). This wonder

Figure 2

Structures of the antiproliferatives taxol and discodermolide, the antibiotic erythromycin A and the marine toxin maitotoxin.

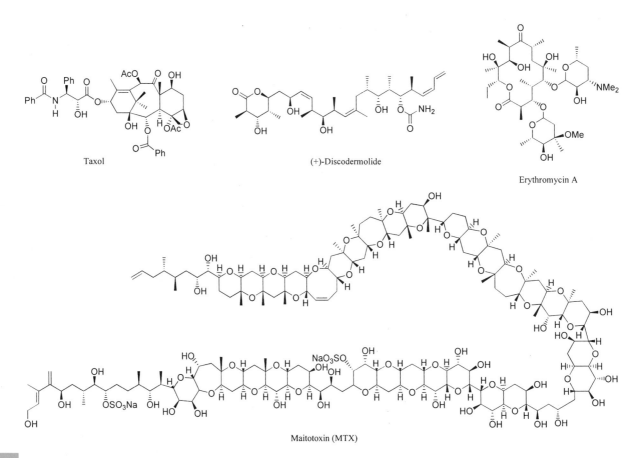

Taxol

(+)-Discodermolide

Erythromycin A

Maitotoxin (MTX)

drug operates by inhibiting the uncontrolled growth of cancerous cells. In order to grow, malignant cells must undergo rapid division, a process that depends on the assembly and disassembly of a protein called tubulin, which is the major component of the cell's internal structural support, the cytoskeleton. Taxol halts mitotic division in a rather counter-intuitive manner: unlike many anti-cancer drugs that act by inhibiting polymerization of microtubules, it actually promotes the formation of stable bundles of microtubules, killing the cells instead by arresting the disassembly of the cytoskeleton.

A significant initial drawback to the use of taxol in chemotherapy was that the yew bark yields only the tiniest amounts of the drug relative to the potential worldwide market, so it appeared that the tree species would have to be sacrificed for the sake of human life. Fortunately, we have discovered two alternative sources of taxol. First, the drug can be chemically synthesized from a precursor of taxol found in needles of European yews. Secondly, two scientists at the University of Montana stumbled upon a taxol-producing fungus that nests in the folds of the yew bark. Crucially, the fungus continues to produce the drug even when it is separated from its host (which is never a given), an observation that suggests that it somehow picked up the taxol-producing genes from its host tree. Speculation aside, the implication is that large-scale fermentation of the fungus could provide us with a relatively cheap and abundant source of taxol.

Marine organisms and micro-organisms also continue to be a fruitful source of interesting drug leads. Like taxol, (+)-discodermolide (Figure 2) is a spectacular anti-mitotic agent. However, unlike its terrestrial counterpart, this antiproliferative is produced in the ocean by the sponge *Discodermia dissoluta*. Despite their disparate origins and lack of apparent structural similarities, the two compounds act in a comparable manner both in cells and biochemically; in fact, discodermolide stabilizes microtubules more potently than does taxol, although its overall cytotoxicity is comparable. On the basis of its mode of action, discodermolide will probably serve as another potent weapon in our ongoing battle with cancer and provides a new structural class for further synthetic elaboration towards new pharmaceuticals.

Not to be outdone, dirt has proven to be a rich source of bioactive natural products. Soil is a thriving microbial community, where thousands of bacterial species battle each other for limited resources and space. One survival strategy is not simply to out-compete your neighbours, but rather to do away with them completely by producing bacteria-killing chemicals, or antibiotics. Although they are obviously generated by bacteria for bacteria, these anti-bacterial chemicals, for example erythromycin A (Figure 2), are some of the most important drugs ever discovered. Despite our attempts to grow more of these soil species, there is still a vast number of bacteria that have resisted culturing and analysis. To this end, some biotechnology firms have been founded to exploit the diversity of chemical structures potentially available from the organisms found in dirt. Instead of attempting to grow the bugs and isolate their metabolites directly, these companies are attempting to clone the genomic DNA that directs antibiotic synthesis. The hope is that one will be able to bypass the difficulties associated with culturing soil micro-organisms by extracting the synthetic pathways and introducing them into more amenable species. The currently favoured hosts are various species of bacteria called *Streptomyces*, which are already sources for many important pharmaceuticals. Although some difficulties remain, this technology holds great promise for further exploiting the wealth of soil antibiotics.

As a final illustration of the amazingly complex structures that await our discovery, a Japanese group recently discovered and characterized a compound called maitotoxin (MTX) (Figure 2), the largest and most toxic secondary metabolite ever described. This gigantic molecule is produced by a marine micro-organism called a dinoflagellate and, although its function for these creatures remains unknown, 1 milligram of the compound is enough to kill 50 000 000 mice! This metabolite clearly demonstrates that we can scarcely predict nature's tremendous synthetic creativity or the potential biological activities – for good and ill – of these remarkable compounds.

Although natural products are a proven source of drug candidates, scientists have begun to pioneer a new method of discovering drugs – genetically engineered biosynthesis. In this radical technology, which is the main subject of this chapter, the enzymes of an existing biosynthetic pathway leading to the production of a natural product are suitably altered by genetic engineering with the aim of redirecting the biosynthesis to form a novel compound. If this approach lives up to its initial promise, it will provide a powerful new source of pharmaceuticals complementing the traditional approach of screening the biosphere to discover organisms that produce hitherto unknown compounds.

The Elucidation of Biosynthetic Pathways

Genetic engineering would not be possible without detailed knowledge of the candidate biosynthetic pathway. Fortunately, efforts towards the structural elucidation of natural products have been matched by enormous progress in investigations into their biosynthesis.

It was first recognized that even complex structures would have to be assembled from smaller molecular building blocks and it was logical to assume that they would be components involved in the primary metabolic processes used by the cell for growth – amino acids, sugars and fatty acids. This conviction was reinforced by the observation that natural products share structural features with primary metabolites and such similarities allowed the natural products to be grouped into families. One such family, a set of alkaloids, is shown in Figure 3.

The structures differ markedly in size and complexity but all contain a 'C_6–C_2' unit comprising a six-membered carbocyclic ring with a two-carbon chain (indicated by heavy lines) attached. This structural feature suggested that this shared residue derives from a common primary source. Obvious candidates for this source metabolite were the amino acid phenylalanine and its close structural relation tyrosine. The idea of a common building block was confirmed by use of radioactively labelled precursors. For example, tyrosine labelled with ^{14}C at C-2 was administered to opium poppy plants at the time they are known to produce papaverine. The labelled precursor found its way into the sap of the plant and so labelled the existing pool of tyrosine. From there, the radioactivity was incorporated into the alkaloid by the various biosynthetic enzymes. As had been expected, radioactive papaverine was isolated from the plants and, more significantly, it was shown that all the radioactivity resided in the two sites marked in Figure 4. The accepted interpretation of this result is that two C_6–C_2 units derived from tyrosine are incorporated into papaverine, which accounts for the origin of all the carbons in the main skeleton. Similar results were obtained for morphine and, furthermore, mescaline was proved to incorporate one C_6–C_2 unit in the predicted way.

This type of experiment identifies the building blocks but does not define the step-by-step processes by which the final structure is reached via a succession of biosynthetic intermediates. The standard approach to gaining these added insights is to postulate a pathway and then synthesise key putative intermediates in radiolabelled forms. In the case of papaverine, for example, it was thought that the simpler benzylisoquinoline, norlaudanosoline (Figure 5), might be an intermediate. Feeding synthetic radiolabelled norlaudanosoline to the opium poppy resulted in the formation of papaverine labelled specifically at the single predicted site (Figure 5). Other inter-

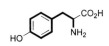

Tyrosine

Mescaline

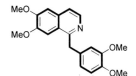

Papaverine

Morphine

Figure 3
Alkaloid natural products that incorporate a C_6–C_2 unit (heavy lines) from tyrosine.

mediates were similarly identified in a subsequent methodical investigation and, as a result, the pathway to papaverine is now known to follow the step-by-step sequence up to the key intermediate norlaudanosoline (Figure 6). The order of the remaining steps has yet to be established.

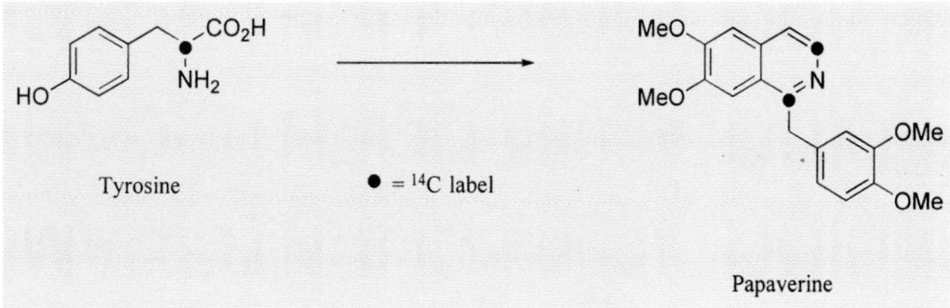

Many other biosynthetic routes have been elucidated in great detail by this experimental approach. Extra confirmatory evidence came in many cases from direct detection of biosynthetic intermediates and other highly sophisticated isotopic labelling studies. In recent years, with the advent of high-field nuclear magnetic resonance (NMR) capable of direct detection of stable isotopes, radioactive isotopes such as ^{14}C and ^{3}H have largely been replaced by the chemically equivalent but stable isotopes ^{13}C and ^{2}H. The advantage of NMR is that it allows direct detection of isotopic labelling sites without resorting to the degradative chemistry necessary to determine the position of radioactive labels.

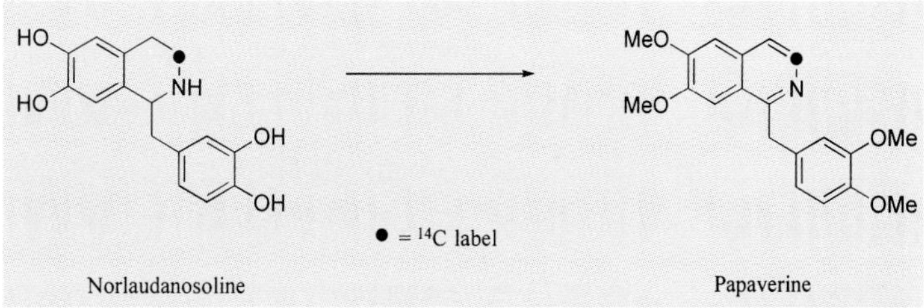

Examples of Biosynthetic Pathways

The penicillins need no introduction as important antibiotics other than to say that they are produced by a number of fungi, not just the one discovered serendipitously by Alexander Fleming in 1928. The pathway to the penicillins begins with three amino acids, two of which, cysteine and valine, are drawn from the pools of amino acids used for protein synthesis. The third precursor, aminoadipic acid, is generated by a dedicated pathway that need not concern us here. In the first steps of the pathway (Figure 7), a tripeptide is formed by linking the three amino acids. This intermediate is sometimes called ACV after the order in which the three amino acids are joined or the Arnstein tripeptide after its discoverer. Very recently it has been shown that the linking reaction is carried out by a member of an extraordinarily complex set of enzymes called non-ribosomal peptide synthases.

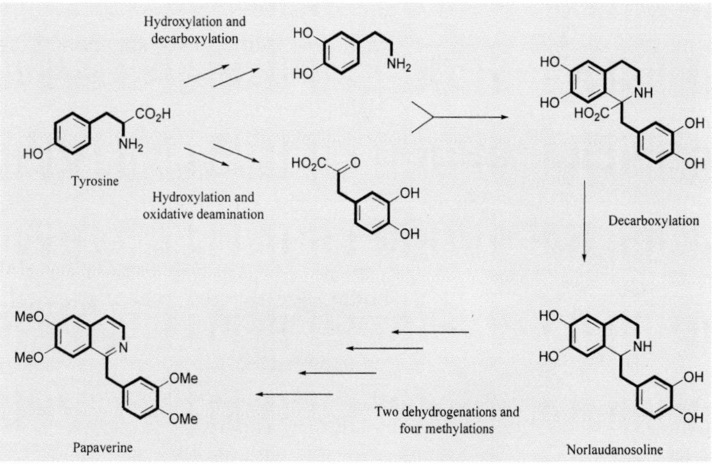

The following step, the synthesis of the bicyclic isopenicillin N, involves the formation of two bonds. Although the reaction is clearly a multi-step process, it is nonetheless carried out by a single enzyme, isopenicillin N synthetase (IPNS), using extraordinary chemical mechanisms. Finally, the aminoadipic acid residue attached to the bicyclic ring is replaced by another carboxylic acid that can vary widely according to the culturing conditions under which the producing organism is grown. This replacement can also be achieved chemically. These final penicillin analogues are the active antibiotics in commercial use.

Knowledge of the biosynthetic pathway has allowed it to be manipulated towards the production of novel penicillins. Jack Baldwin and colleagues in the UK pioneered a method for altering the core portion of the penicillin molecule. Using purified IPNS,

Figure 4
The incorporation of ^{14}C-labelled tyrosine into papaverine.

Figure 5
The incorporation of ^{14}C-labelled norlaudanosoline into papaverine.

Figure 6
An outline of the biosynthetic pathway from tyrosine (two equivalents) to papaverine.

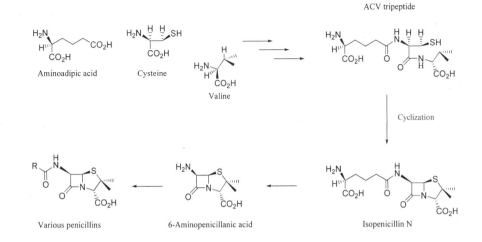

ACV tripeptide

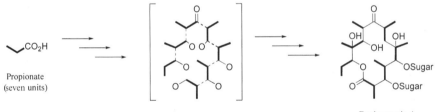

the researchers were able to vary the structure of the nucleus by feeding unnatural synthetic dipeptides and tripeptide substrates. Although none is in commercial use, many of the new penicillins exhibit antibiotic activities comparable to that of the parent compound.

The antibiotic erythromycin A is widely used in medical practice. It belongs to a family of natural products called polyketides because of the common feature of their biosynthesis: their structures are derived formally by repeated joining of small carboxylic acids such as ethanoate or propionate (ketide units). In the case of erythromycin A, this analysis applies to the macrocyclic lactone (macrolide) core, as can be seen in Figure 8.

The idea was confirmed initially by adding propionate that had been radiolabelled in the carboxyl group to the culture medium of the producing organism. This feeding resulted in the production of radioactive erythromycin A (Figure 9(a)). More recently, the experiment was repeated with ^{13}C-labelled propionate. In this case, it was evident from ^{13}C NMR that all the predicted carbons had been enriched in ^{13}C over the level of natural abundance by incorporation of the labelled precursor (Figure 9(a)). Further research revealed that the propionate is incorporated through one unit of propionyl-CoA (termed the starter unit) and six units of methylmalonyl-CoA – termed the extender unit (Figure 9(b)).

In parallel to these studies with isotopically labelled precursors, biosynthetic intermediates were isolated from blocked mutants (bacteria that are unable to produce one of the enzymes catalyzing the conversion of an intermediate in the biosynthetic pathway, so that that metabolite accumulates in sufficient quantities to be characterized). By identifying all of the late intermediates between 6-deoxyerythronolide B and

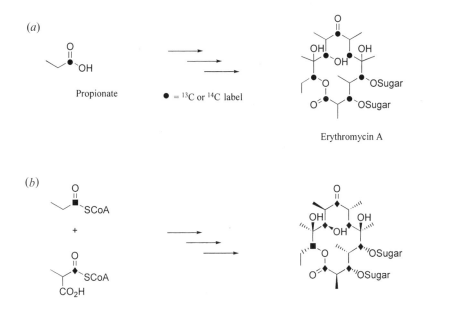

Figure 9
(*a*) The incorporation of ^{13}C- or ^{14}C-labelled propionate into erythromycin A. (*b*) Actual building blocks for the biosynthesis of erythromycin.

erythromycin A (Figure 10), the nature of phase II of the biosynthesis of erythromycin was determined in its entirety.

However, here the chemistry failed: unlike the pathways leading to papaverine and penicillin, that for the biosynthesis of erythromycin could not be fully characterized through isolation of intermediates. No free intermediate simpler than 6-deoxyerythronolide B was ever discovered, so the reactions of phase I remained a 'black box'. The lid was raised only by recent genetic studies.

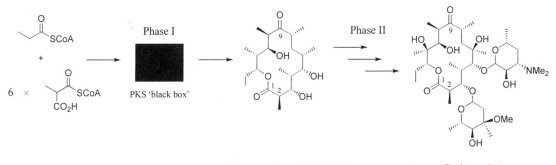

6-Deoxyerythronolide B (6-dEB) Erythromycin A

The Genetics of Erythromycin Biosynthesis

Figure 10
Phases I and II in the biosynthesis of erythromycin A.

Pioneering work in polyketide genetics was carried out by Sir David Hopwood and co-workers at Norwich in the UK. Beginning in the 1980s, as a culmination of a large amount of exploratory work, the group first identified the genes encoding for the biosynthesis of the polyketide actinorhodin. This advance encouraged other groups to locate genes for the biosynthesis of polyketides. The concurrent discovery of the genes for the biosynthesis of erythromycin by Peter Leadlay and co-workers in the UK and Leonard Katz and co-workers in the Abbott Laboratories in the USA was based on the assumption that the biosynthetic genes would be clustered in the vicinity of the

Figure 11

An outline map of the cluster of genes coding for enzymes involved in the biosynthesis of erythromycin.

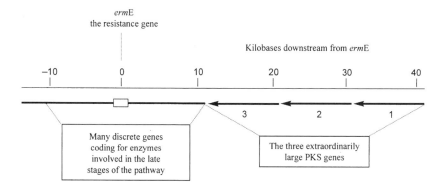

bacterium's own antibiotic self-resistance gene. The structure of the gene for the biosynthesis of erythromycin is outlined in Figure 11. On both sides of the resistance gene are regions of DNA that code for enzymes involved in the late stages of the pathway (phase II) in which, as mentioned previously, 6-deoxyerythronolide B is converted into erythromycin A. Sequencing further away from the resistance gene revealed the genes for the polyketide synthase (PKS) proteins responsible for phase I. From the sequence of the PKS genes the chemistry of the early stages of the biosynthesis of erythromycin immediately became clear.

The PKS genes consist of three large stretches of DNA (called open reading frames), each coding for a large multi-functional protein. Sets of active-site motifs, consisting of 4–6 amino acids, are strategically distributed along the primary sequence of these proteins. Because these motifs are characteristic of known activities involved in the biosynthesis of polyketides, it was reasoned that the protein would fold locally around these amino-acid residues to form a conventional enzyme active site. These localized folded regions were termed domains (Figure 12(a)) and each one was assigned a catalytic activity on the basis of its putative active site. Although they are functionally discrete, the domains remain covalently attached to a neighbour or neighbours by short linker regions of 10–30 unfolded amino-acid residues. One can therefore think of each giant protein as a set of biosynthetic enzymes joined to one another like beads on a string.

The individual domains within the multi-enzymes are further organized into sets of functional units called chain-extension modules, two per protein (Figure 12(b)). Each of the six modules adds a three-carbon unit to the relevant developing chain and produces the appropriate chemical structure at the β position (a keto, a hydroxyl, a double bond or a fully reduced methylene group). The modules also control the three-dimensional arrangement or stereochemistry of these groups. Figure 13 shows in more detail how the PKS genes are arranged on the genome and how the catalytic sites are ordered in the primary structure of the protein in each corresponding set of modules.

In effect, the PKS functions as a molecular assembly line. Each module is a team of workers that collaborate to carry out one cycle of chain extension. The nature of the chemical structure produced during chain extension is therefore dictated by the composition of the workers in the team. The modular arrangement of these activities immediately suggested their suitability for rearrangement: simply changing the combination of the workers in one or more modules can alter the structure of the end product. Alternatively, the chain length can be altered by increasing or decreasing the number of modules in the production line. On a practical level, such manipulation or engineering must necessarily occur on the level of the genes encoding for the PKS. Recent experiments have demonstrated the promise of this genetic-engineering approach towards producing novel erythromycins.

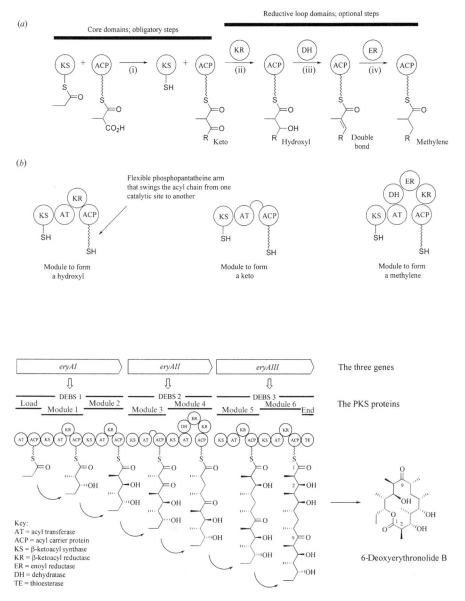

Figure 12
The role of the domains in the erythromycin PKS and the sequence of chain-extension reactions. (*a*) In the first steps (not shown), a unit of the chain extender, methyl malonate, is loaded onto the active thiol of the acyl carrier protein (ACP) by an acyl transferase (AT) domain and the starter acyl is loaded onto the active thiol of the β-ketoacyl synthase (KS), also with the help of an AT. A condensation reaction (i) between the methylmalonate and the acyl group) is then carried out by the KS. These three domains (ACP, AT and KS) are therefore essential for the formation of the new C—C bond. The extended chain remains bound to the thiol of the ACP during the subsequent sequence of optional reductive operations in which the newly formed keto group is appropriately modified; (ii) the ketone is reduced to a hydroxyl by the β-ketoacyl reductase (KR); (iii) dehydratase (DH)-mediated dehydration occurs; and (iv) the double bond is reduced to a saturated methylene group by the enoyl reductase (ER). The extent of reduction in each module depends on the number of domains in the reductive loop. (*b*) The arrangements of the domains in the three types of modules used in the biosynthesis of erythromycin.

Figure 13
The arrangement of the genes and the primary structure of the PKS protein.

Genetic Engineering of Erythromycin Biosynthesis

The erythromycin PKS has been modified in many different ways that have dramatically demonstrated the ability of genetic engineers to produce targetted mutations. Pioneering experiments in this area were carried out by Leonard Katz and co-workers in the Abbott Laboratories in the USA. These researchers disrupted the section of the gene coding for the β-ketoacyl reductase (KR) domain in module 5, to produce a modified protein that lacked this key player. This alteration to the production line resulted in the formation of a macrolide ring bearing a keto group rather than a hydroxyl in the predicted position (Figure 14). Similarly, they also inactivated the enoyl reductase (ER) in module 4 and isolated an analogue of erythromycin with a double bond at the expected position (C-6/C-7) (Figure 14). The formation of these two analogues and the fact that they were correctly cyclized showed that the structure of the growing chain need not play a critical role in the biosynthesis of polyketides and that altered polyketides can be substrates for further chain extension. This result implies that the correct transfer of the growing chain from one module to the next resides not in the conventional substrate specificity of a particular activity but rather in the specific juxtaposition of the various domains. Such substrate tolerance bodes well for the success of rearranging domains within the PKS to produce novel metabolites. It is also significant that the product released from the PKS in both experiments was at least partially processed towards an analogue of erythromycin A, demonstrating that the elaboration enzymes of phase II also exhibit a useful degree of relaxation of their substrate specificity.

A common characteristic of the research just described was the essentially destructive nature of the mutations. Although the experiments demonstrated that individual domains could be inactivated without destroying the overall capacity of the PKS to make macrolides, the range of variation accessible by this strategy is clearly limited. A potentially more versatile approach is to reposition domains within the PKS. However, such rearrangement can work only if the relocated domain is capable of carrying out its normal reaction in its new context. The first demonstration of the feasibility of this approach by Peter Leadlay and co-workers at the University of

Erythromycin A core ring

C-5 keto analogue

C-6, C-7 double bond

C-2 desmethyl analogue

C-10 desmethyl analogue

C-12 desmethyl analogue

Avermectin 'heads' on erythromycin 'bodies'

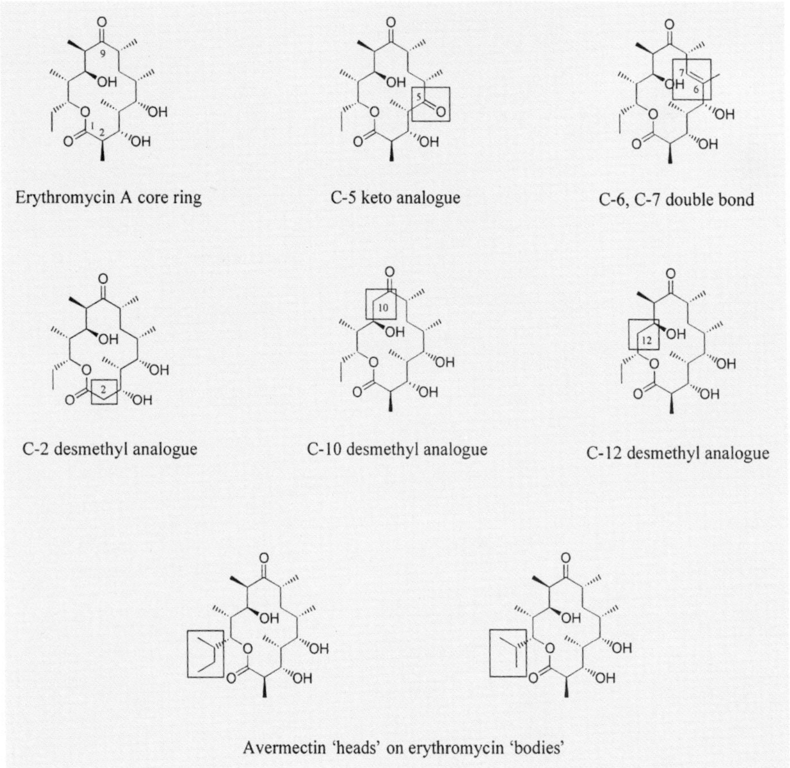

Figure 14
Modified erythromycin cores generated by genetic engineering.

Cambridge was to reposition the thioesterase (TE) domain from the terminus of DEBS 3 to the end of DEBS 1 (Figure 15). In its normal context the TE is responsible for off-loading and cyclizing the fully processed heptaketide chain to produce the macrolide 6-deoxyerythronolide B (Figure 13). It was therefore hoped that the TE would carry out the same functions in its new location, catalysing the release of the growing chain at the triketide stage to produce a δ-lactone (Figure 15). A mutant of the erythromycin-producing bacterium *Saccharopolyspora erythraea* containing the engineered protein (called DEBS 1-TE, Figure 15) did generate the anticipated δ-lactone and another lactone incorporating an ethanoate instead of a propionate starter unit. As expected, the production of erythromycin was completely abolished. To demonstrate that the TE was playing an *active* role in the biosynthesis of the lactones, a further mutant, in which a genetically engineered *inactive* copy of the TE was placed at the end of DEBS 1, was constructed. This control mutant again did not produce erythromycin and generated the two lactones with a greatly reduced yield. These results were supported by contemporary experiments with a very similar construct, DEBS 1 + TE, that also produced the δ-lactones.

In further applications of this relocation strategy by David Cane, Chaitan Khosla and co-workers in the USA, the TE has been moved to the termini both of module 5 and of module 3 (Figure 16). Products consistent with truncation of chain extension at the expected stages were obtained in both cases. The termination at module 5 was particularly significant because it led to release of the hexaketide intermediate as a 12-membered-ring macrolide. The DEBS 1 + module 3 + TE construct produced two

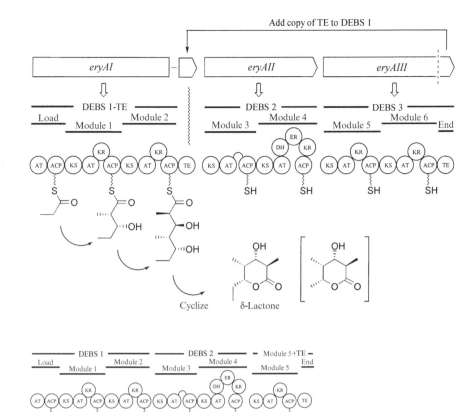

Figure 15
Genetic engineering to generate
DEBS 1-TE.

Figure 16
Further examples of relocating the
TE to produce novel metabolites.

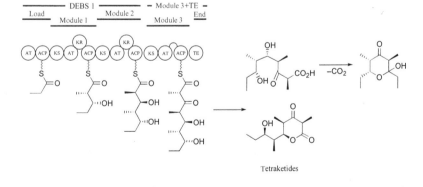

tetraketide products; the first is possibly formed from a keto acid by decarboxylation after its release from the enzyme.

A more versatile strategy for generating novel products is to transfer domains between PKS clusters. Effectively, this amounts to playing 'mix and match' with portions of different natural products. In the first demonstration of this concept, the acyl transferase (AT) of module 1 of DEBS 1-TE was replaced by the AT derived from module 2 of the rapamycin PKS to produce a hybrid multi-enzyme (Figure 17). This

Figure 17
The construction of an
erythromycin–rapamycin hybrid
PKS.

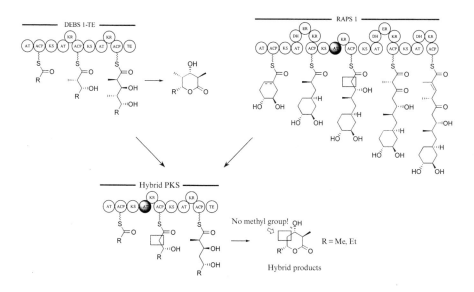

transplanted AT domain is responsible for the incorporation of a malonate chain-extension unit rather than methylmalonate as in erythromycin. As had been expected, the lactone products lacked methyl groups at C-4, but were otherwise normal in structure. The *hybrid* PKS therefore produced the predicted *hybrid* product.

In further swapping experiments, modified erythromycins have been generated by utilizing the AT domains from other PKS systems. For example, the ATs in modules 1, 2 and 6 of the erythromycin PKS were swapped with AT activities that supply malonate. As expected, the resulting macrolide products of these hybrid PKSs lacked the methyl groups normally observed at C_2, C_{10} and C_{12} of the erythromycin core (Figure 14). Although attempts to swap AT domains into modules 3–5 as well were made, they were unsuccessful. It is clear, however, that at least some ATs will function in downstream modules of the erythromycin PKS. Certainly genes for acyl transferases can be included in the large pool or 'toolbox' of PKS genes with which to perform rational engineering.

The swapping approach has also been applied successfully to ketoreductase (KR) domains. Because these activities participate in controlling the configuration of hydroxyl groups on the backbone of erythromycin, swapping KRs is a potential means towards designing stereochemical patterns into novel polyketides. The experiment was carried out with the DEBS 1 + module 3 + TE system described earlier. Exchange of the KR domain in module 2 (KR2) for that of module 5 of DEBS 3 (a reductase domain that, in its native environment, generates the same stereochemistry as KR2) gave the expected products, which were identical to those of DEBS 1 + module 3 + TE (Figure 16). However, when KR2 was replaced by the KR domains from either module 2 or module 4 of the rapamycin PKS, a triketide lactone product was obtained, in which the stereochemistry at C-3 was opposite to that generated by the DEBS 1-TE system, but consistent with that observed for rapamycin KR2 in its normal context (Figure 18). The hydroxyl stereochemistry normally produced by rapamycin KR4 is hidden by a subsequent dehydration, but was revealed in this experiment. It would appear, therefore, that the transplanted KR domains are capable of accepting structurally diverse β-ketothioesters and of reducing them in a stereospecific manner. The reason for the isolation of triketide products rather than the expected tetraketides is not known.

In order to produce modification of oxidation states at the β-keto-derived carbons of the erythromycin structure, the existing catalytic domains must be altered in some manner. In addition to the specific domain inactivations mentioned previously, possible methods that can be utilized include removal, addition and exchange of the optional KR, ER and dehydratase (DH) domains. In the first experiment of this type the KR domain of module 2 of DEBS 1 + TE was replaced by that from module 3 of DEBS 2, which is *non-functional* in its normal context. This produced novel δ-lactones in which the keto group at C-3 generated by the second condensation step survived, with the normal pattern of functionality at all other carbons (Figure 19).

The exchange of *functional* domains was demonstrated to occur when the KR domain of module 2 (KR2) of DEBS 1 was successfully swapped for the reductive domains DH and KR from module 4 of the rapamycin PKS. This swapping experiment was also carried out in the context of the trimodular DEBS 1 + module 3 + TE system. The products of this hybrid PKS contained a double-bond moiety in the place of the normal module-2 hydroxyl functionality (Figure 20). This work was extended by the successful swapping for KR2 of the DH–ER–KR tridomain of module 1 of the rapamycin PKS (Figure 21). In this example, loss of the C-5 hydroxyl group (as a consequence of the substitution with the tridomain) promotes the formation of an unusual eight-membered lactone ring, a result that again demonstrates the remarkable tolerance of the thioesterase domain of erythromycin. The unexpected reduction of the carbonyl group at C-3 is assumed to have occurred as a consequence of an additional metabolic enzyme.

Further evidence for the viability of swapping a *series* of domains came from an experiment in which the propionate-selective loading module (the AT–acyl carrier protein (ACP) didomain) of DEBS 1–TE was replaced by that from the avermectin PKS (for the structure of avermectin, see Figure 1). Results of previous studies had shown that this loading module, which normally recruits branched-chain starter acids, can also accept more than 40 unnatural carboxylic acids under the correct conditions. The products of the fermentation of this mutant strain included two novel lactones as well as the normal products (Figure 22). The new products had starter acyl groups characteristic of the avermectins and therefore can reasonably be viewed as hybrid molecules incorporating elements both of the avermectin structure and of the erythromycin structure. In a subsequent experiment the same loading module was placed at the start of the entire erythromycin PKS in *S. erythraea*. As had been expected, the new macrolides produced by this organism contained starter acids characteristic of avermectin (i.e. they had avermectin 'heads' on erythromycin 'bodies' (Figure 14)). The broad specificity of the loading module of avermectin is well documented and it has already allowed the facile production of many more novel erythromycin analogues through the feeding of the system with exogenous carboxylic acids. It is also expected that this strategy may be applied to other PKSs.

In all the cases shown in Figure 14 we have illustrated not the final isolated natural product, but rather the intermediate product of the modified 'black box'. In many cases the enzymes responsible for phase II of the pathway transformed these intermediates into active antibiotics analogous to erythromycin A. This further processing has caused great excitement in pharmaceutical companies interested in marketing existing

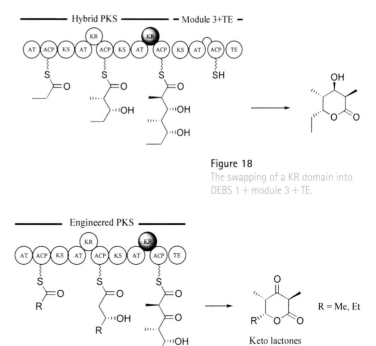

Figure 18
The swapping of a KR domain into DEBS 1 + module 3 + TE.

Figure 19
The swapping of a non-functional KR domain into DEBS 1 + TE.

Keto lactones

R = Me, Et

Figure 20

The swapping of reductive domains into DEBS 1 + module 3 + TE.

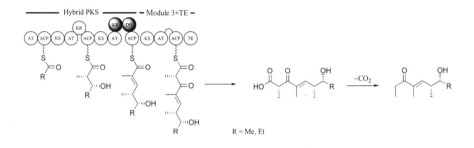

Figure 21

The biosynthesis of an eight-membered ring lactone by reductive domain swapping into DEBS 1 + module 3 + TE.

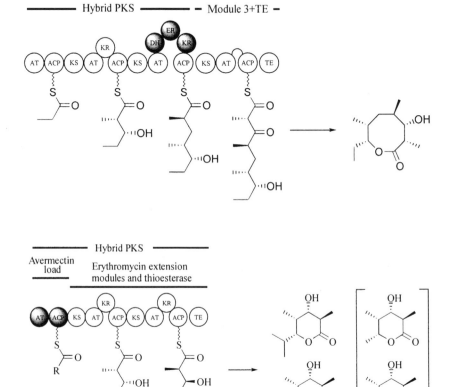

Figure 22

Swapping the loading module into DEBS 1-TE.

polyketide medicines. Obviously this novel technology has enormous potential for producing new candidate drugs.

As any synthetic chemist would know, compounds of this complexity could not be produced efficiently by total synthesis – molecular biology, however, makes their production a practical reality. None of these novel compounds has been discovered in nature and, indeed, they may never have been produced before. Genetic engineering has already begun to take evolution into new areas and, it is to be hoped, will provide us with thousands and perhaps millions of candidate antibiotics and other pharmaceuticals in the near future.

It is important to keep this new technology in perspective, however. Although we now know the structures of many natural products and, in many cases, their biosynthetic pathways, we can still scarcely predict what nature has up her sleeve. Therefore, while genetic engineering may allow us to create in a rational way perhaps thousands or even millions of novel structures, teams of modern medicine men and women in white lab coats will, like their predecessors of old, continue to labour over heating mantles and oil baths to isolate novel natural metabolites from plants and micro-organisms.

Further Reading

1. F. Myers. Surprise! A fungus factory for taxol? *Science* **260**, 154–155 (1993).
2. C. J. Cowden and I. Paterson. Cancer drugs better than taxol? *Nature* **387**, 238–239 (1998).
3. G. A. Cordell. *Introduction to Alkaloids* (John Wiley, New York, 1981).
4. J. Mann. *Secondary Metabolism* (Clarendon Press, Oxford, 1978).
5. J. Staunton and B. Wilkinson. Biosynthesis of erythromycin and rapamycin. *Chem. Rev.* **97**, 2611–2629 (1997) and other articles therein by D. A. Hopwood and J. E. Baldwin, *et al.*
6. Articles within *Science* **264** (1994).

10 From Pharms to Farms

Chemists are extremely influential in the design and synthesis of new medicines for the clinic and in the preparation of materials for enhancement of yield and protection of crops. We shall illustrate these well-defined roles, using selected examples to illustrate the key points. In addition, chemists are also extremely important individuals in the fragrance industry and some of their contributions to this field are noted for comparison. All these areas need and utilize high-value compounds called 'fine chemicals'.

New Drugs, Novel Aids for Agriculture and the Preparation of Refined Fragrances
Stanley Roberts

New Medicines

Historically treatments for various diseases have originated from observations of the actions of various natural compounds on human beings and other animals. For example, the properties of morphine (from the opium poppy) indicated that it could find use as a pain-relief agent. Digitoxin (from the foxglove) came into use in the treatment of heart failure. Scientists use such observations as sensible starting points with the aim of providing pharmaceuticals that retain the beneficial properties of the natural product and eliminate any harmful (toxic) effects.

Plants can also contain materials that, although they are of no biological importance in themselves, can be transmogrified to give compounds of great value. Steroids available in some plants but unsuitable for the treatment of human diseases can be changed into compounds of use in the treatment of inflammation. Thus the development of useful ethical drugs based on the steroid structure and the morphine skeleton provides good starting points for discussion.

Steroids act as hormones in the mammalian system, being released from the adrenal cortex and other depots upon receipt of the appropriate stimulus. These naturally occurring compounds are important in the control of various biological processes, e.g. fertility and inflammation. Although high concentrations of some steroids are found in certain parts of the body (e.g. cholesterol (1) in gall stones) extraction of large quantities of such materials from animal organs and bones is generally impractical. (In fact the major supply of cholesterol is from wool grease, the mixture of compounds obtained on cleansing wool.)

In the 1950s supplies of the steroid nucleus (2) were urgently required as it became clear that application of some steroid derivatives would suppress the inflammation associated with arthritis and severe asthma (i.e. inflammation of the bronchial system). Unfortunately the size and complexity of the molecules meant that they were difficult to synthesize *de novo*.

A remarkable breakthrough came when it was found that large quantities of steroids were present in the Mexican yam *Dioscerea mexicana*. Extraction of the principal steroid (diosgenin) and chemical modification gave rise to progesterone and man-made derivatives that exhibit potent anti-inflammatory properties.

As the price of Mexican yam escalated a new source of steroids from plant materials was sought. The sisal plant from east Africa was found to contain large quantities of the steroid hecogenin and this substance was then used by organic chemists as a

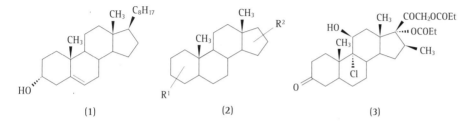

(1)   (2)   (3)

good source of material to prepare anti-flammatory halogeno-steroids such as Betnovate™ and Becotide™ (3).

Research in various laboratories (for example Organon in the Netherlands) focused on the role of oestrogens (such as **4**) in the control of female fertility and the oestrus cycle. Various types of steroids were prepared in chemical laboratories and, after painstaking work, a combination of steroids such as norethynodrel (**5**) and mestranol (**6**) was used in the pill and later in the mini-pill. It is estimated that in 1990 about 100 million women worldwide used steroid contraceptives to control their fertility.

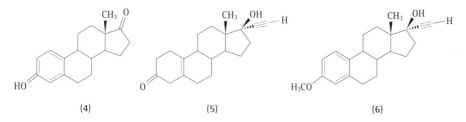

(4)   (5)   (6)

It should be hastily added that not all developments of biologically active natural products result in such a satisfactory conclusion. For instance, prostaglandins and thromboxanes are widely distributed (albeit in minute quantities) in the human body; related compounds are found in coral. The compounds in question, being relatively small and, for the most part, uncomplicated, can be made from simple materials by total synthesis. Prostaglandins and thromboxanes have been likened to biological policemen controlling a vast array of processes *in vivo*, including haemostasis. In the latter connection thromboxane A$_2$ (**7**) causes aggregation of human blood platelets and is pro-thrombotic; prostaglandin I$_2$ (**8**) inhibits this aggregation. In healthy individuals the thromboxane/prostaglandin levels are such that they maintain the appropriate control but when the arteries start to develop lesions, especially in old age, the prostaglandin I$_2$ levels decrease. To counter this effect it would be beneficial to supply the patient with a drug that would counter the effects of the thromboxane (a thromboxane antagonist).

Despite the implication of prostaglandins and thromboxane in a large number of important processes, despite detailed knowledge of the biochemistry of some of these processes (e.g. haemostasis) and despite the huge amounts of high-quality investigative research work carried out in the area, there is not a top-selling drug being marketed that is based on the prostaglandin template.

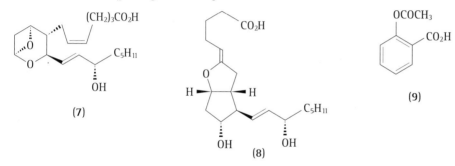

(7)   (8)   (9)

Indeed, the search for a thromboxane antagonist was downgraded when it was found, in clinical studies, that aspirin (9) acts as a preventative medicine for thrombosis by readjusting prostaglandin/thromboxane levels. A cursory comparison of the structures (7)–(9) will identify which compound is the easiest and cheapest to produce!

The effects of opium, containing morphine (10) (9–17%), and the closely related substance codeine (0.3–4%) have been known for many centuries. The former compound is a powerful analgesic (pain-killer) but it has the great disadvantage that it is highly addictive, a condition manifested as an overwhelming craving for the drug. Release from the addiction to morphine is a traumatic experience. Twentieth century research into the biological activity of compounds related to morphine quickly showed that the two key properties (i.e. the pain-killing ability and the addictive nature) were not intimately linked. Closely related substances such as codeine (11) and more distantly related substances in structural terms, such as pethidine (12), were developed from these structure–activity studies. More recently the mechanism by which morphine and analogues give relief from pain has been investigated and is now better understood.

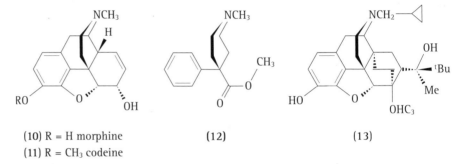

(10) R = H morphine
(11) R = CH$_3$ codeine

(12)

(13)

Morphine interacts with the so-called opioid receptors in the brain. Receptors are complex molecules of high relative molecular mass that are composed of condensed amino acids. They recognize particular substrates (in this case morphine) and, on binding to the small partner, the receptor acts to release a signal that can elicit a physiological effect (Figure 1).

The opioid receptors have been classified into various types, the δ-opioid receptor, the κ-opioid receptor *etc*. These receptors are similar to each other (in that they all 'recognize' morphine) but there are small differences among them. (In other words they are made up of slightly different arrangements of amino acids.) These small differences mean that morphine analogues do not all bind in the same way to this set of receptors and the responses to the various morphine analogues will be different. One company that has undertaken a lot of work in this area is Reckitt and Colman and, through an understanding of the interaction between morphine and its receptors, they were able to develop very powerful and relatively non-addictive agents as pain-killers and sedatives, such as buprenorphine (13), which became increasingly commonly used in hospitals, particularly to treat the pain of carcinoma and following surgery.

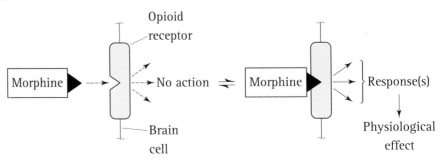

Figure 1
The effect of morphine interacting with one of its receptors.

As a final point, it is of interest to note that morphine is not a substance produced by mammals. Thus morphine is not the 'natural' partner for opioid receptors; in fact, the compounds in the human brain which interact with these receptors are the enkephalins. Comparison of the structures of buprenorphine and a typical enkephalin (14) would lead even a non-expert to deduce that there is little similarity. If the action of morphine had not been known for thousands of years and if its structure had not been elucidated last century then drugs such as buprenorphine would surely not have been designed from the structure of an enkephalin. Moreover, an understanding of the mechanism of action of a compound often allows selectivity of biological action to be obtained.

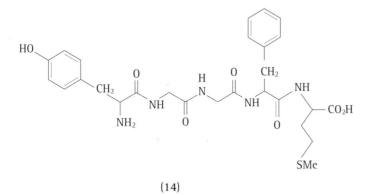

(14)

This final point will become a recurring theme throughout the next part of the text. Putting it another way, the chemist's design and synthesis of highly selective non-toxic medicines for the clinic often follow years of background work by (microbial) biochemists and/or pharmacologists. The development of the anti-asthma compound ventolin and the β-blocker propranolol further illustrates this typical phasing of the work of the different scientific disciplines in the process of discovery.

Asthma is a widespread problem that is particularly prevalent (10%) among children, who can die after a severe attack. Much research was undertaken concerning the biology of asthma, in particular the processes by which the bronchial smooth muscle controls the opening and closing of the airways leading to the area of the lungs in which exchange of oxygen and carbon dioxide occurs. Bronchoconstriction can lead to the asthmatic being unable to empty their lungs properly. The 'bronchial tubes' of those who suffer from asthma should be kept dilated to allow the maximum throughput of air.

The bronchial muscles react to adrenaline and noradrenaline, amongst other things. This is because the muscles have receptors for (nor)adrenaline. When these receptors combine with (nor)adrenaline the bronchial muscles relax, leading to opening of the airways. Is the solution, then, to give the asthma sufferer more adrenaline at the onset of the attack? The answer is no, because adrenaline has many more effects on the body, including causing an increase in heart rate, dilation of the pupils of the eye and constriction of peripheral blood vessels (and hence a consequent increase in blood pressure). Indeed, a burst of adrenaline into the system is the natural way of getting a person ready for 'flight or a fight' in response to an adverserial situation – more air to the lungs, an increase in delivery of oxygen to the muscles, more light to the eye, pallor, etc., etc.

Thus, since *selective* action was required, that is opening of the airways but not the multitude of other effects, a greater understanding of the adrenergic ((nor)adrenaline-utilizing) receptors was needed. This understanding came initially from the work of

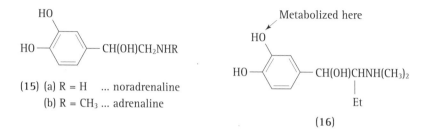

(15) (a) R = H   ... noradrenaline
(b) R = CH₃ ... adrenaline

(16)

Alquist, who classified adrenergic receptors into α-receptors and β-receptors. Alquist found that various analogues of noradrenaline (15) could differentiate among the various tissues and concluded that the receptors were therefore different. Bronchial muscle and heart muscle are similar in that they both possess β-receptors; action of (nor)adrenaline on bronchial β-receptors causes dilation of the airways, whereas the action of (nor)adrenaline on heart muscle increases the rate and force of contraction. Thus there was still a need to differentiate between the bronchial and heart receptors. Following the work of Lands, it was found that the two receptors could be differentiated by the compound isoetharine (16), which is closely related in structure to (nor)adrenaline. The compound was more effective on bronchial receptors (β_2-adrenergic receptors) than it was on heart-muscle receptors (β_1-adrenergic receptors).

Thus compound (16) possessed the selectivity required of an anti-asthma agent. Unfortunately the compound is quickly metabolized in the body as indicated (see formula (16)) to produce a compound that does not activate the β_2-receptors. Scientists in the Allen and Hanburys (now part of Glaxo) laboratories studied this situation carefully. It was obviously necessary to retain the selective action of (16) on β_2-receptors while modifying the position that is prone to metabolization. As a result of the analysis, compound (17) was synthesized and tested. This compound proved to be metabolically stable and highly selective towards the β_2-receptor system (several hundred times more effective than on the β_1-receptor) and the compound was developed to provide the anti-asthmatic drug called salbutamol (Ventolin®) (17). This drug is most often used as an aerosol and was first marketed in 1969. It is now sold worldwide and is the most widely prescribed anti-asthmatic in the world.

Asthma is an inflammatory disease and, in the worst of cases, the patient suffers hypersecretion of mucus, which can lead to complete plugging with mucus of the bronchioles. In this situation the inflammatory has to be cleared up first using an anti-inflammatory steroid such as Becotide® (3) before treatment with the β_2-stimulant begins.

The studies on the classification of adrenergic receptors led to the development of a second major sort of medication. As outlined above, β_1-adrenergic receptors are found on the heart muscle and on relatively few other tissues. Activation of β_1-receptors causes an increase in the rate and force of contraction of the heart muscle. Following the work of Lands, it was realized that blocking the recognition site for (nor)adrenaline on the β_1-receptor could have important implications in the control of some forms of heart disease (Figure 2). In this case it was necessary for the chemists to design and prepare compounds that would act as antagonists to (nor)adrenaline on β_1-receptors

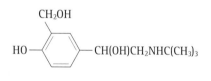

(17)

Figure 2
The mode of action of a β-blocker.

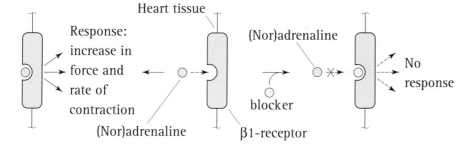

but would not interact with β$_2$-receptors. Thus the compound (a β-blocker) should selectively block β$_1$-receptors. The hunt was on in the ICI laboratories in Cheshire. The compound would probably bear some semblance of similarity to (nor)adrenaline since this was known to fit the receptor. However, the required compound had to bind to the receptor more strongly than (nor)adrenaline (in order to exclude the natural stimulator) and elicit no response. In this way, access to the active site of the receptor by the neurotransmitter (nor)adrenaline would be denied and the activity of the heart reduced. Success was achieved upon the synthesis of the adrenaline analogue (**18**). This proved to be an effective block to (nor)adrenaline and of value in the treatment of hypertension. The compound was given the name propranolol and it and more selective congenors such as atenolol still command a prominent position in the treatment of high blood pressure (hypertension).

An apocryphal story suggests that compound (**18**) was not the primary target compound for the ICI workers. They had planned to append the side chain to the 2 position of the bicyclic ring system. However, the relevant starting material was not available in the laboratory, but 1-naphthol (**19**), which leads in a straightforward manner to propranolol, was in stock, so compound (**18**) was made instead and as a consequence became the parent of a major class of drugs. When, later, the 2-substituted isomer of propranolol was made, it proved to be much less active as a β-blocker. Perhaps if 2-naphthol had been in the stockroom the discovery of the β-blockers may have taken longer.

(18) propranolol

(19)

Sir James Black was a driving force behind the discovery of the important anti-ulcer agent Tagamet™. The search for a drug in this area was stimulated by the difficulty of treating stomach and duodenal ulcers. Such ulcers are caused by the erosion of the membrane lining the stomach and duodenum, exposing underlying layers of tissue. The situation is aggravated by the release of hydrochloric acid in the stomach (this acid is released to help to digest food). As the meal-time approaches, the acid level in the stomach rises (as you think about food now, the acid may well be pouring out of parietal cells in your stomach). The secretion of acid was found to be under the control of a variety of agents (Figure 3), including acetylcholine, gastrin and histamine. Efforts to find the primary control agent were made under the assumption that blocking the action of the principal effector should decrease the secretion of gastric acid, thus allowing the ulcer time to heal. If it were successful, the new treatment would obviate the need for surgery to heal the wound.

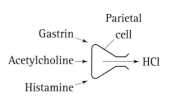

Figure 3
Agents causing the release of HCl into the stomach.

Initial work concentrated on blocking the effects of acetylcholine. This *can* be achieved and release of HCl *can* be reduced; however, acetylcholine is a neurotransmitter that is utilized widely in the mammalian system. Treatment with an acetylcholine antagonist led not only to a decrease in secretion of acid but also to dryness of the mouth, blurred vision and retention of urine. It was concluded by the investigators, probably correctly, that any candidate compound that left a prospective patient parched of mouth, half-blind and with a bladder twice the normal size was not likely to be a successful drug substance!

A different approach concentrated on the action of gastrin. This seemed to be a pretty good bet because gastrin, a polypeptide, is found only in the stomach and hence a selective inhibition of gastrin would have no effect on the rest of the body. In due course gastrin blockers were found but they were not as effective at lowering secretion of HCl as had been hoped. Perhaps gastrin had a more profound influence in man's earlier times but is no longer important.

This left histamine as a prime contender. Once again an understanding of the sub-classification of receptors led to the discovery of a drug. Histamine (20) has more than one receptor in the body but the important receptors on the parietal cells in the stomach are designated H_2 receptors. To reduce the secretion of acid it was necessary to design a compound that could selectively block the action of histamine at H_2 receptors. Histamine was taken as the template since this was known to fit into the active site. The heterocyclic ring was retained while the substitution pattern on the periphery was changed. The long side chain evident in compound (21) was eventually appended and this compound was found to be a potent antagonist of histamine at H_2 receptors. This compound was developed and introduced into medical practice as Tagamet™. The time scale of the whole work carried out in the Smith–Kline–French (now SmithKline Beecham) laboratories in Harlow is interesting and is outlined in Table 1.

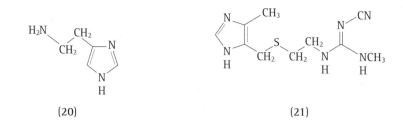

(20)   (21)

The data in Table 1 emphasize how long it takes for a new compound to reach the market as a drug. The intervening toxicity studies and clinical trials are expensive and time consuming. Tagamet has now been joined by a second compound that also commands large sales as an anti-ulcer agent that works by blocking H_2 receptors. Compound (22) is Zantac®, a Glaxo product that notably has a dimethylaminomethyl-furanyl unit replacing the imidazole moiety of histamine (20) and Tagamet (21). Zantac and Tagamet are two of the world's best-selling drugs (Table 2).

Enzymes are similar to receptors in some ways and different in others. Both are naturally occurring macromolecules made up of a particular and specific sequence of condensed amino acids. Both recognize partner, with a lower relative molecular

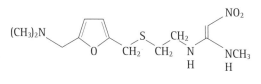

(22)

Table 1 Time scale of events leading to the launch of Tagamet

1964	Start of the project
1968	Discovery of the first lead compound
1972	Discovery of Tagamet
1976	Launch in the UK
1977	Launch in the USA

Table 2 The world's best-selling drugs in 1998

Rank Drug	Category	Company	Sales (10⁶£)
1 Zantac (ranitidine)	H_2 antagonist	Glaxo	2150 A
2 Adalet/procardia (nifedipine)	Ca antagonist	Bayer/Takeda/Pfizer	1095 E
3 Vasotec (enalapril)	ACE inhibitor	Merck & Co	1055 A
4 Capoten (captopril)	ACE inhibitor	BMS	1030 A
5 Mevacor (lovastatin)	Hypolipaemic	Merck & Co	1000 E
6 Carizem/Herbesser (diltiazem)	Ca antagonist	MMD/Tanabe/other	780 E
7 Epogen/Procrit	Erythropoietin	Amgen/J&J/other	780 E
8 Losec (omeprazole)	Anti-ulcer proton-pump inhibitor	Astra	780 A
9 Ventolin/Proventil (salbutamol)	Bronchodilator	Glaxo/Shering-Plough	755 A
10 Ceclor (cefaclor)	Oral/cephalosporin	Lilly/Shionogi	750 E
11 Voltaren (dichlofenac)	NSAID	Ciba-Geigy	750 E
12 Omnipaque (iohexol)	Contrast agent	Nycomed/Sterling	710 A
13 Zovirax (aciclovir)	Antiviral	Wellcome	690 E
14 Tagamet (cimetidine)	H_2 antagonist	SB	670 A
15 Prozac (fluoxetine)	Antidepressant	Lilly	670 E
16 Augmentin (amoxicillin plus clavulanate)	Antibiotic	SB	645 A
17 Tenormin (atenolol)	β-blocker	Zeneca	625 E
18 Naprosyn (naproxen)	NSAID	Syntex	595 E
19 Rocephin (ceftriaxone)	Inj. cephalosporin	Roche	540 E
20 Sandimmun (cyclosporin)	Immuno-suppressant	Sandoz	515 A

Note:
A, Actual; E, Estimate. ACE, angiotensin-converting enzyme; NSAID, non-steroidal anti-inflammatory drug.

mass (or a small family of compounds) through having an active site of a complementary nature to the partner. Interaction of a receptor and its partner leads to a response: the partner leaves the active site unchanged. Interaction of an enzyme and its partner leads to a chemical reaction leaving the enzyme unchanged but the partner modified in some way. The enzyme catalyses a chemical reaction involving the partner (Figure 4).

For an enzyme inhibitor to be an effective medicine, the mammalian enzyme must be over-working in the diseased state *or* the enzyme must be one that is peculiar to a pathogenic organism, i.e. a bacterium, fungus or virus. In the latter case inhibition of the microbial enzyme should have no effect on the host, i.e. the patient.

Figure 4
An enzyme catalysing a chemical reaction involving the partner.

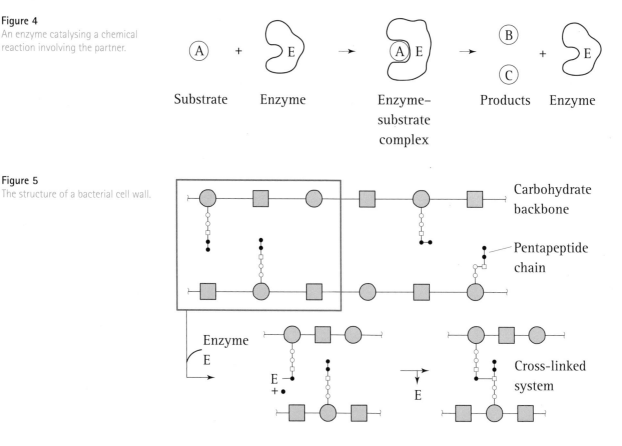

Substrate · Enzyme · Enzyme–substrate complex · Products · Enzyme

Figure 5
The structure of a bacterial cell wall.

Carbohydrate backbone

Pentapeptide chain

Enzyme E

Cross-linked system

Some of the best-known medicines are effective because they act as enzyme inhibitors. Take the penicillins. Penicillins act as anti-microbial agents by interfering with the synthesis of bacterial cell walls. Bacterial cell walls are much more substantial and sturdier than mammalian cell walls, allowing them to survive changes of temperature and osmotic shock. The enzymes that are involved as catalysts in the production of bacterial cell walls have no counterparts in the mammalian system. Figure 5 shows, diagrammatically, the cross-linking of the cell wall of a bacterium. The pentapeptide unit is attacked by the *exo*-peptidase enzyme to give an activated complex which is attacked by one of the constituent parts of an adjacent chain. The enzyme, having performed its job as a catalyst, is regenerated ready to perform the next cross-linking process. The effect of the penicillin is to inhibit the cross-linking enzyme, thus weakening the bacterial cell wall. The effect can be quite dramatic (Figure 6), the osmotic pressure within the bacterial cell causing rupture of the cell wall and damaging loss of cell contents.

The penicillins that were originally obtained, for example penicillin-G (**23**), display only weak antibacterial activity. It was the job of the chemist to modify the natural product (**23**) to produce a more potent antibacterial agent. Retaining the ring systems and changing the side chain emanating from the four-membered β-lactam ring was found to be a good way of increasing the potency of the antibacterials and increasing the range of susceptible bacteria. Thus penicillin-G was degraded to give the parent compound (**24**). Appending moieties such as the one in compound (**25**) produced very active compounds. The 'semi-synthetic' penicillin (**25**) is called amoxicillin and is a medicine that is widely used for the treatment of gonorrhoea and infections of the respiratory and urinary tracts.

Of course, neither the mechanism of action of penicillins nor the concept of making

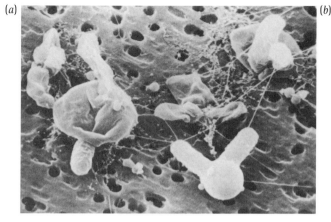
(a)

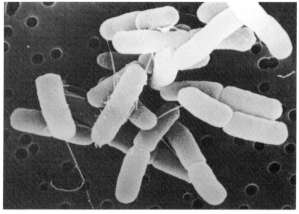
(b)

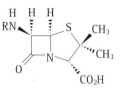

(23) R = PHCH₂CO
(24) R = H

(25) R = HO —⟨ ⟩— CH(NH₂)CO

Figure 6
A natural bacterium (*a*) before and (*b*) after treatment with a β-lactam. (From *Medicinal Chemistry: The Role of Organic Chemistry in Drug Research*, edited by B. J. Price and S. M. Roberts (Academic Press, New York, 1989) pp. 25–6.)

semi-synthetic materials was obvious to Alexander Fleming when he first observed the effect of penicillins in 1928. This crucial observation owed a lot to chance. Fleming had left bacteria to grow on plates near open windows at St Mary's Hospital in London. A small quantity of *Penicillium fungus* was deposited on one of the plates and, in the area around the fungal material, growth of the bacterial lawn was inhibited (diagrammatically shown in Figure 7). Fleming realized that this was a very important observation in that the fungus was exuding an anti-bacterial substance.

It took much effort and many years to isolate the fungal metabolite and to identify its structure. It was only an impetus given by World War II that saw production of material on any scale (bacterial infection of wounds was a dire problem at the battle front). The first treatments of septicaemia by the crude penicillins must have seemed like miracle cures. Patients that would otherwise have died of blood poisoning were saved by the crude drug; an understanding of its action and optimization of the biological activity came later.

Obviously the production of penicillin by the fungus and the biosynthesis of morphine by the poppy are related. Both compounds are complex and are produced by the organism as secondary metabolites. Both compounds have important biological activities. The concept of finding useful materials in the fermentation broths of bacteria and fungi has been taken up and refined by most pharmaceutical companies. Thus many different types of micro-organism have been grown and the broths have been taken at different stages of the growth cycle and the constituents of the broths tested for inhibition against a battery of enzymes. The enzymes are chosen carefully, on the basis of evidence that inhibition of the enzyme would have a clinically beneficial effect. Testing up to a thousand broths a week against a battery of ten enzymes is possible using sophisticated robotics and highly refined analysis techniques.

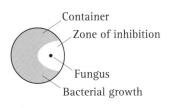

Figure 7
The anti-bacterial action of a fungus.

This 'random screening' of secondary metabolites has certainly paid dividends. For example, a very important hypocholestaemic (cholesterol-level-lowering agent) was discovered using this strategy.

Cholesterol (1) has long been associated with the 'furring' of the arteries in the cardiovascular system which leads to thrombosis. Certainly patients with a certain genetic disorder leading to extraordinarily high levels of cholesterol in the blood stream frequently die of heart attacks at a very early age. There is a body of opinion that suggests that the lowering of cholesterol levels in most individuals is beneficial. Cholesterol is produced in the body from simple molecules by a sequence of reactions catalysed by a portfolio of enzymes (Figure 8). One of these enzymes reduces HMG-coenzyme A; as is customary the enzyme's name reflects its function, i.e. HMG-CoA reductase. Inhibition of this enzyme leads to a down-regulation of the production of cholesterol. Random screening of broths by scientists at Merck in New Jersey led to the discovery of mevinolin (26), one of a group of compounds produced by *Streptomycetes*. A compound of this type has successfully been marketed as the hypocholestaemic agent called lovastatin and is already in the 'top ten' of best-selling drugs (Table 2).

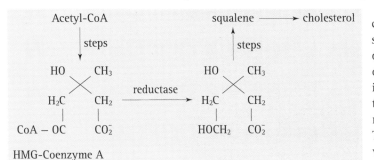

HMG-Coenzyme A

Figure 8
Some detail of the biosynthesis of cholesterol.

Another interesting activity of medicinal chemists can be illustrated using the lovastatin story. Mevinolin (26) is a natural product produced as a secondary metabolite by a microorganism; it is fortuitous that it acts as an inhibitor of HMG-CoA. In such cases it is logical to ask just how much of the natural product is necessary for potent inhibition of the enzyme. The natural product can be simplified by various degradative transformations. In complementary work simpler molecules, which represent part of the whole structure, were made from readily available starting materials. In this way it was found that the monocyclic ring system was the most important feature of the molecule and that synthetic compounds such as (27) act to inhibit HMG-CoA and reduce cholesterol levels *in vivo*.

It is hoped that it is becoming clear to the reader that, before a chemist can begin to design and synthesize a drug, a great deal of biological data has to be gathered. Key enzymes have to be identified and characterized, important receptors must be classified and mechanism(s) of action have to be understood. As the biological sciences become more sophisticated and analytical techniques improve, so the process becomes less 'hit and miss'. The advances in biotechnology have made a major impact. The ability to make substantial quantities of pure protein (e.g. an enzyme), by cloning and expressing the relevant gene, gives material for structural studies and for screening for inhibition of enzymes. Site-directed mutagenesis has made it much easier to decide which amino-acid residues are important in the active site of an enzyme. Thus, if one amino

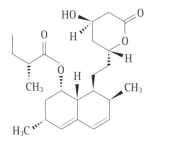

(26)

Ar = relatively simple aromatic moiety

(27)

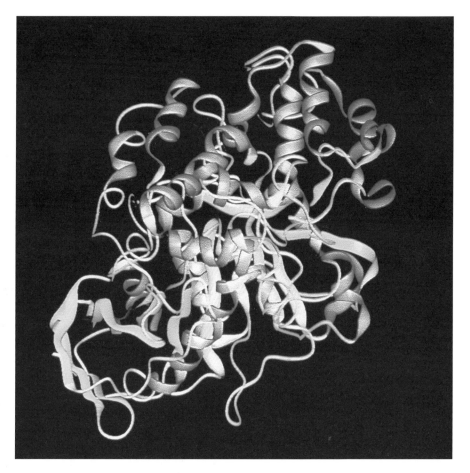

Figure 9
The crystal structure of *Candida* sp. lipase.

acid is removed from or substituted in an enzyme and the resultant protein is inactive as a catalyst, then it can be concluded that the amino acid in question must play a key role in the function of the enzyme.

Various analytical techniques have made more and more impact in medicinal chemistry, none more so than protein crystallography. If crystalline enzymes that inhibit the same catalytic activity are available from, say, a pathogen and man, then the subtle differences in structure can be observed and, where differences are readily apparent, small molecules that might interact and interfere with the foreign protein but not with the host enzyme can be designed and prepared. Figure 9 shows the structure of a fungal lipase. Detailed examination of such proteins from pathogenic organisms, in the presence of the appropriate substrates, using computer graphics to visualize the macromolecules and to allow rotation to obtain various views of the active site, allows one to design molecules that will fit into the active site of the foreign enzyme. These molecules can then inactivate the protein by a Trojan-horse strategy, i.e. using a mimic of the natural substrate of the enzyme that, on bonding to the active site, will reveal a reactive unit (a warhead) to interfere with the active site.

Thus modern approaches to drug design often incorporate programmes of protein-structure determination (by X-ray crystallography) and modelling (using computer graphics). One of the first success stories in modern drug discovery involving these techniques was the development of inhibitors of the enzyme angiotensin-converting enzyme (ACE) which (as the name might suggest) converts angiotensin I into angiotensin II. Angiotensin I is biologically benign whereas angiotensin II elevates blood pressure by a variety of mechanisms. Inhibiting ACE reduces hypertension by moderating the level of angiotensin II.

It was known that ACE was inhibited by components of the venom of the Brazilian viper *Bothrops jararaca*. However, vipers, just like leeches, are not widely used in modern medicine and the peptidic components of the venom are too unstable for use in the clinic. Nevertheless, using the structures of the active peptides as templates, gathering knowledge of ACE and understanding of its relationship to other similar crystalline enzymes allowed progress to be made. The breakthrough came by incorporating a sulfur atom into the peptide-like ACE inhibitors in the knowledge that a sulfur-loving zinc atom was present in the enzyme's active site. The affinity of zinc for sulfur means that compounds such as captopril stick avidly to the surface of ACE in, or close to, the active site such that angiotensin I is excluded and does not undergo conversion. Captopril (**28**), developed in the Squibb laboratories, and its congeners make up an important class of drugs for controlling the commonly encountered forms of hypertension.

As biochemical systems become better understood, so better medicines can be designed. There is much work still to be done in the area of anti-viral chemotherapy and in delineating new treatments for the various forms of cancer.

It is not difficult to understand why these diseases cause so many problems. Most viruses, unlike bacteria, cannot reproduce on their own; instead they infect a host (e.g. a human cell) and use the machinery of that cell to replicate. Thus the human immunodeficiency virus (HIV) seeks and infects cells of the human immune system. Therein the virus replicates, causing damage to the host cell with the result that the affected person has a weakened immune response and often dies of an opportunistic infection or a breakaway cancer (acquired immune deficiency syndrome – AIDS).

Much effort has been expended on finding out which biochemical processes are crucial for HIV. The enzymes that manufacture DNA for the virus are certainly different from mammalian systems, yet the overall process is so similar to that employed by the host that a really good chemotherapy for AIDS has not been found. One of the few compounds used in the clinic is azidothymidine (AZT) (**29**). The compound does marginally increase the life-expectancy of the HIV-infected person but it also produces side effects such as suppression of bone marrow, which results in under-production of blood cells. Interfering with production of virally coded RNA and DNA and the transcription of viral DNA into protein is not easy.

This is the reason why cancer is still a frightening problem. In this situation the medicines are known as 'cytotoxic drugs' because in almost all cases they function by causing cell death. The first chemotherapeutic agent used clinically was nitrogen mustard. This compound was derived from the mustard gas used in World War I, during which, among other unpleasant effects, it inhibited the production of blood cells in the bone marrow of its victims. It was reasoned that these agents may be of value in the treatment of leukaemia, a malignant disorder of the blood that results in over-production of white blood cells. Nitrogen mustard (**30**) appeared to kill cells by attach-

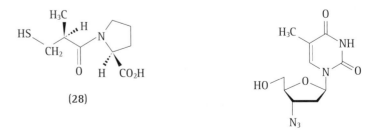

(28)

(29)

ment of each of the chlorine-containing side chains to molecules located on parallel strands of DNA. In chemical terms this is known as an alkylation reaction, whence the name 'alkylating agents' for the class of medicines with this mechanism of action. The alkylation of parallel DNA strands produces a cross-link that destabilizes DNA and results in breakages of DNA strands that are lethal to the cell. Unfortunately, nitrogen mustard exhibited no selectivity between normal cells and cancer cells so that its toxicity to healthy tissues was exceptionally high.

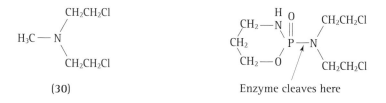

(30)　　　　　　　　　　　　(31)

The search for alkylating agents with selective activity against cancer cells led, in the 1960s, to the synthesis of cyclophosphamide (31), one of the world's first 'designer drugs'. It was thought (erroneously as it happened) that cancer cells contain large quantities of the enzyme phosphoramidase. Cyclophosphamide comprised a nitrogen-mustard molecule that was chemically attached to an inactive carrier molecule. The rationale behind the design of cyclophosphamide was that, on entering cancer cells, the bond between the nitrogen mustard and the carrier molecule would be cleaved by phosphoramidase to release the nitrogen mustard which would destroy the cancer cells. It was hoped that, in cells of normal tissues, which were thought to contain little phosphoramidase enzyme, the active nitrogen mustard would not be liberated, so that the drug would be selectively toxic to cancer cells. Unfortunately, it was found that all cells contain phosphoramidase and, although cyclophosphamide required cleavage to make it active (in other words it was a pro-drug), it was found that cleavage occurred in the liver. Production of the active molecule in the liver meant that cyclophosphamide was equally toxic to normal and tumour tissues. Nevertheless, cyclophosphamide and a more recently produced derivative, ifosfamide, are valuable anti-cancer drugs that are used in the treatment of leukaemic tumours, lymphomas and solid tumours of the breast, lung, cervix and ovary. The main toxicity of cyclophosphamide and ifosfamide arises from depression of the production of bone marrow.

The platinum-containing drugs exhibit a similar mechanism of action to that of the alkylating agents. The first platinum drug, cisplatin (32), was marketed in the late 1970s, when it made an immediate impact on the treatment of ovarian cancer and, in particular, testicular cancer. In combination with radiotherapy and other drugs, the introduction of cisplatin raised the cure rate for testicular cancer to over 80%. The main toxicity of the platinum drugs includes depression of the production of bone marrow, damage to the kidneys and severe nausea and vomiting. The latter form of toxicity has become much less of a problem recently with the introduction of the powerful anti-emetic drugs ondansetron and granisitron.

The 'anti-metabolite' class of anti-cancer agents works by a totally different mechanism of action. These drugs are designed to imitate closely the chemical structures of vital precursor molecules that are the 'building blocks' used by cells to make the genetic

(32)

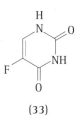

(33)

material, DNA. The anti-metabolites are therefore readily taken up by cells and incorporated into biochemical pathways. However, subtle differences in the anti-metabolite molecule cause it to bind tightly to enzymes involved in assembling the 'building blocks' into the final genetic material. The bound enzymes are inactivated and so production of DNA is shut down, resulting in prevention of the division of cells and (through other complex mechanisms) the death of cells. An example of an anti-metabolite drug is 5-fluorouracil (33), an analogue of uracil. 5-Fluorouracil is used to treat a variety of cancers. The principal toxicities of such an anti-metabolite agent include depression of production of bone marrow and irritation to the gastro-intestinal tract.

A third important category of anti-cancer drugs comprises the 'anti-tumour antibiotics'. This group of drugs (which should not be confused with anti-microbial antibiotics such as pencillin) includes the relatively complex molecules doxorubicin, epirubicin and mitozantrone. Two main mechanisms of action have been proposed for the anti-tumour antibiotic drugs. The first mechanism involves the specific inhibition of an enzyme known as topoisomerase II, the enzyme responsible for momentarily breaking, and then re-joining, the strands of DNA so that the DNA molecule can unwind to facilitate replication of the genetic template and cell division. Inhibition of topoisomerase II by the anti-tumour antibiotic drugs prevents the re-joining of broken DNA strands, thus preventing replication of cells and causing cell death. The second mechanism is dependent upon the flat, planar structure of the antibiotic molecule which allows it to fit between or intercalate the DNA base-pair molecules that bound the two strands of DNA helix together. This 'foreign implant' in the DNA structure again inhibits the process of replication.

The anti-tumour antibiotics are used to treat a wide range of leukaemias and cancers, usually in combination with other anti-cancer drugs. Apart from depression of production of bone marrow and nausea and vomiting, the anti-tumour antibiotics exhibit a unique toxicity to the heart, which limits the total amount of drug which may be administered. The cardiotoxicity is thought to be associated with the glycoside (sugar-like) portion of the drug molecule which closely resembles the molecular structure of cardiac glycosides used to treat heart failure (e.g. digoxin).

In addition to the three major categories of anti-tumour agent there are several other cytotoxic drugs that have alternative mechanisms of action. The most important of these is the group of drugs known as the 'vinca alkaloids', which are derived from *vinca rosacea*, the periwinkle. Among drugs in this category, vincristine, vinblastine and vindesine are used clinically to treat a wide range of solid tumour cancers and leukaemias.

Taxol, another natural-product anti-cancer drug, is derived from the bark of the Pacific Yew tree. Taxol and taxotere (a semi-synthetic taxol (cf. the story of penicillin)) have similar actions to those of the vinca alkaloids. Taxol has recently received a product licence for use in the treatment of resistant ovarian tumours and clinical studies to evaluate the activity of this drug for treating breast cancer have been started. In addition to depression of production of marrow, one of the main problems associated with Taxol is that it is only very slightly soluble in water, making it difficult to formulate as an injection. The commercial formulation contains, of necessity, a solubilizing agent that can cause hypersensitivity reactions when it is administered to humans. It is therefore necessary to administer steroids and antihistamines to patients before Taxol treatment in order to reduce the incidence and severity of this type of adverse reaction.

The scarcity of Taxol coupled with the desire for the preservation of yew trees has made this molecule an important target for synthetic organic chemists. Despite the molecular complexity of Taxol (34), total syntheses of the material were reported early in 1994. This feat by synthetic organic chemists (Robert Holton at Florida State University and K. C. Nicolaou (see Chapter 8) at the Scripps Institute, San Diego) allows a more fundamental study of structure–activity relationships to be undertaken.

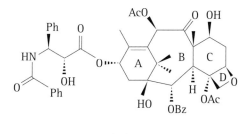

(34)

The cytokine group of drugs, which includes interferon alpha and interleukin II, is an example of the new types of medicine produced by biotechnology. The cytokines are polypeptides. Cytokines have no direct cytotoxic action on cancer cells but instead activate cells of a special type present in the human immune system known as 'natural killer' cells. These recognize and attach to cancer cells, allowing the immune system to identify and destroy the 'invading' or 'foreign' cancer cells. Small quantities of cytokines are produced naturally by the body in response to a stimulus such as an infection or a malignant disease. However, it was thought that, by administering additional cytokines from an external source, it would be possible to augment the immune response against cancer cells. With the exception of certain types of leukaemia, the cytotoxics have been disappointing in clinical practice when they have been used as single agents. However, in combination with other drugs the cytokines have been useful in the treatment of malignant melanoma and renal-cell carcinoma tumours that are normally very unresponsive to drug treatment.

The above discussion highlights the range of different cancers that can appear and the problems associated with the treatment. Indeed, there are many reasons why chemotherapy may fail to cure cancers and leukaemias completely. Sometimes the treatment is too toxic for the host tissue (the patient) or alternatively the drug does not penetrate into the core of the tumour. One of the most common reasons for failure of treatment however, is the development by tumour cells of resistance to a drug. Of particular concern is the development of multiple drug resistance, whereby the tumour becomes resistant not only to the drug or drugs that have been used to treat it but also to other quite different drugs, to which the tumour has not previously been exposed. With the lack of an effective treatment for many types of cancer and the emergence of drug-resistant tumour cells, it is clear that the research efforts directed towards discovering anti-cancer drugs must be maintained and that new drug targets in cancer cells must be identified and the appropriate cancer chemotherapeutics designed and synthesized by chemists.

The Chemistry of Fragrances

The production of fragrances often involves the chemistry of materials of small relative molecular masses such as terpenes. Terpenes occur naturally as highly complex mixtures in a number of fragrant plants, a typical example of which is lavender. The oil obtained from this plant is rich in linalool and linalyl ethanoate, which account for its characteristic fragrance. The relative amounts of the oil constituents are not uniform and vary among different varieties of the same plant. For example, English lavender has an odour distinct from those of other lavenders, owing to its low ester content.

The essential oils were originally separated from the plant tissue by gentle heating and collection of the crude distillates. Today, several isolation techniques are employed. Lavandin oil is obtained by steam distillation from hybrid lavenders – a process that is economical and allows production of large quantities of the oil.

Jasmine oil is isolated from mature jasmine flowers by extraction with petroleum ether. Removal of the solvent leaves a waxy residue called a concrete, which on washing with warm alcohol, gives jasmine absolute, a concentrated mixture of the odorous constituents.

A quite different process is utilized for flowers that continue to give off perfume oil long after harvesting. This process is called enfleurage. Since it is costly it is little used nowadays, solvent extraction being less expensive. However, tuberose absolute, obtained from a strain of polyanthus, is isolated in this way. During the process, the flowers are laid on a layer of fat, which absorbs the fragrant oils as they are exuded. Once the fat has become saturated, it is treated as a concrete and the oils are extracted from it with alcohol.

Citrus oils are taken from lemon, lime, orange and bergamot fruit, usually from the peel, from which they are separated by expression. This technique physically crushes the walls of the oil-bearing cells, thereby releasing the fragrant materials.

Since the volatile components of the essential oils were beyond early techniques of separation, the distillates were used as mixtures. For example, rose oil is compounded of geraniol, citronellol, phenylethanol and rose oxide.

Early perfumers were faced with the problem of the variance of olfactory homogeneity. The advent of chemical analysis by gas chromatography, together with improved, efficient distillation techniques, overcame the problems of separating the oil's constituents. Perfumes manufactured today can be compounded to exact proportions, if necessary.

The demand for perfumery chemicals is high, since most household products – soaps, detergents, polishes, disinfectants, plastics, rubbers – are artificially scented, either to mask malodorous base ingredients, or just to make them more attractive to the customer. Disadvantages of using essential oils include the relatively high cost – for instance, 1 tonne of jasmine blossom, which comprises 6–8 million hand-picked flowers, yields only 1.4 kg of jasmine absolute – and variability of supply and quality, which are dependent on climatic conditions. In view of this, synthesis and modification of terpenes are of great importance to the perfumery industry, since synthetic terpenes are cheaper, more easily available and uniform in odour.

Two main routes to monoterpenoid perfumery materials exist at present. The first involves non-terpenoid starting materials and relies on building the skeleton from smaller units.

The second route leads to perfumery materials derived from pinene. Pinene itself is isolated from turpentine, whose production is about ten times greater than that of all other essential oils combined. The composition of turpentine from different parts of the world varies considerably; however, α-pinene (**35**) and β-pinene (**36**) constitute the major part of the oil. These are extensively utilized in the preparation of fragrance chemicals. For example, α-pinene can be hydrated, under acidic conditions, to furnish

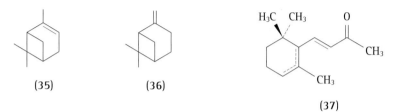

(35)   (36)   (37)

an important perfumery compound called α-terpineol, whilst similar conversions of oxygenated pinene derivatives have led to synthetic menthols, carveols and carvone.

Hydrogenation affords pinane, which can be readily oxidized, via the corresponding hydroperoxide, to pinanol. Pyrolysis of this yields linalool, a key terpene alcohol, which is widely employed because of its floral fragrance. A rearrangement of the hydroxyl group produces nerol–geraniol mixtures, which have pleasing rosacious odours and also find many applications in the fragrance industry. Oxidation of these alcohols leads to a mixture of E- and Z-citrals, which are intermediates for the synthesis of ionone (37) and its derivatives.

Hydrogenation of the conjugated double bond of citral gives citronellal, which, albeit not a perfumery chemical in itself, is a precursor for menthol, a flavour chemical. Citronellol, a versatile fragrance chemical, with a mild floral odour, is the reduction product of citronellal, which completes the circuit.

Further Reading

1. Case Histories: C. R. Ganellin and S. M. Roberts (editors), *The Role of Organic Chemistry in Drug Research* (Academic Press, London, 1992).
2. Issues of Scale-up Chemistry: K. G. Gadamasetti, *Process Chemistry in the Pharmaceutical Industry* (Marcel Dekker, USA).
3. Medicinal Chemistry: T. Nogrady, *A Biochemical Approach*, 2nd edition (Oxford University Press, USA, 1998).
4. Medicinal Chemistry: F. D. King (editor), *Principles and Practice* (CRC Press Inc., USA, 1994).

Discovering Pesticides

Bhupinder P. S. Khambay and Richard H. Bromilow

Introduction

Ever since Man's first attempts at cultivation to secure adequate and reliable food supplies, he has had to suffer competition from pests, diseases and weeds. One of the main reasons for our failure to find a permanent solution to this problem lies in the ability of our biological competitors to adapt to almost any change in their environment.

At first, insects were the major source of agricultural problems (Figure 10) and thus methods for controlling them have been known for many centuries. However, over the past 200 years, a shift towards intensive single-crop farming practices has led to fungi and weeds becoming increasingly important problems (Figures 11 and 12). In 1993, of the total world-wide pesticide market (£15 billion), insecticides accounted for 30%, fungicides 18% and herbicides 46%, although the proportions vary with geographical location and climatic conditions. For example, whereas herbicides dominate the market in Europe, insecticides are the most important class in Africa.

Historically, an increase in the use of pesticides can be traced back to the mid-1800s, when demand on food production led to an increase in monocultural farming practices. Organic products, especially plant-based insecticides, were traditionally used but these were in limited supply and often rather ineffective; this led to the widespread use of inorganic compounds, which trend has been reversed only during the past 60 years. However, even though many effective synthetic organic pesticides have been introduced since around 1940, their repeated, and sometimes indiscriminate, use in the field has already led to the development of resistance (i.e. a decrease in susceptibility to the pesticide).

Some aspects of the design of pesticides parallel those of the design of drugs, though others differ, as summarized in Table 1. Drugs may be considered to fall into two classes. Those designed to affect physiological processes, mostly in human beings, usually act at a specific target site (e.g. an enzyme). In contrast, drugs controlling infections and also pesticides are aimed at a wide range of species and may operate at one or more sites; for pesticides in particular, the varied requirements for delivery for different pests (e.g. volatility for insecticides to control soil pests) can lead to each class spanning a wide range of physicochemical properties. Another factor to be considered for pesticides is their potential impact on the environment, including effects on non-target species and the fate of the pesticide. For example, a highly effective and selective

Figure 10

Damage to cabbages by caterpillars of the diamondback moth (*Plutella xylostella*): (*a*) untreated control and (*b*) treated with insecticide. (Photographs courtesy of Zeneca Limited.)

(*a*)

(*b*)

Figure 11
Competition with grapevines by Bermuda grass (*Cynodon dactylon*): (*a*) untreated and (*b*) sprayed with herbicide. (Photographs courtesy of Zeneca Limited.)

Figure 12
Competition with field crops by weeds: (*a*) poppies in spring rape and (*b*) charlock in spring barley. (Photographs courtesy of P. J. W. Lutman, IACR-Rothamsted)

pesticide may nonetheless fail to pass the registration process if its physicochemical properties and persistence are such that it is leached into ground water.

From a commercial point of view, drugs, being delivered directly to the organism (e.g. by injection), are used in small amounts and can command a high price in human healthcare (Table 1). In contrast, pesticides cannot be delivered precisely (e.g. they are delivered by crop spraying). Consequently, even modern highly active pesticides need to be used in relatively large quantities (10–50 g ha^{-1}), yet they must be produced cheaply relative to the value of the crop. Chemists involved in pesticide research today have to face the challenge of finding new pesticides that are both highly effective and with a potential market large enough to bear their high development costs. In addition, they must ensure that such pesticides satisfy all the safety criteria for mammalian toxicology and environmental behaviour.

A further consideration in the design of drugs and pesticides is the occurrence of resistance of the target organism. Resistance usually arises either from changes in susceptibility of the target site or from a decrease in availability of the compound at the site; the latter can arise from a restriction of movement of the compound into the organism or from an increase in breakdown *en route* due to an enhancement of enzymatic

Table 1 Contrasting aspects of the discovery and use of drugs and pesticides

Aspect	Drugs	Pesticides
Spectrum of activity	Narrow	Wide
Target	Specific receptors (in humans) and whole organisms (e.g. parasites)	Whole organisms
Delivery to target	Precise (e.g. injection)	Imprecise (e.g. spraying)
Hazard to environment	Not usually an issue	Important to minimize it
Requirements on stability	Metabolic	Photostability, chemical and metabolic
Major source of lead compounds	Natural products	Random
Screening methods	Emphasis on receptor assays	Traditionally whole-organism
Price	can be $>£100 \text{ g}^{-1}$	$<£2 \text{ g}^{-1}$

Table 2 Biochemical processes affected by major classes of pesticides

Process affected	Insecticides	Herbicides	Fungicides
Transmission of nerve signals	✓		
Hormonal processes	✓	✓	
Respiration/production of energy	✓	✓	✓
Biosynthesis of lipids/sterols		✓	✓
Synthesis of amino acids		✓	
Photosynthesis		✓	
Production of RNA			✓
Formation of microtubules		✓	✓

activity. In the past, industry has responded by introducing new classes of pesticides with different modes of action. However, given the ever-increasing cost of development, more emphasis is now being placed on resistance-management strategies.

In the development of pesticides one often tries to exploit biochemical processes (Table 2) that are absent from mammals (e.g. moulting of insects, photosynthesis in plants) and this is facilitated by the ready availability of test organisms for bio-assays. However, the evaluation of drugs for use on the main target species – Man – is possible only during the final stages of development; therefore, not surprisingly, folklore-based and, more recently, biochemically directed research has provided the major source of lead structures for drugs. In contrast, leads provided by random screening have been of most importance in pesticide research and consequently many pesticides have been commercialized even before their mode of action has been understood fully.

This chapter is not intended to be an exhaustive overview but aims to illustrate the process of discovery with the aid of selected examples. It does not detail the use of toxin-producing bacteria, e.g. *Bacillus thuringiensis* (commonly known as BT), genetically engineered organisms and living organisms (e.g. viruses and pathogenic fungi).

Insecticides

There are many similarities between the physiological make-up of vertebrates and that of invertebrates, so it is not surprising that some insecticides exhibit a degree of tox-

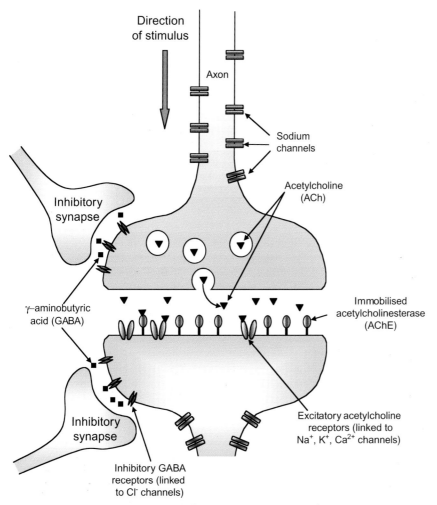

Figure 13
Schematic diagram of an excitatory nerve junction and its modulating neural connections.

icity towards non-target organisms (e.g. birds and mammals). Selectivity is achieved either by exploiting differences between the rates of detoxification or by targeting mechanisms that are absent from vertebrates.

Traditionally, plant extracts have provided the most important means of controlling insects. Pyrethrum and derris continue to be used today in selected applications such as indoors, although their lack of stability in the field and their limited availability impose severe constraints on their use. Consequently, inorganic compounds, such as copper acetarsenite (Paris Green) and lead arsenate, many of which are highly toxic, were introduced in the mid-1800s, but are little used today. In the late 1940s, DDT dominated the world market for insecticides. Currently organophosphorus compounds have assumed this position with, until recently, the remainder split between carbamates and pyrethroids. However, newer compounds such as the chloronicotinyls are now rapidly increasing in importance.

Disrupters of the Nervous System

Three of the major classes of established insecticides act on the central nervous system. They act at two distinct sites (Figure 13), either at the nerve junctions (synapses), so disrupting signal transmission, or on individual cells along the nerve membrane by distrupting the balance of ions within them (mainly modulated via sodium and chloride channels). In both cases, lack of coordination between the brain and the rest of the body leads to death.

The activity of DDT (**38**) was discovered by Müller in 1939 whilst he was employed by Ciba Geigy in Switzerland. This discovery, for which Müller received the Nobel Prize for Medicine and Physiology in 1948, marked one of the most significant events in the history of mankind. DDT brought untold benefits in all aspects of insect control, especially in agriculture and human welfare.

At the time, DDT was the most effective, easy-to-produce, broad-spectrum insecticide with low mammalian toxicity and adequate persistence for use in the field. It was extensively used to disinfest soldiers' clothing during World War II to prevent typhus and was responsible for the almost complete eradication of malaria from many countries. However, two of the properties of DDT that were originally seen as advantageous prompted eventual restrictions on its use. Firstly, the broad spectrum of activity extended to many beneficial insects. Secondly, its long persistence resulted in bioaccumulation throughout the food chain, which affected wildlife, especially birds. Adverse effects on human health are less clear and are a current subject of debate.

The discovery of DDT prompted intensive research which led to the discovery of non-aromatic polychlorinated cyclic insecticides such as hexachlorocyclohexane (HCH). Their mode of action, different from that of DDT which acts on the sodium channels, involves binding to post-synaptic γ-aminobutyric acid (GABA) receptors (Figure 13), and this blocks chloride channels. Their volatility being greater than that of DDT allows their use in a much wider range of applications, e.g. seed dressing to combat soil insects.

Of the nine possible isomers of HCH, only the γ-isomer, commonly known as lindane (**39**), is active. Of the fused polycyclic compounds, the most important groups contain either three or four fused five-membered rings; many compounds in this class, such as dieldrin and endrin, have now been withdrawn because of their high mammalian toxicity, extreme persistence and toxicity to beneficial organisms. However, less persistent derivatives, e.g. endosulfan (**40**), are still used.

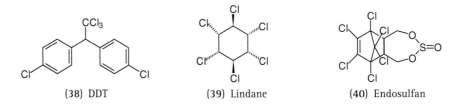

(38) DDT   (39) Lindane   (40) Endosulfan

Organophosphorus compounds (OPs) are the most important class of commercial insecticides and their discovery provides an excellent example of the application of chemistry to the investigation of the dependence of biological activity on structure. The highly toxic nerve gases such as sarin (**41**) and dimefox (**42**), developed during World War II, were recognized also to be powerful insecticides. Despite this apparently alarming starting point, studies on structure–activity relationships quickly led to the discovery of insecticides that had acceptable safety margins.

OPs act by mimicking acetylcholine (**43**), which is present both in vertebrates and in invertebrates and is the 'carrier' of nerve impulses across synapses (see Figure 13). Once the surge of acetylcholine has crossed the synapse to stimulate the receptor on the second nerve, the acetylcholine is hydrolysed by the enzyme acetylcholinesterase (AChE) so that it does not continue to cause stimulation. The first step in this reaction involves the formation of a complex between acetylcholine and a serine-OH residue at

(41) Sarin   (42) Dimefox   (43) Acetylcholine

the active site of the enzyme. This is followed by a two-stage hydrolysis to choline, ethanoate and the regenerated enzyme. OPs function by reacting irreversibly with the serine-OH residue of the AChE, a leaving group on the OP being displaced to form a stable phosphorylated enzyme whose active site is now effectively blocked. This leads to the accumulation of acetylcholine, causing disruption of nervous activity, with consequent loss of muscular coordination, convulsions and ultimately death. In the early 1940s, the general structure (44) for insecticidal activity was recognized. The effectiveness of OPs is primarily dependent on the efficiency of the leaving group Y, which in turn is influenced by the three other groups. R^1 and R^2 are generally alkoxy or alkylthio groups, CH_3O- and CH_3S- being preferred for lower mammalian toxicity. A high intrinsic activity can be achieved only when X is oxygen but these OPs (e.g. phosphates) exhibit high mammalian toxicity. However, this can be selectively lowered by replacing oxygen by sulfur such as in parathion (45) and chlorpyrifos (46). These compounds act as 'pro-insecticides', relying on enzymatic oxidation *in vivo* to the active oxygen analogue. Whereas in insects this process occurs in the fat body close to the target site, in mammals the corresponding process occurs primarily in the liver where further efficient metabolism and excretion prevent significant levels of the toxic compound from reaching susceptible sites.

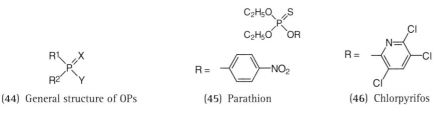

(44) General structure of OPs   (45) Parathion   (46) Chlorpyrifos

The most important class of commercial OPs is that of the phosphorodithioates. As well as two sulfur atoms they usually incorporate an additional metabolizable moiety in the leaving group. Dimethoate (47), which was discovered in 1951, is still one of the most widely used. Another important class is constituted by those incorporating a sulfur atom in the phosphoramides, e.g. methamidophos (48). Such compounds are believed to be activated by sulfoxidation, the resultant CH_3SO- being a better leaving group. Mammalian toxicity is further reduced by introducing an additional metabolizable group, e.g. by acylation of the amine group such as in acephate (49).

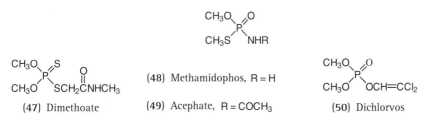

(47) Dimethoate   (48) Methamidophos, R = H   (50) Dichlorvos

(49) Acephate, R = COCH₃

OPs have dominated the insecticide market since the 1940s. The main reasons for their continued success are the low cost of their syntheses and the wide range of possible structural variations, which can provide OPs having properties suitable for a wide range of applications. For example, parathion (45) and chlorpyrifos (46) are contact insecticides with poor solubilities in water. In contrast, dimethoate (47) and acephate (48) are water soluble and act systemically, i.e. when they are sprayed on the crop plant, they become absorbed into the vascular system and translocated such that insects feeding on new unsprayed leaves receive a toxic dose. Other compounds such as dichlorvos (50) are sufficiently volatile for use as fumigants.

The alkaloid physostigmine (51) is the toxic principle of calabar bean (*Physostigma venenosum*), which was formerly used in West Africa in 'trials by ordeal'. This compound was known to act by inhibiting AChE. When it was established that the OP insecticides also acted on AChE, physostigmine was tested on insects but found not to be insecticidal by topical application. This was attributed to its poor penetration of insect cuticle, probably owing to the alkaloid amine moiety being protonated at physiological pH with the resulting polar cation penetrating cuticle only very slowly. During the 1950s studies of structure–activity relationships led to the discovery of the insecticide carbaryl (**52**), an un-ionized analogue that retained the toxic *N*-methylcarbamate group and was sufficiently lipophilic to penetrate insect cuticle. The discovery of insecticidal carbamates thus provided an excellent example of using an established class of drugs as a starting point.

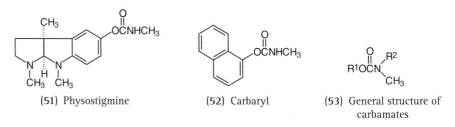

(51) Physostigmine   (52) Carbaryl   (53) General structure of carbamates

The general structure (**53**) for insecticidal carbamates was soon established and the importance of structural resemblance to acetylcholine (**43**) was recognized. The nature and stereochemistry of the leaving group R^1O- was shown to influence both biological (e.g. insect selectivity) and physical (e.g. volatility and water solubility) properties. R^2 is usually H but occasionally CH_3.

Replacement of the naphthyl ring in carbaryl (**52**) by the more polar pyrimidine ring led to the discovery of pirimicarb (**54**), a highly effective and systemic *N,N*-dimethylcarbamate. A further development was the replacement of the aromatic ring by an acyclic chain containing an oxime group, exemplified by aldicarb (**55**).

(54) Pirimicarb   (55) Aldicarb

This modification retained useful systemic activity and also increased the spectrum of activity to cover mites and nematodes. Unfortunately, such compounds are rather toxic to mammals. Also, their high polarity, especially after oxidation of compounds such as aldicarb to the sulfoxides and sulfones, leads to low sorption by soil and hence the potential for leaching into ground water. More recently, carbamates based on the pro-insecticide approach towards reducing toxicity to mammals have been introduced. For example, substituting the H in the carbamate moiety of carbofuran (**56**) by a thiosulfenyl group led to the discovery of carbosulfan (**57**). This compound is metabolized back to the active carbofuran in insects whereas the major detoxification pathway in mammals involves cleavage of the O–C bond, leading to the formation of non-toxic fragments.

Resistance to carbamates and OPs can develop in several ways. A major mechanism involves a mutational change in AChE such that certain compounds no longer bind so well to the enzyme. This change can in some instances differentiate between the two classes of insecticides and even among chemicals in the same class, so that cross-

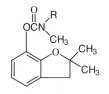

(56) Carbofuran R = H

(57) Carbosulfan R = SN((CH₂)₃CH₃)₂

resistance may be very restricted. For example, some strains of cotton aphids that are strongly resistant to pirimicarb (54) remain fully susceptible to other carbamates and OPs. This mechanism of resistance is often accompanied by others involving enhancement of detoxification of insecticides, e.g. through increases in production of the metabolic enzymes, such as esterases, monooxygenases and glutathione transferases. These mechanisms are less specific and can give broader cross-resistance, often extending to insecticides such as the pyrethroids which have an unrelated mode of action.

Pyrethroids, like DDT, affect sodium channels, but it is not known whether they act at the same site. Pyrethroids for use in agriculture were introduced during the 1970s and to date remain the most effective, selective and environmentally benign class of insecticides. They have low mammalian toxicity and are degraded quite rapidly in the field.

Modern pyrethroid insecticides are synthetic analogues of the natural compounds known as pyrethrins, the insecticidal constituents found in flowers of the pyrethum daisy (*Tanacetum cinerariaefolium*). The insecticidal activity of this plant, which is still utilized for controlling indoor pests and pests in stored grain, had long been known and the structures of the active compounds were established during the period 1900–1966. The natural pyrethrins are more active than DDT but their use in agriculture is restricted by their instability in light. Studies of their structure–activity relationships by Michael Elliott at Rothamsted over 30 years showed the way to better pyrethroids. Using pyrethrin I as the lead structure, he divided the molecule conceptually into seven segments (Figure 14). Each segment

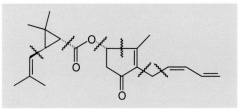

Figure 14
Segmentation scheme of pyrethrin I.

proved amenable to substitutive replacements as long as the overall shape of the molecule was retained. The essential features for binding appear to be unsaturation at each end of the molecule and the presence of a central *gem*-dimethyl (or equivalent) group. In 1965, Elliott discovered the commercially important compound resmethrin (58); this has the cyclopentenone ring, considered as a spacer group, replaced by a furylmethyl group and the terminal diene side chain by a benzyl group to retain unsaturation. Although resmethrin is more active than pyrethrin I, unfortunately it is also photolabile and therefore its use is mainly restricted to indoor applications.

In 1973, photostability was finally achieved by introducing two further changes. The dimethylvinyl group was replaced by a dihalovinyl group and the furyl by a benzyl group, yielding the commercial compound permethrin (59). The insecticidal activity was further increased, especially for those compounds containing a *cis*-substituted cyclopropane ring, by the introduction of an α-CN group (such as in deltamethrin (60)). Because the activity of a pyrethroid molecule depends on its overall shape, the absolute configuration at each of the three chiral centres is important. Deltamethrin, which is still one of the most effective commercial insecticides, is produced as a single isomer of the eight possible. The spectrum of activity was increased to include mites by

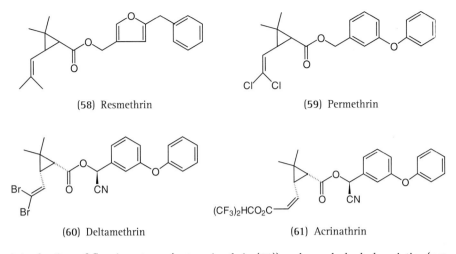

(58) Resmethrin

(59) Permethrin

(60) Deltamethrin

(61) Acrinathrin

introduction of fluorine atoms (e.g. acrinathrin (**61**)) and novel alcohol moieties (e.g. bifenthrin (**62**)).

Independent studies in Japan led to active non-cyclopropane analogues, e.g. fenvalerate (**63**). The next advance came in 1987 with the introduction of the non-ester compound etofenprox (**64**) which was based on fenvalerate but contains changes in all seven segments of the lead compound, pyrethrin I. A key property of the non-ester pyrethroids, in contrast to the ester analogues, is their low toxicity to fish, allowing their use in rice paddy fields.

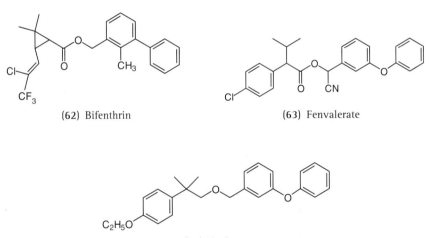

(62) Bifenthrin

(63) Fenvalerate

(64) Etofenprox

The tobacco plant (*Nicotiana tabacum*) and related species have a long history of use as insecticides. Although the active principle, nicotine (**65**), has long been known, extensive structure–activity studies have not produced useful alternatives, primarily due to their high toxicity to mammals.

In contrast to the OPs, nicotine mimics acetylcholine (**43**) by binding (in the protonated form) to the acetylcholine receptor on the post-synaptic nerve, which in turn opens a sodium channel (Figure 13). Imidacloprid (**66**), reported in 1990, is a member of a new class of commercial compounds, often referred to as the 'chloronicotinyl'. Despite the structural similarities to nicotine, it was developed from a synthetic lead compound, nithiazine (**67**), which was reported in 1960 but not commercialized due to its instability under field conditions. Imidacloprid overcame these problems and is a systemic compound that is most effective through ingestion.

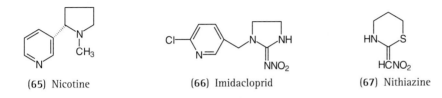

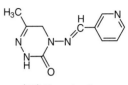

(65) Nicotine   (66) Imidacloprid   (67) Nithiazine   (68) Pymetrozine

In contrast to most established insecticides, pymetrozine (68) was developed from a lead compound identified from a programme driven by investigation of novel heterocyclic chemistry. Pymetrozine has a unique mode of action, probably acting on the nervous system, causing malfunction of the salivary gland and leading to irreversible blocking of feeding.

The ease of generating and processing secondary metabolites from microbes in fermentation broths has moved the emphasis during the past 20–30 years from plants to microbes as sources of novel lead compounds. For example, highly active macrocyclic compounds, the avermectins and the related milbemycins, have been isolated from soil-dwelling *Streptomyces* spp. The two series differ mainly by virtue of the presence of an oleandrose sugar at position 13 in avermectins (Figure 15). To date, their main commercial impact has been on animal and human health, with several products available for controlling parasites. Although avermectins bind to the GABA receptor, as do the cyclopolychlorinated compounds and pyrazoles (see below), they act by activating rather than blocking the chloride channels and so cause the insects to die from muscular paralysis. In agriculture their use has been limited by their poor photostability, relatively narrow spectrum of activity and high costs of production. The major commercial agrochemical in this class, abamectin, a mixture comprised mainly of avermectins B1$_a$, is used against mites. The search for semi-synthetic analogues with better biological spectra of activity has led to the discovery of several useful analogues, e.g. emamectin, in which the oleandrose −OH of avermectin B1$_a$ at the 4″ position is replaced by −NHCH$_3$ with inversion. This compound is nearly 400 times more active towards lepidopterous pests than is abamectin, but it is less active against mites and aphids.

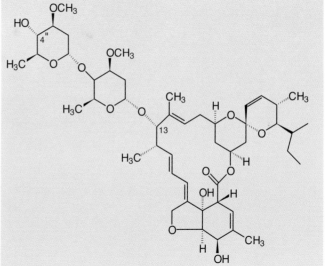

Figure 15
Structure of natural avermectin B1$_a$.

This approach of exploiting semi-synthetic derivates has been applied successfully to several natural macromolecules, e.g. azadirachtin from the neem tree, *Azadirachta indica*. However, it has not been successful with rotenone, the active principle of derris dust.

The recent class of pyrazoles, exemplified by fipronil (69), was developed through a conscious effort to investigate new areas of relatively simple heterocyclic chemistry. It also acts by blocking chloride channels. Although these compounds bind at a different site on the GABA receptors from that for polychlorinated compounds, nonetheless cross-resistance occurs.

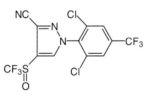

(69) Fipronil

Insect Growth Regulators

Insects can be divided into two broad classes in terms of their life cycles – those that emerge from eggs directly as miniature adults and those that develop through several larval stages leading to a pupa that undergoes metamorphosis into an adult. The process involved for the latter class, which is regulated by two series of hormones, is unique to insects and so presents an ideal target for the aim of developing novel insecticides with low mammalian toxicity. The insect is maintained in the larval stages by the presence of the natural juvenile hormones (JHs), e.g. JHIII (70). The formation of the cuticle during moulting is dependent on the presence of another hormone, 20-hydroxyecdysterone (71), derived from ecdysone.

Methoprene (72) was designed to imitate the structure of natural JHs and did indeed

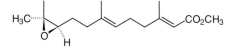

(70) Juvenile hormone JHIII

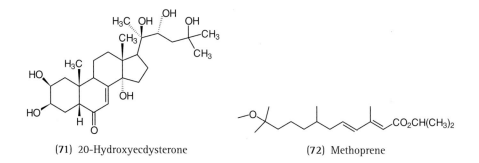

(71) 20-Hydroxyecdysterone

(72) Methoprene

mimic their activity, thereby preventing metamorphosis to the adult. It was the first example of a 'biorationally' designed insecticide. Poor photostability has been the major limitation of this class of compounds; this has been overcome in recent analogues such as pyriproxyfen. Compounds of another group inhibit the synthesis of chitin and so kill the insect during moulting. The first of these, diflubenzuron (73), which was introduced in 1976, was developed from a simple benzoylurea identified during routine screening of synthetic compounds from a herbicide synthesis programme. A recent class based on hydrazine, e.g. tebufenozide (74), acts by mimicking 20-hydroxyecdysterone, so inducing premature and incomplete moulting.

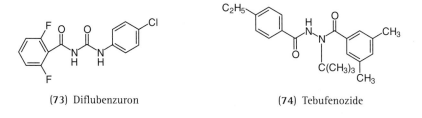

(73) Diflubenzuron

(74) Tebufenozide

Respiration Inhibitors

Respiration is essential in order for cells to generate and store energy. In respiration, the oxidation of NADH and succinate in the mitochondrion matrix with the concomitant reduction of oxygen to water powers the creation of a proton gradient across the inner

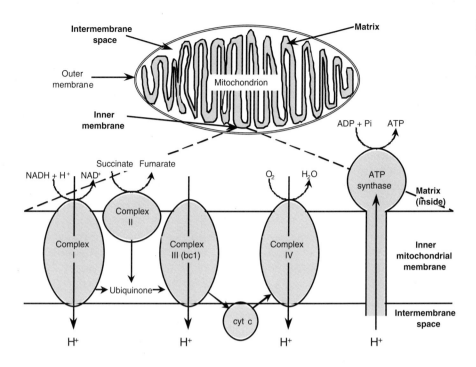

Figure 16
Mitochondrial respiration, with
detail of the processes occurring
across the inner membrane
bounding the matrix.

membrane; this gradient then drives the production of the 'high-energy' ATP inside the matrix by ATP synthase. The two main mechanisms of some recent insecticides are the uncoupling of phosphorylation from oxidation and the blocking of electron transport during the process of respiration (Figure 16). Because respiration occurs both in vertebrates and in invertebrates, commercial compounds are generally pro-insecticides that exploit differences between the mechanisms of detoxification for the two classes.

Dioxapyrrolomycin (**75**), a fungal metabolite from *Streptomyces fumanus*, was the lead structure in the development of chlorfenapyr (**76**). The presence of an ethoxymethylene group on the nitrogen atom reduces both toxicity to mammals and phytotoxicity without reducing the insecticidal activity. It is removed *in situ* by oxidative metabolism. Chlorfenapyr acts by uncoupling oxidative phosphorylation.

Diafenthiuron (**77**) is also a pro-insecticide, being oxidatively metabolized within the insect to the active carbodiimide (**78**). It is thought to act by blocking the mitochondrial ATP synthase. Other examples of insecticides/acaricides that interfere with mitochondrial electron transport include rotenone and fenzaquin (acting at complex I) and acequinocyl (acting at complex III).

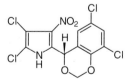

(**75**) Dioxapyrrolomycin

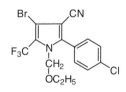

(**76**) Chlorfenapyr

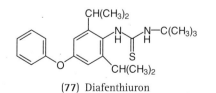

(**77**) Diafenthiuron

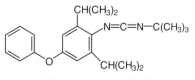

(**78**) Derived carbodiimide

Herbicides

Weeds can cause serious loss of crop yield, usually simply by competition with the crop, but also, mainly in more tropical areas, by direct damage to the crop plants from parasitic weeds such as *Striga*. Traditionally, mechanical methods of weed control such as hoeing have been practised but, though these can be scaled up to tractor-drawn implements for certain situations, most modern agriculture relies heavily on the use of herbicides.

Plants are so different from animals in terms of their biochemistry that most herbicides have little toxicity towards animals. However, the main task for herbicides in agriculture is to kill weeds in the presence of crop plants, so we require selective herbicides. Selectivity is usually achieved in one of two ways. Firstly, the herbicide may be physically kept away from the crop, e.g. simazine, even though it is toxic to bean plants, will kill shallowly rooted weeds via uptake by their roots when it is sprayed onto the soil's surface but will not usually be leached down to the deeper-rooted beans. Secondly, there may be biochemical selectivity between the crop and weed plants, either due to differing sensitivities of the target sites in different plant species or, more commonly, due to rapid metabolism of the herbicide in the crop plant, e.g. the rapid breakdown by hydroxylation of atrazine in maize. Changes in these biochemical processes in weed populations treated repeatedly with pesticides may result in the evolution of weeds resistant to certain herbicides.

Many modes of action are utilized by herbicides and different chemical classes of herbicide may act at different points in the same biochemical pathway. The following description covers the main modes of action together with the discovery and roles of herbicides active on that pathway.

Photosynthesis Disrupters

The process of photosynthesis combines water and carbon dioxide in the presence of light to produce sugars. These reactions all take place in the chloroplasts, which contain the green pigment chlorophyll. The processes involved are complex and still not fully understood. In outline, two photosystems, PSI and PSII, each comprising a protein complex including the pigment proteins, capture the energy of light and use it to remove electrons from water; these electrons pass along a series of carriers, including plastoquinone and cytochrome *f*, and ultimately provide the driving force to convert adenosine diphosphate (ADP) to the 'high-energy' triphosphate (ATP). This energy then drives the fixation of carbon dioxide into carbohydrate. The process of the photo-oxidation of water to evolve oxygen during photosynthesis is generally known as the Hill reaction.

ICI (now Zeneca) discovered the bipyridinium herbicide paraquat (**79**) by random

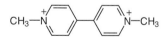

(**79**) Paraquat

screening in the late 1950s. Paraquat and related compounds are non-selective herbicides that, in the presence of light and oxygen, cause rapid death of plant leaves following their application by spraying. These dicationic compounds appear to divert electrons from the iron–sulfur centres in photosystem I. It seems that the reduced paraquat species reacts with oxygen to give the superoxide anion, O_2^-, which generates hydroxyl radicals either directly or via a hydrogen peroxide intermediate. These highly reactive radicals attack membranes and induce breakdown of cells in the sprayed tissue, though untreated and non-photosynthetic organs such as roots remain unaffected and their ability to regrow is not impaired.

Photosystem II consists of two regions, these being a reaction centre and a group of pigment proteins. The reaction centre contains two proteins and it is one of these, of relative molecular mass 32 kDalton and comprising 353 amino acids, that is the target for phenylurea, triazine, uracil and a number of other herbicides. This leads to inhibition of the Hill reaction in the chloroplasts and ultimately to the death of the plant. Most of these herbicides were developed during the period 1950–1970 and, although they are less active (application rates 1–3 kg ha^{-1}) than many modern herbicides, they are still widely used as soil-applied compounds. Examples of the phenylureas, triazines and uracils are (80), (81) and (82), respectively; these compounds share the same site of action, whereas other compounds such as hydroxybenzonitriles and dinitrophenols also act at photosystem II but their inhibition exhibits different characteristics.

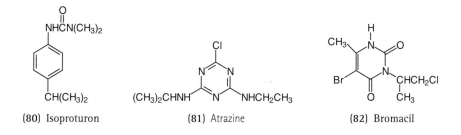

(80) Isoproturon   (81) Atrazine   (82) Bromacil

Porphyrins are based on the tetrapyrrole nucleus and the best known of them is protoporphyrin IX (82), called *haem* when it is complexed with an iron(II) ion. Such compounds play important roles as coenzymes in animals and plants; haem for instance is a constituent of haemoglobin, which is the oxygen-transporting protein in blood, and serves as a coenzyme for the cytochromes, a class of oxidation–reduction proteins.

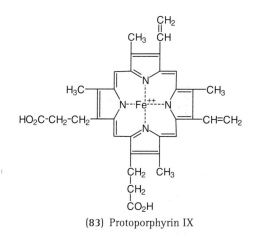

(83) Protoporphyrin IX

Protoporphyrin IX (83) is biosynthesized by oxidation of protoporphyrinogen IX, this precursor having the same carbon skeleton as that of protoporphyrin IX but with the four pyrrole rings not linked by conjugation. Several herbicides inhibit the enzyme protoporphyrinogen-IX oxidase, among which the best studied class is that of the diphenyl ether herbicides such as nitrofen (84). Such inhibition leads to the accumulation of protoporphyrinogen IX and it is thought that this leaks from the cells and undergoes non-enzymatic oxidation to protoporphyrin IX. This highly reactive compound, in the presence of light and in the absence of the functioning of the control mechanisms that operate at its normal site, then causes peroxidative degradation of membranes, leading to bleaching of green plant tissue.

The diphenyl ethers were introduced in 1962, but recently there has been a lot of research interest in the area of inhibitors of protoporphyrinogen-IX oxidase. One such new compound is thidiazimin (85), a representative of the imino-1,2,4-thiadiazoles, which, applied as a contact herbicide at 20–40 g ha^{-1}, controls broad-leaved weeds in cereals.

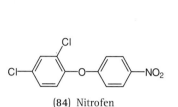

(84) Nitrofen

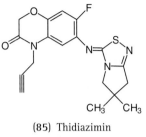

(85) Thidiazimin

Hormone mimics

Hormone mimics that influence the response to auxin were discovered during the 1940s and can be considered the first modern herbicides. They were the first to provide efficient control of broad-leaved weeds in cereals and some of the early compounds are still widely used today. The endogenous hormone auxin is indol-3-ylacetic acid (86) and it controls the growth of plants. Unlike auxin itself, the hormone mimics, mostly based on phenoxyalkanoic acids (e.g. 2,4-D, 87) and benzoic acids (e.g. dicamba, 88), are metabolically stable and so disrupt the growth of the plant; we can often see this effect on treated lawns on which broad-leaved weeds such as dandelions and buttercups grow tall and twisted before collapsing.

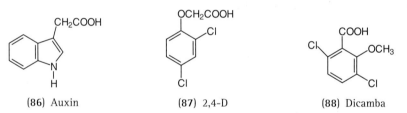

(86) Auxin   (87) 2,4-D   (88) Dicamba

The effect of stereochemistry may be noted for the enantiomers of mecoprop (89 and 90), of which only (R)-mecoprop (89, sold as mecoprop-P) is active. Thus, the site of action clearly has chiral requirements, even though auxin (86), the natural hormone, is not a chiral compound.

(89) (R)-Mecoprop   (90) (S)-Mecoprop

Inhibitors of amino acid synthesis

The shikimate pathway is very important in plants, for it produces chorismate which is the precursor for a large number of aromatic products, including the amino acids phenylalanine, tyrosine and tryptophan. About 20% of the carbon flux is processed via this pathway, other important products being ubiquinone, folic acid and lignin. The start of the shikimate pathway (Figure 17) is the reaction of phosphoenolpyruvate (91) with erythrose-4-phosphate, which after several steps gives shikimate itself. This is then phosphorylated by the enzyme shikimate kinase and the product is then combined

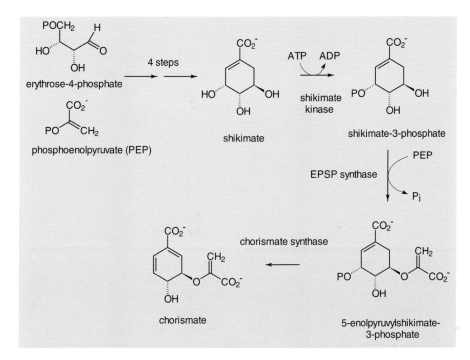

Figure 17
Synthesis of chorismate via the shikimate pathway.

with phosphoenolpyruvate in a reaction catalysed by 5-enolpyruvylshikimate-3-phosphate synthase (EPSP synthase) to give the chorismate skeleton.

Glyphosate (**92**) inhibits EPSP synthase in a complex fashion, one aspect of which is that it acts as a reversible and competitive inhibitor of the above reaction with respect to the structurally similar phosphoenolpyruvate (**91**). By so doing, glyphosate is a potent, non-selective herbicide. Glyphosate has additional favourable features in that it is translocated in plant phloem (i.e. with the photosynthates such as sucrose);

$$HOOCCOP(OH)_2 \atop \underset{CH_2}{\overset{O}{||}}$$

(**91**) Phosphoenolpyruvate

$$HOOCCH_2NHCH_2\overset{O}{\overset{||}{P}}(OH)_2$$

(**92**) Glyphosate

hence foliar applications result in glyphosate being translocated to roots and so preventing regrowth. Furthermore, glyphosate is quickly and strongly sorbed by soil and so crops can be sown immediately after spraying to kill weeds in stubbles. Such desirable properties have led to much effort to produce analogues with the same properties, but even slight changes to this simple molecule cause substantial loss of activity.

Bilanophos is the alanylalanyl derivative of glufosinate (**93**), and was isolated from

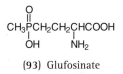

$$CH_3\overset{O}{\overset{||}{P}}CH_2CH_2\underset{NH_2}{\overset{|}{C}}HCOOH \atop OH$$

(**93**) Glufosinate

the soil micro-organisms *Streptomyces viridochromogenens* and *S. hygroscopicum*. This dipeptide derivative is herbicidal, being rapidly hydrolysed in plants to give the active principle glufosinate. Glufosinate inhibits the enzyme glutamine synthase, apparently by mimicking the transition state for the precursor glutamate reacting with ammonia and adenosine triphosphate. As a consequence of this inhibition, free ammonia accumulates and this kills the treated tissues. Though glufosinate is generally a non-selective herbicide, genetic engineering has been used to create crop plants that metabolize it rapidly and so can tolerate applications of glufosinate to kill weeds. These transformed crop plants contain a gene from a strain of *Streptomyces* that codes for an acetyltransferase enzyme, which converts glufosinate into its inactive *N*-acetyl derivative.

The synthesis of acetolactate is the first step in plants in the making of branched-chain amino acids such as valine, leucine and isoleucine. Two classes of herbicide, the sulfonylureas and imidazolinones, inhibit the enzyme acetolactate synthase. This inhibition is due not to their binding at the active site but to allosteric binding, i.e. binding away from the active site that in this case distorts the shape of the enzyme and so renders it disfunctional. It is believed that these two classes of herbicide bind at different allosteric sites on the enzyme. The development of these herbicides began in the 1970s and their effectiveness at low-to-very-low rates of application, together with their wide range of selectivities, makes them an important group of herbicides. More recently, other classes of herbicides that inhibit this enzyme have been discovered and these include the triazolopyrimidine sulfonamides and the pyrimidinyl oxobenzoic acids.

The simple phenylurea herbicides already discussed encouraged chemists at Du Pont to study the related compounds having an SO_2 group between the ring and the urea group. The reaction of benzenesulfonylisocyanate with 4-chloroformanilide led to the *N*-formylsulfonylurea (**94**) but this compound, first prepared in 1957, did not exhibit any interesting activity and so this area of chemistry was not then pursued further. However, 16 years later, in 1973, what proved to be a great stroke of luck intervened. In a new screening for chemosterilant activity against mites, this *N*-formylsulfonylurea was tested and found to have some activity. One of the analogues, **95**, proved to be a weak retardant of plant growth.

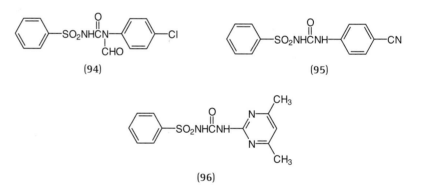

(94)   (95)

(96)

George Levitt then started a programme of synthesizing heterocyclic derivatives of **95** and, in 1975, a key pyrimidine derivative (**96**) was found to be a potent herbicide. The synthesis of further analogues led to the discovery of the first two commercial compounds, chlorsulfuron and metsulfuron-methyl (**97** and **98**). These compounds selectively kill broad-leaved weeds in cereals and were found to be active at astonishingly low rates of application. Rates of 7.5–30 g ha^{-1} are effective, i.e. these sulfonyl-

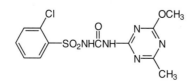

(97) Chlorsulfuron

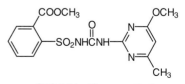

(98) Metsulfuron-methyl

ureas are about 100 times more active than the phenylurea herbicides such as monuron! The selectivity of the sulfonylureas is due to the fact that they are more rapidly metabolized in the crop plants than they are in the weeds.

Many thousands of analogues have now been synthesized and the structural requirements are well defined. A heterocyclic ring, usually pyrimidine or 1,3,5-triazine, substituted in the 4- and 6-positions is attached to the urea; the group attached to the sulfur is usually a 2-substituted phenyl group, but may be a 2-substituted thiophene or even an aliphatic group such as that in amidosulfuron (**99**). Varying the structure has led to further types of selectivity being found; for example, although sugar beet is very sensitive to chlorsulfuron, altering the substituents to give triflusulfuron-methyl (**100**) has allowed the latter compound to be introduced for the control of broad-leaved weeds in sugar beet.

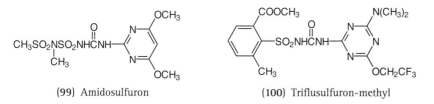

(**99**) Amidosulfuron   (**100**) Triflusulfuron-methyl

The development of the sulfonylureas, a highly successful group of herbicides, well illustrates the role in the discovery of pesticides of an initial serendipitous observation of activity followed by the thorough elucidation of structure–activity relationships, in this example leading to compounds 1000 times more active than the lead compound.

The story of the beginnings of the development of the imidazolinone herbicides is as strange as that of the initial work on the sulfonylureas. The Cyanamid company prepared the phthalimide (**101**) as an anticonvulsant, but, from a random screening test in 1972, some herbicidal activity was noted. This prompted further synthesis in this area, from which it was found that the chloro analogue (**102**) had interesting effects on plant growth even though it was essentially without herbicidal activity. Subsequently (**103**) was prepared and this exerted a strong effect on plant growth akin to that of gibberellic acid.

(**101**) X = H   (**102**) X = Cl   (**103**)

Further work by Marius Los led to the cyclization (Figure 18) of the phthalimide (**104**) to the imidazoisoindole (**105**), which proved to be a broad-spectrum, non-selective herbicide. The isoindole ring in this compound was in turn opened by methoxide in methanol to give the imidazolinone (**106**), the first member of this class of herbicides to be prepared. This and related compounds were herbicidally active (the free carboxylic acid is the active molecule); changing the two alkyl substituents on the imidazolinone ring and introducing substitution in the phenyl ring generally led to a reduction of their activity. However, introducing a methyl group into (**106**) either at position 3 or at position 4 led to a mixture that had good selective activity against wild oats, blackgrass and mustards in wheat and barley. This mixture has now been commercialized.

The final story with these compounds was the replacement of the benzene ring by a

Figure 18
Discovery of imidazolinone
herbicides.

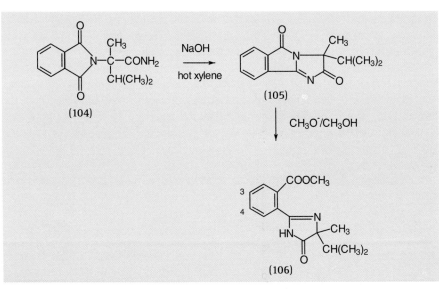

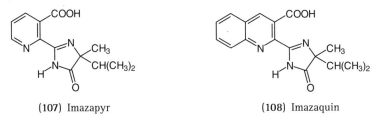

pyridine ring. Again, fortune played a part, for, of the four positions at which the pyridine nitrogen could be placed, the first isomer synthesized (imazapyr, **107**) proved to be a highly active non-selective herbicide. Further synthesis led to compounds with differing selectivities, such as imazaquin (**108**), which is selective for use with soybeans.

(107) Imazapyr   (108) Imazaquin

Inhibitors of lipid synthesis

Lipids play an important role in the life of plants; phosphoglycerides and glycosylglycerides are the main such components in membranes, the triacylglycerols being the major storage lipids. The surface layers of plants consist of wax, cutin or suberin structures and many lipids contribute to this cuticular layer. Finally, the carotenoid pigments assist in photosynthesis.

Several classes of herbicides affect the production of fatty acids and we will first consider the biosynthetic pathway for the synthesis of simple fatty acids (Figure 19). The initial step is the reaction of the acetyl-coenzyme A complex (acetyl-CoA) with bicarbonate, catalysed by acetyl-CoA carboxylase, to give malonyl-CoA, and a second acetyl-CoA becomes attached to an acyl carrier protein (ACP); these two molecules react together to give a β-ketoacyl-ACP. This in turn undergoes a cycle of reduction of the keto group, dehydration and reduction of the resultant alkene bond to give a new acyl-ACP, each such cycle adding two carbon atoms to the chain length. Until the formation of the final products containing 16 carbon atoms, the condensations are catalysed by the enzyme β-ketoacyl-ACP synthetase I. Further enzymes can extend the chain length to 18 carbons, such as in stearate, and modify the chain by desaturation and/or attach other groups such as glycerol-3-phosphate.

Aryloxyphenoxypropionates and hydroxycyclohexenones (cyclohexanediones) were introduced as herbicides in the late 1970s and are important because they selectively control grass weeds in crops. The structural variation of the herbicides within each of

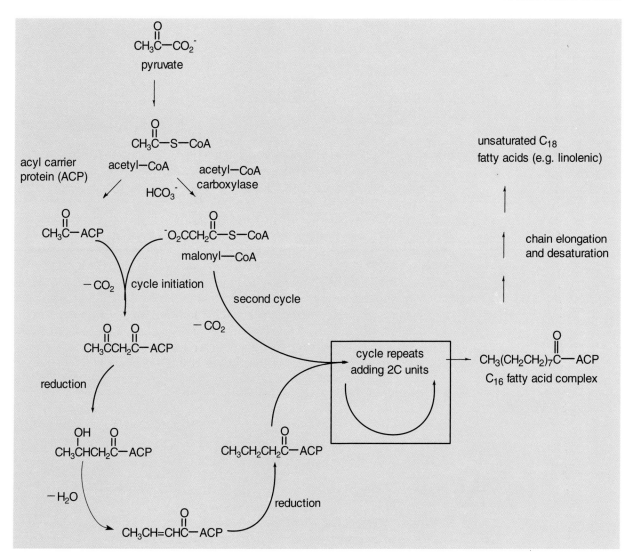

Figure 19

Schematic outline of the biosynthesis of fatty acids.

these classes is rather small. Aryloxyphenoxypropionates such as diclofop (109) vary only in the nature of their terminal aryl moiety, whilst the hydroxycyclohexenones such as sethoxydim (110) all have the central skeleton and alkoxime group but with varying substituents on the ring and alkoxime fragment.

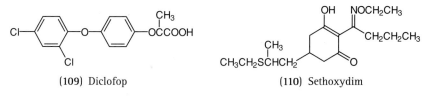

(109) Diclofop   (110) Sethoxydim

The development of the hydroxycyclohexenones is another example in which the initial area of activity being investigated was not that finally obtained. Scientists at Nippon Soda had introduced benzoximate (111) as an acaricide in 1971 and efforts to improve its activity against mites led to analogues such as (112) that were found to kill grass weeds. Further work showed that modifying the ring structure to the hydroxycyclohexenone skeleton improved the activity and this led to the commercial herbicide alloxydim (113). This herbicide was more effective against annual grasses than

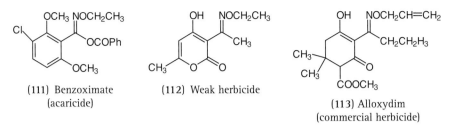

(111) Benzoximate
(acaricide)

(112) Weak herbicide

(113) Alloxydim
(commercial herbicide)

it was against perennial grasses such as Johnson grass, *Sorghum halepense*, which is a serious weed problem in the southern USA; varying the substituents led to compounds such as sethoxydim (110) that control both annual and perennial grasses effectively.

These herbicides inhibit acetyl-CoA carboxylase and so stop the first committed step in the synthesis of fatty acids rather than influencing the subsequent chain-elongation cycle. The selectivity in the killing of grasses apparently arises through more than one cause: the acetyl-CoA carboxylase from grasses appears to be much more sensitive to the hydroxycyclohexenones than does that from other plants; on the other hand, wheat is tolerant of diclofop even though its acetyl-CoA carboxylase is sensitive, indicating that rapid metabolism is conferring selectivity for this species.

Thiocarbamates, such as triallate (114), are normally applied to the seed bed before the emergence of the seedlings, controlling grasses and some broad-leaved weeds in many crops including maize, cotton and soyabean. They inhibit the production of the long-chain fatty acids and waxes and so are believed to act on the elongase enzymes that elongate the base 16-carbon chain.

(114) Triallate

(115) Norflurazon

The pyridazinones such as norflurazon (115) were developed initially as inhibitors of photosynthesis, but were found also to decrease levels of lipids, including that of linolenic acid, $CH_3(CH_2CH=CH)_3(CH_2)_7CO_2H$. There is evidence that their mode of action is inhibition of the desaturation of precursors such as linoleic acid, $CH_3(CH_2)_3(CH_2CH=CH)_2(CH_2)_7CO_2H$, and some are known also to prevent the production of carotenoids by inhibiting the enzymes that increase the degree of unsaturation of the precursor phytoene.

Fungicides

Most crop plants are susceptible to serious fungal diseases and such diseases have in the past limited the geographical distribution of certain crops, e.g. the growing of cereals in the wetter western part of the UK can be difficult without fungicides. One of the first effective fungicides was the copper-based 'Bordeaux mixture', which was first used during the nineteenth century, primarily for the control of downy mildew on grape vines.

The first generation of fungicides exerted only a protectant rather than a curative effect. These materials, such as sulfur powder and the *bis*-dithiocarbamates, will, when they are sprayed onto the surface of a leaf, inhibit the germination of fungal spores and so prevent the consequent attack. Nevertheless, diseases already on the plant will not be cured and the protection of new growth requires repeated spraying. However, in the early 1970s, the first systemic fungicides appeared, i.e. compounds such as carboxin, ethirimol and benomyl that move within the plant and exert a curative effect.

The long-standing use of elemental sulfur and of copper compounds has already been mentioned. Fungicidal compounds were amongst the earliest synthetic organic pesticides, the dithiocarbamates being discovered in 1934 and the bis-dithiocarbamates in 1943, exemplified by thiram (116) and maneb (117), respectively. These compounds exert protectant activity against a broad spectrum of fungi, although they are ineffective against powdery mildews, and are believed to act as multi-site inhibitors in fungal cells.

(116) Thiram   (117) Maneb

Membrane Disrupters

Compounds such as quintozene (pentachloronitrobenzene, 118) have been used for many years in the control of seed- and soil-borne diseases; these compounds are moderately volatile and so can be redistributed via the vapour phase, e.g. amongst potatoes in storage. Diphenyl (119) has similar physical properties and is used to treat

(118) Quintozene   (119) Diphenyl

wrappings to maintain stored fruit in good condition. The use of such compounds has declined in recent years, owing both to the development of resistance by some fungi and to the introduction of modern compounds with greater activity. The aromatic hydrocarbons interfere with cytochrome *c* reductase and the free radicals thus produced react with unsaturated fatty acids in the membranes and so damage the mitochondrial and other membranes.

The first and simplest member of the dicarboximide series, dimetachlon (120), was introduced in Japan around 1971 for the control of fungi in rice and other crops. A series of closely related and very active compounds was subsequently developed, all containing the *N*-(3,5-dichlorophenyl)dicarboximide moiety, exemplified by iprodione (121). These compounds are of particular value because they can control both grey mould (*Botrytis cinerea*) and *Sclerotinia* species that have become resistant to the benz-

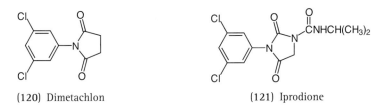

(120) Dimetachlon   (121) Iprodione

imidazole fungicides. Their mode of action is also inhibition of the flavin enzyme cytochrome *c* reductase, leading to damage to membranes in the mitochondria; indeed, fungal strains resistant to the dicarboximides generally exhibit cross-resistance to the aromatic fungicides described above.

Inhibitors of Energy Production

The discovery of the carboxamides in 1966 yielded the first systemic as opposed to protectant fungicides; within the subsequent decade, several other classes of systemic and curative fungicides were found. Carboxin (122) is the best known among the carboxamide fungicides and it is interesting to note that the structurally similar compound salicylanilide (123) was the first completely organic compound with fungicidal action; after its discovery in 1930, it was used to protect materials and textiles against *Cladosporium fulvum*.

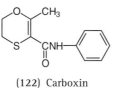

(122) Carboxin

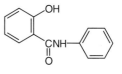

(123) Salicylanilide

The carboxamides are especially active against basidiomycetes such as rust, smut and bunt fungi in cereals. They act by inhibiting mitochondrial respiration (Figure 17) in a complex fashion, primarily by limiting the oxidation of succinate.

A family of fungicidal natural products that contain the (*E*)-methyl β-methoxyacrylate group was first reported by Steglich and co-workers in 1977. Most of these compounds are obtained from the mycelia of basidiomycete fungi. The simplest members are the strobilurins A and B (124 and 125) together with oudemansin A (126). All these compounds have been shown to exert their fungicidal action by binding to a site on cytochrome *b*, thus blocking the transfer of electrons between cytochromes *b* and c_1.

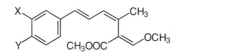

(124) X = Y = H Strobilurin A
(125) X = CH_3O, Y = Cl Strobilurin B

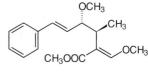

(126) Oudemansin A

This novel mode of action indicated the possibility that these compounds could be developed to control fungi that had become resistant to established fungicides, so their potential as lead structures for synthesis programmes was investigated by agrochemical companies. Several of the natural compounds were moderately active in bioassays *in vivo*, but, disappointingly, the simplest compound, strobilurin A, was active only in tests *in vitro*, not in the greenhouse. It was soon recognized that losses of this compound by rapid photodegradation and also by volatilization were negating its intrinsic activity. Because the β-methoxyacrylate group is present in virtually all of the natural products, it seemed unlikely that this group was the cause of the low photostability, for compounds such as oudemansin A were moderately active *in vivo*. Thus, modifying the other half of the strobilurin A molecule seemed the way forwards.

At Zeneca, the approach taken was that of fixing the geometry of the (*Z*) double bond of strobilurin A by encompassing it within a benzene ring. The resultant stilbene retained activity and had better photostability, but was still insufficiently photostable in field tests. To overcome this, the stilbene moiety was replaced by various aryloxy groups, which culminated in the synthesis of azoxystrobin (127). The chemists at BASF followed a not dissimilar route, but with the additional feature of replacing the β-carbon of the methoxyacrylate by a nitrogen to give the fungicidal methoximino derivative kresoxim-methyl (128). These two compounds are now being developed commercially and should provide a new weapon in the fight against a wide range of ascomycetes, basidiomycetes, deuteromycetes and oomycetes that attack agricultural crops.

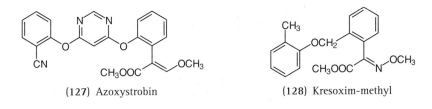

(127) Azoxystrobin   (128) Kresoxim-methyl

Inhibitors of Cell Division

Interest in the benzimidazole series began with long-chain compounds such as 1-methyl-3-dodecylbenzimidazolium chloride (129), which was developed during the 1960s and had a curative effect on apple scab. By the end of that decade, scientists at Du Pont had discovered carbendazim (130) and its derivative benomyl (131). Benomyl breaks down rapidly in plants to give carbendazim, which is probably the main fungitoxic agent. A related compound, thiabendazole (132), was known by 1962 to be an anthelminthic that is effective against internal parasites and it was subsequently discovered to control phytopathogenic fungi such as storage moulds on fruit and potatoes.

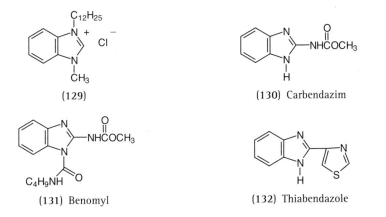

(129)   (130) Carbendazim

(131) Benomyl   (132) Thiabendazole

These compounds act by binding to the β-monomer of tubulin, which is a subunit of the microtubules that comprise a major constituent of the cytoskeleton. This binding results in failure to assemble the microtubules and this damages functions of the cell. Such formation of microtubules is essential during cell division and also occurs in plants; indeed, dinitroaniline herbicides (not discussed in this chapter) such as trifluralin act by inhibiting the formation of microtubules, but the benzimidazole fungicides affect this process only in fungi, not in plants.

Phosphorylated pyrimidines analogous to the insecticide diazinon (133) were found to be active against powdery mildews and structure–activity studies led to the development of three compounds exemplified by ethirimol (134). This class of compounds introduced during the early 1970s was one of the first to exercise systemic and curative activity on plants. They appear to act by interfering with the metabolism of purine by inhibiting adenosine deaminase, the enzyme which catalyses the hydrolytic deamination of adenosine to inosine. This enzyme occurs in fungi but not generally in plants.

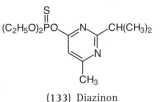

(133) Diazinon

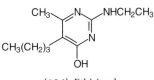

(134) Ethirimol

The class of phenylamides embraces several important fungicides, comprising acyl-anilines, butyrolactones and oxazolidinones, exemplified by metalaxyl (135), ofurace (136) and oxadixyl (137), respectively. Their discovery arose from the observation of fungicidal activity among the chloracetanilide herbicides and it proved possible to retain and improve this activity whilst minimizing the herbicidal effects. These compounds were commercialized during the late 1970s and are systemic fungicides active against pathogens in the order Peronosporales (e.g. downy mildew on grapes). Metalaxyl has been studied the most and it apparently acts by inhibiting the incorporation of uridine into RNA.

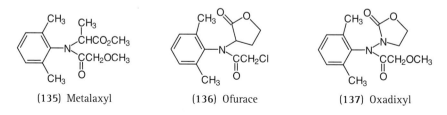

(135) Metalaxyl (136) Ofurace (137) Oxadixyl

Inhibitors of Sterol Synthesis

Sterols, especially ergosterol, are important components of the membranes of fungal cells. Two classes of fungicides act by inhibiting their synthesis by the pathway outlined in Figure 20.

Figure 20.
Biosynthesis of ergosterol in fungi.

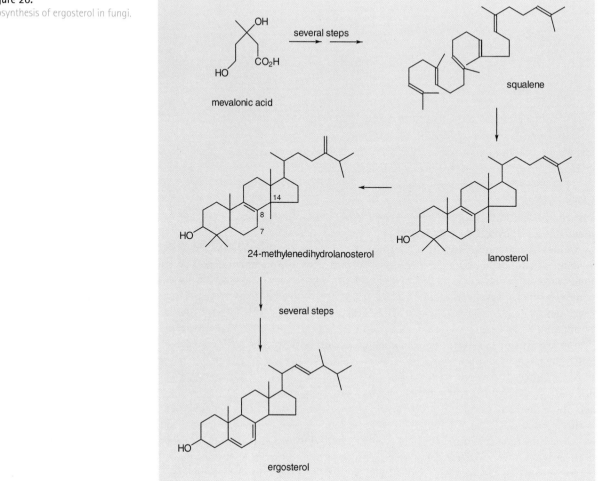

The fungicide triarimol (138) was reported by Eli Lilley in 1970 and proved to be the first of a very diverse range of compounds that stop the production of ergosterol and related sterols by inhibiting the demethylation at C-14. The inhibition involves binding with a cytochrome P-450 enzyme involved in the demethylation, this reaction involving several steps. The presence of a heterocyclic ring able to complex to the iron atom in P-450 is a requirement for activity. Many of the compounds developed later on, e.g. triadimenol (139), utilize a 1,2,4-triazole moiety to perform this function. The range of permitted structures is such that even a compound derived from silicon chemistry, the substituted silane flusilazole (140), has now been commercialized. All these compounds control a wide range of fungal diseases in crops.

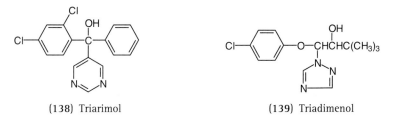

(138) Triarimol   (139) Triadimenol

Compounds of a second class, the morpholine fungicides, also inhibit the biosynthesis of sterols, but by a different mode of action. They inhibit the isomerization of the double bond at C8 to its position at C7 in ergosterol. These morpholine fungicides, such as tridemorph (141), all have a large alkyl group attached to the morpholine nitrogen.

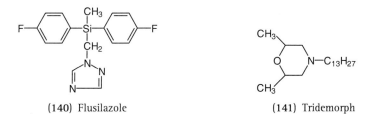

(140) Flusilazole   (141) Tridemorph

Conclusions

Pesticides have been developed largely via two routes, namely either from natural product leads or by a chance finding of some biological activity, in each case followed by studies of structure–activity relationships and a large programme of synthesizing analogues in order to optimize activity. In many areas, good fortune has played a major role; for example, insecticides have been developed from a herbicide programme and potent herbicides from a chemosterilant of mites! There have been attempts to design pesticides on the basis of knowledge of candidate target sites but, although some active compounds have been produced by this approach, to the best of our knowledge no commercial compound except for methoprene has this background.

A wide range of pesticides now exists and most pests and diseases can be controlled to some degree. The search for new and better compounds thus becomes increasingly difficult; it is estimated that typically 50000 compounds must now be screened for activity to produce just one commercial compound. Clearly the financial commitment is enormous and this has led to many mergers of agrochemical companies in an effort to be big enough to minimize lean spells between introductions of new pesticides. Part of the driving force for producing new compounds is the development of resistance to older compounds, together with the increasingly stringent testing of environmental effects and toxicity, for which the older compounds will not necessarily come up to the current standards.

Pesticides are generally used in conjunction with other means (e.g. breeding plants that are resistant to pests and diseases, agronomic practices and biological control) to give an integrated approach and to minimize the usage of pesticides. Nonetheless, pesticides are believed to safeguard approximately 30% of total crop production in the world. As such, pesticides are likely to be with us for the foreseeable future, and so it is incumbent on us to use them wisely.

Further Reading

1. L. G. Copping and H. G. Hewitt, *Chemistry and Mode of Action of Crop Protection Agents* (Royal Society of Chemistry, Cambridge, 1998).
2. W. Draber and T. Fujita (editors), *Rational Approaches to Stucture, Activity and Ecotoxicology of Agrochemicals* (CRC Press, Boca Raton, 1992).
3. I. Ishaya and D. Degheele (editor), *Insecticides with Novel Modes of Action* (Springer, London, 1998).
4. R. C. Kirkwood (editor), *Target Sites for Herbicide Action* (Plenum Press, New York, 1991).
5. W. Köller (editor), *Target Sites for Fungicide Action* (CRC Press, Boca Raton, 1992).
6. Gy. Matolcsy, M. Nadasy and V. Andriska, *Pesticide Chemistry* (Elsevier, Amsterdam, 1988).
7. C. D. S. Tomlin (editor), *The Pesticide Manual*, 11th edition (British Crop Protection Council, Farnham, 1997).

The Inorganic Chemistry of Life

Robert J. P. Williams

11

Introduction

It is generally agreed that the evolution of living systems, no matter how they began some 4×10^9 years ago, has been in the direction of increasing their capability of survival through increasing the sophistication of their chemistry. Since all chemicals are made from the limited number of stable elements in the Periodic Table, Figure 1, that is fewer than 100, we can easily accept that the increase in sophistication of the chemistry of life may well manifest itself in the increasing sophistication of the specific uses (fitness) of many of the elements individually. Bio-inorganic chemistry is an approach to living systems that follows this point wherever it may lead, looking first at analysis of living systems in terms of all of the possible 90 elements, Figure 1, and then seeking their functional values. It often uses information from model substances, which has supplemented the knowledge gained from previous examinations of the inorganic chemistry of elements of the Periodic Table from a purely non-biological interest. All these non-biological studies have shown elements within different groups, Figure 1, to be very different in physico-chemical character. It is reasonable to assume that living systems, as they evolved, had to utilize and develop these differences effectively as aids to survival. Part of the test of fitness for a living organism is therefore to examine the use of the elements within it, taking into account both the availability of each element, since extracting an element costs energy, and its known chemical as well as biological functional value. All three features are based on chemical properties of course. In fact the availability and biological functional value of an element are associated with the strengths of its binding to inorganic constructions in the mineral world on the one hand and to organic constructions in biology on the other, whereas in chemistry both types of constructions are analysed functionally without an immediate view of their biological significance.

Since substitutions of similar elements for one another in inorganic chemical and in extracted biochemical systems have often been shown to be ineffective for generating a given function, it must be the case that biological systems have selected elements on the basis of the product of the availability and the functional value. All would be well in this analysis if the availability of each element had remained fixed, but it has not. However, even in an evolving environment, there is not the opportunity for biology to develop species *de novo*. Change must be gradual. We shall see in fact that

IA	IIA	IIIB	IVB	VB	VIB	VIIB	\multicolumn{3}{c}{VIII}	IB	IIB	IIIA	IVA	VA	VIA	VIIA	0
H																	He
Li	Be											B	C	N	O	F	Ne
Na	Mg											Al	Si	P	S	Cl	Ar
K	Ca	Sc	Ti	V	Cr	Mn	Fe	Co	Ni	Cu	Zn	Ga	Ge	As	Se	Br	Kr
Rb	Sr	Y	Zr	Nb	Mo	Tc	Ru	Rh	Pd	Ag	Cd	In	Sn	Sb	Te	I	Xe
Cs	Ba	Ln	Hf	Ta	W	Re	Os	Ir	Pt	Au	Hg	Tl	Pb	Bi	Po	At	Rn
Fr	Ra	Ac	Th	Pa	U												

- ● Bulk biological elements
- Trace elements believed to be essential for plants and animals
- Possibly essential trace elements

Figure 1
The Periodic Table of the chemical elements, showing those with functional presence in biological systems.

biology has undergone a series of modifications whereby organisms have had to compensate for the loss or gain of availability of an element within a pre-existing system after relatively minor adjustment. For example, often use of the elements lost required the development of scavengers for them, while the elements gained by the environment required internal cellular protective devices. The protective devices could then evolve to generate novel functions and eventually new species. In part, of course, biological activity itself is responsible for changing the availabilities of elements. In order to appreciate the roles of the elements in biology we obviously need first to have a very simplified view of a typical biological cell and its essential activities in mind. The description to be given will apply to all cells from bacterial cells to those in higher animals so that its features will have remained common throughout evolution. This means that we use for illustrative purposes at first a single containing membrane and no extracellular matrix linking cells together, Figure 2.

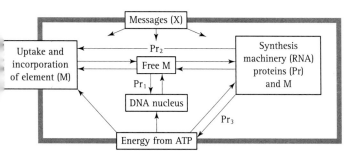

Figure 2
A schematic representation of general features of all cells. Pr is a protein. The important interactive nature of all the units shown is stressed in the text, especially the role of elements, M, in connecting parts together.

A Typical Cell

Every cell has several features in common, Figure 2. Cells need to take in material describable in terms of elements (M); they need energy from an external source; and they receive information (X). All three reach the interior through a membrane that retains an aqueous cytoplasmic solution and in which the elements M, energy and information are distributed. The distribution is to synthetic protein-based machinery for the manufacture of all the cell's components, including the coded DNA which instructs the cell. Thus a cell leads a responsive existence and it can also reproduce itself. Different cells have different major functions. It is seen immediately that there is a basic material limitation to life – the supply of certain 'inorganic' chemical elements (M).

Now, the properties of proteins are central to all activity. They are synthesized from a very few of these elements, H, C, N, O and S, and dominate the activities of all other elements, although by themselves proteins cannot function in cells. Before we can see the essential role of the rarer elements we must tackle in outline the dominant chemical activity, namely the required synthesis, in relation to the contents of the typical cell. The major synthetic pathways of cells involve the acid–base, especially condensation, reactions and the redox reactions of the common non-metal elements H, C, N. O, P and S, Figure 3, to form the polymers DNA, RNA, proteins and polysaccharides as well as the lipids. The reactions require internal carriers of materials and energy as well as catalysts and controls. Here we cannot describe the multitude of biological pathways known to occur in cells, such as the glycolytic pathway, the citric-acid cycle, synthesis of amino acids, fixation of nitrogen, transduction of energy and assimilation of carbon. However, these pathways must all be linked to one another to maintain balance in the cells, Figures 2 and 3. Thus the supply of major elements and energy must be in a feedback loop connected to synthesis and the production of the catalysts for that synthesis. The same limitation will be seen to apply to all the elements (M). In effect the cell has to be seen to be maintaining some steady state, homeostasis, in all activities and this steady state relates to all its component chemicals, which can be broken down to the level of the steady states of elements (M) in the cell. We illustrate this condition by reference to the two major reaction systems of cells, namely acid–base reactions, with special reference to synthesis by condensation, and redox reactions, with special reference to the capture of energy. This will show the ways in which several *major* inorganic elements, e.g. P, Mg, Ca, Fe and K, apart from H, C, N, O and S are required for essential functions in all cells before we tackle the reasons for the presence of several minor but also essential elements.

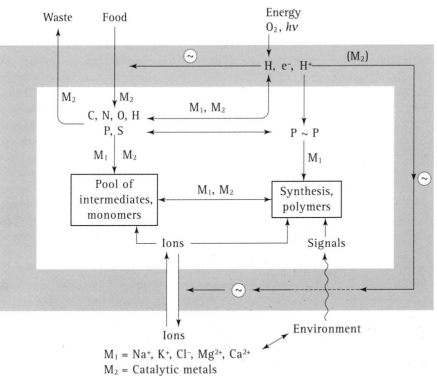

Figure 3
A second schematic diagram of the way in which various elements, M, are locked into the activities of cells, which may but need not be aerobic.

All major synthesis of polymers in biology is by condensation, namely the removal of water between monomers. The condensation reagent is adenosine triphosphate (ATP), see Figure 5 later, which is a compound of pyrophosphate. In effect pyrophosphate's action as a soluble condensation agent is really just the last step in the action of P_2O_5 as a drying agent in organic chemistry

$$P_2O_5 + 2H_2O \rightarrow P_2O_7^{4-} + 4H^+$$

$$P_2O_7^{4-} + H_2O \rightarrow 2HPO_4^{2-} (2P)$$

$$ATP + H_2O \rightarrow ADP + P$$

Thus ATP is a reagent that has stored chemical energy so that we write $-\Delta G_1$, its free-energy capacity to drive reactions, as

$$ATP + H_2O \rightarrow ADP + P - \Delta G_1$$

Now, any condensation reaction requires energy $(+\Delta G_2)$:

$$X-OH + HY \rightarrow X-Y + H_2O + \Delta G_2$$

Adding the two reactions together, we have

$$X-OH + HY + ATP \rightarrow X-Y + ADP + P - (\Delta G_1 - \Delta G_2)$$

Since $\Delta G_1 > \Delta G_2$, this is the way ATP, see Figure 2, drives much of the synthesis in cells.

As inorganic chemists we need to see why such phosphate compounds are so essential. The answer is simple. Although from C, H, N and O very kinetically stable organic polymers can be built, i.e. their acid–base reactions are intrinsically slow, they are not thermodynamically stable. Thus energy is needed to construct them, requiring a chemical condensation system which is also somewhat kinetically stable and yet carries energy. This chemical has to be somewhat easier to make, more reactive than the polymers and it is also desirable that it is safe from redox reactions (which is not true of C, H, N, O and S chemistry). Such kinetic control and redox properties are possible neither for compounds of metals of the Periodic Table in water nor for compounds of extreme non-metals, as simple inspection of their chemistry shows, Figure 4. This narrows the choice of the element to use for driving condensation almost directly to phosphate compounds, though certain sulfur compounds can be used. Especially pyrophosphate has a unique combination of acid/base, kinetic and energy-carrying properties with no redox properties, making phosphorus the very fittest element to be at the heart of condensation polymerization and of acid/base homeostasis.

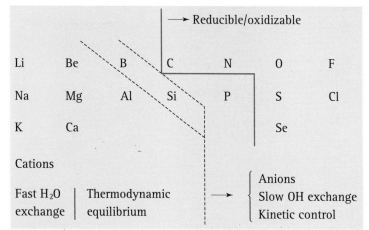

Figure 4
The early part of the Periodic Table, showing how the elements may be classified as cations (acids) or anions (bases), as redox active or inert; and as in fast exchange under thermodynamic control or in slow exchange under kinetic control. Phosphorus, as phosphate, has a special kinetic role very different from that of sulfur, see the text.

In fact ATP (together with other nucleotide triphosphates) is the major chemical for general *energy distribution* as well as chemical condensation in cells. Almost obviously it must therefore be used simultaneously with ADP and P as a *messenger* of the energy state of the cell, i.e. in control, Figure 5, since syntheses or degradations of proteins, DNA and RNA etc. must not occur independently of one another and cannot be independent of energy considerations. (There always must be a link between synthesis and energy and this is clearly most effective if they are based on the same message control unit.) If there is high [ATP]/([ADP][P]) then the cell progresses towards division but if the ratio is low the cell rests. Thus ATP, ADP and P must act in feedback loops upon enzymes:

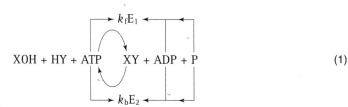

$$XOH + HY + ATP \quad XY + ADP + P \tag{1}$$

where ATP, ADP and P bind to the enzymes E_1 and E_2, to control the rate constants of the reaction, k, but they also control the rates of many other reactions. The control obviously extends to the pH through acid–base equilibrium, so the homeostases of ATP and H^+ are linked and are ultimately linked to all acid–base homeostasis and hence, through condensation reactions, to all polymer synthesis and energy transfer in biology.

To summarize: the fitness of phosphate for all these chemical reactions and message systems is now clear. It is the best element for carrying energy for all purposes, for signalling the energy status to catalytic systems and using condensation energy in the production of polymers due to its relatively easy hydrolysis kinetics and its lack of redox reactions. It is just this fitness for biological functions which inorganic biochemistry should demonstrate element by element. Note that, for every element incorporated, there should be a fit function and a signalling system controlling it.

We are treating phosphate biochemistry as part of bio-inorganic chemistry in this article and four other factors strengthen this approach, Figure 5.

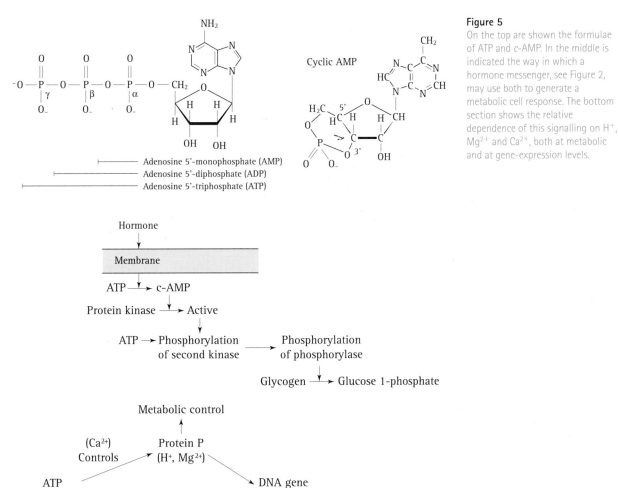

Figure 5
On the top are shown the formulae of ATP and c-AMP. In the middle is indicated the way in which a hormone messenger, see Figure 2, may use both to generate a metabolic cell response. The bottom section shows the relative dependence of this signalling on H^+, Mg^{2+} and Ca^{2+}, both at metabolic and at gene-expression levels.

(1) ATP and ADP are largely present as MgATP and MgADP, hence the presence of Mg^{2+} is part of the internal message and control over acid/base metabolism in biology.

(2) In the presence of equimolar amounts of Mg^{2+}, ATP, ADP and P can bind protons to some degree. The result is that there is a network of compounds all of which can act as messengers based on the three elements Mg^{2+}, H^+ and P^{2-} and two compounds, ADP and ATP. The separation of Mg^{2+} and H^+ from P messages is made, however, in the additional use as a messenger of cyclic AMP (c-AMP), which does not bind Mg^{2+} or H^+ until below pH 1, Figure 5.

(3) ATP is made into a complex part of messages in cells through its ability to make not only c-AMP but also protein-P, which together are major separate factors acting on the synthesis of proteins at the DNA level. They will be shown later to be in a large part dependent on the input of calcium to eukaryote cells for the initiation of a response to external events, Figures 2 and 6, though this is not the case in prokaryotes.

(4) To complete the feedback ATP is used in pumps to control the levels of Mg^{2+}, H^+ and Ca^{2+} as well as those of several other elements in cells.

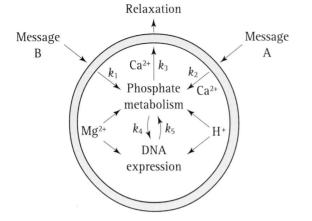

The complexity of the message systems for general acid/base metabolism in cells based on the elements H^+, P^{2-}, Mg^{2+} and also Ca^{2+} cannot be overstressed and involves a wide range of enzymes such as (a) kinases giving phosphorylation of proteins, phosphorylation of substrates and so on; (b) phosphatases that reverse the phosphorylation; (c) cyclases that produce c-AMP from ATP; and (d) esterases that hydrolyse c-AMP to AMP. Some of the network of connections of calcium to these enzymes is shown in Figure 6. The choice of these elements in the control of acid/base cell activity is obviously based on numerous features of their inorganic chemistry, which we analyse below.

The Capture of Energy and Redox Homeostasis

The ultimate source of energy for life is light from the sun. The light, $h\nu$, is absorbed by photopigments, chlorophyll in parts of plants, the chloroplast, and, much as in a photocell, it creates a separation of charge, Figure 7. In plant cells the separation of charge is across a membrane on one side of which the electron goes to an organic molecule, shown as Q for quinone, via iron complexes and reduces it to QH_2. Further reaction of QH_2 can be used to reduce many organic compounds, e.g. CO_2, so assisting in the assimilation of carbon in sugars, while removing protons from one aqueous phase. Meanwhile the positive hole created by the charge separation of Figure 7 is filled by taking an electron from water on the other side of the membrane. When this is performed four times water is converted to oxygen and protons are generated. The reaction requires manganese. There are then two gradients, one of pH and the other of redox potential, between sugar and oxygen. In a different part of a plant or in an animal there are other particles, mitochondria, which can bind oxygen on one side of a membrane and sugar or rather a reduced compound made from it on the other. The reduced compound gives electrons to the membrane, making protons, gradients and CO_2 from the sugar. The electrons pass through the membrane to the oxygen, giving water on arrival of protons there. Iron and copper are required, see Figure 21 later. Both chloroplasts and mitochondria use redox reactions to make proton gradients across membranes. The proton gradient next makes ATP by utilizing a synthetic machine in the membrane of mitochondria or chloroplasts that allows only the protons to flow through it while it forces ADP and P to condense to ATP and H_2O. This scheme was put forward by myself and later Peter Mitchell in 1961. (It was quickly realized that this synthesis of ATP from a proton gradient required a mechanical device but it was not until 1996 that this protein machinery was shown to be a cyclic set of physical changes in proteins related to the flagellum motor of bacteria. The major contributors to this discovery were J. Walker and P. Boyer.)

In this series of reactions there is a requirement for the elements Mn, Fe and Cu as well as those mentioned above. Their fitness for this function will be examined below. First we must extend our brief examination of redox reactions to the general metabolism of the cell as opposed to their role in energy production. Here we find that, just as phosphate compounds are peculiarly suited to carrying energy for condensation reactions, so sulfur compounds are ideal for carrying oxidation/reduction equivalents to the less reactive H, C, N and O substrates due to their kinetic competence. In fact the greater reactivities of P and S are related to their being in the second rather than the first row of the Periodic Table, see Figure 1. The major sulfur reaction, where RSH is glutathione, is

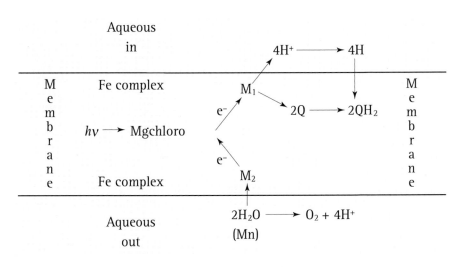

Figure 7
The basic step in the production of ATP is the generation of a proton gradient that acts on an ATP synthetase, see Figure 19. The gradient can be generated by light or by a fuel-cell reaction (Figure 21) but in all cases it involves directly or in controls both iron and manganese, usually to transfer electrons.

$$2RSH \rightleftharpoons R-S-S-R + 2H^+ + 2e$$

whereby, for example, a ketone can be reduced:

$$>C=O + 2H^+ + 2e \rightarrow >CHOH.$$

We make one further point before we leave the brief description of a biological cell's main activities. Given the machinery of Figures 2 and 3, it is necessary to transfer the energy, the condensing ability and the redox equivalents with special molecules such as ATP and glutathione. It is also necessary to carry chemical fragments bearing H, C and N, e.g. CH_3CO^-, amino acids and nucleotides, for synthesis. The coenzymes, namely the carrier molecules used, are given in Table 1. Just because of their somewhat greater kinetic labilities, there are many derivatives of phosphorus and sulfur in the list. It is this fitness for functions to which we must return time and again in assessing the need for elements in biology. To keep the whole working and in an integrated fashion there must be also a balance of synthetic products, so that links, message systems based on coenzymes, to control (catalytic activity) and regulation (transcription of DNA) are both necessary.

The above very brief general account of biological activity is given in order to indicate the absolute need for certain elements in particular activities, including not only common elements but also some rarer elements. It is now that we must return to our introduction and describe the more general reasons for the functional value of some 15 specific inorganic elements. The first consideration is their availability as inorganic elements in the biosphere.

Availability

It is the case that the strongest binding for most elements in the presence of excess oxygen has been shown by chemical studies to be in oxides. Some obvious examples are H_2O, CO_2, SO_3, Fe_2O_3 and ZnO. Thus, on the basis of the affinities of elements for one another, most elements, excluding nitrogen and the halogens, should occur on the earth as oxygen compounds. This did not happen during the formation of the earth owing to the abundance in the universe of the sum of all elements being greater than that of oxygen and to peculiarities of inorganic reactions. Instead of an oxidizing, oxygen-rich universe giving rise to the planet earth, it was an oxygen-depleted earth that was generated. As the universe cooled the elements with lower affinities for oxygen were forced to combine together, e.g. in metal sulfides, or even found themselves without partners, e.g. the iron–nickel core of the earth and the noble metals, Au, Pt etc. Life evolved, then, in a reducing chemical system in the presence of an

Table 1	Some basic element-flow units of biology	
Metabolite (element)	Reactant and control carriers (coenzymes)	Protein-catalysed reactions
Electron (e⁻)	Metal ions, glutathione	Many, e.g. energy conversion, ribose reductase
Hydride (H)	Pyridine nucleotide (NAD)	Many, e.g. energy conversion
Methyl (C)	Methionine	Methylation
Acetate (C) (and other carbon fragments)	Coenzyme As	Fatty-acid synthesis, etc.
Phosphate (P)	Adenosine (guanosine) triphosphate (ATP)	Many, e.g. glycolysis and oxidative phosphorylation, synthesis of DNA and RNA
Nitrogen (N)	Pyridoxal phosphate	Exchange of NH_3
Amino acid (N)	Adenosine (guanosine) monophosphate (AMP)	Ribosomal synthesis of proteins
Sugars (C, H, O)	Uridine diphosphate, dolicol phosphate	Many, e.g. formation of glycoproteins
Nucleotides (C, N, O)	Nucleotide phosphates	Synthesis of RNA/DNA in the nucleus
Hydride/electron (H⁻/e)	Coenzyme Q (quinone)	Energy conversion (membrane)
Sulfate (S)	Adenosine phosphate sulfate (APS)	Sulfonation
Energy (ΔG)	Triphosphates	Utilization of energy
Light quanta (hv)	Chlorophyll	Capture of energy (membrane)

oxide/sulfide surface layer, since the residual pure metals settled through gravity to the earth's centre. The distinction between those elements which are to be found combined with oxygen (hard) and those to be found combined with sulfur or free (soft) is shown in Figure 8 using a simple classification based on the size and affinity for electrons of each ion. In essence, large heavy elements that prefer to form covalent bonds, i.e. those towards the bottom right-hand side of the Periodic Table, Figure 1, have relatively weak affinities for oxygen and those towards the left-hand side and top prefer oxygen more strongly. Some elements, N and the halogens, bind both O and S very weakly. (Such rules apply equally in mineral and in biological chemistry. The interior of cells especially is still an oxygen-depleted zone and remains reducing, although the present-day atmosphere is oxidizing for reasons that are outlined below.)

Before we examine further the availability of elements and restrictions on their entering biology we must note a second factor affecting the former. Whereas elements such as Na, K, Mg, Ca, B, Si and P have a generally fixed chemical character of one oxidation (or valence) state in nature, that is as Na^+, K^+, Mg^{2+} and Ca^{2+} cations or as BO_3^{3-}, SiO_4^{4-} and PO_4^{3-} anions, there are other elements such as S, Fe and Cu that exist in two or more ionic states, HS^-/SO_4^{2-}, Fe^{2+}/Fe^{3+} and Cu^+/Cu^{2+}, and some non-metals exist in series of neutral mixed oxidation states with bound oxygen/hydrogen atoms, e.g. carbon which gives the basis of organic chemistry. The combination of all these features provides competitive conditions and extra selectivity so that much iron is found in minerals as Fe^{2+}, in FeS_2 (pyrite), and much as Fe^{3+}, in Fe_2O_3 (haematite), and carbon occurs both as CO_2 and CO as well as in coal (C) and hydrides (CH_4). The dominant oxidation state of an element affects its availability.

Putting all the chemistry of availability together, we see that the non-metals H, C,

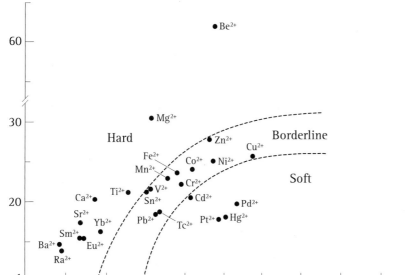

Figure 8
A plot of the reciprocal of the radius *r* for a cation of charge *z* (here 2+), against its ionization energy, I_2. The separation of elements preferring N and S as partners (softer) as opposed to those preferring O donors (harder) is shown.

Table 2	Available free-ion concentrations in the sea	
Metal ion	Original conditions (M)	Aerobic conditions (M)
Na^+	$> 10^{-1}$	$> 10^{-1}$
K^+	$\simeq 10^{-2}$	$\simeq 10^{-2}$
Mg^{2+}	$\simeq 10^{-2}$	$\simeq 10^{-2}$
Ca^{2+}	$\simeq 10^{-3}$	$\simeq 10^{-3}$
Mn^{2+}	$\simeq 10^{-6}$	$\simeq 10^{-6}$
Fe	$\simeq 10^{-7}$ (Fe(II))	$\simeq 10^{-17}$ (Fe(III))
Co^{2+}	Low	Low
Ni^{2+}	$< 10^{-9}$	$< 10^{-9}$
Cu	$< 10^{-20}$	$\simeq 10^{-10}$ (Cu(II))
Zn^{2+}	$< 10^{-12}$	$\simeq 10^{-8}$
Mo	$\simeq 10^{-9}$ (MoS_4^{2-})	10^{-8} (MoO_4^{2-})

N, O, P, S and Cl have always been readily available ever since the earth formed, Table 2. They occur as gases or in soluble salts (in the sea), as do the metals Na, K, Mg, Ca and Mn. In the early dioxygen-free atmosphere other elements such as Fe and Ni were somewhat available as soluble divalent cations or colloidal sulfides from volcanoes. Thus these were the dominant elements in gases or solution when life began. If we suppose that, inevitably to some degree, just these elements formed the primaeval soup of mixed organic and inorganic character, then we are led to the proposal that it was some fluctuation within their abiotic chemistry, that is not a necessary logical progression based on organic synthesis, that created life. The nature of the fluctuation was just some particular combination of circumstances and many candidates have been proposed, but none has become generally accepted. Let us leave this question to one side since it is sufficient for our purposes to use the facts that all these elements are present in what we think of as primitive life forms today and, for all we know, always were present in the most primitive forms of life and that each and every one of these elements is now certainly known to be essential. There is no *organic* chemistry of life known to us that is independent of the inorganic elements above, Figure 1. A much less speculative question to ask is that of how the system evolved in terms of these elements individually and how evolution affected the availability and use of other elements. I describe changes in availability first.

Changes of Availability with Time

The evolutionary development I shall describe here is an interactive one of a pre-existing non-biological inorganic environmental undergoing abiotic changes with the development of life's bio-inorganic chemistry, which now concerns all the elements in living systems locked into evolution, Figure 9. The earth itself evolves very slowly and with many fluctuations and this all-embracing, geo-chemical planetary evolution has constantly re-organized large masses of material over thousands of millions of years. The surface and the atmosphere of the earth are a part of this evolution; for example, at least 90% of the initial CO_2 has been lost. Availabilities of elements have also been affected strongly by living systems, which evolve substantially only at about the same rate as that of geological change,

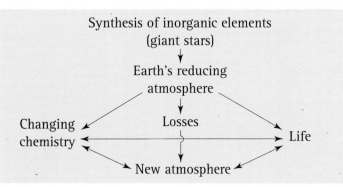

Figure 9
The scheme of the development of the functions of elements on an abiotic or a biotic earth.

i.e. over billions of years. The major elements of the atmosphere are the same as those turned over by life, i.e. H, C, N and O, so that the present atmosphere is now to a fair extent a biochemical product from the primordial atmosphere. On the other hand the sea and the mineral surface of the earth and its waters are still largely dominated by geological rather than biological influences, though this does not hold for the trace-element composition. The balance among the waters, the minerals and the air is constantly undergoing slow change due to biology and erosion but is open to violent readjustment from the core (or even by man). The major adjustment, no matter what the cause, to the atmosphere and thence to the sea, starting some two or three billion years ago, is largely characterized by the change from a dioxygen-depleted to a dioxygen-rich atmosphere over a period of at least a billion years, Table 2. Availability has probably remained approximately constant over the last 7×10^8 years. All carbon chemistry is very unstable in the presence of dioxygen, as are the chemistries of many other elements. This arrival of such an unstable chemical, dioxygen, relative to a pre-existing biological and geological system has had enormous consequences for life and is largely *sui generis*, Figure 10.

First we must ask what difference the excess of dioxygen made to the availabilities and the states of chemicals. The switches in the chemistry of the elements important for life can be shown in part by using oxidation-state diagrams, Figures 11 and 12. As dioxygen accumulated, so the states of elements in Figures 11 and 12 moved progressively from positions nearer the bottom left-hand side towards the top right-hand side. For the non-metals the shifts are from compounds of C/H to CO_2, from N/H to NO_2, from S/H to SO_3, from halides to halogens and so on. The metal elements shifted from low to higher cationic states, e.g. Fe^{2+} to Fe^{3+} and Cu^+ to Cu^{2+}, while the non-metals shifted to covalent higher oxides and their anions. At the same time combinations between elements changed. Thus the availability of the elements slowly altered, either in the nature of the oxidation state or in quantity. H_2S disappeared from the oceans while SO_4^{2-} appeared, NH_3 went to N_2, some insoluble sulfide minerals gave rise to more soluble sulfates or oxides so that zinc, copper and cadmium became available as free aquo-ions in waters, Figure 13, while this oxidative activity reduced the availabilities of some metals, e.g. Fe^{2+} went to insoluble $Fe(OH)_3$ and also Ni^{2+} may have become of

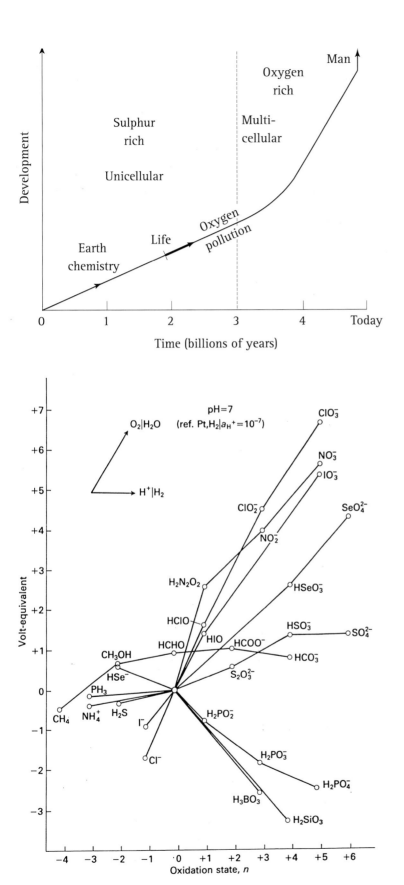

Figure 10
The change of redox conditions on
the earth's surface with time.

Figure 11
An oxidation-state diagram for the
non-metals in water (pH 7). The
redox potential between any two
states is given by the slope of a line
joining them. The relative oxidizing
power is clear.

Figure 12
A similar plot to Figure 11, for the
metal elements in water, pH 7.

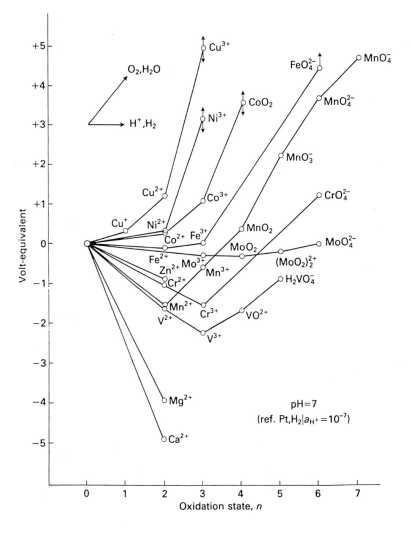

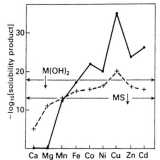

Figure 13
The solubility products of divalent
metal hydroxides (- - -) and sulfides
(—), which have been of extreme
importance in evolution. The
horizontal lines indicate a solubility
product that will give a precipitate
at pH 7 and for $[M^{2+}] = 10^{-4}$ M. The
hydroxide of Fe^{3+}, not shown, is
extremely insoluble at this pH.

somewhat lower availability. This evolving chemistry outside life controlled the possibilities available to the evolving chemistry of biological systems and hence has dominated evolution. Biological evolution is not just random searching within DNA, though this was associated with it, but is a modification directed by functional fitness to cope with environmental change, whenever and wherever this took place. It did not occur in anaerobic zones, for example, where primitive systems are still to be found, allowing us to follow evolution. Moreover, we can see evolution as a gradual timed sequence since it is clear that it had to be that the initially large amounts of reduced non-metals such as carbon (CO), sulfur (sulfides) and nitrogen (ammonia) and reduced iron would mop up and buffer the slow initial build-up of dioxygen. The earliest changes to CO_2 and N_2 need not have had a major influence since they effectively removed O_2 without altering the availability of these elements. They just became more difficult to convert to useful forms. The removal of sulfide was different. Sulphide/sulfur organisms evolved to sulfide/sulfate organisms as the partial pressure of O_2 reached considerable (10^{-8} atm) though still very small levels *before* true aerobes or nitrifying organisms evolved. Molybdenum chemistry had to change from MoS_4^{2-} to MoO_4^{2-} before Fe^{2+} went to Fe^{3+} and before CuS went to CuO, and that is before true modern aerobes evolved around a billion years ago. The slow progression of availabilities of elements, in terms of bound states and amounts, therefore dominated the slow

Table 3	Functional values of the elements
Osmotic controls: transfer and storage of information (electrical)	Na^+, K^+, Cl^-, H^+, Mg^{2+}, Ca^{2+}, HPO_4^{2-}
Chemical–mechanical transmission	Mg^{2+}, Ca^{2+}, HPO_4^{2-}
Acid-base catalysis	Non-metals and divalent and trivalent ions, e.g. Zn^{2+}, Fe^{3+}, H^+, N and S
Redox catalysis	Transition metal ions and some non-metals, Cu, Fe, Mn, Co, Mo, Se and S
Structural roles (excluding the organic polymers)	Si, B, P, S, Ca, Mg (Zn)
Transmission and storage of chemical energy	P, S (C)

Note:
The functional values of the elements arise partly from the chemistry and partly from the input of energy to that chemistry. The input of energy can be to the gradient of the free concentration of an element, to a chemical bond of given kinetic stability or to the synthesis of a polymer in a compartment that acts as a trap.

progression of life, as we can see in the geological records of life and minerals on the earth. Surface geochemistry and life's evolution are one topic.

The Functional Values of Elements

It was necessary to describe the earth's chemistry in terms of availability and its changes before we analyse further the functional values of elements in living systems since we must not be confused by the state of the earth today. Against the background of modern chemistry and biological chemistry, described in the first sections of this chapter, it is easy to recognize the differences among the functional values of elements within compounds, Table 3, within the limitation that the chemistry of life is peculiar in that it is permitted by the only naturally available solvent, water. Life is a process in aqueous solution that is limited grossly by the properties of the elements H and O. It is functions (uses by man or biology) in *aqueous chemistry* that concern us here. We note the following points.

(1) The covalent, yet kinetically somewhat unstable, non-metal chemistry of hydrogen, oxygen, nitrogen and phosphorus (see above) with carbon allows the construction of ordered linear polymers: nylon, polyesters and polythene and fats, proteins, polysaccharides, DNA and RNA. The polymers and related small molecules are made soluble in water usually through the incorporation of charged groups and here, in biology, phosphate is valuable. The fats, which are much less soluble than other polymers, naturally form the membranes which contain many of the essential polymers in cells but are used as detergent films by man.

(2) There is a special role for phosphate in condensation and another for sulfur in redox chemistry for kinetic reasons, see above.

(3) The simple ionic aqueous solution chemistry of Na^+, K^+ and Cl^- allows them to be present selectively with the organic molecules yet mostly not interacting with them. These ions are especially valuable for the control of bulk properties such as osmotic pressure and electrical neutrality since they are very available and of course they carry currents in electrolytic cells in biology or man's apparatus.

(4) The aqueous ions Mg^{2+} and Ca^{2+} bind well to the organic anions and so cause

hardness of water (precipitating detergent anions) and help the organization of DNA and RNA (Mg^{2+}) and of cell walls (Mg^{2+} and Ca^{2+}). They equally well support man's material constructs, such as cements and plasters, and the formation of biominerals (Ca^{2+}). They also act as weak catalysts for phosphate transfer in biology, see above.

(5) The trace elements V, Mn, Fe, Co, Ni and Cu, which bind strongly to nitrogen and sulfur centres in minerals and proteins (see Figure 8), together with Mo and the non-metal Se, undergo redox reactions, since they have more than one oxidation state and are therefore excellent catalysts for redox changes of molecules containing H, C, N, O and S. These features in industry and biology have been studied extensively, see above. Although man can use extra ligands such as phosphines to hold metals in organometallic compounds, biology cannot.

(6) Of the non-metals, sulfur plays a special role not only in cross-linking, in the rubber of motor-car tyres and in proteins, but also by virtue of its activity as a catalytic base and redox centre. No other non-metal is as versatile except hydrogen. The redox chemistry of these two elements may well have been linked with that of iron even in the earliest forms of life, e.g. in its capture of energy, see above. Of course, man uses sulfides in many catalysts.

(7) Finally, the trace element zinc is the only metal which is a really good Lewis-acid catalyst, is not redox active and yet has been available, albeit only at modest levels, since the beginning of life but increasingly so with increasing levels of dioxygen. It aids hydrolysis in organic chemistry and of proteins and peptides in biology.

(8) Although man can make and use metals and alloys, biology cannot.

This quick summary of general functional values is given in order to illustrate that man and biological organisms have uncovered the main ways of using chemical elements. However, biology had to refine these uses in limiting and changing circumstances if evolution was to occur. In this chapter we shall no longer look at man's chemistry but rather at the detailed uses of individual elements in biology.

Fitnesses of Elements for Biological Functions

The next step forwards we have to make is in understanding the fitness for a biological function within the limits of availability. We shall now take it that availability is a well-understood feature of the chemical elements and that we understand in simple outline the peculiar general chemical functional value of any element. Fitness is not just a matter of choosing a particular element, of course, but also of choosing its coordination partners, its whereabouts in a cell and its dynamics among sites. All of these separate features can be understood from chemistry in part but there is a very different feature of biological systems. Living is a process, not a static structure, so any activity within it has to be seen within a context of the flow and turnover of material and energy in the whole organism, Figures 2–5. Looked at in this way, we need an integrated or holistic view of biological systems rather than one based on the separation of components, or a reductionist approach. This is not to say that the reductive approach does not help – it is certainly necessary for understanding – but it will always miss the essence of a cellular system. A simple point is that any catalyst could be refined to be as fast as possible but this could be undesirable in biology in that it forces the product to be in much higher concentration than the reactant in a steady state. A cell may need both for synthesis. Catalysts must be designed to enhance rates but they must also have a *selective and controlled* or at least *controllable rate* in tune with the fitness of the whole system. The amount and the way in which an element is incorporated into a biological system is connected to this fitness; see earlier. Thus the uptake and rejection of all elements from cells must be managed.

Now, there is another way of managing cellular systems, which is to limit the amount of protein introduced into a system. All use of elements is linked to proteins. The synthesis, or expression, of the proteins (e.g. catalysts in cells) is through the coded series of preparation

$$DNA \rightarrow RNA \rightarrow protein \ (catalyst)$$

It follows that the *regulation* of DNA limits the transcription of RNA and that the regulation of RNA limits the concentration of protein, or translation. We have to examine the regulation of the outputs of RNA and proteins by inorganic elements as well as the other functions of elements since proteins plus inorganic elements, not proteins alone, are the functional units of interest here, i.e. to bio-inorganic chemistry. It is necessary for us to see how a cell can manage to couple the production of proteins with the uptake of inorganic ions, leaving no unwanted excesses of useless units, neither apoproteins nor inorganic ions, see Figure 2, and hence using feedback connections among their activities.

Finally fitness has to do with the transmission of information, messages, within the whole organism, since it is necessary to maintain communication in a system such as a cell in which all components are in balance but all are constantly changing their partners. The continuous change of a cell is central to living since all materials in cells are in unstable states and have to be remade as they decay as well as made for reproduction. These points were all introduced under the biological chemistry of phosphorus, see Figure 5.

I turn now to a more detailed look at the various fitnesses mentioned, that is which inorganic ions have been allocated to which activity and how a particular ion has been conditioned for its task as a catalyst, structural unit, messenger, or whatever function it performs. We are looking in detail for a discussion parallel to that outlined at the beginning of the chapter concerning the functions of especially phosphorus, sulfur and hydrogen.

Chemical–Site Fitness for Catalysis

The refinement of a chemical site to functional perfection was termed by Vallee and Williams the development of an entatic (constrained) state[†]. It is easy to appreciate the basic idea by starting from the Haldane–Pauling postulate for how enzymes act. This postulate is that the substrate of an enzyme, on becoming bound, is forced into the transition state for reaction by the use of its binding energy. In effect the enzyme's surface is a mould of the transition state of the substrate. Now consider a catalytic centre containing some (inorganic) atom. The catalyst is not changed in a full reaction cycle, of course, but it must cycle through a series of states while the substrate goes to the product on the surface. Just like in the case of a substrate S going to products P through the transition state S*, we can write that the catalyst C, on going through a cycle back to C, passes through a transitory state C* that is best for catalysis. One function of the protein could then be to hold the metal close to the state C*. A real case is the following. The state of copper in electron-transfer catalysis involves the uptake by Cu(II) of an electron, giving Cu(I), before returning to Cu(II). Electron transfer is the most well-understood reaction path. We know that its rate is reduced by the lowering of relaxation energy in the steps of reaction which are the changes in conformational

---

[†] The adjective 'entatic' for a protein-induced state of a metal ion or another group in a protein can be replaced by the word 'constrained', see H. B. Gray, B. G. Malmström and R. J. P. Williams, *Biol. Inorg. Chem.* 2000, in press.

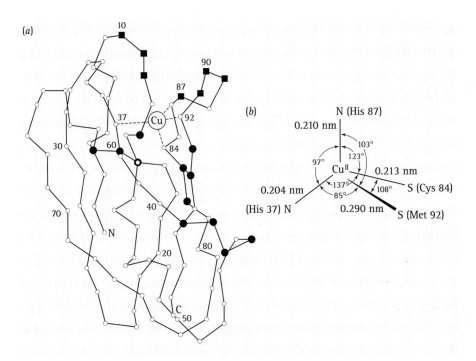

Figure 14
(*a*) A line drawing of the fold of a blue copper protein – note that it is a barrel of β-strands. (*b*) The site of the copper atom which is independent of the oxidation state.

energy of the catalytic electron-transfer agent, here copper in a protein in its two states. The reduction of this relaxation energy requires that the geometry around the metal site both for Cu(I) and for Cu(II) forms be fixed and remain largely the same. This site fitness is observed in many copper proteins, Figure 14, and the site does not have to be either the geometry most liked by Cu(I) or that most liked by Cu(II). In fact more than one rigid site is observed in proteins with the site structure maintained by the rigidity of the protein-fold, a β-barrel. Removal of the copper does not change the fold. The most usual entatic site is closer to that expected for Cu(I) so that there is a greater strain (less optimal binding) for the Cu(II) form and thence the redox potential is elevated. Fixing the redox potential is a second requirement for functional fitness. To keep bond lengths as well as bond angles fixed, it is desirable to have metal ligands that allow covalency, e.g. thiolate. Overall, then, the transition state for an electron-transfer reaction, C*, should require as little change as possible from either the Cu(I) or the Cu(II) state, which is found to be the case.

Although the above is the simplest example, more complex examples such as catalysis of atom transfer demand a comparable but different site fitness. For the attack of an acid on an organic molecule catalysed by a metal ion, the metal ion works best if it has an open coordination site, such as in haemoglobin, Figure 15. For attack by a base the metal ion can hold a base such as OH^- in its coordination sphere at pH 7, whereas in its normal coordination chemistry the OH^- would have remained as H_2O, e.g. zinc in carbonic anhydrase. Site fitness of this kind for catalysis is found for metal and non-metal atoms and ions in most enzymes, but in each case one given element provides the best features for attack on a particular kind of substrate, Table 4. This emphasis on the use of particular metals is based on the differing inorganic chemistries of the elements in the Periodic Table and refined by their incorporation into proteins. For example, inorganic chemists know that, owing to the nature of electronic states, SeO is naturally a two-electron O-atom-transfer agent, as is MoO, whereas the use of FeO for the same step introduces a greater risk of one-electron steps involving radicals, which sometimes should be avoided. However, FeO is of much higher redox potential and is a more powerful oxidant. The following features of inorganic chem-

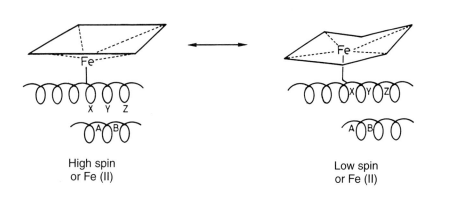

High spin
or Fe (II)

Low spin
or Fe (II)

Table 4	Some specific metal-ion catalyses	
Small-molecule reactant	**Metal ion**	**Examples**
Glycols, ribose	Co in B_{12}	Rearrangements, reduction
CO_2, H_2O	Zn	Carbonic anhydrase, hydration
Phosphate esters	Zn, Mg	Alkaline phosphatase
RNA	Fe, Mn	Acid phosphatase
N_2	Mo (Fe)	Nitrogenase
NO_3^-	Mo	Nitrate reductase
SO_4^{2-}	Mo	Sulfate reductase
CH_3, H_2	Ni (Fe)	Methanogenesis, hydrogenase
$O_2 \rightarrow H_2O$, NO, N_2O	Fe	Cytochrome oxidase
Insertion of oxygen (high redox potential)	Fe	Cytochrome P-450
SO_3^{2-}, NO_2^-	Fe	Reductase
$H_2O \rightarrow O_2$	Mn	Oxygen-generating system of plants
H_2O_2/Cl^-, Br^-, I^-	Fe (Se)	Catalase, peroxidase
H_2O/urea, CH_3CO^-	Ni	Urease

istry in biology are therefore consistent with chemical good sense. MoO is an intermediate in the reactions of $NO_3^- \rightarrow NO_2^-$ and $SO_4^{2-} \rightarrow SO_3^{2-}$, O-atom transfer at a relatively low potential. SeO is an intermediate in ROOH $\rightarrow$ ROH + H_2O and again an O atom is transferred in detoxification without the risk of free radicals being formed. The catalyst intermediate is here at a low redox potential and removes the risks associated with peroxides of high potential. FeO is an intermediate in many hydroxylations, in which the path is FeO(IV)$\rightarrow$FeOH(IV)$\rightarrow$Fe(III) and the group R to be hydroxylated goes through two radical one-electron steps RH$\rightarrow$R$^\bullet$$\rightarrow$ROH, while it is trapped at the site. H$^\bullet$ is transferred to FeO and Fe acts twice as a one-electron-transfer site. The reaction requires a high one-electron redox potential to make the radical R$^\bullet$ and hence needed a much more powerful reagent to catalyse the change. In enzymes all these features of inorganic chemistry are retained but refined by the protein in quite remarkable ways. Reference to the known enzyme sites together with knowledge of kinetics makes it clear that many, if not all, catalytic sites are idiosyncratically refined in entatic ways, Table 4. The particular properties of a given element, which we note in the Periodic Table, are utilized in enhanced and very selected ways, see Table 5 for an example using copper.

Copper protein	State of copper
Table 5 Functional values and constrained states of copper in proteins	
Blue (type I) proteins. Ranging from azurin to laccase	Distorted tetrahedral or trigonal. $E° = 300-800$ mV. Thiolate and three histidines. Electron transfer. β-sheet.
Superoxide dismutase	Distorted tetragonal, open-sided. $E° = 280$ mV. Four histidines. Superoxide reactions. β-sheet.
Dopamine β-hydroxylase. Peptide glycine α-hydroxylase (2 Cu)	Three-coordinate open. $E° > 350$ mV. Methionine and two histidines. β-sheet. Three-coordinate (open). $E°$ unknown. Three histidines. Activation of O_2 via H_2O_2. Activation of the substrate. β-sheet.
Laccase (type II)	Distorted tetragonal. $E° = 500$ mV. Three histidines. Reactions converting O_2 into H_2O. β-sheet.
Copper ATPase	Linear. $E° = 500$ mV. Two thiolates. Copper(I) pump. β-sheet/α-helix.
Laccase (type III)	Distorted tetragonal. $E° = 400$ mV? Three histidines. Reactions converting O_2 into H_2O. β-sheet.
The copper(A) pair of cytochrome oxidase	Distorted trigonal. $E° \approx 400$ mV. Electron transfer. β-sheet.
Haemocyanin	Two trigonal copper atoms (Cu^+). $E° = 700$ mV. Three histidines per Cu. O_2 carrier. α-helical.
Galactose oxidase	Distorted square pyramidal. $E° = 200$ mV. Two histidines and two tyrosines. Oxidation. β-sheet.
Metallothionein	Tetrahedral Cu(I). $E° = 300$ mV. Thiolates. Copper buffering. The apoprotein has a 'random' structure.
Phytochelatins	Unknown. $E°$ is low and unknown. Thiolates. Buffering and transport of copper. The apopeptide has a 'random' structure.
Albumin	Tetragonal. $E° = 200$ mV. Histidines and peptides. Carrier for copper(II). The nature of the fold is not known.

The Self-Assembly of Chemical Partners

Under the rubric of site fitness we have already illustrated element-selection fitness for reaction but we must also look at the way elements can be bound selectively and retained in proteins using at first only thermodynamic (equilibrium) methods and later element-transfer kinetics, see below. It is this selection of the synthesis of suitable proteins for different metals which allows the positioning of elements in cells for functional advantage.

There is no space here to give more than the briefest outline of the factors leading to the matching of the chemical donor character of a site with the metal's binding strength. The important features of physical–chemical fitting to a site are very similar to those controlling availability, see Figure 8.

(1) Size, e.g. $K^+ > Na^+$, $Ca^{2+} > Mg^{2+}$ and $Mn^{2+} > Zn^{2+}$. (In electronegativity-based considerations of 'hard' and 'soft' this size term is unfortunately forgotten by chemists. It is very important in selectivity.)

(2) The electron affinity (EA) of an ion, or the ionization potential (IP) for the change to an ionized state, e.g. for divalent ions the order of the IPs is $Cu > Zn > Ni > Co > Fe > Mn > Mg > Ca$. This series reflects electronegativity or soft/hard scales, Figure 8. For a fixed size it is the dominant factor. Note also that $F^- > Cl^- > Br^- > I^-$ (the order of the IPs of the ions).

(3) The electron configuration and core-polarization energy, e.g. in an octahedral field $Ni > Co > Mn$ and Zn for divalent ions. This is secondary to (1) and (2) above and is called the ligand-field term. It includes the manipulation of spin states of ions.

The way in which the matching of protein structure, including the positioning in space of between two and six coordinating donor non-metal centres, allows much selective incorporation of elements using the above factors is analysed in reference 1 at the end of this chapter. An example of the matching of particular binding to a special function is given in Table 5. The idea of evolved fitness in biology, selective binding to a protein, here is in close accord with chemical complex-ion equilibria of small molecules. Further selection is heavily biased by the selective transfer of particular elements into coenzymes and proteins and of particular proteins into separated compartments and by kinetically controlled insertion, see below. Analytical chemists use very much the same manipulations of chemicals as do organisms.

Redox–Potential Fitness

The desired matching of a redox potential to a function is very dependent on the reaction in question. In order to attack a very stable organic molecule, say a hydrocarbon, using chemical elements in proteins, it is necessary to have a high redox potential since the removal of $H^\bullet$ or an electron, e, from the molecule can itself be a step with a redox potential approaching 1.0 V. From Figure 13 it is clear that simple pairs of inorganic ions M^+/M^{2+} or M^{2+}/M^{3+} do not fall within this range in water. Many are of too high a potential (manganese, iron, cobalt and nickel) and some of too low a potential (molybdenum, tungsten and copper). The protein binding must create fitness for the task of matching redox potentials of the attacking metals to the high potentials of the relatively inert substrates O_2, NO_2^-, Cl_2 and so on, Figure 12. This is managed by the relative stabilizing of states of elements $M(n)/M(n+1)$ where n is a given oxidation state. The equation

$$E = E^0 + \frac{RT}{nF} \ln\left(\frac{K_{n+1}}{K_n}\right)$$

where E is the desired (fittest) redox potential, E^0 is the potential of the hydrated ion in Figure 13, K is a binding constant and n is the oxidation state of M, gives this stabilization. For example, in the case of haem iron the Fe(II)/Fe(III) redox potentials range from -0.5 to $+0.4$, dependent on the protein which holds the haem and 'designed' to fulfil a specific function. Particularly interesting are the redox potentials of iron in the cytochrome chain (see Figure 21 later), which fit the steps of the reaction in such a way as to optimize the conservation of energy in the oxidation of substrates. Another example among copper proteins is given in Table 5. Once again, using models, the chemistry observed in biology has been shown to be not so much mysterious as extremely refined. The relative binding strengths of various oxidation states are made fit by the appropriate evolution of protein binding sites.

Spatial Fitness and Protein Structure

There are two distinct considerations here. The first is the position of an element's site within a protein, namely its depth below the surface, and the second is the position of the protein within an organization. The metalloproteins which form the electron-conduction paths of biology provide a good example. Since proteins are insulators the metal sites have to be placed at distances around 15–20 nm apart to get (1) adequate rates of electron transfer while generating distance across a space; (2) protection of sites from adventitious reactions; and (3) a protein surface for recognition, Figure 16. It is now clear that through-matrix electron transfer which connections of 1.5–2.0 nm distance is similar for many matrices, which are best treated as bulk dielectrics with a dielectric constant of 5–10. The exact amino-acid composition of a protein is of little concern here. Many electron-transfer rates are then not the fastest possible but the fastest commensurate with competent and correct self-assembly and transfer-path lengths. It is then necessary to have in addition to a fast rate switches, controls and limitations on synthesis over the 'inorganic' assembly.

A second example of spatial fitness is the depth of an open site in a channel of a protein containing a highly reactive atom or ion that is wanted selectively for a reaction involving small substrates only. Examples are catalase, superoxide dismutase and carbonic anhydrase, in which the metal is deeply buried and the small substrates and products, H_2O_2, O_2, H_2O, CO_2, HCO_3^- and H^+, reach the site by passing down very tightly constrained channels in the protein, see Figure 16. Larger substrates cannot reach the catalytic sites. Enzymes with similar active sites but that have the function of attacking larger organic molecules allows the organic molecules to enter the mouth of the tunnel at least, e.g. peroxidases and cytochrome P-450. Such channels are also common in membrane pumps for the selective uptake of elements.

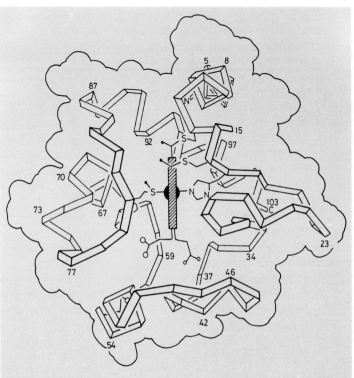

Figure 16

A fold diagram of cytochrome c, an iron haem protein, in which, as described in the text, the iron (black circle) is hidden deeply in the fold. The very specific surface is shown. It controls self-assembly.

Compartmental Fitness

Biological space is far from homogeneous. There are easily discernible regions (extracellular, periplasmic, vesicular, cytoplasmic, membranous and organelle spaces) and there is also even divided organ space in higher plants and animals. Elements are moved into these spaces to fixed concentrations by means described below and this must be done in such a way as to optimize operations of the organization if the organism is to be of the fittest. It is not generally wise to use reactive high-redox-potential reagents such as H_2O_2 close to DNA, for example. The best position for production and use of H_2O_2 is either in peroxyzomes (special vesicles) or outside cells (especially plant cells), Figure 17. In fact, it is precisely in these parts of space that catalases and peroxidases (Fe haem proteins) and some high-potential copper proteins are found. Apart from producing oxidized organic molecules and so acting in defence these vesicles produce a vast range of messenger hormones, including sterols, adrenaline and thyroxine (note the use of inorganic iodide), usually in vesicles. Even the cytochrome P-450 enzymes, which activate O_2, for the removal of dangerous chemicals, detoxification, in the cytopasm are held on membranes away from the DNA in its nuclear compartment.

There are many examples of the fitness of elements in extracellular space. Special notice must be taken of the findings that most, if not all, copper oxidases are outside the cytoplasm. Thus we need a careful analysis of the spatial distribution since this is clearly of the essence of biological organization. It is related also to the evolution of the environment and hence to the evolution of living forms.

A more obvious compartmental division than that of trace-metal catalysts is that of the concentrations of the simpler ions Na^+, K^+, Mg^{2+}, Ca^{2+}, Cl^-, SO_4^{2-} and HPO_4^{2-}, whereas HCO_3^- and Mg^{2+} are evenly distributed to a large degree (but the validity even of this statement depends on the environment). The other simple ions fall into groups in different parts of space, see for example Figure 18,

Na^+, Ca^{2+} in vesicles or extracellular space,
K^+ (Mg^{2+}) in cytoplasm,
Cl^- in extracellular space,
HPO_4^{2-} in cytoplasm and
SO_4^{2-} in the Golgi apparatus.

These separations have many links to fitness. Thus the Na^+/K^+ and Cl^-/HPO_4^{2-} divisions allow osmotic balance and electrical charge balance and, later, the conduction of ions in nerves. The rejection of calcium gives, in addition to the possible use of this ion as a messenger, Figure 6, extracellular structural support and the production of many biominerals, see Table 7 later, while protecting the internal polymers from cross-linking. The situation of sulfate allows the sulfation of outgoing polysaccharides. Such separation requires selective pumping through protein channels in membranes since pumps are deeply connected to securing biological needs regarding the capture and use of energy. Now, many of the above cases are of kinetic trapping behind physical barriers, but we must also consider chemical-kinetic traps.

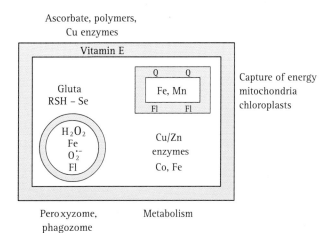

Ascorbate, polymers, Cu enzymes

Capture of energy mitochondria chloroplasts

Peroxyzome, phagozome

Metabolism

Figure 17
A schematic example of compartments in a cell. Notice how different elements are placed in different compartments. Membranes are shaded.

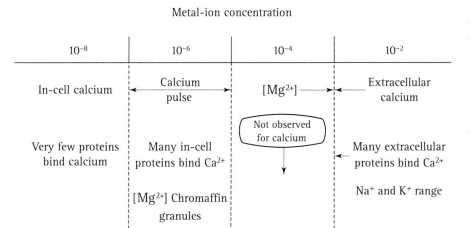

Metal-ion concentration

Figure 18
The concentration levels of some simple ions in and around cells, noting the switch for a calcium signal.

Table 6	Some elements acting in dual kinetic capacities
Element	Forms
H	NADP, NADPH
P	ATP, c-AMP
Fe	Fe^{2+}, Fe haem
Ni	Ni^{2+}, F-430
Mg	Mg^{2+}, Mg chlorophyll
S	Glutathione, coenzyme A

Coenzymes and their Fitnesses

There is one further series of reactions that we have to include before we can attempt to integrate the disparate activites of elements described so far. Although it is true that many metal ions are bound in sites due to their affinities of binding in a general sense or in a localized space into which they have been pumped, this description does not explain the appearance of many metals and non-metals in a considerable number of different sites. The most obvious example is the use of the porphyrin ring of uroporphyrin. Four different metals are placed specifically in six or more rings. Although there is only one for Ni and Co, there are two or more for Mg and certainly more than three for Fe, all used differently. If we consider any one ring chelate as L and any one metal as M, then the insertion necessary has to be under kinetic control in order to gain selectivity. The control rests in the recognition of the metal or a metal complex by a 'chelatase' enzyme, so we have

$$M + L \xrightarrow[\text{chelatase}]{} ML$$

ML is effectively not open to dissociation and so it can act as a free agent as a catalyst and also provide a signal different from that of M or L alone. (In fact, it is known that free haem acts as a cofactor of a transcription factor and that haem in a receptor acts as an agent responding to the messenger NO, quite disconnected from the activity of Fe^{2+}).

In effect the irreversible transfer of M into a given non-exchanging complex is the creation of a new 'element' within biology, Table 6, just as there are diverse ways of using phosphate in covalent non-exchanging compounds such as ATP and *c*-AMP, Figure 5; and note the coenzymes of Table 1. These cofactors then undergo the same selection for function as that which we have seen for the simple ions, i.e. their chemistry is adjusted for catalysis (the constrained state), they are selectively bound, they are placed inside cellular constructs and their free diffusion allows them to become signalling factors. Thus the fitness of iron is in numerous kinetically separate compounds and compartments and a similar description applies to Ni, Mg and Mo. Clearly covalence and steric trapping allow the selection of mostly heavy metals including Cu and Zn.

Now organization also needs a flow of information among the compartments. Which elements are most suitable, fittest, for the transfer of information? Time and again the catalytic ion or element, free or in a coenzyme, is used as a carrier of information, material or energy, which is obviously so desirable if action, control and regulation are to be linked. Fitness for function is then not just the adjustment of binding but also the possibility of use in conveying information. Thus, in the above haem is mentioned as a messenger but Ni, Co and Mo are rarely, if ever, so used.

Table 7	Regulation and control ions		
Ion	Binding constant	Association rate	Dissociation rate
H^+	$> 10^7$	10^{11}	$< 10^4$
Na^+, K^+	< 10	10^9	10^8
Ca^{2+}	10^6	10^9	10^3
Mg^{2+}	10^3	10^6	10^3
Zn^{2+}	10^{12}	10^8	10^{-4}
Mn^{2+}	10^{10}	10^9	10^{-1}
Fe^{2+}	10^{10}	10^9	10^{-1}
Ni^{2+}	10^{12}	10^5	10^{-7}
Cu^+	10^{15}	10^9	10^{-6}
Cu^{2+} (cluster)[a]	10^{10}	10^8	10^{-2}
Cu^{2+} (enzymes)	10^{15}	10^8	10^{-7}

Note:
[a] Not present in cells.

Rates of Exchange and Messengers

Even though catalysis is essential, it is not the dominant feature of biological construction. It is organization that dominates biology and this has two factors, controlled structure and, often, energized flow. Thus, it is not just the nature of binding of the substrates, coenzymes and ions and of their transition states which is dominant but rather the rates of binding and dissociation. In a biological context we do not look at catalysis or equilibria in a thermodynamic sense but at each step or along each pathway, which has to be examined for the choice of its rate control, i.e. the on/off rate constants. In the present chapter we are not especially concerned with transformations of substrates but rather with the variety of values of ions and elements within biology. The on/off rates of ions are very specific, Table 7, which allows their differential use as messengers.

Diffusion and Binding Kinetics

There are many ways of passing a message. Man's favourites are light and electronic devices. They are the fastest but are little used by biology, which is very slow in comparison. It is difficult, if not impossible, for biology to make metal wires for fast electronic conduction. It does make molecular light pipes for transferring photons to reaction centres by adjusting the absorption spectra of chlorophyll molecules with bound magnesium in series. This is comparable with the adjustment of redox potential of units for electron transfer, see Figure 21 later, but both act over very short ranges and, even in a series, 10 nm of biological 'wire' is rarely, if ever, exceeded.

The most frequently used long-range means of transmitting messages in biology is electrolytic; the migration or flow of ions down electrical potential gradients. The fastest diffusing ions at pH 7 are Na^+, K^+ and Cl^- since they do not bind to biological surfaces. They are primary message carriers. (At pH 7 H^+, although it is the most mobile one in water, has a low carrier concentration.) The selective movement of $Na^+_{(IN)}$ and $K^+_{(OUT)}$ exchange across cell membranes generates the best (fastest) current-carrier system we know that is based on ions. It made possible both nerves and brains but these ions can do no more than carry current since they do not bind to proteins and hence cannot change activities. Their use had to await the arrival of elongated cells in multicellular organisms.

Calcium is a fast second messenger. It is held outside cells at around 10^{-3} M and inside cells at 10^{-8} M, Figure 18. The depolarization of a cell membrane by the Na^+/K^+ message

R. J. P. Williams

Figure 19

A schematic diagram of an ion pump. The movement of helices is due to the conformational energy imparted from the β-sheet ATPase to the rotation/translation of helices. Similar mechanical devices are used in the reverse flow of protons to make ATP.

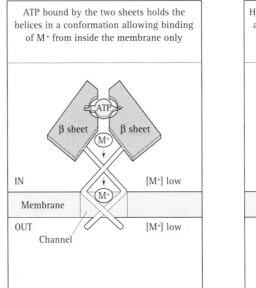

ATP bound by the two sheets holds the helices in a conformation allowing binding of M⁺ from inside the membrane only

β sheet   β sheet

M^+

IN   [M⁺] low

Membrane

OUT   [M⁺] low

Channel

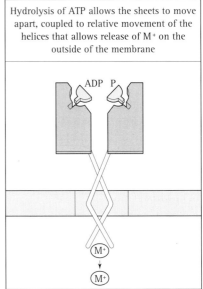

Hydrolysis of ATP allows the sheets to move apart, coupled to relative movement of the helices that allows release of M⁺ on the outside of the membrane

ADP P

M^+

can also open channels for calcium to enter. Now, at 10^{-6} M, after its entry into the cytoplasm, the calcium is recognized by selective binding, even in the presence of 10^{-3} M Mg^{2+}, by a series of special selected proteins. On binding calcium these proteins change shape in such a way as to generate *mechanical* signals in cells, Figure 6. Proteins that are components of mechanical devices must have conformational mobility, just like man's mechanical devices made from levers. They are 'designed' from bundles of helices in which the helices can move over one another and, indeed, the helices can move over β-sheets. Thus proteins designed on the basis of helices are built for dynamic mechanical functions, pumps and triggers, so we expect them to be associated with allosteric binding of ions (and molecules), which is useful in signalling. (Other examples of the movement of helices are in actions such as the cooperative uptake of O_2 such as by haemoglobin, Figure 15, and in compulsory-order enzymes, such as in P-450 and in the transduction of energy, Figure 19). Their value is especially associated with the transmission of messages for this requires mechanical devices, often based on movements of Ca^{2+} (channels and pumps) subsequent to flow of Na^+/K^+. The sustained calcium pulse (or even quite a fast pulse) does not just communicate with contractile systems in cells, as mentioned above. It also affects metabolism, generally, and it does so through a connection with phosphate metabolism in the cell, see Figure 5. This requires more detailed examination since the on/off rate of the calcium, given a fixed binding constant, controls the duration for which the pulse can be sustained. Different biological activities are best suited by different pulse lengths. All these expectations increase our belief that fitness for function lies behind evolution, here of α-helical proteins, designed to match the possibilities associated with particular metal ions, whereby controlled rates and concentrations, Lewis-acid strengths and redox potentials are adjusted by proteins, in such a way as to generate a holistic organized activity.

(The contrast with β-sheet structures such as that described for the copper proteins, Figure 14, is striking. A β-barrel is immobile and excellent for recognition by antibody binding fragments and at dominating small bound units stereochemically. We know from models that rigidity is useful for *selective* recognition whereas mobility is valuable for rapid action. Thus, different proteins are evolved to enhance and control the activities of inorganic elements differentially, choosing the inorganic ion apposite for the required function, i.e. for its intrinsic fitness.)

Slow Controls and Regulators

The binding constants for Ca^{2+} rarely exceed 10^6–10^7 and, given a diffusion-limited on rate constant of 10^9, this means that the relaxation or off rate is relatively fast, occurring within, say 10^{-3} s. This speed is common to fast muscle action. Now, there are also hormones, e.g. adrenaline, which bind at cell-surface receptors and work by letting Ca^{2+} into cells. They commonly have a sustained action, lasting for, say, 10 s, though usually they are not long-term effectors. They may well hold calcium channels in an oscillating, open/closed, condition for as long as 1 min. The value of this pulse will be considered below. The binding constants for Fe^{2+}, haem Fe, Mn^{2+} and Zn^{2+} are many orders of magnitude larger than those of Ca^{2+}, say from 10^{10} to 10^{13}, Table 6. If we assume that their on rate is at least 10^8 then the off rate or duration for relaxation is 10^2–10^5 s. Thus the messages involving Na^+/K^+ and Ca^{2+} cause activation and relaxation without affecting the bindings of these ions. The slower relaxation rates of hours or days of transition metal ions and in particular of zinc can have quite different functions, now separate from those of calcium.

Perhaps the best example is the control and regulation exerted by free Fe^{2+} in aerobic prokaryote cells. The cells need iron for very much of their redox metabolism but there is little iron about in the environment. It is essential that, when the organism runs short of iron, it switches off the (redox) metabolism for growth and devotes its synthetic machinery to scavenging for iron. It does so as illustrated in Figure 20. The machinery which is switched from growth to scavenging is part of the citric-acid (Krebs) cycle for the production of protein intermediates and ATP. This machinery is directed by lack of iron, which dissociates from some critical enzymes, away from the production of ATP towards the synthesis of scavenging chelating agents, ferroxamins (L), which are exported to the environment. The switch from one condition to the other is dependent on the iron-dependent Fe protein, FUR (ferric uptake regulator). When iron is in excess in the environment and adequate in the cell FUR·Fe^{2+} binds to DNA and switches off production of the enzymes for synthesis of ferroxamins while the iron-dependent normal metabolism of the cell, E_1 and E_2, continues. When the amount of iron in the cell is limited, FUR (with no Fe^{2+} bound) switches on the scavenger system while the lack of iron acts as a limiting control of all oxidative activity. Note the dual function of iron in control of catalysis and regulation of the production of protein. Control is generally much faster than regulation. There is a further slow regulatory switch, called FNR, which depends on the association/dissociation of an Fe/S unit to switch the cell from aerobic to anaerobic metabolism.

In higher organisms there is a parallel system using transferrin as the synthesized scavenger, now a protein, and the internal signal is again an iron/sulfur protein, an aconitase. In all cases the overall rates of response to the levels of free Fe^{2+} in the cell are in terms of minutes, hours or days. There are likely to be several other controls based on Fe^{2+} levels in advanced and primitive cells.

The peptide hormones which were also developed in multicellular organisms act in a different way. They eventually effect the phosphorylation of internal proteins at the membrane, which diffuse and then bind to the nucleus and alter the synthesis of proteins. The phosphorylations are now maintained for long periods. This is made to occur by the intermediate release of Ca^{2+} into cells by the peptide binding on the surface, Figure 6, possibly in long sustained series of short pulses so that calcium is involved in transmission over a variety of durations depending on the signalling device. Interestingly the synthesis, release and destruction of the peptides is copper- and zinc-dependent, indicating their recent appearance in evolution, see below.

We can see how message systems can be extended to much longer periods of time by looking at the rates of on/off reactions for series of ions. The rates of exchange in water are

Figure 20

The homeostasis of iron. Iron has functions in enzymes that are controlled by the levels of iron and protein in the cell. The two are produced, protein by DNA and Fe by uptake, so as to match. If Fe goes short then the cell produces the scavenger chelates (ferroxamins) shown below. When iron is in excess more proteins are produced for growth.

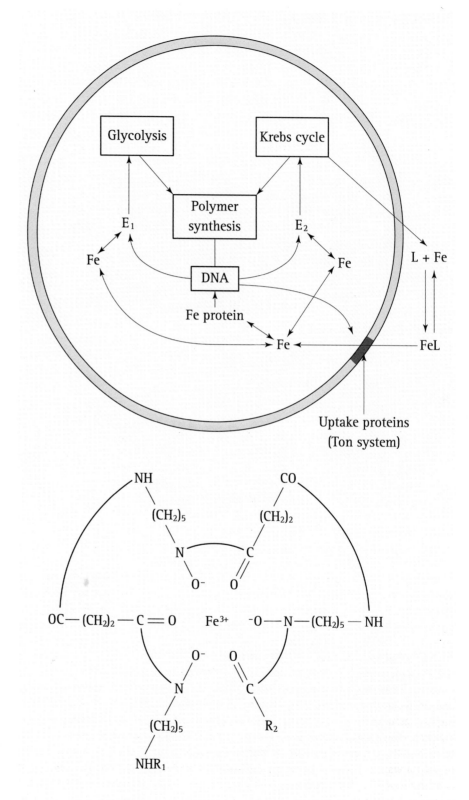

$$Na^+, K^+, Cl^- > Ca^{2+} > Fe^{2+} > Zn^{2+}$$
$$10^9 \qquad 10^8 \quad 10^8 \quad 10^8$$

and the binding constants are roughly

$$Na^+, K^+, Cl^- > Ca^{2+} > Fe^{2+} > Zn^{2+} > (Cu^{2+}, Mo) > haem$$
$$0 \qquad 10^6 \quad 10^9 \quad 10^{13}$$

Thus the speed of relaxation, given equilibrium exchange, gives the off rate as

$$Na^+, K^+, Cl^- > Ca^{2+} > Fe^{2+} > Zn^{2+} > (Cu^{2+}, Mo, haem)$$
$$10^9 \qquad 10^2 \quad 10^{-1} \quad 10^{-5}$$

Now there is a group of organic hormone regulators that bind to receptors either in the membrane or deep inside advanced cells, causing differentiation. They are the peptide, steroid and similar hormones. The steroids commonly cause the protein to which they bind in the cytoplasm to migrate into the nucleus and alter the synthesis of proteins. The hormones' binding constants are $>10^{12}$ and they may have slow on rates or associated rates of formation $<10^8 \text{ s}^{-1}$. Thus, like Fe^{2+} and Zn^{2+}, they have long relaxation or residence times, bound to proteins or associated with DNA, of hours or days. In fact the protein sterol receptors which bind DNA are in large part zinc fingers. The hormones have similar on/off rates to the rate of exchange of zinc, which suggests that zinc could act as an internal cytoplasmic messenger in advanced cells responsive to these hormones of multicellular eukaryotes. Thus, whereas Fe^{2+} is known to act as a similar (somewhat faster) transmission cofactor inside all cells, zinc, which is not redox-active, may dominate direct connection of DNA in advanced eukaryotic cells internally. The evolutionary features will be described below. We shall see how a new organic chemical set of signals evolved with the new chemistry of elements in the presence of dioxygen. Individually the same criteria for fitness as those for individual elements described above must be used. (Note that while all prokaryote signals are fast, eukaryotes may have signalling rates of days, months or even years).

To summarize these rate-of-exchange data, much as the inorganic and organic chemistry come together in biology to make fit constructions, catalysts and osmotic and electrical controls, so they associate in matched fitness for messages:

inorganic	Na^+, K^+, Cl^-	Ca^{2+}		$Fe^{2+}, Zn^{2+}, Cu^{2+}$
organic	acetylcholine	c-AMP, adrenaline		sterols, peptides

In the end, however, we have to see how the value of one element or organic compound is knitted into the structure of the whole system rather than just into these bits and pieces. This is seen too in the continuously developing change of homeostasis as the availability and compartmentalization of elements evolved.

A Summary of In-Cell Fitnesses

In this chapter we are trying to demonstrate the fitness of elements for biological functions. So far we have shown (i) the suitability, after adjustment of the stereochemistry and electronic states by proteins, of many elements for the catalytic action with which they are associated; (ii) the way in which thermodynamic binding and kinetic controls allow specific associations of particular elements with special organic groups and proteins; and (iii) the use of pumping energy makes it possible to place elements in space in such a way as to optimize their value, which implies control over free and bound concentrations in compartments. Furthermore, in the last few paragraphs we have shown (iv) how a combination of binding strengths, relaxation rates and diffusion allows a series of independent non-interacting in-cell message systems based on inorganic chemistry and with very diverse rate constants to be devised. Having examined all these separate uses of the elements we must see how they are put together to make a cooperative whole. (Eventually we will need to show how much of this activity is connected to the genome but we cannot do this yet (2000)).

The Cooperative Action of Elements

When we described the condensation reactions driven by ATP as an illustration of the fitness of phosphate for driving these reactions (no other element can replace it), we observed that its functions were inseparable from the uses of H^+ and Mg^{2+}, for example. We then connected H^+ to redox chemistry involving sulfur chemistry. We also suggested that these activites are linked to the whole of biological synthesis of polymers, Figure 5. To complete the loop of dependences we mentioned, and will now describe in somewhat more detail, that the formation of ATP itself is largely dependent upon energy-capture redox chemistry via proton gradients, Figure 7. This redox chemistry takes place in membranes and requires the use, because of their functional fitness set out above, of a further set of elements. Thus we start to build up the requirement for mutual homeostasis of as many as 15–20 elements.

Chloroplasts and Mitochrondria

The ultimate needs for biological synthesis are the intake of material, some 15–20 inorganic elements, and energy, Figures 2 and 3. While H (H_2O), C (CO_2), N (N_2), O (H_2O), S (H_2S) and Si ($Si(OH)_4$) can enter freely by diffusion, most other elements have to be obtained from water initially by forced (pumped) movement across membranes and then distribution of ionic or complexed-ion forms. These include Na^+, K^+, Mg^{2+}, Ca^{2+}, HPO_4^{2-}, Cl^-, Mn^{2+}, Fe^{2+}, Co^{2+}, Ni^{2+}, Cu^{2+}, Zn^{2+}, SeO_4^{2-}, MoO_4^{2-}, Br^- and I^-. All these elements, with one or two exceptions, are essential to all life, including man, and below we shall illustrate why this is so. To retain the elements they have to be incorporated into covalent structures and/or pumped across further membranes using chemistry plus molecular mechanical devices (pumps) using energy. The covalent incorporation (of C, H, N, O, H, P and S) is initially in mobile coenzymes, Table 1, and then in the synthesized polymers, requiring material and energy. Simultaneously there are feedback messenger molecules to keep synthesis and degradation in harmony with one another and with the energy supply. All the syntheses and pumps use ATP and we must see how bio-inorganic devices lead to the formation of the necessary trapped pyrophosphate, ATP (1), starting from light ($h\nu$) or energy released from fuel cells (sugars + O_2) (3). Of course we have to include O_2 and the production of sugars (2) in the energy path from the essential light:

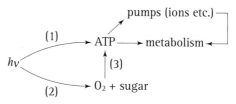

In this diagram we wish to place some 15 elements in cooperative activity. The first very special inorganic feature of this set of activities is the difference between photosystems (I) and (II), Figure 21. In the latter Mn is the unique oxidizing centre giving O_2 and as yet the centre is of unknown structure. On the reductive side there is a series of iron complexes leading to the ferredoxins (Fe) for the hydride reactions. Most of the photon-capture system is made from Mg complexes, chlorophylls. When O_2 and sugars form two ends of a fuel cell, (2) above, the whole chain of energy transduction is dependent on more than 20 iron atoms in three trans-membrane units and several more in succinate dehydrogenase, Figure 22. The terminal oxidase, cytochrome oxidase, has one magne-

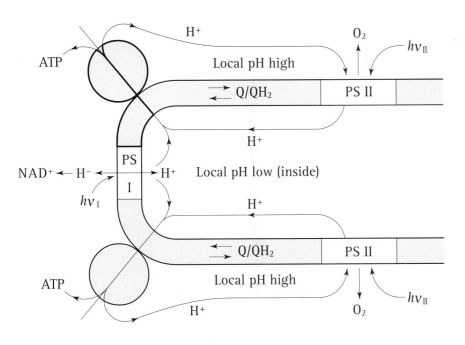

Figure 21
The membrane of the light-capture and energy-producing (ATP) membrane of plants. Note how the elements are arranged along the membrane, e.g. photosystem II has manganese, chloride and calcium dependences but photosystem I does not. The ATP synthetase is shown schematically in Figure 19.

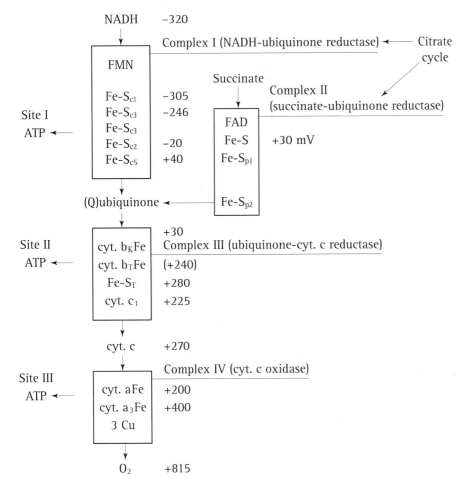

Figure 22
The oxidative energy-producing membrane of animals and plants. Note how the elements copper and iron are arranged in the membrane.

sium, three copper, one zinc and two haem iron centres. The central particle has two haems (Fe) and one Fe–S complex and connects to the first via cytochrome c (a haem protein). (This is in contrast to one connection between the two photosystems, which depends on a mobile copper protein in aqueous solution.) The first complex has several Fe–S proteins, perhaps six. If we treat the proteins and the membrane as an organizing uniform dielectric matrix for the essential metal atoms then we have a picture of much of the electron-transfer chain of the capture of energy for oxidative phosphorylation. We have to observe the essential use of iron (particularly), manganese and copper. They are the fittest elements for the necessary electron-transfer reactions since they and only they generate the possibility of electron transfer, O_2 chemistry and redox potentials.

As pointed out by the present author in 1961, oxidative phosphorylation and photophosphorylation chains produce ATP from these electron flows and with the help of the intermediate proton gradient, Figures 6, 21 and 22, and the basic inorganic chemistry is just oxidation, $H^- \rightarrow H^{\bullet} \rightarrow H^+$. Now the H^+ gradient can also be used to pump other ions or compounds by exchange when we have

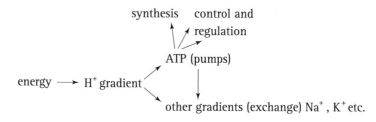

The basis of the transduction of energy is therefore the appropriate use (fitness) of a series of inorganic elements for different tasks, i.e. Na, K, Cl, Ca, Mg, Mn, Fe, Cu, H, P and S all associated with proteins, which like DNA and RNA are made from the essential elements H, C, N, O, S and P. (Notice that the proteins plus special elements are placed in space. It is this organization which creates the flow so indicative of life.) Each and every stage must have control based on the mobile elements incorporated into it. Thus immediately there is no biology without at least 15 elements. The fitness of the electronic, the electrolytic, the structural and the kinetic characters of elements is wonderfully illustrated in the transduction of energy. In fact *all* cellular activities are knit together by the selective activity of very many elements.

We now have two distinct tasks. The first is to show that, given a certain availability of elements, the original forms of life would be driven by the existing environmental chemistry to form a network within life that utilized in some optimal way the energy then available. The second task is to show how, when availability of elements changed, the life forms changed to take advantage of the new chemical possibilities and new regimes of fitness evolved.

Integration

A cell is an integrated chemical system using elements in selected ways for a multitude of purposes, Figure 2. We must see that the transfer of material and energy in metabolism is constantly monitored in a network, see Figure 30 later, to keep its composition and structural stability. Unlike a man-made computer network of feedback the medium for carrying the information cannot be wires, which are separate channels for discrete bits of information. Instead there is one general continuous medium for transport, the aqueous space, divided only by membranes, with channels and pumps. Confusion is avoided since messages are carried by a multiplicity of selective messengers, which can diffuse and are received by receptors in limited ways, though generally in water. We turn then to the development of these systems during the evolution of constructs and catalysts within whole cells, starting from the most primitive cells we know.

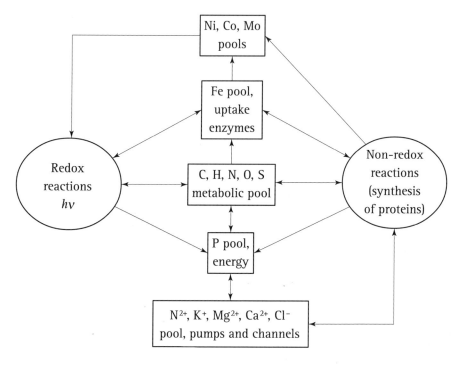

Figure 23
A diagram illustrating the cross-talk relationship of the elements in a primitive cell (with no Cu and little Zn).

Primitive Organization

We described earlier the basic requirements of cellular activity in terms of the synthesis and degradation of C, H, N, O, P and S compounds. All cells maintain this cytoplasmic metabolism in the range from -0.5 to 0.0 V. The most primitive organisms we know, the anaerobes which are often called archaebacteria, work in this redox range and maintain the catalytic elements such as Mn, Fe and haem Fe, Co (in vitamin B_{12}), Ni and F430 Ni (Zn?), Mo and W (as their pterin complexes) at fixed levels both as free and as bound ions, Figure 23. The feedback controls which maintain this homeostasis have been described in the introduction to this chapter as controlling cytoplasmic activity common to *all* cells. At the same time the levels of Na^+, K^+, Cl^-, Mg^{2+} and Ca^{2+} were controlled as described above. We do not know about the internal controls inside the archaebacteria but in general all cells have been shown to have controls involving phosphorylation and c-AMP (phosphorus), methylation and acetylation (carbon), redox reactions (hydrogen and sulfur) and probably general electron-transfer reactions (iron). This includes regulation of expression at DNA. We have described these systems only in briefest outline but they are known to relate to the switching on and off of the major pathways of the uptake of elements, metabolism and cycling of cells. In essence we can understand these internal systems and their interactive feedback. We now wish to look more closely at the evolutionary role of the elements in higher organisms.

We saw that elements are selected for their functional value within organizations. One heavy restriction for biology was the lack of accessibility (availability) of elements, including the form in which any element was available. The availability has changed continuously over some five billion years, which has grossly affected the way in which higher organisms could evolve and will continue to change with or without man's input, see Figure 10. The changes follow changes in the pH and the redox potential, the loss of gases (CO_2) and changes of temperature. In this chapter we have only enough space for an illustrative example – the coming of dioxygen, a change in the available form of the element oxygen from that in H_2O, which led to the introduction of truly multicellular life based on new chemistry *external* to cells while most of the chemistry that had been going on in cells was retained. We use it as an illustration of the total dependence of higher life forms on a combination of organic and inorganic elements.

Multicellular Organisms

Many of the new features of cells after the first three billion years concerned the external matrix which stabilized the extracellular matrix. This matrix had to be fairly rigid in part but an open mesh to allow the transport of material to the cells. The rigidity is brought about by *copper* cross-linking oxidase systems involved in large part in the production of chitins and collagen, Figure 24. Copper became available only after the partial pressure of dioxygen had increased and simultaneously iron became less available due to its precipitation as $Fe(OH)_3$. (There are virtually no simple iron enzymes outside the cells of higher animal and plant organisms, simply because the loss of availability or iron made it less easy to employ it in new possibilities, much though it had to be spared for metabolism inside cells.) The *fitness* of *copper* as a factor in external oxidases was then brought about by the environmental change owing to which iron could not be used externally. Copper is of course an excellent redox catalyst. The change in inorganic chemistry, Table 2, changed evolution interactively.

The additionally required open-mesh polymers outside cells in animals are based on sulfated polysaccharides formed in the apparatus of Golgi. (In plants, polysaccharides, e.g. cellulose, give a parallel matrix.) Their fitness is due to their lack of binding to almost any kind of ion or molecule and the fact that their anionic surfaces repel, creating space for free diffusion. This oxidized-sulfur chemistry is possible only outside cells in oxidizing media, see Figure 11. Note that it was necessary to evolve a suitable catalyst to handle sulfate (and nitrate later), which needed a two-electron, atom-transfer centre. This was provided by a *molybdenum* cofactor and is a modification of primitive aldehyde oxido-reductases, found in cells of anaerobes. We have described above the fitness of molybdenum for such reactions insofar as it readily carries out O-atom-transfer reactions at low potentials.

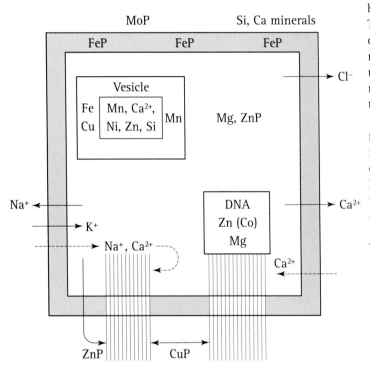

Figure 24

A diagram illustrating the relationship of compartment activities in a modern cellular system, showing the connection of copper and zinc to the extracellular network of filaments.

Within the external, more rigid, matrix of collagen or chitin, cells have to divide and grow so that the matrix must be constantly cut and then re-assembled. The cutting is done by a set of enzymes, collagenases, elastases etc. in animals, chitinases in plants and insects. They are all metal-based. Those in the first group are hydrolytic ones dependent on zinc, see Figures 26–29 later; those in the second group depend on iron and manganese using peroxide. None of these systems could have existed before the advent of dioxygen, since one required the release of zinc from sulfides and the other the metabolism of dioxygen itself. The hormones controlling these systems include many produced in the presence of dioxygen by iron (plant and insect ectozomes), by zinc (peptide hormones) and by copper (the amidated peptides and adrenaline). There is, then, a new network of messages linked to the new extracellular matrix, Figure 25, related to, but almost independent of, the cytoplasmic systems of Figure 24. However, the new messengers must also have new proteins linking them to the regulation of

Table 8	Examples of regulatory DNA-binding metalloproteins
DNA-binding protein	Enzyme or protein expressed
ACE 1 (Cu(I) cluster)	Metallothionein Cu, Zn Superoxide dismutase Cu, Zn
FUR (Fe(II))	TON B receptor system (Fe, Co) Enzymes for synthesizing siderophor Superoxide dismutase Fe, Mn
IRE (RNA) (Fe)	Transferrin Transferrin receptor Ferritin
Vitamin-D receptor (Zn)	Alkaline phosphatase (Zn) Calbindin (Ca) Osteocalcin (Ca)
Glucocorticoid receptor (Zn)	Na, K pumps?

DNA, Table 8. In fact copper and zinc are linked to a new set of thiolate-containing proteins, the metallothioneins, which control their levels in cells, while the synthesis of signal peptides and other hormones is largely confined to vesicles, separating it from primary (primitive) and basic reactions of the cytoplasm.

Of course, we must also enquire about the origin of the steroid and similar hormones which carry messages between cells located now in extracellular matrices and here we return to the in-cell secondary metabolism based on new iron–oxygen chemistry, much like that which we have seen in the iron-uptake scavenger chemistry of aerobic prokaryotes, which also developed with the advent of dioxygen, Figure 20. The iron enzymes *in cells* of animals and plants were adopted to new oxidizing conditions by placing them in association with membranes both of the plasma and of vesicles, Figure 25. (In plants and bacteria as haem enzymes they are often outside cells.) Thus they do not interfere with primary metabolism or its signals, though they must be interwoven with it. These oxidized or even iodinated (thyroxine) molecules then became the long-term hormones, stimulating differentiation. (Note that iodination requires the oxidation of available iodide by peroxide.) The hormones also regulate many of the essential proteins which came to be used in control of long-term calcium signalling, see Figures 26–29.

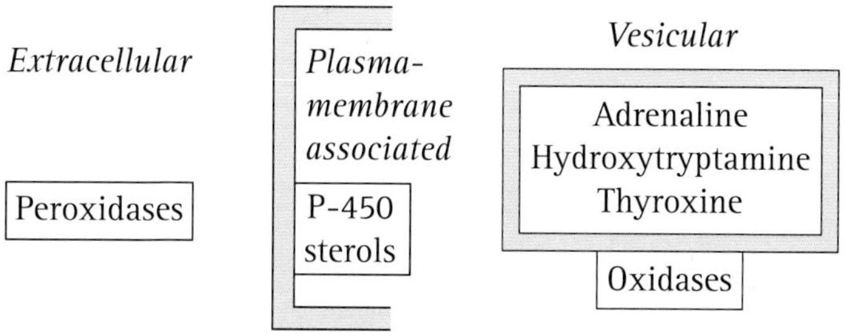

As we described for simple single cells, the uptake, incorporation and then use of any element are integrated and the integration extends to all other elements through cross connections. The case of iron was illustrative, see Figure 20. In multicellular systems the same integration is, if anything, more important. The structure of an extra-cellular matrix depends on relationships between zinc and copper but also on iron and calcium, as shown in Figure 26, the last acting not just as a messenger but also as a structural unit binding to surfaces, e.g. collagen, produced by iron-, copper- and zinc-catalysed metabolism. We need to note the cross connections to the development of extracellular structures and the proliferation of cells and it is this complication which demands the cross connection of messages based on the activities of at least Cu, Zn, Fe and Ca. One illustrative example will serve here.

The sterol hormones are clearly implicated in growth controls. As stated above,

Figure 25

A variety of iron enzymes produces hormones. In animals these proteins are mainly associated with the cytoplasmic faces of membranes whereas in plants they can also be extracellular.

Figure 26

A scheme to illustrate how several metals and phosphate can be linked in the activities of a cell. See the description in the text for the way energy, ATP, is also involved. Single-headed arrows show unidirectional action while double-headed arrows show exchange.

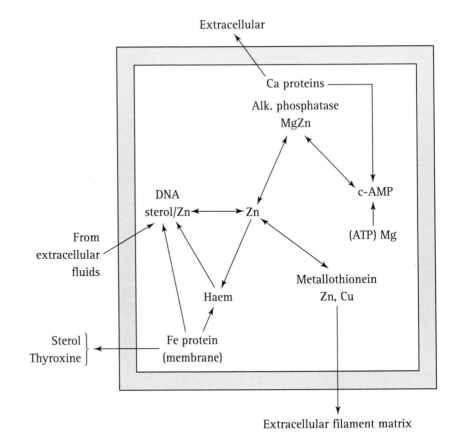

Figure 27

A brief outline of the cross-talk of Fe/Cu/Zn with hormones and connective tissue including calcium and phosphate in bone.

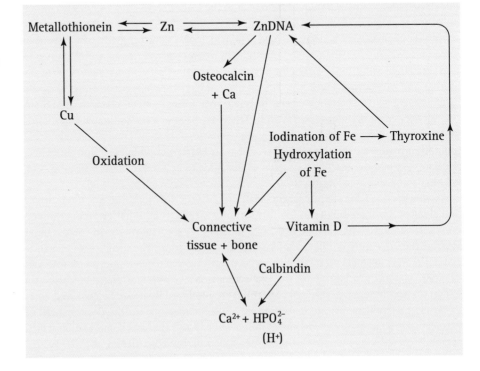

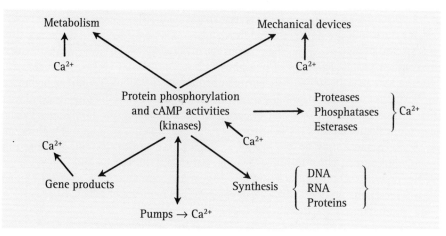

Figure 28
Gene regulation by calcium, which itself controls many functions, probably occurs indirectly through phosphorylation (kinases), dephosphorylation, phosphatases, and through cyclic AMP. The homeostasis of calcium in an advanced multicellular system produces a network leading to the formation of connective tissue and bone.

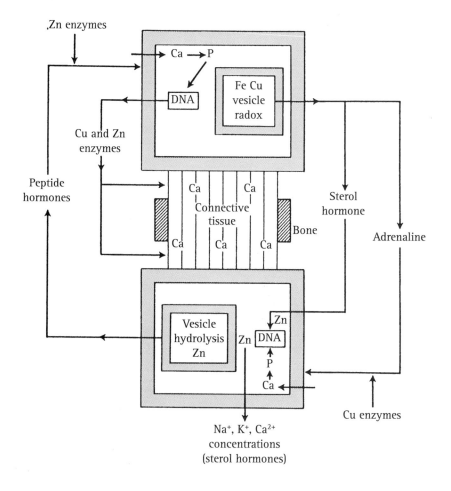

Figure 29
The relationship of two cells from different organs of the body indicating their relationship through hormones and connective tissue and some of the roles of inorganic elements in the various connections.

sterol hormones are produced by oxidative metabolism using Fe enzymes. The receptors for the sterols are Zn fingers and thus an Fe–Zn system regulating the production of RNA and protein generates the proteins for many cellular activities, some of which are coupled to the regulation of phosphorylation, i.e. internal events, see Figure 29. Here we note that the sterols also generate the proteins for mineral controls, including the Ca-uptake proteins of epithilial cells (calbindin), those controlling the levels of Na^+ and K^+ (the glucocorticoids) and the proteins for controlling the formation of bone material (calcium phosphates of osteoblast cells, osteocalcin), see Figure 27. Thus, Ca, Na, K, P, Fe and Zn are linked. Insofar as Zn is linked to Cu and both are linked to peptide hormones, which in turn are connected to activities of Ca and P in cells, we have a network of cross-interacting messengers and synthetic paths, Figures 27 and 28 and see Figure 30 later. This was an essential pre-requisite of advanced cells.

To summarize, we see links of the type

$$\left.\begin{array}{l} \text{Fe (hormones, I)} \\ \text{Cu(hormones)} \end{array}\right\} \rightarrow Zn \rightarrow Ca \rightarrow P \rightarrow Mg, H \text{ and so on}$$

where the first two or three are relatively recent in functional use, as has just been described, whereas the last three are primitive. Functional calcium arrived as eukaryotes appeared at an intermediate stage of evolution.

We see all these activities of inorganic elements as guiding the pathways of metabolism of non-metals, i.e. they act as the fields and switches in a computer-like circuitry of cell metabolism, see Figure 30 later. To complete this account of interwoven fitness, we give a brief description of the peculiarities of the biological fitnesses of minerals that support the extracellular constructs.

Biominerals and Extracellular Space

The fitness of biomineralization of extracellular tissue, Table 9, for its purpose cannot be contradicted. Early in evolution, minerals coated single cells, giving protection, but even here the structures were irregular, allowing pores to form, since access of food through the mineral coat was essential. Sophistication grew as the two materials which had concentrations close to their solubility products, $CaCO_3$ and $Si(OH)_4$, were manipulated on the surface of proteins and then using internal filaments and vesicles. The mineral structures incorporated polymers on the outside of the organisms, whence sophisticated forms evolved and even irregular colonies such as corals and sponges. The incorporation of polymers conferred greater physical strength on early composites.

Once the partial pressure of dioxygen had increased and cross-linked external polymers could be made, the biologically directed mineralization increased again in sophistication. Shells of $CaCO_3$ and surfaces of opal-included form were generated by animals and plants, respectively. The shell now grew as the multicellular organism grew and the growth pattern, logarithmic growth as a cone or a twisted cone, is admirable for retaining protection as the size of the organism increases.

There is no space to elaborate here, but, at some stage in evolution, the animals invaded fresh water, where the concentrations of $Si(OH)_4$ and Ca^{2+} are low. Mineralisation required a new procedure. The growth of animals had always demanded a high supply of phosphate and a high level of external calcium around internal cells. This was managed by pumping in phosphate and calcium so that the levels in the extracellular fluids approached 10^{-3} M. This level of calcium did not approach the solubility limit of $CaCO_3$ but could be pushed to that for $Ca_2(OH)PO_4$ – bone appeared. Now, this particular form of calcium phosphate is extremely interesting since it is not a conventional crystalline solid. It is a piezo-electrical. This makes it extremely fit for its function *inside* a growing organism since the required growth of the bone is stress-dependent, i.e. field-dependent. The fields are stresses of any kind that are

Table 9 The main inorganic solids in biological systems

Cation	Anion	Formula	Crystal	Occurrence	Function
Calcium	Carbonate	$CaCO_3$	Calcite	Widespread in	Exoskeleton
			Aragonite	animals and plants	Gravity, Ca store
			Vaterite		Eye lens
	Phosphate	$Ca_{10}(PO_4)_6(OH)_2$	Hydroxyapatite	Shells, some bacteria, bones and teeth	Skeletal, Ca store (piezoelectric)
	Oxalate	$Ca(COO)_2 \cdot H_2O$	Whewellite	Insects' eggs Stones in vertebrates	Deterrent Cytoskeleton
		$Ca(COO)_2 \cdot 2H_2O$	Weddellite	Abundant in plants	Ca store
	Sulfate	$CaSO_4 \cdot 2H_2O$	Gypsum	Statocysts in coelenterates	Gravity
				Plants	S store, Ca store
	Silicate			Phytoliths	
Iron	Oxide	Fe_3O_4	Magnetite	Bacteria, teeth of animals	Magnetic device
		$FeO(OH)$	Ferritin	Widespread	Iron store
Silicon	Oxide	SiO_2	Amorphous (opaline)	Sponges, protozoa Abundant in plants	Skeletal Deterrent
Magnesium	Carbonate	$MgCO_3$	Magnesite	Reef corals	Skeletal

created by the functional demands of other parts of the body – e.g. muscles and sinews. The bone is incorporated inside these structures using special proteins and cells that monitor the surface. This is not possible for $CaCO_3$, though hydrated silica, which is amorphous, can be and is manipulated in plants. Bone matures with growth.

The bone growth is therefore integrated with the whole body through three major constituents – H^+, HPO_4^{2-} and Ca^{2+}. These ions are each in a fixed relationship with a solubility product, hence extracellular fluids of bony animals are well controlled, Figures 26 and 28. The control does demand a constant supply of these elements and the appropriate proteins from outside. It is also the case that external H^+, Ca^{2+} and HPO_4^{2-} are critical to internal cell functions. Thus a link from extracellular uptake and circulation to intracellular conditions is needed and the reverse for the supply of protein. It is this total integration of inorganic and organic functions which we must tackle finally. It is extremely well illustrated by the activities of calcium, Figure 28. It is present in every cell, in every organ and in every organism. Though there is a variety of combinations of elements in different organisms, there is much that is common, Figure 2.

Note that with the coming of multicellular organisms a new internal mineral evolved, bone, which is a living material as much as DNA or a protein since it responds through the piezo-electrical effect to pressure. Thus a new connection, to a vastly superior material within the whole homeostasis of the elements was made.

The Holistic Nature of Biological Chemistry

This chapter was written to generate a view of biological systems that is not based on the traditional description of so-called 'organic' chemical metabolism. Starting from the Periodic Table, it sets out to describe the common way in which elements are used in biology. This can be called at first a routine inorganic analytical approach to finding a minimum set of elements on which every cell is based. There is no escape from the first conclusion: the following elements, put in the order of the Periodic Table, are essential for all life:

H	C	N	O	Na	Mg	P	S	Cl	K	Ca
(Mn, Fe)	Co	Ni	Zn	Se	(Mo)					

Many organisms utilize a secondary set of elements but as yet it is not known how extensive the requirements are. These elements are

B	F	Si	V	(Cr)	Cu	As	Br	Sr	I
Ba	W								

The first task for the future is that of gaining a better knowledge of which elements are used.

There is then a second task, which is to show the ways in which specific elements are selected, incorporated (including located) and given local functional uses in biology. We desire to know how and why iron, copper, carbon and nitrogen, among other elements, are of significance in life at the level of the constructed molecules in which we see them. The finesse with which biological systems operate in all these respects is becoming more obvious as chemists compare simple models with proteins, RNA and DNA in which the above-mentioned essential minimal 15 or so elements reside. (Make no mistake, each of these polymers is tuned to the presence of many other elements than H, C, N and O.)

The above tasks, which are far from having been completed, have on top of them a higher level of kinetic consideration, which concerns how a cell works as an organization of the 15–20 elements, Figures 29 and 30. This is a very difficult step beyond the consideration of the construction of cellular components. Any machinery made by man

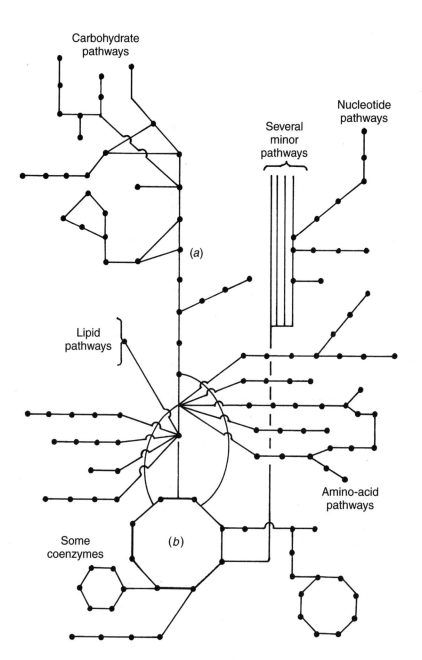

Carbohydrate
pathways

Several
minor
pathways

Nucleotide
pathways

(a)

Lipid
pathways

Amino-acid
pathways

Some
coenzymes

(b)

Figure 30
In the future it will be necessary to draw up circuit diagrams that parallel computer circuits in order to show the intricate way in which metabolic pathways are linked in order to provide a holistic view of a cell's activity. The way in which the inorganic elements are involved in every branch of the circuit and even in almost every step will have to be considered. Here is a scheme for carbon (after W. Kaufman).

or biology is not to be understood in terms of a structural description of idle components. We required further knowledge about the application of energy to cause a desired flow. It is the very essence of life that we are approaching – the organization of 15–20 elements in a flowing system. Much of the flow is not of material for synthesis but rather is of energy and of information, feedback communication, from transfer of material utilizing energy. The elements which are heavily involved are H, Na, K, Ca and P at one, relatively fast, level and at least Fe and Zn at a slower level. The communication network of the machinery is dependent on the selective input of energy, which, as in all machines, is controlled by feedback too. Thus we come to have an holistic view of life based only in part on traditional 'organic chemistry' within biology, which applies within single cells, organs and organisms and even to ecosystems.

Immediately this creates a problem in any consideration of evolution. Whatever part inorganic chemicals play in biology today, they came before organic chemicals on the earth. Water is the solvent and has always had not just CO_2 and N_2 but considerable quantities of Na, K, Mg, Cl, S, P and trace elements in it. They were present from the beginning. However, evolution of the planet and of life brought about changes in the distributions of most other elements within aqueous media and none of them is distributed in true equilibrium to this day. The system continues to evolve. A major contributor to change in the past was the photosynthetic introduction of dioxygen by algae. There followed a succession of changes in life:

new poisonous chemical→new protection→new function

In this chapter we have shown how it is new inorganic chemistry, different from that of the primitive earth, which directed evolutionary possibilities in this process. The examination of these views has hardly begun.

Of the 90 stable elements about 60 are not used. They can be used, as can the other 30, in a variety of combinations in medicines, drugs and poisons. Some known examples of medical use include Li, Al, Sb, Bi, Pt and Au. A huge range of apparent possibilities has not been explored by biology, as we know it, or by man.

We must also reflect further on the use of elements by man. We do not have a good assessment of the impact of the introduction and long-term use of elements in industry. All of this subject material falls under the title of *inorganic biochemistry* and in the first place it demands analysis in order to tackle the task of relating the environment to living systems.

The final point therefore clearly concerns the role of man himself. Man, like all advanced multicellular organisms, is not an independent living organism but rather is dependent on an ecosystem. Any ecosystem is a combination of living and non-living chemicals that are differentiated only by the way in which energy is applied to them. Man, by applying energy in new ways to external materials, makes them interact with himself and all biological systems in new ways. He is the 'photosynthetic–oxygen-producing' algae of today. Like them he puts everything in the melting pot as he plays with the 90 stable elements of the Periodic Table. It is a part of biological inorganic chemistry to look at the developing situation. Change, flatteringly called progress, is inevitable but unpredictable unless great care is taken with biological inorganic chemistry.

[It maybe that some readers will be worried that there is so little description of the role of genes in this account. The major reason for leaving DNA to one side is that it is only representative of the living system in coded form. It is not active except in relationship to regulation. Understanding life is dependent upon the insight we have into *systems* not individual molecules.]

Further Reading

1. J. J. R. Fraústo da Silva and R. J. P. Williams, *The Biological Chemistry of the Elements* (Oxford University Press, Oxford, 1994).
2. I. Bertini, H. B. Gray, S. J. Lippard and J. S. Valentine (editors), *Bio-Inorganic Chemistry* (University Science Books, Mill Valley, CA, 1994).
3. W. Kaim and B. Schwederski, *Bio-inorganic Chemistry: Inorganic Elements in the Chemistry of Life* (Wiley, Chichester, 1994).
4. R. J. P. Williams and J. J. R. Fraústo da Silva, *The Natural Selection of the Chemical Elements* (Oxford University Press, Oxford, 1996).
5. D. F. Schriver, P. W. Atkins and C. H. Langford, *Inorganic Chemistry* (Oxford University Press, Oxford, 1995).

6 to 9 relate to energy transduction.

6. R. J. P. Williams, *J. Theoret. Biol.* **1**, 1–13 (1961).

7. P. Mitchell, *Nature* **191**, 144–148 (1961).

8. F. M. Harold, *The Vital Force – A Study of Bio-energetics* (W. H. Freeman & Co., New York, 1986).

9. *Les Prix Nobel (Chemistry)* (Almquist and Wiksell International, Stockholm, 1998).

10. Review papers in *J. Biol. Inorg. Chem.* **3**, 372–404 (1997) on bioinorganic chemistry give details of catalytic processes.

11 to 16 give views on the evolution of biological organisms, biochemicals and of Earth.

11. J. Maynard Smith and Szathmáry, *The Major Transitions in Evolution* (W. H. Freeman/Specktrum, New York, 1995).

12. L. G. Harrison, *Kinetic Theory of Living Pattern* (Cambridge University Press, Cambridge, 1993).

13. L. Margulis and K. V. Schwartz, *Five Kingdoms* (W. H. Freeman, New York, 1996).

14. J. E. Lovelock, *The Ages of Gaia* (Oxford University Press, Oxford, 1998).

15. J. Press and R. Siever, *Understanding Earth* (W. H. Freeman, New York, 1994).

16. L. Stryer, *Biochemistry*, fourth edition (W. H. Freeman, San Francisco, 1995).

17. S. Mann, J. Webb and R. J. P. Williams (editors), *Biomineralization* (VCH Press, New York, 1989).

18. R. J. P. Williams and J. R. R. Frausto da Silva, *Bringing Chemistry to Life* (Oxford University Press, Oxford, 1999).

19. J. L. R. Chandler and G. Van de Vijver (editors), *Closure: Emergent Organisations and their Dynamics* (Annals New York Academy of Science, New York 2000).

These publications give an approach to biological organisms involving the elements and avoid much of the relationship with genes.

12 Supramolecular Chemistry

Jean-Marie Lehn and Philip Ball

Introduction

When in 1944 the physicist Erwin Schrödinger asked 'what is life?', he was neither the first nor the last to ponder this mystery. (Not all, however, have shown the wisdom of the British biologist J. B. S. Haldane, who, in an essay of the same title just 5 years later began by confessing 'I am not going to answer this question'.) Yet, Schrödinger and his contemporaries had the decided advantage over previous generations that they knew pretty well where to look. For by that time, the puzzle of life had been passed (and not without reluctance on each occasion) from philosophers to theologians to zoologists and, finally, to chemists. Ultimately, it is on the stage of chemistry that most of life's fundamental dramas are played out.

That the raw material of life is the product of chemical processes alone has been in little doubt since Friedrich Wöhler's synthesis of urea – a compound previously associated only with living creatures – from inorganic materials (ammonium cyanate) in 1828. Wöhler's discovery sounded the death knell for the idea of vitalism, which ascribed some ethereal, non-chemical 'vital force' to living matter. By the time Schrödinger's book appeared, the focus had come to rest on proteins as the stuff of life, but 9 years later his guess that the genetic material was composed of an 'aperiodic crystal' was borne out beautifully when James Watson and Francis Crick deduced the double-helical structure of deoxyribonucleic acid (DNA), the gene library. In that same year of 1953, Harold Urey and Stanley Miller in Chicago showed that the primary building blocks of proteins – amino acids – could be synthesized by the crudest of procedures and from the simplest of raw materials: hydrogen, ammonia, methane and water (the compounds that were then thought to have comprised the earth's early atmosphere), subjected to the miniature lightning bolts of electrical discharges.

Yet the idea that life is 'just chemistry' does not necessarily simplify matters. In fact, it demands that we regard chemistry from an entirely new perspective. We will learn a great deal about the way the body functions by, for example, elucidating the various steps of the Krebs cycle of metabolism, or painstakingly decoding the molecular sequence of each gene in the human body, as molecular biologists are now doing. However, life is not to be reduced to a list of constituents, like the parts of an engine dismantled and laid out all oiled and greasy before us. Neither can it be summed up as a miasma of simultaneous chemical reactions – some making carbohydrates, others amino acids, others proteins and so forth. Rather, these parts, these reactions, must be orchestrated more precisely than clockwork and must be acutely responsive to one another. The molecules of life somehow have to interact, to get organized, to engage in cross-talk. Writing in 1812, the Swedish chemist Jöns Berzelius realized this:

> This power to live belongs not to the constituent parts of our bodies, . . . but [is] the result of the mutual operation of the instruments and rudiments on one another.

It was this 'mutual operation', not the generation, of the body's constituents that so confounded Berzelius and his contemporaries. However, today's chemists are able to understand and to make practical use of the mutual operations of molecules both natural and synthetic. To do so, they have had to expand their horizons, to see chemistry not just as the science of individual molecules but also as an investigation of how molecules come together and interact in groups – in pairs, in small aggregates or in vast throngs. This is the business of supramolecular chemistry – the chemistry beyond the molecule, the study of ensembles of molecules working together. Only by taking a perspective this broad can chemists hope to understand life's molecular complexity.

Yet that will be but a by-product of the supramolecular chemist's craft. For this chemistry 'beyond the molecule' is demonstrating that chemistry itself has a vaster potential than any scientist of Schrödinger's generation would have guessed, a potential whose realization will demand not just technical aptitude but also creative imagination. It is as if the brick-makers have suddenly realized that their products need not be an end in themselves but provide a means for them to become architects.

Lines of Communication

Schrödinger's perusal of life's mystery led him to appreciate that living organisms find a way to keep ahead of the continual drift towards disorder, towards increasing entropy. This prompted him to suggest that organisms are somehow able to extract 'negative entropy' from their surroundings, a proposal that sounds suspiciously like vitalism dressed as thermodynamics. The battle against entropy is fought in our social structures too. Civilization combats entropy through a network of information exchange. (Information was made formally the opposite of entropy in Claude Shannon's information theory in the 1940s.) We talk to each other, we send letters, faxes and electronic mail, we write things down and store them in libraries where others can look them up. We pass on this information from generation to generation – and, because it comes mixed with a dash of inevitable disorder, it changes slowly in the process.

When molecules need to get organized, they adopt analogous strategies. This is why the key concepts of supramolecular chemistry embrace not just those of traditional molecular chemistry – structure and energy – but also a third, information. We can regard supramolecular chemistry as a kind of molecular sociology, wherein the behaviour of the collective results from the nature of the individuals and the relations among them. The components of supramolecular chemistry communicate, they form associations, they have preferences and aversions, they follow instructions and pass on information. Central to these exchanges is the idea of molecular recognition, whereby one molecule is able to distinguish another by its shape or properties.

The structures and the energetics of supramolecular systems differ in significant ways from those of molecular chemistry; it is these differences that allow information to be passed around. At the structural level, the difference is not so much one of scale as it is one of organization. Supramolecular systems – groups of two or more discrete molecules that interact in specific ways to form an organized assembly – can be smaller than some of the exquisite single molecules that organic chemists have now synthesized (the latter can easily contain many hundreds of atoms); but within them one can identify several levels of structural order. Whereas molecules consist of a single, continuous network of atoms linked together by covalent bonds, 'supermolecules' in the strict sense are generally comprised of individual ions or molecules united with one another by weaker bonds into a discrete unit with a well-defined structure and dynamics. Furthermore, both molecules and supermolecules can be brought together in organized supramolecular arrays that are more or less extended in space, amplifying the scale of the chemist's creativity from the scale of tenths of nanometres to dimensions of many nanometres. This hierarchy can be carried right up to the scale of bulk materials organized in three dimensions by interactions of the same kind as those that hold together a single supermolecule (Figure 1). A hierarchy of structural levels, from the molecular to the supramolecular and beyond, is commonly found in the complex systems of nature, such as viruses, cells and tissue fibres.

The secret of this 'modular' approach to the synthesis of complex chemical systems

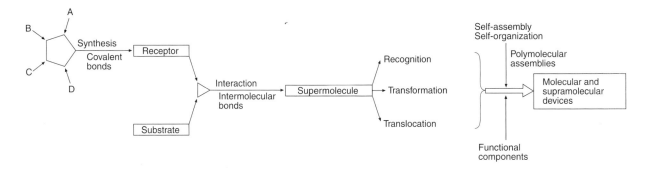

Figure 1

From molecular to supramolecular chemistry.

lies with the energetics. It is here that we find the most telling distinction between the living and inanimate worlds – in the choice not of bricks but rather of mortar. The forces that hold together supramolecular systems are not strong covalent or ionic bonds, but interactions that are weaker, commonly by an order of magnitude in binding energy. Hydrogen bonds, donor–acceptor interactions, metal-ion coordination, hydrophobic interactions – these are the glues that stick supermolecules together. This subtle attraction means that molecules can communicate without becoming locked into barren unions.

In biology, covalent bonds define the composition of the fundamental materials – proteins, nucleic acids, lipids, hormones – but non-covalent bonds play a critical role in determining their functions. It is through the use of weak interactions that enzymes are able not only to bind but also to release their target molecules, that the DNA double helix can be unzipped and copied, that nerve cells can send each other signals. Likewise, it is by exploiting this kind of chemistry that supramolecular scientists can become technologists, making molecular devices that will perform specific functions.

Living organisms provide abundant and inspirational examples of 'proofs of existence' for the kind of things that chemists might one day be able to achieve by working at the supramolecular scale. Life has its wonderful machinery and it is likely that the best way to build artificial analogues is not via the top-down approach of the mechanical engineer but by the bottom-up strategy of the molecular engineer – the same way, indeed, that such devices are made in nature. Take the flagella of the gut bacterium *Escherichia coli*, for instance. These protein threads, 15 nm thick and 10 μm long, extend from the bacterium's cell wall and their corkscrew motion propels the organism through its fluid medium. This motion is driven by a molecular motor, a rotary device embedded in the cell membrane that has all the characteristics of the best human-made motors but is just 25 nm or so across. It can be regarded as a highly sophisticated supramolecular assembly, with protein rotors that spin on protein bearings (Figure 2). At the base of the rotor sit other proteins that can switch the direction of rotation. The motor is driven electrochemically, by the electromotive force of a hydrogen-ion imbalance (a difference in pH) across the membrane, and it spins at a rate of around 60 revolutions per second.

To be able to arrange molecules with this kind of precision and extract such an ingenious and efficient function from them is the kind of goal that supramolecular chemists can now regard as realistic.

The Origins of Supramolecular Chemistry

In any field of science, novelty is linked to the past. Central to supramolecular chemistry is the business of making unions between molecules. That fleeting molecular liasons take place in nature became apparent to biochemists at the end of the nineteenth century: in 1890, Cornelius O'Sullivan and Frederick Tompson suggested that the enzyme invertase, which converts cane sugar into its 'inverted' form, forms a temporary bound complex with its sugar substrate. When the German chemist Emil Fischer observed how selective certain enzymes are in inverting sugar derivatives that differ only minutely in structure, he was led in 1894 to propose the metaphor that anchors our understanding of molecular recognition today: the lock and key. His idea was that the enzyme, the 'lock', has a molecular shape that is precisely complementary to that of the substrate which the enzyme subjects to a chemical transformation.

This concept of geometrical complementarity (modified in some important ways) continues to underpin much of modern research into the modes of action of enzymes. Yet binding alone is not the end of the story. Enzymes are catalysts, which means that they facilitate a chemical reaction of the substrate and are recovered unchanged at the end. Thus, while selective binding is a pre-requisite, the protein must also stabilize the ephemeral transition state of the reaction it catalyses, thereby lowering the activation energy (the energy hill that must be overcome) and increasing the rate of the reaction, and it must let go of the reaction products afterwards. Finally, complementarity between an enzyme's binding cavity and its substrate may be expressed not so much in terms of overall shape as in terms of the location of interacting chemical groups – the donor and acceptor components of a hydrogen bond, for example, must sit opposite one another. This complementarity of interacting groups is critical: there would not be much advantage in having geometrical complementarity if the substrate and binding site repelled each other!

The principles of molecular recognition and complementarity resonate throughout the whole of biochemistry. Complementarity between the hydrogen-bonding groups on the twin strands of DNA is what makes the double helix stable. Intercellular communication is commonly effected by the extracellular transfer of a signalling molecule, which is recognized and bound by proteins dangling from the cell's outer surface. These cell-surface recognition processes are extremely selective; the lymphocyte cells that patrol the immune system, for example, recognize the blood type of red blood cells by the difference in a single sugar group on a cell-surface glycoprotein. The idea of selective binding introduces the concept of a receptor, a molecule with a binding site of a well-defined geometry and disposition of binding groups. A trial-and-error process of identifying 'substrate mimics' that would bind to a specific receptor in preference to the true substrate was the typical *modus operandi* of the pharmacological industry that sprang from Paul Ehrlich's discovery of the first synthetic drug in 1909.

At much the same time as these biochemical insights were emerging, the existence of 'loose' molecular associations was becoming recognized. In 1935 Klaus Wolf introduced the term *Übermoleküle* to describe molecular associations, such as the acetic-acid dimer, that were clearly not covalently bound. Yet it was not until the 1980s that these elements merged into the coherent framework of supramolecular chemistry. Why this delay?

One reason is concrete enough: only in the past few decades have the powerful physical methods that are needed to analyse the structures and properties of

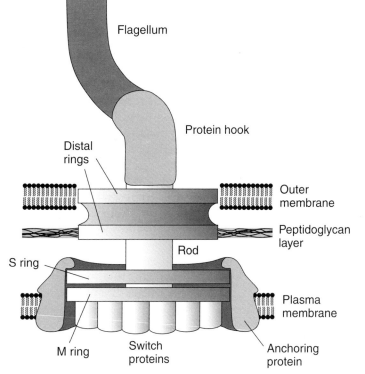

Figure 2

The flagellum of the bacterium *E. coli*, a molecular-scale rotary motor.

supramolecular assemblies existed. Spectroscopies, particularly nuclear magnetic resonance, have been refined to a degree that allows the structure and dynamics of these systems to be studied in great detail in solution, while X-ray diffraction techniques can now routinely reveal their complex crystal structures. Equally important for a field to ripen and mature is a receptive and fertile scientific scene – one that is able and prepared to embrace ideas from diverse disciplines and to recognize connections between them; and that has the vision to set new agendas and objectives without any guarantee that they will be attained or that they will in the end have justified the effort. From molecular biology come not only inspirational examples of supramolecular chemistry in action (and an array of machinery that can on occasion be commandeered for new uses) but also a demand for new molecules with finely tuned recognition capabilities, to act as drugs, inhibitors or gene-targetting agents. From engineering technology comes a call for ever smaller devices and materials-patterning procedures, to the extent that a bottom-up approach begins to look more attractive than the tricky business of carving out components with ion beams – especially when complex architectures can be made by 'self-assembly'. From information technology comes a conceptual framework that allows us to look at chemistry not as cookery but as a means of storing and transmitting data on the smallest scales imaginable.

Every new field struggles for a time to establish a useful vocabulary. Perhaps the most tortured terms in supramolecular chemistry are those that seek to convey the autonomy and 'programmed' nature of its building blocks: 'self-assembly' and 'self-organization'. There is an increasing tendency both to use these interchangeably and to apply them to any molecular process that is not somehow guided by the researcher's manipulations. Here we try to refine these terms with the following definitions.

- *Self-assembly* is the spontaneous association of several (more than two) molecular components into a discrete, non-covalently bound aggregate with a well-defined structure. This will generally involve more than one kinetically distinct step. Self-assembly involves molecular-recognition processes – binding events, but not 'mere' binding. Rather, one may say that recognition is binding with a purpose.
- *Self-organization* is the spontaneous ordering of molecular or supramolecular units into a higher-order non-covalent structure characterized by some degree of spatial and/or temporal order or design – by correlations between remote regions. A self-organized system may be either at equilibrium or in a dynamic state characterized by several stable configurations; and it will exhibit collective (and often non-linear) behaviour. Such a definition does not (and need not) exclude crystallization and related ordering phenomena such as liquid-crystallinity.

As is ever the case, these terms should become more sharply focused in their application than they are in their definition.

Perhaps the most useful way to build a lexicon is not by sterile definition but by analogy. Drawing a parallel with languages, we might say that atoms are chemistry's letters: fundamental and distinct building blocks, but in themselves generally devoid of meaning. The molecule is the word (and hence already contains some information!). So supramolecular chemistry represents the beginning of an attempt to develop chemistry into a literature: the supermolecule is the sentence; and, beyond that, we can hope to put together (probably by self-organization) books of polymolecular assemblies; and in that way to start telling stories.

Crown Ethers and Cryptands

Supramolecular chemistry's first synthetic receptors recognized the simplest of substrates – individual ions of alkali metals, good approximations to uniform charged spheres. All that is needed in such a receptor is a 'hole' into which the 'ball' fits snugly.

Receptors that recognize these ions already exist in nature. Many of the cell's biochemical processes are regulated by the flow of ions across the cell membrane; nerve signals, for instance, correspond to the propagation of a wave of varying electrical potential along an axon, as sodium, potassium and calcium ions are pumped one way or another across the nerve cell's walls. Some ions are transported through protein pores or 'channels' in membranes; others are ferried across by molecules called ionophores, which surround the water-soluble ions with a hydrophobic coat that allows them to pass amongst the fatty tails of the membrane's lipid molecules. An exemplary ionophore is valinomycin, a cyclic peptide that effects the transport of potassium ions (Figure 3). The ring contains carbonyl groups whose oxygen atoms point inwards, so that they can interact electrostatically with a positively charged metal ion in the centre.

Valinomycin's ability to recognize and bind potassium ions selectively is a consequence of its geometrical complementarity – the hole is the right size. Sodium ions are less tightly bound because they are smaller – crudely speaking, they rattle around in the hole. (More accurately, the oxygen atoms cannot get close enough to compensate in binding energy for that which is lost by stripping away the sodium ion's hydration sphere of water molecules.)

In addition to their complementarity to the shape and size of the substrate, ionophores like valinomycin exemplify several more of the principles that make for a good receptor. For example, they have several interaction sites, so that strong binding is possible even though the individual interactions at each site are relatively weak. Furthermore, for binding small, compact substrates, a concave receptor cavity is usually the ideal design, with binding groups that converge towards the bound species.

These characteristics provided clues for what one should look for in a synthetic analogue of natural ionophores. It was the search for such molecules that led one of us (J.-M. L.) to begin work on molecular recognition of metal ions in the mid-1960s. A minimal model of valinomycin might contain oxygen atoms incorporated into a cyclic structure with a central cavity about the right size to fit the target ion. Molecules that fitted this prescription were cyclic ethers: chains of oxygen atoms and methylene (CH_2) groups linked together in rings (Figure 4(a)). During the 1960s Charles Pedersen, investigating additives for rubber products, found serendipitously that related cyclic ethers will bind metal ions. The ring then adopts a corrugated conformation reminiscent of a crown, so Pedersen christened the receptors crown ethers (Figure 4 (b)). The molecule in Figure 4 is called 18-crown-6, indicating that there are 18 atoms in total in the ring, of which six are oxygen atoms.

Pedersen's work led Lehn to realize that crown ethers could serve as mimics of valinomycin and other ionophores, with the advantage that the ether rings would be less reactive (and thus less susceptible to chemical attack) than the peptide rings of the natural compounds. He suspected that modifications of the cyclic molecules might offer stronger and more selective metal binding. In 1967–68, Lehn and his co-workers synthesized bicyclic molecules in which nitrogen atoms unite three ether chains (Figure 5). These receptor molecules are three-dimensional receptors with roughly spherical cavities that bind alkali metal ions several orders of magnitude more tightly than do the single-ring crown ethers. Because of their ability to bind and 'hide' a metal ion within their internal cavity, Lehn called these molecules cryptands and the bound metal complexes were then cryptates.

The metal-ion-binding ability of cryptands is selective – they exhibit molecular recognition – but the size of the cavity is only a crude guide to this selectivity. The preference for a particular metal ion depends on a rather subtle balance between the influences of the binding energy (enthalpy) and the gain or loss of configurational

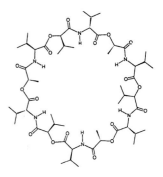

Figure 3
Valinomycin is a cyclic peptide molecule that binds potassium ions in its central cavity.

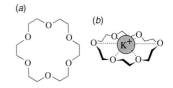

Figure 4
Crown ethers (a) are cyclic molecules that bind metal ions via the lone pairs of electrons on the oxygen atoms. In the metal complex, the ring adopts a puckered configuration (b).

Figure 5
Cryptands have three-dimensional central cavities for binding metal ions.

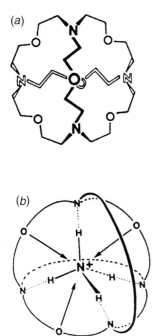

(a)

(b)

Figure 6
Tricyclic cryptands (*a*) exhibit strong binding and high selectivity for various metal ions. The nitrogen atoms at the threefold junctions are well situated to interact with the hydrogens of a bound ammonium ion by hydrogen bonding (*b*).

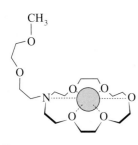

Figure 7
Lariat crown ethers contain a flexible arm, which allows strong binding while at the same time permitting rapid release of the guest.

freedom (entropy) when the solvent cage of the metal ion is replaced by the ether cage. This is characteristic of all supramolecular complexes, whose geometrical preferences should not blind us to the importance of entropic effects.

If two rings bind more tightly than one, what will three rings do? In 1975 Lehn and colleagues synthesized the tricyclic receptor shown in Figure 6(*a*), which will bind large metal ions such as caesium in its cavity. This cavity has tetrahedral symmetry – the four nitrogen atoms sit at the corners of a tetrahedron. So the molecule has binding groups that are ideally placed for grasping the tetrahedral ammonium cation (NH_4^+). The four hydrogen atoms of the ammonium ion point towards the nitrogen atoms in the cryptand, forming hydrogen bonds with the lone pairs on the latter, while the lone pairs on the oxygen atoms of the ether chains point towards the central, positively charged nitrogen atom of the ion (Figure 6(*b*)). This is an example of molecular recognition guided by complementarity not just of size but of shape too.

It is straightforward to convert these tricyclic cryptands to receptors for neutral molecules and negatively charged ions (anions) too. In acidic solution, all of the nitrogen atoms in the receptor can be protonated – they acquire a hydrogen ion, giving the nitrogen atoms a positive charge and a hydrogen substituent that points towards the centre of the cavity. This highly positively charged cage is very attractive to anions such as chloride. In somewhat less acidic solution, the cryptand may be given just two protonated nitrogen atoms, whereupon it becomes an ideal receptor both in size and shape for a water molecule. Thus the tricyclic cryptand behaves as a sort of molecular chameleon, responding to changes in pH of the surrounding medium by becoming selective to different kinds of substrate.

With the bicyclic and tricyclic cryptands one can attain greater binding strengths than are provided by monocyclic crown ethers; but, because the binding cavity is more enveloping, the substrate takes longer to get in and out. This is a potential drawback for, say, receptor-mediated membrane transport of ions, in which the release of the ion may be the rate-limiting step. George Gokel and co-workers at the University of Miami devised a kind of crown ether that retains some of the binding ability of cryptands while offering the rapid binding and release characteristic of monocyclic crown ethers. These molecules have a single ether ring, to which is attached a flexible ether-chain arm that can swing round to afford additional binding capability for a captured ion (Figure 7). They are a kind of molecular lasso for cations; Gokel called them lariat ethers (from the Spanish *la reata*, 'the rope').

Engineered Binding

The crown ethers and cryptands set the agenda for subsequent work on molecular recognition, both by providing a set of versatile constructional elements (ether chains and rings, proton-binding groups, nitrogen junctions in so-called aza-crowns) and by illustrating the features that would ensure good recognition of a substrate by a receptor. They show, for instance, how increasingly strong and selective binding results from an increasing degree of pre-organization of binding groups in the receptor. In the cryptands, pre-organization is fully three-dimensional.

The exploitation of pre-organization is taken a stage further in the spherands of Donald Cram's group at the University of California at Los Angeles (Figure 8(*a*)): macrocyclic receptors in which the ether binding groups are attached to benzene rings. Whereas crown ethers are rather floppy in the uncomplexed state, like a rubber band, the spherand rings are more rigid – like a washer – because the benzene groups cannot easily bend out of the plane of the ring. So the circular cavity in the centre of the spherands is already in place before a metal ion is bound and hence the binding event is not penalized by the enthalpic and entropic cost of having to organize a floppy loop into a circular ring. Cram's group has also made hybrids of cryptands and spherands, such as the cryptaspherands (Figure 8(*b*)), which combine the benefits of both.

Receptor molecules with relatively rigid cavities are nothing new, however. Since the 1950s a strand of molecular-recognition studies was unwinding essentially independently of the work on crown ethers. It focused on a class of naturally occuring carbohydrates called cyclodextrins, which contain six glucose rings linked head to tail in a ring (Figure 9(*a*)). These molecules have a cylindrical cavity running through their centre, into which small molecules can fit. The outer face of a cyclodextrin bristles with hydroxyl groups with hydrogen-bonding ability, so the molecule is soluble in water; but the inner cavity is lined mainly with carbon–carbon and carbon–hydrogen groups, so it is hydrophobic and able to receive hydrophobic molecules such as benzene. In the 1950s and 1960s Friedrich Cramer studied the formation of inclusion complexes of cyclodextrins, in which small molecules like benzene and phenyl methyl ether are bound within the cavity (Figure 9(*b*)). Within the paradigm of supramolecular chemistry, these complexes can now be regarded as supermolecules in which complementarity of size and affinity (hydrophobicity) between the cavity and the guest governs the molecular-recognition event. During the 1960s and 1970s several groups, particularly those of Myron Bender at Northwestern University in Illinois and of Ronald Breslow at Columbia University in New York, built on Cramer's investigations of cyclodextrins as enzyme mimics that enhance the selectivity of a chemical transformation carried out on the bound substrate. Breslow found in 1969 that the hydrogen atoms of the benzene ring of phenyl methyl ether could be substituted by chlorine specifically in the *para* position (opposite the methyl ether group) and in no other, when the molecule was bound within a cyclodextrin – because only this part of the molecule, protruding from the bottom of the cavity, was exposed to chemical attack (Figure 9(*b*)).

Yet another pathway to the rigid-cavity receptor was paved by C. David Gutsche of the Texas Christian University. In the 1970s Gutsche studied a cyclic oligomer of formaldehyde and phenol derivates that formed during the manufacture of a petroleum demulsifier (Figure 10(*a*)). He realized that this cyclic molecule would adopt a cup-like shape (Figure 10(*b*)), which reminded him of a type of Greek vase called a calix crater. He called the cup-shaped molecule a calixarene (the suffix 'arene' denoting the presence of the benzene 'aryl' groups). During the 1970s synthetic methods for making a wide range of calixarenes with different numbers of phenolic groups in the rings were devised (the molecule in Figure 10 is called a calix[4]arene).

The lining of benzene rings in the bowl-shaped cavities of calixarenes allows them to act as 'molecular baskets' for hydrophobic substrates like toluene, benzene itself and xylene. The strength with which the substrate is bound again depends on complementarity of size: the large cavity of a calix[8]arene holds toluene only loosely, but this substrate becomes so firmly embedded in the bowl of a calix[4]arene that it is hard to extract. In 1994 three groups in the USA, Japan and the Netherlands independently found that a calix[7]arene will exhibit molecular recognition towards the carbon-cage fullerene molecules C_{60} and C_{70}: the former, with a soccer-ball shape, is bound tightly in the cavity, whereas the latter, with a rugby-ball shape, is not. This provides a means of separating the two molecules, which are generally formed in an intimate mixture.

The cup shape of calixarenes is not rigid – the benzene rings can rotate outwards so

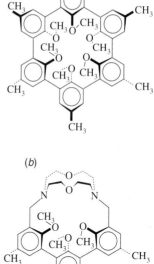

(*a*)

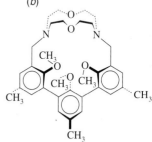

(*b*)

Figure 8
Spherands (*a*) have relatively rigid, pre-organized binding cavities. Cryptaspherands (*b*) combine the properties of spherands and cryptands.

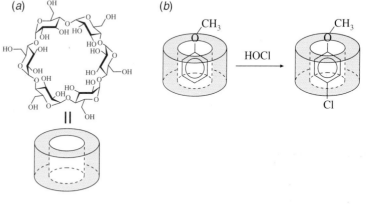

(*a*) (*b*)

Figure 9
Cyclodextrins (*a*) are naturally occurring, cyclic carbohydrates with a cylindrical central cavity that will bind small hydrophobic molecules (*b*). Phenyl methyl ether bound within α-cyclodextrin can be selectively chlorinated at the *para* position, which is exposed at the bottom of the ring (*b*).

(a)

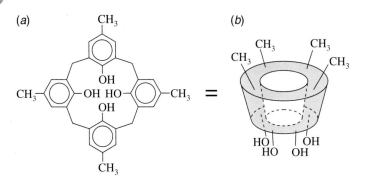

(b)

Figure 10
Calixarenes will accommodate guest molcules in a cup-shaped cavity.

(a)

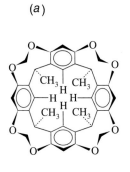

(b)

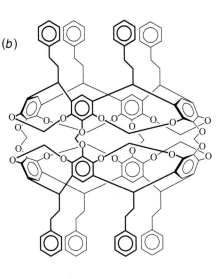

Figure 11
Cavitands (a) are rigid, bowl-like molecules. When they are joined in a rim-to-rim fashion, they make roughly spherical cage-like molecules called carcerands (b).

that the cup becomes partly inverted. These inversions are happening all the time in solution. To suppress them and make a more rigid cavity, Donald Cram's group modified a calixarene-like molecule called a resorcinarene by linking adjacent benzene rings via an ether chain. The 'hard' bowl-shaped receptors that resulted were christened cavitands (Figure 11(a)). Cram realized that the ultimate pre-organized cage-like receptor could be fashioned by joining two cavitands together around their upper rims. This yielded a carcerand, a molecular prison for small substrates (Figure 11(b)). The carcerand's cage is so firm and all-encompassing that small molecules trapped inside when it is formed cannot get out. Supramolecular complexes of this sort are true compounds, in the sense that the constituents parts, although they are not bound together by chemical bonds, cannot be separated without breaking bonds. They are united mechanically.

There are advantages to being incarcerated – you are safe, for instance, from the ravages of the world outside. With this in mind, Cram and co-workers have shown that carcerands can provide a protective environment for chemical species too unstable to survive outside the cage. They have made a carcerand with an entrance – a gap in the wall, through which small molecules can pass if the temperature is high enough. This slightly incomplete cage is called a hemicarcerand. They allowed a molecule of α-pyrone to pass into the cage. This molecule then underwent a photochemical reaction to yield cyclobutadiene – a highly reactive species that nevertheless remained stable inside the cage for a long time because no other molecules could get at it (Figure 12).

As we saw with the tricyclic cryptands, receptor molecules can be designed to bind a particular substrate not simply by virtue of having a cavity of about the right size but rather through the careful placement of binding groups to complement those of the substrate. Crown-ether chemistry has now yielded all manner of 'sculpted' cavities that will bind substrates that are rather more complex than mere ions. Amine receptors can be devised using the same principles as those that allow cryptands to bind an ammonium ion: the oxygen atoms of a crown-ether ring form hydrogen bonds with a protonated amine. A six-oxygen crown will bind a protonated primary amine (in which some substituent R replaces one of the protons of the ammonium ion) (Figure 13). The side-arms X that we have shown here attached to the crown-ether ring can be chosen to interact favourably with the R group of the amine, providing additional binding; for example, donor–acceptor or hydrogen-bonding functionalities of the R and X groups can be matched up.

Ithmar Willner at the Hebrew University in Jerusalem has made an intriguing receptor that binds both metal ions and small organic molecules, by combining crown-ether chemistry with cyclodextrins. He has attached an aza-crown ring to the mouth of a cyclodextrin molecule: this receptor will bind both the metal cation and the organic anion of alkali-metal *p*-nitrophenolate salts (Figure 14).

Hydrogen bonds are one of the supramolecular chemist's favourite glues. Their big attraction is that (unlike electrostatic or donor–acceptor interactions) they are directional in space, which means that precise geometrical matching of the receptor's and substrate's binding groups is possible. George Whitesides and co-workers at Harvard University have investigated the complementarity between hydrogen-bonding units on cyanuric acid and melamine in order to make a large family of increasingly complex supramolecular assemblies. Both of these molecules contain a central six-membered ring from which hydrogen-bonding groups extend with threefold symmetry in the plane of the ring

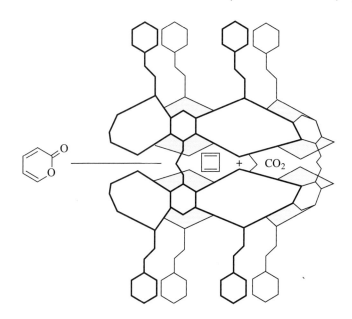

Figure 12
Reactive molecules, such as cycloybutadiene, remain stable if they are formed within the protective environment of a hemicarcerand.

(Figure 15(*a*)). Whitesides' supermolecules are based on the 3:3 motif in which three isocyanurate units (derived from cyanuric acid) alternate with three melamine units (Figure 15(*b*)). Although this structure cannot be formed from the six discrete components on their own, it self-assembles when the three melamines are pre-organized by joining them together with flexible arms that meet in a central 'hub' (Figure 15(*c*)). By adding another layer of isocyanurate groups to the three that are bound by the first melamine trimer, a second one of these trimers can be attached below (Figure 15(*d*)). This generates a supramolecular assembly of five components with a relative molecular mass of 5519 – larger than that of most covalently bound synthetic molecules. We will see later how these principles can be used to generate supramolecular arrays that extend indefinitely, like polymers, in one dimension or more.

Threading the Needle

Although they do not share the directional properties of hydrogen bonds, donor–acceptor interactions have proved to be a highly versatile means of sticking supermolecules together. Fraser Stoddart and co-workers have used them to make some of the most complex supermolecules yet created. While he was at the University of Birmingham (and previously Sheffield) Stoddart focused his attention on assemblies in which the molecular components are intertwined, with one threaded through another like the beads of a necklace or the links of a chain. Such molecular assemblies were first created in the 1960s from hydrocarbon chains and rings, without the assistance of molecular recognition. An assembly in which one ring is linked through another is called a catenane, from the Latin *catena*, 'chain' (Figure 16(*a*)). A ring molecule threaded on a linear one, with bulky caps on the thread to prevent the ring from falling off, is a rotaxane (from the Latin *rota*, 'wheel', and *axis*, 'axle') (Figure 16(*b*)).

While they were working on the cyclodextrins in the late 1950s, Friedrich Cramer and others realized that molecular recognition might be used to guide the threading of a linear molecule through a cyclic one. Stoddart and colleagues have exploited this idea to make catenanes and rotaxanes. A ring containing electron acceptors in its cavity will bind a molecule containing an electron-donor unit. If the donor group has a long arm attached at both ends, it sits in the cavity with the arms poking out each side of the ring (Figure 17(*a*)). For the acceptor unit, Stoddart's group mainly used molecules related to paraquat, containing two positively charged pyridine rings linked

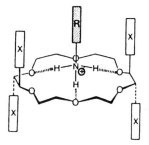

Figure 13
A receptor designed to bind to a protonated primary amine. The side-arms on the crown ether chain (X) are chosen to interact favourably with the R group of the amine.

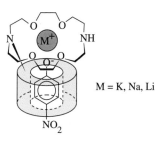

Figure 14
A receptor that binds both the anion and the cation of organic salts.

(a)

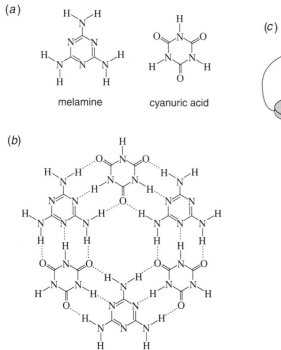

melamine   cyanuric acid

(b)

II

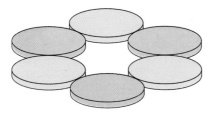

(c)

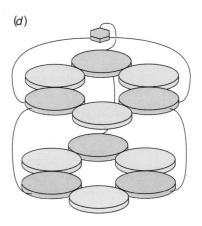

+3

(d)

(e)

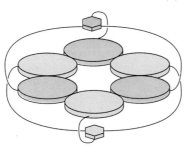

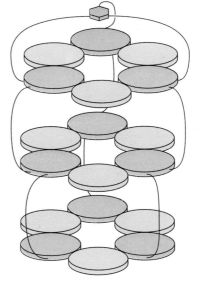

Figure 15
(a) Melamine and cyanuric acid are complementary hydrogen-bonding partners. (b) These two molecules can link up into 3:3 hexagonal groupings. (c) The hexagonal complex self-assembles when three melamines are pre-organized by linking them to a central hub. (d) Further tiers of this basic assembly can be added. (e) These same principles can be extended to make organized assemblies of almost arbitrary complicity.

tail to tail. For the donors, they selected molecules based on hydroquinone, in which two hydroxyl groups on opposite sides of a benzene ring push negative charge towards the ring and make it electron-rich.

Stoddart's group found that a crown-ether ring containing two hydroquinone units will spontaneously thread itself onto a paraquat-based thread, so that the acceptor and donor units stack face to face. The same is true if the acceptor groups constitute the ring and the donors make up the thread. When the ends of the thread are capped, the assembly becomes a rotaxane – a supermolecule held together mechanically (Figure 17(*b*)). The favourable interactions between the ring and the thread are here supplemented by π–π stacking of the aromatic groups.

In 1989 Stoddart and colleagues linked up the two ends of a rotaxane's thread to form a catenane (Figure 18(*a*)). If the thread is made long enough, two rings can be threaded – one on either end, where the acceptor groups sit. In this way, Stoddart and colleagues have made a three-link catenane, denoted a [3]catenane (Figure 18(*b*)). By judicious choice of thread lengths and capping sequences, they have even built a

(*a*)   (*b*)

Figure 16

(*a*) Cyclic molecules joined into a two-link catenane. (*b*) A rotaxane is a linear molecule threaded through a cyclic molecule. Unthreading is prevented by subsequently adding bulky end caps.

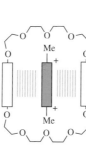

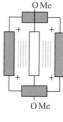
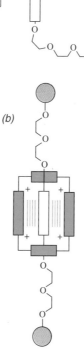

(*a*)   (*b*)

Figure 17

(*a*) Linear donor molecules will spontaneously thread through cyclic acceptors in solution and vice versa. (*b*) Adding bulky groups to the threaded molecule prevents unthreading, creating a rotaxane.

Figure 18
(a) A catenane can be made by linking together the ends of a threaded molecule. (b) A larger loop provides access to a three-link supermolecule: a [3]catenane. (c) A [5]catenane that has been christened 'olympiadane'.

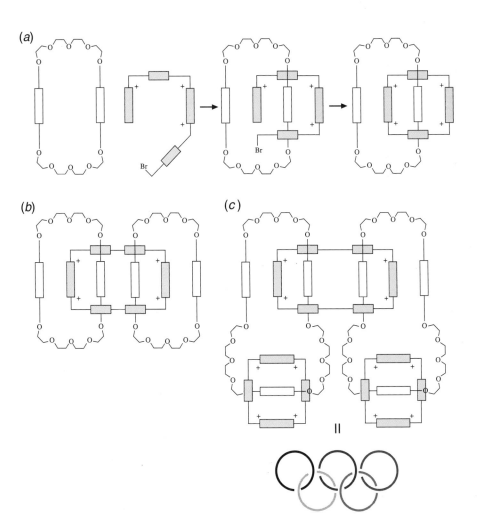

[5]catenane, which Stoddart named 'olympiadane' – for reasons that the structure makes clear (Figure 18(c)). Synthesizing these assemblies could not even be contemplated without the aid of molecular-recognition processes to guide the elements into place.

More recently, Stoddart's group has employed hydrogen bonding to guide the threading of rotaxanes. The hydroquinone/ether macrocycles can be threaded onto dialkylammonium axles owing to interactions between the lone pairs on oxygen of the former and the NH groups of the latter (Figure 19(a)). A rotaxane results when the ends of the axle are 'stoppered' with bulky groups. Three-armed variants of the axle unit give rise to a 'supramolecular cage' in which two tripods capture three rings (Figure 19(b)).

In 1990 David Lawrence at the State University of New York and colleagues reported a rotaxane made from a threaded cyclodextrin. Akira Harada and co-workers at Osaka University in Japan have found that up to 23 cyclodextrin 'pearls' will become threaded spontaneously on the polymer polyethene glycol to form a 'molecular necklace' (Figure 20). In 1993 Harada's team created a 'molecular tube' by linking the threaded molecules with rim-to-rim covalent bridging units (Figure 20). This shows how supramolecular assemblies can be used to organize molecular components for subsequent synthesis of covalently bound products – a trick that Fraser Stoddart calls 'supramolecular assistance to molecular synthesis'. The polymer thread in this example can be regarded as a mould or template around which the molecular tube is formed.

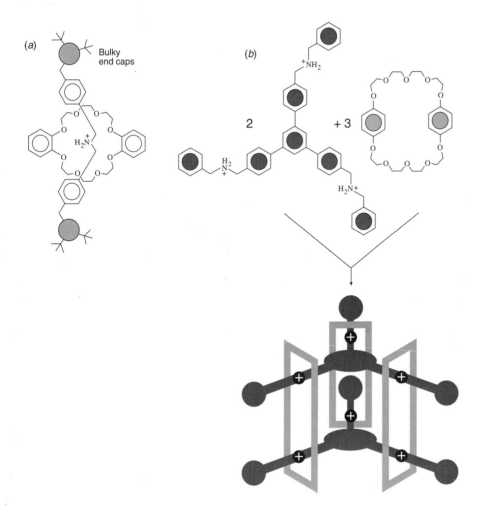

Figure 19
(*a*) A rotaxane whose threading is guided by hydrogen bonding. (*b*) The same principles assist the self-assembly of this five-component supermolecule.

Gerhard Wenz and co-workers at the Max Planck Institute for Polymer Research in Mainz have made polyrotaxanes from cyclodextrins and polymer chains containing alternating hydrophobic and hydrophilic units. The cyclodextrins are thought to encircle the hydrophobic segments, while the hydrophilic parts render these large molecular assemblies water-soluble even with as many as 37 beads on the thread. Harry Gibson and colleagues at the Virginia Polytechnic Institute have generated polyrotaxanes from crown ethers threaded on polymers.

Molecular and Supramolecular Devices

One of the goals of supramolecular chemistry is to build synthetic systems that carry out some of the functions of biochemical systems – trans-membrane ion transport, perhaps, or enzyme-like catalysis. In short, the aim is not merely to achieve binding and (self-)assembly, but also to develop functionality. This drive towards functional supramolecular systems is motivated not only by a wish to emulate nature but also by a growing awareness that chemistry might have something to offer technology. Microelectronics engineers, for example, want ever smaller conduits for channelling electrons and smaller devices that perform switching operations; and analytical chemists, drug manufacturers and biotechnologists need chemical sensors with high sensitivity and selectivity.

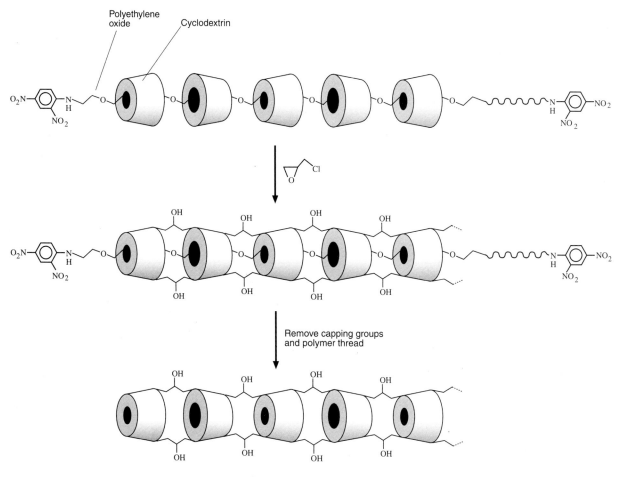

Polyethylene oxide

Cyclodextrin

Remove capping groups and polymer thread

Molecular tube

Figure 20

A molecular tube can be synthesized by threading several cyclodextrin molecules on a linear polymer thread to form a polyrotaxane (after end-capping), then covalently linking the 'beads' and removing the thread.

Rites of Passage

The first supramolecular devices were concerned with transport – of ions, molecules, electrons and energy. The key to much of the communication – that is to say, the transfer of information – that takes place at the level of the cell is trans-membrane exchange: the ferrying of charged or neutral species across cell membranes. The membrane itself is designed to inhibit such processes: its interior consists of the closely packed water-insoluble (hydrophobic) parts of the constituent lipid molecules, an environment in which anything that is soluble in the watery medium to either side of the membrane will not be welcomed. One way to counter this hydrophobic barrier to the passage of water-soluble species is to bind them inside ionophores, carrier molecules with a hydrophobic coat. Another option is to perforate the membrane with channels whose interior is water-loving (hydrophilic) and whose outer face is hydrophobic. In nature these channels are made from proteins that span the cell wall. Protein channels are generally highly selective for a particular ion or small molecule and are often 'gated' so that they open or close in response to chemical signals.

Supramolecular systems that mimic both of these kinds of transport have been created. Crown ethers and cryptands will ferry metal ions selectively through a layer of organic liquid sandwiched between two aqueous phases (a crude analogue of a biological membrane). The highest rates of transport are obtained for cryptands that bind their substrates neither too weakly to achieve significant uptake nor too strongly to let them go again. Selective metal-ion-transport systems like these might be useful not only as models of biological systems but also for removal of toxic heavy-metal ions and the recovery of valuable trace metals from ground water or industrial effluent.

David Reinhoudt of the University of Twente and co-workers have designed a macrocyclic receptor that performs the trick of binding both a positive and a negative ion selectively and ferrying them across a liquid membrane. The carrier molecule is based on a calix[4]arene, which has a fragment of a crown-ether ring attached to one rim and a uranyl (UO_2) group appended, via amido linkages, to the other (Figure 21). The ether chain binds caesium ions, while the uranyl group acts as a receptor site for several anions, including chloride and nitrate.

Trans-membrane transport in biological systems commonly takes place against a concentration gradient – that is, the substrate is ferried across the membrane even though there is already a higher concentration of the substrate on the far side. Like transporting water uphill, this requires a pump. Trans-membrane biological pumps have to be driven by some energy source, such as light. For example, the membrane protein bacteriorhodopsin, which pumps hydrogen ions across the plasma membrane of salt-loving bacteria, is photochemically driven. Synthetic supramolecular ion pumps can be created by coupling transport processes so that the pumping of ions in one direction is accompanied by (and driven by) the pumping of some other species in either the same or the opposite direction.

For example, John Grimaldi and Jean-Marie Lehn pumped potassium ions across a membrane by coupling their transport (mediated by a crown-ether-type carrier) to the simultaneous transport of electrons. The latter was achieved by having an oxidation–reduction reaction take place across the membrane. The electron carrier within the membrane was a nickel complex (Figure 22), which was reduced (given an electron) by a reducing agent on one side of the membrane and oxidized (relieved of an electron) by an oxidizing agent on the other side. This provided a kind of electrochemical battery that pumped potassium ions in the same direction as the electrons, maintaining neutrality of charge on either side.

Light-coupled transport processes offer the prospect of mimicking photosynthesis: capturing the energy of sunlight by using it to drive the transport of ions. When ions of opposite charges are pulled apart, an electrochemical potential is formed, just like when a battery is charged. For example, Tom Moore and co-workers at Arizona State University have demonstrated the light-driven transport of hydrogen ions and electrons across lipid membranes, curled up into artificial cell-like structures called liposomes or vesicles. When the central component of the triad, a porphyrin, absorbs light, the energy is used to pass an electron from the innermost (donor) to the outermost (acceptor) components of the triad. This electron is then passed on to a mobile shuttle molecule sequestered within the liposome membrane. The shuttle also picks up a hydrogen ion from the fluid outside the membrane, and diffuses to the inner surface. Here it returns the electron to the donor group of the triad, and releases the hydrogen ion inside the liposome (Figure 23).

So, when the liposomes are illuminated, the concentration of hydrogen ions inside them goes up – their interiors become more acidic – and a 'proton potential' is established. A similar proton potential is developed across the membranes of a leaf's chloroplast, the photosynthesizing organelle, and is tapped to allow the chloroplast to generate the energy-storage molecule adenosine triphosphate (ATP) from adenosine

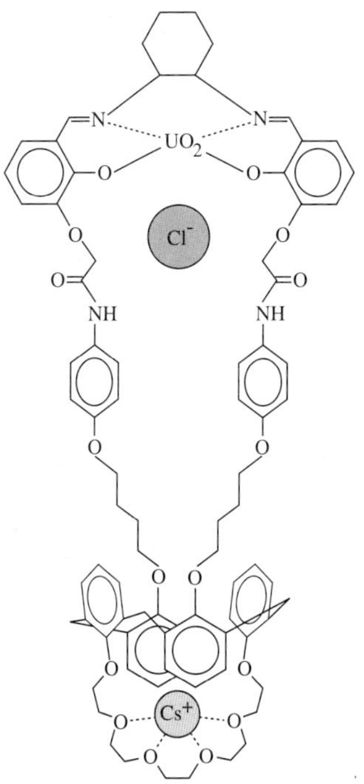

Figure 21
This bifunctional receptor will bind both cations (caesium) and anions (chloride) and will transport them across a liquid membrane.

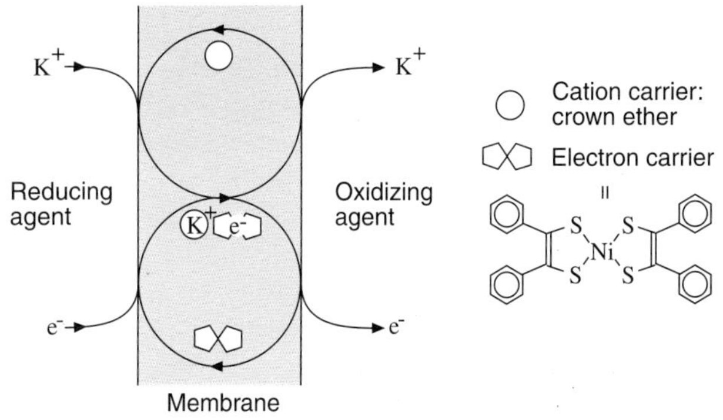

Figure 22
Coupled transport of metal (potassium) cations and electrons across a membrane has been achieved by using two membrane-localized transport agents, a nickel complex for the electrons and an 18-crown-6 ether for the cations. The transport is driven by a redox gradient across the membrane.

Cation carrier: crown ether

Electron carrier

Figure 23

Light-driven electron transport across the wall of a liposome can be coupled to proton transport, decreasing the pH inside the liposome and creating a proton-motive force. This can then be coupled to the synthesis of energy-rich ATP by the enzyme ATP synthase. The overall process is the conversion of sunlight to chemical energy.

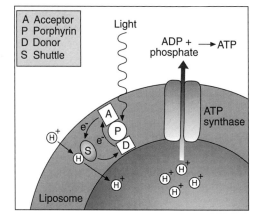

diphosphate (ADP). Moore and colleagues were able to mimic this process, by peppering the liposome walls with the enzyme ATP synthase. This enzyme allows protons to pass back to the outside of the liposome and at the same time taps this 'proton-motive force' to drive the conversion of ADP to ATP (Figure 23). Thus the liposome-based system is truly an organized, multicomponent supramolecular assembly – a supramolecular device, if you will – that executes the conversion of energy.

Artificial ion channels have been constructed from all manner of macrocyclic molecules stacked together to form a tunnel of hoops through a membrane. The stacking is usually driven mainly by the energetic need for efficient molecular packing and the resulting, partially ordered, materials are liquid crystals (discussed later). Reza Ghadiri and colleagues at the Scripps Research Institute in California have made macrocycles whose stacking into channels leaves less to chance. They synthesized cyclic peptide molecules, made from eight or more amino acids (Figure 24(*a*)). These peptide rings have hydrogen-bonding groups pointing up and down from the plane of the rings, so that the rings can stick to others above and below via hydrogen bonds. Ghadiri found that these cyclic peptides will crystallize in long cylindrical channels about 0.8 nm in diameter (Figure 24(*b*)). Hydrocarbon substituents appended to the outer rims of the hoops render them compatible with lipid membranes. The peptide channels have very high rates of trans-membrane transport for sodium and potassium ions – about three times higher than those of the natural channel-forming protein gramicidin A, which is used as an antibiotic.

Another way of encouraging macrocycles to stack into channels is to link them together, for example in a string or a rack. Two aza-crown ethers linked by a flexible chain, called boland molecules, will fold up to form a two-ring stack in membranes (Figure 25(*a*)). These molecules are similar to the bipolar lipids (which have charged head groups at both ends) found in the cell walls of some very ancient species of bacteria. Several crown-ether cycles can be joined together to form a tunnel of hoops (Figure 25(*b*)).

Hoops are not the only kind of channel-forming element. Rod-like molecules can form a cylindrical bundle, such as the protein-based connexons that allow the cells of the heart, liver, kidneys and brain to synchronize their electrical behaviour by exchanging ions. Lehn has made artificial ion channels based on the 'bundle' motif, gathered at the mid-point of the rods (Figure 25(*c*)). These so-called bouquet molecules will enhance the permeability of artificial membranes to sodium and lithium ions.

Molecules capable of transporting DNA into cells ('gene vectors') are of great interest for gene therapy. Positively charged lipid molecules, which will adhere to the negatively charged DNA, show some promise, but the vectors most favoured in current gene-therapy trials are modified viruses, which are particularly adept at breaching the walls of the cells.

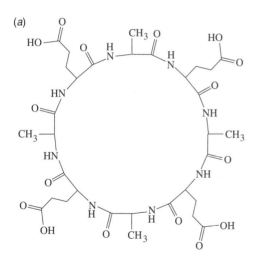

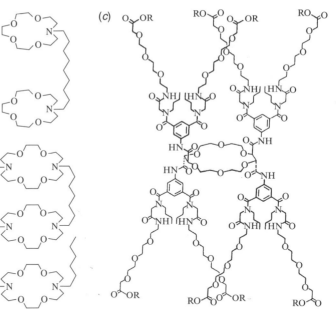

Figure 24
Cyclic peptides (*a*) can be designed to assemble into tubules (*b*) by hydrogen bonding.

Molecular Wires

While the transport of ions or molecules is commonly associated with biology, the transport of electrons is more familiar within electronic technology. With the continuing miniaturization of electronic circuitry, electrons are being passed down wires of ever narrower dimensions. At the last count, conducting channels less than 200 nm wide can be etched into silicon wafers; and, if current trends persist, the size of electronic components on microchips will reach that of individual molecules in less than three decades. This has stimulated interest in the development of molecular electronics, in which individual molecules or supermolecules perform the functions of today's semiconductor diodes and transistors.

One of the fundamental components of a molecular-electronic technology is the molecular wire – a conducting pathway supplied by a single molecule. Molecules like this are found in nature, in the form of the carotenoid pigment molecules that complement the light-absorbing properties of chlorophyll in the reaction centre of the photosynthetic apparatus. These linear molecules possess conjugated systems of double and single bonds along their backbones, which can act as conducting pathways for electrons.

Synthetic molecules based on carotenoids, with polar head groups at each end (Figure 26(*a*)), will transport electrons through lipid membranes, allowing a reducing species outside a lipid vesicle to reduce an oxidizing agent within the vesicle. The development over the past decade of a wide range of conducting organic polymers has broadened the pool of potential molecular wires. The conducting backbone of carotenoids is essentially a short stretch of the conducting polymer polyacetylene (polyethyne), while wires made from oligothiophenes (Figure 26(*b*)), oligoanilines and related carbon backbones have also been devised (here 'oligo' denotes a short polymer chain). David Allara, Paul Weiss, James Tour and their co-workers have shown that short conjugated molecular wires can pass about a trillion electrons per second between a gold surface and the needle tip of a scanning tunnelling microscope. Some of the hollow tubular carbon molecules called carbon nanotubes, discovered in

Figure 25
(*a*) Boland molecules contain aza-crown-ether hoops at both ends, linked by a membrane-compatible strand. (*b*) Several such hoops will stack spontaneously in a lipid membrane to form a tubular channel. (*c*) Bouquet molecules provide synthetic bundle-type molecular channels, which can increase the permeability of lipid bilayers to sodium and lithium ions.

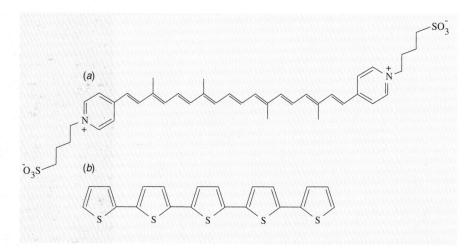

1991, can conduct electricity and have been used as single-molecule interconnects between microfabricated electrodes on a silicon chip.

Switching On

Building a real molecular electronics from these wires will require many more advances, not the least of which will be gaining the ability to position and connect them into circuit patterns and to control the flow of electrons. The tentative beginning of such a technology may be seen in the development of molecular wires that can be switched on or off photochemically (by incorporating optically responsive groups such as bistable chromophores) or electrochemically (by incorporating redox-active groups, for example metal complexes such as ferrocene).

Arieh Aviram and Mark Ratner gave molecular electronics perhaps its first concrete realization in 1974 when they proposed that donor and acceptor groups placed at either end of a linear molecule might turn it into a molecular rectifier, allowing a current to pass one way but not the other. In 1988 Aviram postulated a 'junction box' between two molecular wires that could provide a voltage-switchable bistable device. James Tour's group at the University of South Carolina has now synthesized these right-angled molecular junction switches (Figure 27).

There are many ways to flip a switch and many reasons for wanting to do so. Information is stored on magnetic tape by flipping the orientation of atomic magnets in the coating of iron or chromium oxide particles on the tape's surface; and it is processed in computers by switching operations carried out by transistors. Our bodies make use of all kinds of switches – mechanical (such as when an ion channel in the auditory system is pulled open by vibration), ionic (in nerve cells), optical (in the retina) and molecular (all manner of cell-surface signalling processes). These biological examples offer electronic engineers the hope that switching operations for computation and data storage might be carried out by single molecules, allowing phenomenal amounts of processing power or memory space to be packed onto 'chemical chips'.

In its minimal form, switching involves the conversion of a system from one stable state to another. In 1991, Fraser Stoddart and co-workers constructed a 'molecular shuttle' in which the bead of a rotaxane flipped back and forth between two possible docking positions on the axle (Figure 28). Because the two sites were equivalent, this was not a very useful switch, because one state could not be distinguished from the other; but the fact that the shuttling was taking place was apparent from nuclear magnetic resonance spectroscopy. At room temperature the bead shuttled back and forth so quickly that the two ends of the rotaxane looked the same, whereas at -50 °C the bead spent more time in one position than it did in the other and the two ends then gave distinct NMR signals.

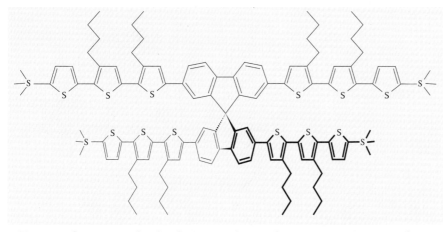

To use such a supramolecular device as a binary data-storage unit, it must be possible to control the switching between the two states. Angel Kaifer and co-workers at the University of Miami introduced controllability in 1994 when, in collaboration with Stoddart, they made a shuttle with two distinct states that could be interconverted electrochemically. The axle of the rotaxane contained two stations, rendered distinct by virtue of the incorporation of a redox-active group into just one of the molecule's end caps. The threaded hoop was sent to one station when the supermolecule was oxidized and to the other when it was reduced. Abraham Shanzer and colleagues at the Weizmann Institute of Science in Israel have shown that metal ions can be controllably switched between two different sites in a supramolecular assembly by changing their oxidation state and thus their preferred coordination geometry.

Flipping a switch between two positions is one of the simplest of the logic operations involved in computing. If molecular computing is ever to become a reality, more complicated logic gates will be needed. Stoddart and his colleague Steven Langford, working with Italian collaborators Alberto Credi and Vincenzo Balzani at the University of Bologna, have developed one such, called an XOR gate. This is a binary logic device with two input channels and one output; the output signal is 1 if the inputs are different (0,1 or 1,0) and 0 if they are the same (0,0 or 1,1). This kind of behaviour was realized in a pseudo-rotaxane – a hoop-and-spindle molecular assembly with no end stoppers, so that the hoop can slip off the spindle (Figure 29). The isolated spindle emits strong fluorescence (output 1), but this is suppressed (output 0) when the hoop threads itself. The input signals correspond to the addition of an acid or an amine base. If they are left to their own devices (inputs 0,0), the two molecules form the pseudo-rotaxane (output 0); but, if either an acid or an amine base is added (input 1,0 or 0,1), the components became unthreaded (output 1), either because the hoop is protonated and prevented from threading or because the amine molecules become complexed with the spindle and obstruct the hoop. If both the acid and amine are added together, however (input 1,1), the pseudo-rotaxane is regenerated because the acid and amine tie one another up in a complex of their own. So the appearance and disappearance of the fluorescence signal is triggered in a manner corresponding to an XOR operation.

All switching processes are subject to the randomizing influence of thermal fluctuations. The fidelity of data stored in bistable memory elements therefore depends on the relative thermal stabilities of the two states. This applies to any data-storage technology, be it magnetic, optical or molecular. It means that, in general, the smaller the storage element the smaller the thermal fluctuation sufficient to erase the data and hence the more the device has to be cooled to retain its memory. It may be that different electronic, rather than conformational, states of molecular assemblies will provide the most stable memory elements; indeed, 'molecular' data-storage devices that work

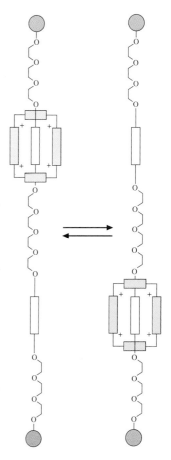

Figure 28
A molecular shuttle based on a rotaxane supermolecule. Above −50°C, the 'bead' will jump back and forth between the two docking points on the thread (where it is stabilized by donor–acceptor interactions).

Figure 29

A molecular-logic operation is demonstrated by this pseudo-rotaxane (a rotaxane with no end caps, so that the threading is reversible). Addition of acid (H^+) or base (B) triggers unthreading and allows the thread to fluoresce; but, when both are added together, the hoop remains threaded and fluorescence is quenched.

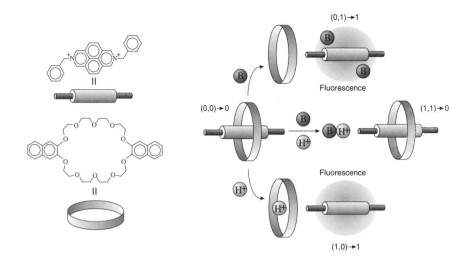

at room temperature have been devised by using the light-induced interconversion of electronic states of the protein bacteriorhodopsin.

To achieve the maximum storage density from such devices one requires the ability to address individual molecules – to write a bit of data onto them and to read it out again. It is conceivable that the scanning probe microscopes such as the scanning tunnelling microscope and the atomic-force microscope, in which an extremely fine needle-like tip can be used to detect or manipulate single molecules or atoms, might provide the means for this. Optical addressing using highly localized light signals is another possibility: already it is possible to make spectral measurements on individual molecules. For molecule-based signal processing more generally, optical interconnects have the advantage that they need not require precise juxtapositioning of molecules in the way that electronic molecular wires do – a spatially distributed optical signal could be converted into a localized, channelled energy packet by a suitable antenna unit. However, these speculative technologies will have to offer truly spectacular benefits in terms of speed, cost and reliability if they are ever going to be competitive with well-established, conventional data-storage and data-processing technologies – and we advise the reader to be wary of overenthusiastic forecasts regarding the advent of 'molecular computers'.

Sensors

Sherwood Rowland, who was awarded the 1995 Nobel Prize for chemistry for his work on the depletion of the stratospheric ozone layer, has confessed that atmospheric chemists only really became alerted to hazards such as this with the advent of techniques for detecting trace constituents of the atmosphere at the level of a few parts per billion – that is, for distinguishing one molecule from a billion others. Toxicologists will agree that there is plenty to fear from such apparently trivial concentrations: just three millionths of a gram of the marine toxin palytoxin can be enough to kill a person. Sensitive and selective methods for determining the concentrations of chemical compounds are in demand from all quarters of analytical science, from the food and drug industries to air-monitoring facilities.

Supramolecular chemistry has much to offer analytical chemistry, because its forte is devising molecular systems that can pick out one particular kind of molecule from amongst hordes of others. However, to turn a molecular-recognition event into a sensing event, binding must trigger some kind of externally detectable signal. Thus a molecular receptor can be made into a molecular sensor by coupling the binding event to a signalling event. Our sense of smell arises from just such coupling. The binding of odorant molecules by a family of proteins embedded in the olfactory membrane triggers changes in conformation at the nether ends of the proteins, which initiate within the cell events that culminate in a neural impulse.

The earliest artificial sensors that exploited molecular recognition were deveoped in the 1960s and took advantage of the selectivity of proteins. They contained the enzyme glucose oxidase immobilized within a permeable plastic membrane. When the enzyme recognized, bound and oxidized glucose molecules in the surrounding fluid, the change in concentration of dissolved oxygen that this engendered was detected electrochemically by a metal electrode. Devices that employ biomolecules such as enzymes for chemical sensing, called biosensors, are now used routinely for clinical diagnosis of body fluids, particularly for monitoring the blood sugar levels of diabetics. Modern glucose biosensors can fit into a pen- or wristwatch-sized device and provide diabetics with an instant analysis of their glucose levels from a drop of blood.

The selectivity of some supramolecular interactions may be exploited to good effect in electrochemical sensors that use chemically modified electrodes. For example, molecular receptors immobilized (say, by covalently bound linking groups) on a conducting electrode's surface may induce a change in the electrode potential when they bind the analyte. In this way, the presence of the analyte may be registered by standard voltammetric techniques. Cyclodextrin molecules attached to an electrode surface have been used in this way to distinguish between the *ortho* and *para* isomers of nitrophenol, owing to the better fit of the latter in the cyclodextrin's bucket-like cavity. Ion-selective electrodes may be obtained by coating a metal electrode with self-organizing supramolecular thin films (see page 343) of molecules that exhibit selective metal-ion binding.

David Reinhoudt and co-workers have married molecular-recognition processes with conventional silicon-based microelectronics to prepare field-effect transistors (FETs) that register a recognition event as a change in current through the device. They exploit the fact that the conductance through a FET, conventionally controlled by the electrical potential of the gate terminal between the source (input) terminal and the drain (output) terminal, is influenced by the electrical potential at the interface of the gate and an aqueous solution with which it is in contact. The top layer of the gate is typically fabricated from silica and the interfacial potential may be altered by protonation or deprotonation of the silanol (SiOH) groups at the surface. Reinhoudt and colleagues have shown that the potential can alternatively be varied by grafting to the silica surface receptor molecules, hybrids of calixarenes and crown ethers, that selectively bind metal ions. These chemically sensitive field-effect transistors (CHEMFETs) are able to measure potassium concentrations in biological fluids in the millimolar range.

Increasingly, electrical signalling is being replaced by optical methods in chemical sensors and biosensors. Here the binding of the substrate being analysed (the analyte) triggers a change in the optical properties of the receptor, which can be detected by, for example, monitoring its fluorescence spectrum. This approach has the advantages that it does not rely on a line of electrical communication between the receptor and the signal-processing apparatus, and that advances in fibre-optical technology allow fluorescence sensors to be very small. One way of coupling an optical response to a recognition-and-binding event is to exploit a transfer of excitation energy between donor and acceptor groups on the bound species. Lehn has proposed that the photosensitive cryptate complex shown in Figure 30(a) could be used as an optical signalling element in an immunoassay, which monitors the level of a species that is recognized by an antibody. The cryptate contains a bound europium ion, which emits strong fluorescence at visible wavelengths. The bipyridine groups of the cryptate will themselves absorb light strongly in the ultraviolet and can transfer the absorbed energy to the metal ion, stimulating it into an excited electronic state from which it decays

Figure 30

(a) A photo-active cryptate complex like this might provide a component in an antibody-based sensing system (an immunoassay). Excitation of the bipyridine group by light is followed by energy transfer (ET) to the europium ion, which then fluoresces at a characteristic wavelength. (b) In the immunoassay, the cryptate and the acceptor group A are linked to two antibodies that bind to different regions of the molecule to be sensed (the antigen). The cryptate and the acceptor are brought together in the presence of the antigen and energy is transferred from the former to the latter, instead of to the bound metal ion. So the europium fluorescence is then quenched.

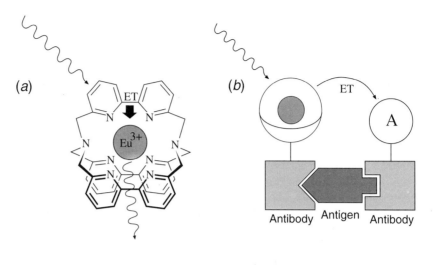

with visible fluorescent emission. If the excited cryptate were to be brought close to another acceptor group, then instead of decaying by fluorescence it could transfer the energy over a relatively long distance to 'switch on' the acceptor. The fluorescent emission would then be at a wavelength characteristic of the acceptor rather than of the europium ion. This kind of long-range energy transfer might be used to signify a recognition event by attaching the acceptor and the cryptate to two separate antibodies that bind specifically to two different parts of a substrate molecule (the antigen) (Figure 30(b)). The three-component supermolecule then acts as a fluorescence sensor for the antigen. This concept has been transformed by Gérard Mathis into a novel method for medical diagnostics.

Other groups have exploited the change in conformation that a receptor experiences when it binds its substrate to develop fluorescent supramolecular sensors. Seiji Shinkai and co-workers at ERATO, the Research Development Corporation of Japan have constructed a receptor for sugar molecules, based on boronic acid, which has two linked naphthalene groups attached. The receptor is chiral – it has two possible structures that differ only in that one is the mirror image of the other – and so, like a glove fitting a hand, it can bind selectively to one particular 'handedness' of a chiral sugar (Figure 31). The naphthalene groups are potentially fluorescent, but the light emission is blocked ('quenched') by adjacent nitrogen atoms which transfer electrons to the naphthalenes. When the chiral sugar is bound, this transfer is inhibited, and the sensor molecule lights up with fluorescent emission. Selective detection of chiral molecules is extremely important to the drugs industry, for a pharmaceutical molecule of one handedness and its mirror image can have very different physiological effects.

Our state of health is signified by the fluctuations in concentration of a fantastic number of chemical species, so assessing the progression of a specific illness might require the simultaneous monitoring of several analytes in the body fluids. Biomedical sensors that are now being developed can provide this kind of information – for example, by immobilizing several different fluorescent receptors on different regions of an optical sensor probe, each served by a separate optical fibre no thicker than a human hair. As the range of supramolecular sensors expands, it might become feasible to think of generating a full molecular body map for a patient: a snapshot of the complete molecular constitution of the bloodstream, or perhaps even a complete spatial picture of how that constitution varies throughout the body. Such a picture might allow doctors to know at a glance exactly where, when and how much medication is needed. A marked reduction in health costs might result.

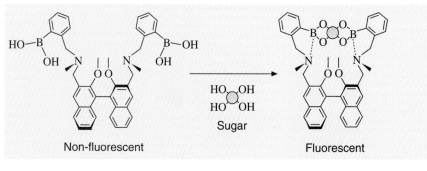

Figure 31
This receptor molecule can sense the difference between two enantiomers of a chiral sugar molecule, fluorescing in the presence of one but not the other.

Non-fluorescent

Sugar

Fluorescent

Catalysis

The chemistry of life needs a helping hand. There is scarcely a biochemical reaction that does not require a catalyst for it to proceed efficiently towards the desired product. Nature's catalysts are enzymes and the way in which they act has the characteristics of a supramolecular process: as Emil Fischer realized, it involves the highly specific binding of a substrate to a receptor site in the enzyme molecule. In his Nobel award lecture of 1902, Fischer said 'I can foresee a time in which physiological chemistry will not only make greater use of natural enzymes but will actually resort to creating synthetic ones'. The tremendous prescience of this remark is attenuated only by the emphasis on physiological chemistry, because today natural enzymes are used not only in medical chemistry but also in industrial processes that synthesize products ranging from foodstuffs to biodegradable plastics. One of the major goals of supramolecular chemistry is to devise new catalysts – artificial enzymes, if you will – that will expand and augment this range.

Just about every catalytic process involves three steps: binding of the substrate, followed by its chemical transformation into the product and finally the release of the product to regenerate the active catalyst. Supramolecular catalysts may be distinguished from traditional industrial catalysts like metal surfaces by their use of molecular recognition both to accommodate the substrate at a reactive site and to carry out 'molecular surgery' to transform it into a particular product.

The trick that all catalysts perform is to reduce the amount of energy needed to transform the substrate to the specified product. At the crest of the energetic hurdle separating reactant from product sits an ephemeral, energetic species called the transition state. The function of a catalyst is to lower the energy of this transition state relative to that of the initial substrate. For supramolecular catalysts, what this means is that the transition state must be more strongly bound (more stabilized) than the substrate – if this condition is not satisfied, the reaction is not assisted no matter how efficiently the catalyst recognizes the substrate.

We mentioned earlier how cyclodextrins can be considered crude enzyme mimics in that they act as non-covalent catalysts, facilitating transformations of a bound molecule without the formation of a covalently bound intermediate. Myron Bender found that a cyclodextrin molecule would accelerate the rate at which a phenyl ester was hydrolysed (split into phenol and a carboxylic acid) by a factor of 12 (Figure 32(*a*)). Bender and Ron Breslow went on to show that the efficiency and selectivity of such catalytic reactions could be increased by attaching substituent groups to the rim of the cyclodextrin's cavity that directed the reaction along a given pathway. For example, a cyclodextrin catalyst devised by Breslow will convert the cyclic phosphate in Figure 32(*b*) into just one of the two products that are generated when the molecule is hydrolysed without the catalyst. This reaction is similar to that carried out by the enzyme ribonuclease A. There is arguably some justification for calling supramolecular catalysts that are designed to produce a specific product 'artificial enzymes'; but the rate

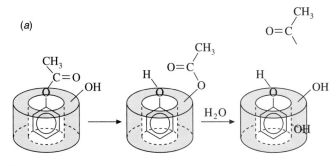

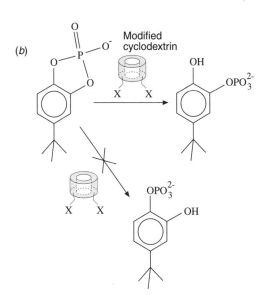

Figure 32

Catalysis by cyclodextrins. (*a*) The rate of hydrolysis of phenyl esters can be accelerated by complexation with α-cyclodextrin. (*b*) A cyclodextrin modified to mimic the enzyme ribonuclease A will selectively catalyse the hydrolysis of a cyclic phosphate to the *m*-phosphate product in preference to the *p*-phosphate.

enhancements that they provide are (at least at present) far less than those offered by real enzymes.

Most of the current work on supramolecular catalysts involves the rational design of the catalytic receptor, with the idea that careful placement of substituents around the binding site will create the right disposition of chemical 'tools' to effect the required surgery on the substrate. The range of catalysts that build on this principle is too vast to do it justice here. To select just one example, Andrew Hamilton (now at Yale University) and co-workers have developed catalytic receptors based on the delicate disposition of hydrogen-bonding groups. In one such case, the rate of the intramolecular Diels–Alder reaction in Figure 33(*a*) was accelerated by the molecule shown in Figure 33(*b*). The receptor seems to hold the substrate in the appropriate conformation for the intramolecular rearrangement to occur, stabilizing the transition state within a supramolecular complex bound by hydrogen bonds (Figure 33(*c*)).

In recent years there has emerged a new philosophy that holds the promise not just of being able to generate novel catalysts but also of yielding clues about how living systems evolved their own impressive array of supramolecular catalysts. This approach is a profligate one, insofar as it generates at random vast numbers of potential receptor molecules in the hope that one of them will just happen to have the structure required to effect good catalysis. It is rather as if one were letting a monkey (these days a computer would do just as well) type characters at random in the hope that somewhere amongst the mess will be, if not a Shakespearian tragedy, then at least a respectable stanza.

The approach works because of the astronomical numbers of molecules involved. Some chemists are accumulating random molecular libraries by using the techniques of polymer chemistry to synthesize random polymers from mixtures of constituent parts, but some of the best results so far have come from techniques that employ the molecular machinery of the immune system to put the libraries together. The immune system tags foreign particles (antigens) for destruction by lymphocytes by producing hordes of antibodies until one happens to provide a good fit. However, if the antibodies are raised against an antigen that closely resembles the transition state for a chemical reaction, they may also act as catalysts, since they might also bind and stabilize the true transition state. In this way, catalytic antibodies that facilitate a wide range of organic reactions have been produced. This technique was pioneered by the groups of Richard Lerner at the Scripps Research Institute and Peter Schultz of the University

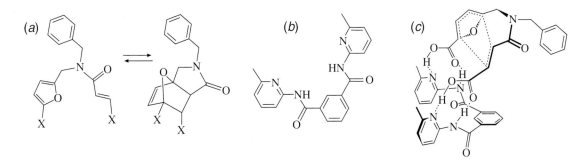

Figure 33
The intramolecular Diels–Alder reaction in (*a*) can be accelerated by the hydrogen-bonding receptor molecule in (*b*), probably via the formation of the complex in (*c*), which stabilizes the transition state.

of California at Berkeley. Catalytic antibodies have not yet achieved the kind of rate accelerations typical of enzymes and it remains unknown whether they ever will – at least without additional restructuring of the protein molecules 'by hand' using standard biotechnological methods. Nonetheless, because the relationships among structure, binding and catalysis can be investigated at various stages as the antibody 'matures' to its final form, they offer the prospect of finding answers to questions about how the pathways of enzymatic catalysis have evolved.

The development of methods for manipulating nucleic acid molecules (DNA and RNA) for biotechnology has given rise to a method for generating both receptors and catalysts from these biological molecules too – even though in nature their primary role is rather as data banks. This approach, called *in vitro* evolution, has been used to create a variety of RNA-based catalysts.

Catalytic RNA molecules do exist in biology: they are called ribozymes and were discovered in the 1980s by Thomas Cech and Sidney Altman and their co-workers. It is believed that ribozymes may have been central to the evolution of living organisms before nucleic acids became able to generate proteins to do the catalytic jobs instead. In support of this idea, Jack Szostak's group at the Harvard Medical School has used *in vitro* evolution to create RNA molecules that can catalyse the formation of peptide bonds, which join amino acids together in proteins. David Bartel at the Whitehead Institute in Massachusetts has reported an artificially evolved ribozyme that will construct a nucleotide – the basic unit of RNA and DNA – from its component parts. Both of these transformations show how RNA-type molecules might have been the engineers that forged a path on the early earth from relatively simple molecules to the complex biomolecules on which all life depends today. This scenario of an RNA-catalysed advent of the earliest life is called the RNA 'world'.

Self-assembly and Self-organization

Many of the supramolecular catalysts developed so far have performed the task of breaking their substrates apart. Other catalysts are more creative: they put new molecules together from their component parts, rather than breaking them down. In this way, catalysis becomes a tool for synthesis, chemistry's creative aspect.

Bringing molecules together is intrinsic to another major concern within supramolecular chemistry: templating. Supermolecules may assist synthesis by acting as templates for the organized assembly of components in a manner that may but need not be catalytic – in some cases, the final assembly might be detached from the template, leaving the latter free to perform the task again, whereas in others the templates may remain part of the final assembly. We have mentioned one such example already: the templating of a 'molecular tube' via the formation of a polyrotaxane.

Metal ions can serve as templates that guide ligands into a particular arrangement and can thereby act as organizing centres for supramolecular synthesis. Because the bonding orbitals of transition metals point in well-defined directions in space, these metals generally exhibit more or less pronounced preferences for certain geometrical

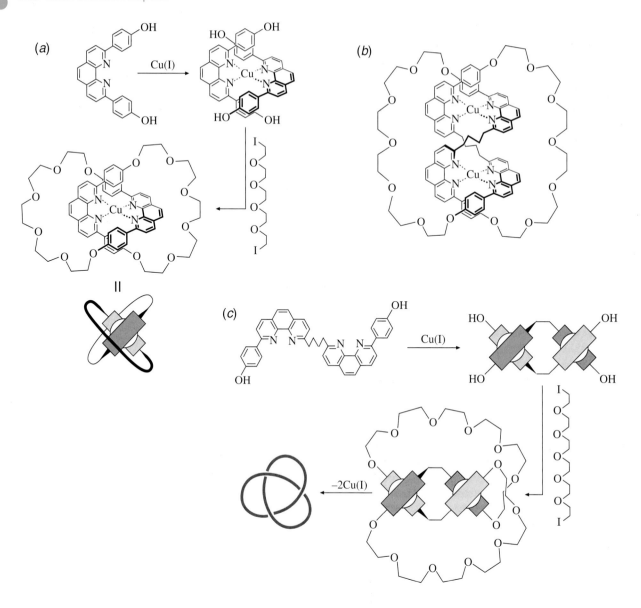

Figure 34

(a) The formation of a catenane using metal-ion templating to pre-organize the link. (b) The same approach can be used to fashion topologically complex knotted catenanes. (c) The molecular trefoil knot, a single strand joined end to end in a threaded knot.

arrangements of ligands, which depend on the nature of the metal and on its charge. Copper(II) (Cu^{2+}) ions, for instance, have a tendency to adopt square-planar (four-coordinate) or square-pyramidal (five-coordinate) geometries, whereas copper(I) ions (Cu^+) prefer a tetrahedral arrangement.

Jean-Pierre Sauvage and co-workers at the Université Louis Pasteur in Strasbourg have used these principles to synthesize topologically complex molecules, in which chains and hoops are interwoven in convoluted ways. Sauvage's group has synthesized catenanes – interlinked rings – by using copper ions to arrange two strands in advance such that, when their ends are joined together via covalent bonds, the loops pass through one another (Figure 34(a)). The chances of capturing one loop on another by closing the rings in the absence of the metal ion are minimal; but, because the ion holds the two strands together by coordination bonds to the nitrogen atoms, closure is guaranteed to form the catenane. This approach can be used to make knotted molecules – Sauvage and colleagues have devised knotted catenanes (Figure 34(b)) and a trefoil knot (Figure 34(c)), a single strand joined end to end whose convolutions are guided into place by the coordination chemistry of the metal-ion template.

The fact that differently charged (oxidation) states of the same metal have different coordination preferences means that the geometry of metal-containing supramolecular assemblies can be altered by changing this charge state through reduction–oxidation (redox) chemistry – by, for example, allowing the metals to pick up or lose electrons at charged electrodes, or by bringing them into contact with electron donors or acceptors. Sauvage's group had made a kind of molecular switching mechanism that works on this principle. They synthesized, using copper(I) ions as templates, a catenane in which one of the rings contained two metal-binding sites; one with three nitrogen atoms and one with two. The copper(I) ions prefer to sit at the two-coordinate sites, where they can adopt a near-tetrahedral geometry; but when the ions are oxidized electrochemically to copper(II), they acquire a taste for a higher coordination number, so the ring flips to bring the three-coordinate site around the metal (Figure 35). Reduction back to copper(I) flips the ring over again. Because the two metal ions absorb light at different wavelengths, this switching process is visible to the eye as a change of the solution's colour from green to pale yellow.

Neutral molecules can also be used as templates for self-assembly processes. This is exemplified in the self-assembling molecular capsules designed by Julius Rebek and Robert Meissner at the Massachusetts Institute of Technology, in collaboration with Javier de Mendoza at the Universidad Autonoma de Madrid. They have developed a new route to molecular encapsulation by synthesizing curved molecules that interlock like the two halves of a tennis ball (Figure 36), held together by hydrogen bonds. In chloroform or xylene solution these molecules form a disordered gel-like state through intermolecular hydrogen bonding. However, when a neutral guest molecule of the right size and shape (such as an adamantane derivative) is added to the solution, the molecules pair up to form a pseudo-spherical host with the guest molecule inside. Rebek has shown that these capsules can accelerate a bimolecular chemical reaction by holding two encapsulated reactants in close proximity.

Figure 35

Switching behaviour can be introduced into metal-templated catenanes by using redox chemistry to change the coordination geometry of the chelated metal ion between the two loops. Oxidation of the copper(I) ion to copper(II) induces a preference for higher coordination, whereupon one ring flips to offer a tridentate ligand instead of the bidentate one.

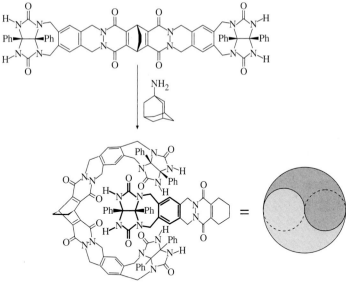

Figure 36

A self-assembling sphere: the two components are held together by hydrogen bonds and dimerize in the presence of a suitable guest molecule. The latter acts as a template for this self-assembly process. (Note that only half of the 'horizontal' molecule in the dimer is shown; the other half is curled over behind it and has been omitted for clarity.)

Replication

A template carries encoded information, which is read during the process of assembly. In its most sophisticated form this information is a unique encoding of the product molecule (or molecular assembly) that is assembled on the template. We can see this most clearly in examples of biological templating – in the synthesis of proteins and nucleic acids. Ultimately the information for making a protein is encoded in the DNA molecule, in a four-letter code that is stored as a sequence of four different nucleotide units along the DNA chain. To convert this information from its nucleic-acid form to

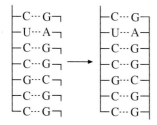

Figure 37

Short nucleic-acid polymers – oligonucleotides – can act as templates for the formation of complementary strands from their monomers, without help from enzymes. Here A, G, C and U denote the nucleic-acid bases of RNA: adenine, guanine, cytosine and uracil. The nucleotides here are modified variants of those in natural RNA with 'activating groups' to promote polymerization. The backbone is shown schematically and the copying mechanism is highly simplified in this representation.

its protein form, the DNA first acts as a template for a messenger RNA molecule, which is assembled on a single DNA strand by binding of each of the RNA's nucleotides to its complementary partner on DNA.

To thrive – to grow and to reproduce – an organism must also be able to carry out replication of its genetic material. That is, the DNA must be able to make copies of itself. Molecular replication implies a particular kind of catalytic process, called auto-catalysis: the substrate multiplies itself by catalysing its own formation. The replication of DNA is again a kind of templated catalysis: a single strand of DNA acts as a template for the assembly of a complementary strand, and then these two separate and each templates the formation of its own complement. Self-templating becomes auto-catalysis, however, only when the product can be separated from the template so that both can effect further templated assembly.

In living cells these templating processes are mediated by enzymes, which ensure amongst other things that the fidelity of the assembly process remains high – that is, that the information in the template is not often misread, which leads to errors in copying and replication. However, when life on earth began, the earliest replicating molecules would not have had an army of enzymes to assist them. In the hope of understanding how replication could have come about in these 'pre-biotic' chemical systems, several researchers have investigated templating and autocatalysis of short-chain polymers (oligomers) made up of similar components to those in RNA and DNA. In the 1980s Leslie Orgel and co-workers at the Salk Institute in San Diego showed that nucleic-acid oligomers can act as templates for the assembly of complementary oli-gomers from the constituent nucleotides, without the assistance of enzymes (Figure 37). However, the nucleotides must be 'primed' with activating groups that help the phosphodiester bonds of the backbone form at the right location on the adjoining units. Moreover, the templated products are typically mixtures of oligomers of various lengths; and, as the strands get longer than a dozen or so nucleotide units, copying mistakes start to creep in.

In the 1980s, Günter von Kiedrowski at the University of Freiburg investigated whether such a templating process can give rise to replication if the template's com-plementary partner is palindromic (identical to the template when read in reverse). If strands assembled in this way can detach themselves from the template and make further copies, the number of copies should increase exponentially over time. (In prac-tice the increase is generally slower than this, since not all the product molecules are readily detached.) Von Kiedrowski and colleagues found that a six-component DNA-based oligomer (Figure 38) performs replication. In 1993, both von Kiedrowski and K. C. Nicolaou at the Scripps Research Institute in California reported that replication of longer nucleic-acid oligomers could be achieved by the templating of non-palindromic partners: each of the initial templates catalysed the assembly of a complementary strand and these then separated and templated the assembly of copies of the original strands.

Does replication have to be based on nucleic acids? Or can we envisage other kinds of 'genetic' molecular systems, not found in nature, which are programmed with the information that allows them to reproduce? Julius Rebek's group showed in 1991 that molecular replication is not a unique property of nucleic acids but can be performed by molecules whose supramolecular templating ability stems from a different kind of chemistry. The requirements of such molecules are easy to enumerate: they must act as receptors that selectively bind their component parts via non-covalent interactions, in a geometry that allows the bound components to react and link up to form a copy of the template. To ensure that autocatalytic replication can occur, the copy must not be so tightly bound as to remain dimerized. This process is shown schematically in Figure 39.

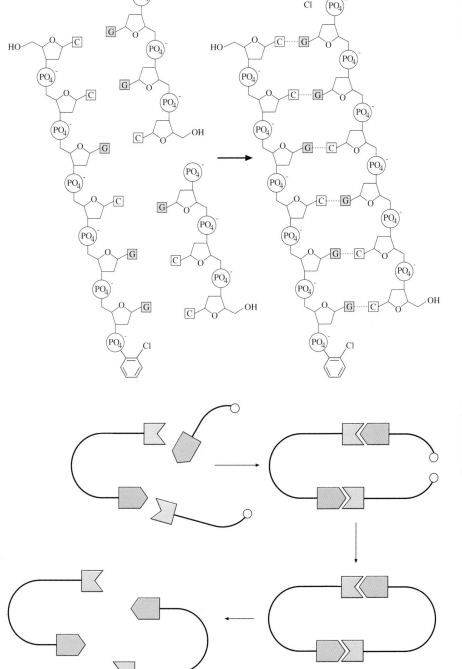

Figure 39
Templated molecular replication requires self-complementarity, so that one molecule can catalyse the assembly of a copy by molecular recognition.

Rebek's synthetic replicating molecules turn back on themselves in a J shape (Figure 40). The binding sites for two separate fragments of the copy sit at the ends of the arms of the J and these fragments are held such that their unbound ends are in close proximity, assisting the formation of the amide linkage that completes the copying. The kink is essential; it ensures that the template's binding sites point in the same direction, so that the fragments of the copy are bound side by side.

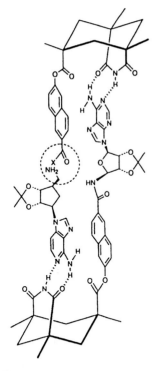

Figure 40

This J-shaped receptor molecule will facilitate the formation of a copy from its two component parts by holding them in close proximity through hydrogen bonding. Linking up of the groups enclosed in the dashed circle produces the copy.

Rebek has demonstrated the generality of this principle for replication by synthesizing other J-shaped molecules that will also template the assembly of copies. He has also been able to mimic evolutionary competition by mixing all four components of two of the replicating molecules and leaving them to compete for parts with which to make copies. As well as the two 'pure' replicators, two hybrid molecules were formed, containing one component of each 'species'. One of these hybrids is a more efficient replicator than is either of the pure systems and therefore becomes the predominant product.

One of the reasons why the 'nucleic-acid-first' RNA world is a favoured scenario for the chemical origin of life is that in all living cells today it is only nucleic acids, not proteins, that are capable of molecular replication. However, Reza Ghadiri and colleagues have shown that this specialization is not fundamental. They have developed small protein-like peptide molecules that will act as templates for the assembly of copies via non-covalent interactions (Figure 41). The peptides are rod-like 'coiled coils', which template the dimerization of their two halves by the juxtaposition of water-insoluble amino-acid groups on the reactants and the template: such groups have a tendency to aggregate in water. So even proteins can replicate through supramolecular interactions if the design is right.

These are primitive steps towards supramolecular systems that exhibit some characteristics of living organisms. However, there is certainly more to life than reproduction. A living organism is self-sustaining (it metabolizes the sources of energy in its environment); it is bounded, generally by a permeable membrane; and it is able to adapt. This means that it carries genetic information that can be passed on and that is susceptible to mutation and Darwinian selection.

Self-assembly through Metal-ion Coordination

Metal-ion templates can be used to guide the synthesis of large supramolecular assemblies from many component parts. One can consider this assembly process as being pre-programmed, in the sense that the information required to build the assembly is encoded in the constituents in the form of their preference to bind certain other components in specific geometries. A series of templating steps leading to an organized supramolecular structure can be considered a genuine process of self-assembly. Self-assembly processes generally produce the thermodynamically most stable entity, but they can also be under kinetic control. This is often the case in biological systems. The use of metal ions for this kind of self-assembly is attractive not only because they provide well-defined coordination geometries but also because they present a range of binding strengths (from weak to virtually covalent) and because their properties can be modified by light (photochemically) or electrons (electrochemically).

Lehn and co-workers have used metal ions with tetrahedral coordination geometries as the organizing centres for the multi-step assembly of double helices (helicates) in which the two strands contain metal-binding bipyridine (bipy) units linked together (Figure 42). Even when strands of different lengths are mixed together with the metal ions, only the 'pure' helicates in which each of the two strands has the same length are formed. Multiple complexation of a single ion by the same strand is prevented by the design of the ligands, which cannot readily bend back on themselves; and the efficiency of the self-assembly process is the result of positive cooperativity, whereby the binding of one metal ion pre-organizes the ligands to facilitate binding of the second. In other words, the assembly process gets easier as it progresses. The fact that imperfect complexes are not formed from mixtures of strands of different lengths is a consequence of the greater thermodynamic stability of the pure complexes, which have no dangling ends. Because of the relative weakness of the coordination bonds holding the assembly together, they can 'anneal' into the pure structures. We can contrast this

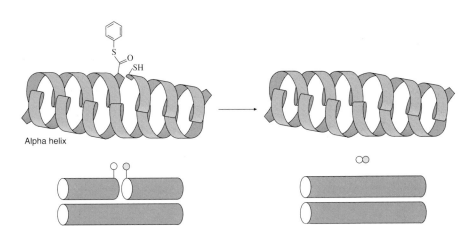

Alpha helix

Template

Figure 41
Replication in peptides. The 'coiled-coil' template has hydrophobic and hydrophilic regions on its surface, which promote the binding of the two components needed to make a copy.

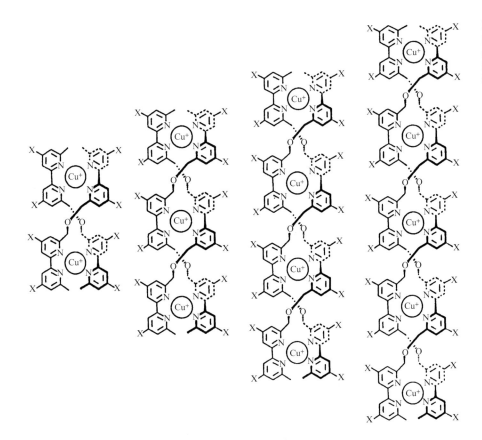

Figure 42
Double helicates are supermolecules organized around metal ions. The two ligands twist around several ions in a double-helical fashion.

with the formation of covalent polymers from a mixture of monomers, in which random polymerization will lead to a diverse mixture of chain lengths and compositions because there is no option of making changes once a bond has been formed.

There is an obvious similarity between the structure of these helicates and the double helix of DNA. The analogy can be made more concrete by attaching nucleosides – the structural elements of DNA, without the phosphate groups – to the helicate backbone (Figure 43). These deoxyribonucleohelicates (DNHs) are like inside-out nucleic acids, held together by metal ions and with their hydrogen-bonding substituents on the outside. However, whereas DNA is negatively charged (because of the ionized phosphate groups), DNHs are positively charged. Because of this and the hydrogen-bonding

Figure 43
DNH is a double helicate with
nucleosides attached to the
backbone.

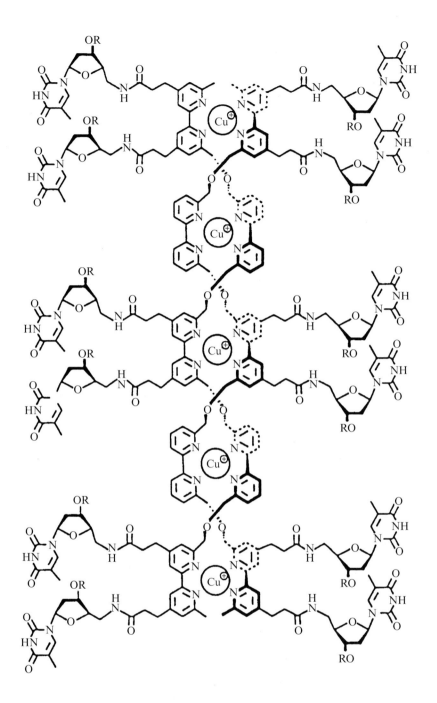

capacity of the nucleosides in DNHs, they will bind to DNA. Compounds that do this
are of considerable pharmaceutical interest, for they offer the prospect of interfering
with and regulating the activity of genes.

The scope of coordination chemistry to provide organized supramolecular assemblies is limited only by our imagination. One can conceive of a set of molecular components that can be welded together by metal ions into all sorts of superstructures. If
the components and ions are well chosen, these assemblies will come together spontaneously in high yield, with no need for the laborious step-by-step procedures of traditional synthetic chemistry. An example is the assembly of rigid components shown
in Figure 44(*a*), in which four flat ligands and three rigid (bipy)$_4$ rods are joined by

(a)

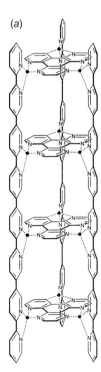

(b)

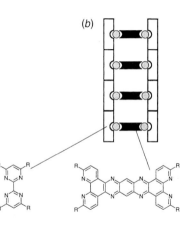

(c)

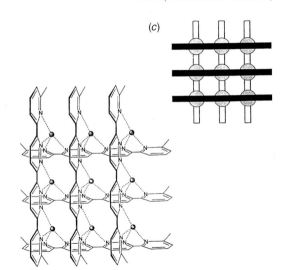

silver ions into a cylindrical complex with no fewer than 19 parts. One possible goal of this kind of chemistry is the creation of molecular grids, which might have useful optical, electronic or mechanical properties. Racks or ladders (Figure 44(b)) can be considered one-dimensional grids, while rod-like ligands can be criss-crossed with metal ions at the vertices to form two-dimensional grids (Figure 44(c)). These supramolecular grids might ultimately prove useful as addressable arrays for very-high-density data storage – information could be read in and out by using light or fine electrochemical probes to alter or monitor the oxidation state of specific metal ions.

Makoto Fujita at the Institute for Molecular Science in Okazaki has pioneered the use of metal-ion coordination chemistry to link organic ligands into complex topologies, such as the cage structures shown in Figure 45. The smaller of these, composed of four triangular ligands and six palladium ions, can encapsulate four bulky adamantane molecules. The larger cage, which self-assemblies in solution from 24 component parts, has an internal volume of 9 nm³, which is capacious enough in principle to accommodate a C_{60} molecule.

This assembly of organized supramolecular arrays via coordination chemistry can be generalized to extended ('infinite') lattices, providing new bulk materials. That is to say, one can make crystals whose molecular frameworks are dictated by the nature of the constituent ions and molecules. Richard Robson and colleagues at the University of Melbourne have made molecular crystals with a channelled framework structure from square porphyrin units linked at their corners by copper(I) ions (Figure 46). The tetrahedral coordination of the metal ions, coupled with the square shape of the ligands, creates rectangular channels of dimensions about 1.5 nm by 2 nm that run through the material. In these particular crystals the channels are filled with solvent molecules and the crystal structure collapses if the solvent is removed by heating. However, this is not a fundamental limitation, as we shall see.

Figure 44

Grids and ladders from metal-ion self-assembly. (a) A self-assembled stack of 19 components. (Here phenyl and methyl substituents on the ligands have been omitted for clarity.) (b) A ladder or one-dimensional grid can be assembled from the ligands shown on the right-hand side, which serve as the rungs. (c) A two-dimensional grid results from tetrahedrally coordinated metal ions (such as silver(I)) and linear rod-like ligands.

Figure 45

Large cage-like structures that self-assemble through metal-ion coordination chemistry.

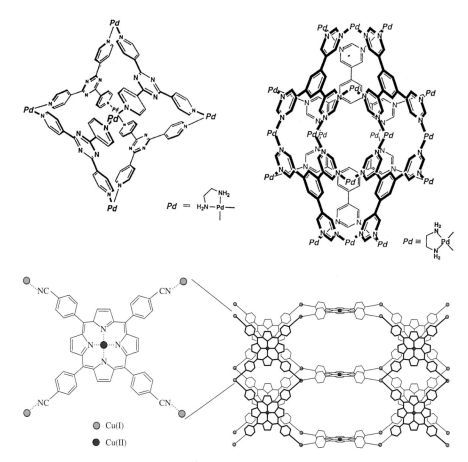

Figure 46

This extended metal-coordination compound contains large channels.

Self–assembly through Hydrogen Bonding

Hydrogen bonds, which also exhibit strong directional preferences, can also be used in the rational design and self-assembly of extended arrays in one, two or three dimensions. This approach has been pioneered by Margaret Etter of the University of Minnesota. A molecular crystal might be considered an infinite supermolecule, and 'crystal engineering' – the control of the crystal structure by manipulation of intermolecular interactions – has become an active area of research. One aim of this field is fundamental: to realize the long-sought goal of being able to predict crystal-packing arrangements *a priori*. Others motivations stem from at practical goals, such as the selective disposition and orientation in space of light-responsive molecular units to produce materials with useful non-linear optical properties, and the synthesis of new 'molecular sieves' with uniform, ordered molecular-scale pores.

The hydrogen-bonded assemblies of melamine and barbituric or cyanuric acid discussed on page 309 are in fact fragments of a hypothetical, extended sheet-like array in which the two complementary molecules alternate on a hexagonal lattice (Figure 47(*a*)). (This structure is hypothetical, because the two molecules do not actually form well ordered co-crystals.) By blocking off the hydrogen-bonding capacity on some sides of the molecules, one can make infinite 'strips' of this sheet-like structure, such as linear or crinkled tapes (Figure 47(*b*)). These supramolecular arrays are not uniquely specified by the structures of the components, however, so the products of their association are mixtures that are hard to characterize. In contrast, the hybrid molecule shown in Figure 48, which contains both melamine-like and barbituric acid-like parts,

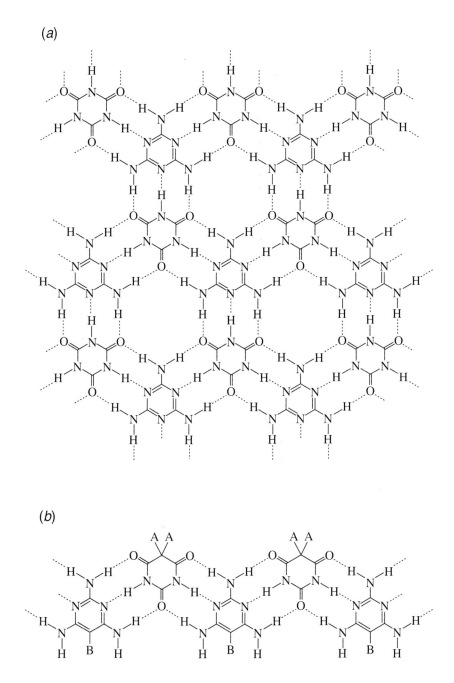

(a)

(b)

Figure 47
(*a*) An equimolar mixture of melamine and cyanuric acid may form an extended, sheet-like hydrogen-bonded lattice.
(*b*) Modifying the molecules to remove some of the propensity for hydrogen-bonding can result in the formation of tape-like structures.

can self-associate only into a specific ribbon structure. This can be considered a kind of supramolecular polymer, held together by hydrogen bonds. Fraser Stoddart and co-workers have reported a clever way of linking hydrogen-bonded tapes into two-dimensional sheet-like arrays by forging rotaxane-like crosslinks between the tapes via hoops that capture adjacent side-arms (Figure 49).

This game can be played in three dimensions too. James Wuest at the University of Montréal and co-workers have made molecular building blocks in which hydrogen-bonding groups sit at the ends of four tetrahedral arms (Figure 50). Wuest proposes the term 'tecton', derived from the Greek word for 'to build', for these building blocks. They will self-assemble into crystals with the same network connectivity as that which links

Figure 48
Molecular design based on the
motifs in the previous figure can
allow specific assemblies to be
engineered by self-recognition.

Figure 49
Hydrogen bonding is used here both
to allow self-assembly of rotaxane
units and to link them into extended
sheets (via the dashed lines).

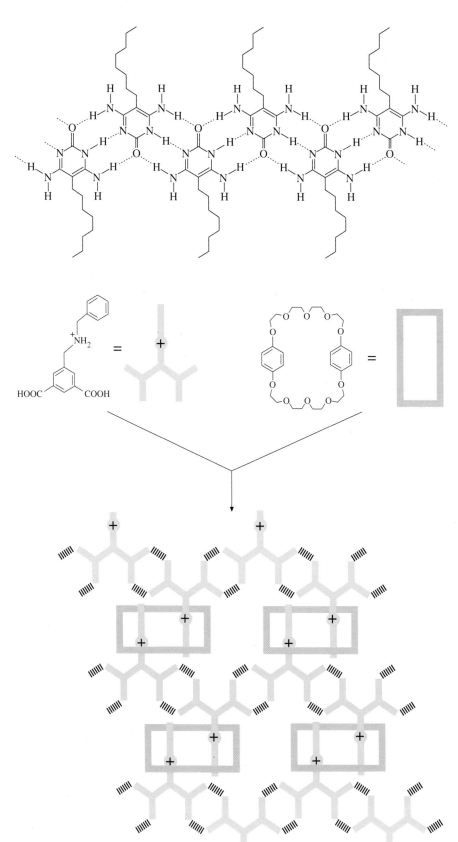

the carbon atoms in diamond. This 'diamondoid' structure contains large voids – so much so that there is room to interweave a second diamondoid lattice with the first. Even with these two interpenetrating networks of tectons, the crystals retain a lot of open pore space. Again, solvent molecules fill the pores of this open-framework material. By multiplying the number of hydrogen bonds holding the tectons together, Wuest's group has achieved the long-sought goal of making an open framework that survives removal of most of these solvent molecules.

Nadrian Seeman and co-workers at New York University have fashioned complex supramolecular structures and arrays using the highly selective hydrogen-bonding capacity of DNA itself. Their framework structures have struts fabricated from the famous double helix, while the junctions between struts consist of several interwoven strands (Figure 51(a)). One of the great attractions of using DNA as a construction material for supramolecular engineering is that it can be programmed to interact very selectively (and with rules that are well known from molecular biology) through the choice of nucleotide sequence along each strand. A second advantage is that nature provides us with molecular tools for cutting and pasting the building blocks, in the form of enzymes that do these jobs in the cell. Seeman's strategy is to create double-helical fragments of the eventual assembly in which short stretches of single strands are left dangling at each end. These provide 'sticky ends', which will bind to those on other fragments via hydrogen bonding – but only if the sequences of the single-stranded portions match up. Once these sticky ends have been joined together, they can be converted into robust, covalently bound connections by using ligation enzymes to link up the helical backbones. In this way, Seeman's group has created topologically complex DNA molecules such as a cube and truncated icosahedron (Figures 51(b) and (c)) and also extended arrays of different ring-like components in a specified tiling pattern (Figure 51(d)). Seeman proposes that these DNA frameworks might provide the scaffolding for the assembly of other structures. By incorporating into them groups that will bind to inorganic materials, for instance, it might be possible to fabricate an ordered array of nanometre-sized semiconductor crystals – an array of 'quantum dots' with possible applications in information processing and storage. Günter von Kiedrowski has investigated this possibility using DNA-bound gold particles.

Liquid Crystals and Supramolecular Polymers

Molecular crystals can be regarded as infinitely extended supramolecular structures in which intermolecular interactions direct the packing of the components into periodic arrays. However, there is an ambiguous middle ground between the extremes of a crystalline solid and a disordered liquid: in this no-man's land we encounter molecular arrays that lack true three-dimensional order, but nevertheless have a sufficient degree of organization to qualify as a kind of supramolecular structure.

It is in this middle ground that we find liquid crystals. While they are able to flow more or less like normal liquids, liquid crystals nevertheless retain some degree of structural order after the manner of crystals. Their constituent molecules are aniso-tropic – generally they have rod-like shapes and in the liquid-crystal state the rods have, on average, a preferential direction of alignment. So even though, unlike a true crystal, the liquid crystal has no strict regularity of molecular positions (indeed, the molecules remain mobile), it retains orientational ordering. In so-called thermotropic liquid crystals, the alignment arises solely from packing effects – the molecules all tend to point in the same direction because they can pack more efficiently that way, just as it is easier to stand up than to lie down in a dense and busy crowd of people. The ran-domizing effect of thermal fluctuations opposes this orientational organization, so that, when they are heated sufficiently, thermotropic liquid crystals lose their order and degenerate into a truly isotropic fluid, in which the rod-like molecules no longer point

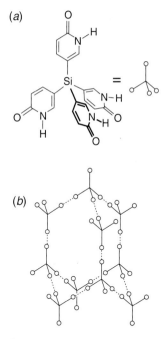

Figure 50
Tetrahedral molecular 'tectons' with hydrogen-bonding extremities (a) will crystallize into an open framework with the topology of the diamond lattice (b).

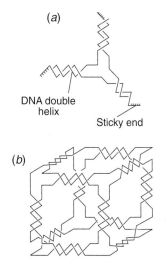

(a)

DNA double
helix

Sticky end

(b)

(c)

Figure 51

Molecular assemblies with complex polyhedral topologies can be formed from double-helical strands of DNA linked into threefold vertices. (a) Enzyme-mediated ligation of the 'sticky ends' of these strands permits the assembly of a cube (b) and a truncated icosahedron (c). The same principles have been used to make sheets of DNA linked into interlocking hoops, visible as a repetitive pattern under the scanning tunnelling microscope (d).

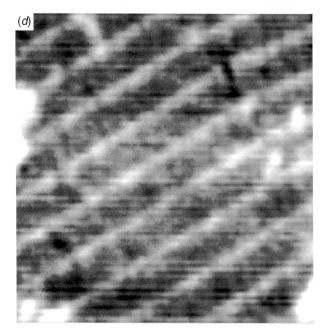

(d)

in a preferred direction. So liquid-crystal phases occur between the fully ordered crystalline phase at low temperatures and the disordered liquid phase at high temperatures. For this reason they are called mesophases – 'middle' phases – and their molecular components are called mesogens – generators of mesophases.

There are many different kinds of mesophase (Figure 52). Liquid-crystal mesophases are also formed by flat, disc-like molecules (Figure 52(c)); these are called discotic phases and the molecular arrangement is akin to a pile of coins. Discotic liquid crystals can form columnar phases in which the 'coins' are stacked into loosely defined columns (Figure 52(d)). A columnar arrangement of disc-like molecules is a common motif in supramolecular chemistry (see Figure 24, for example).

(a)

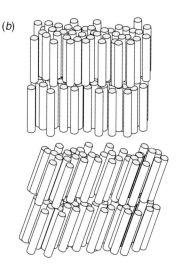

(b)

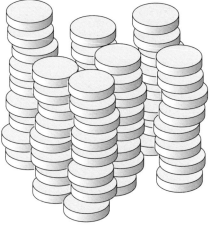

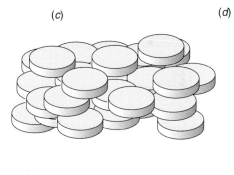

(c)

(d)

Figure 52
Liquid-crystal mesophases. Rod-like molecules may form a nematic phase, in which the molecules have orientational order but no positional order (a), as well as smectic phases (b) in which there is positional order in one dimension (that is, ordering in layers). Disc-shaped molecules also form an orientationally ordered nematic phase (c) and a stack-like columnar phase (d).

The spontaneous ordering that occurs in liquid crystals can be used to 'switch on' a kind of hierarchical organization in supramolecular systems. The association of the two molecules in Figure 53(a) by hydrogen bonding leads to the formation of a columnar liquid-crystal phase, whereas neither of the molecules on their own will form liquid crystals. In Figure 53(b) this organizational principle is extended to a three-component supermolecule: hydrogen bonds hold three identical wedge-shaped molecules in a disc-like assembly that then stacks in a columnar discotic phase. Giovanni Gottarelli of the University of Bologna and co-workers have prepared alkylated derivatives of the DNA nucleoside deoxyguanosine that aggregate into four-component assemblies that again become organized into a columnar structure.

The formation of thermotropic liquid-crystal phases from components that self-assemble into mesogens affords a way of transferring information from molecules to molecular aggregates. For example, chiral mesogenic building blocks with different handednesses may separate spontaneously into liquid-crystal-like superhelices with left- and right-handed threads. In a sense, geometrical (chiral) information is here transferred from the molecular level to micrometre-sized structures.

Biopolymers such as proteins and nucleic acids may self-assemble into well-defined shapes via intramolecular non-covalent interactions. Transfer RNA molecules, for example, adopt a characteristic four-lobed 'cloverleaf' shape through hydrogen-bonding of the base pairs. In proteins the folding of the polypeptide chain into the

(a)

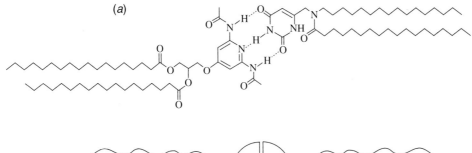

(b)

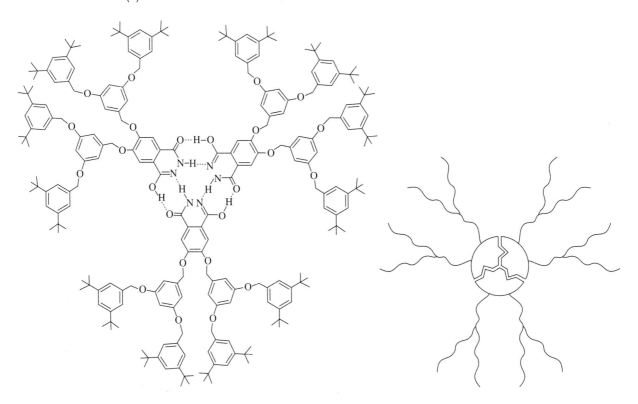

Figure 53

These complementary hydrogen-bonders form columnar liquid-crystalline phases when they are associated. In (a) the central unit is a disc-like structure from which the hydrocarbon tails radiate out. In (b) the three self-complementary components are wedge-shaped.

correct three-dimensional shape is critical to its biochemical function and this folding is guided by a variety of non-covalent forces, primarily hydrogen bonds and hydrophobic forces arising from the insolubility of some parts of the chain in water. The folded structure is generally anchored by covalent disulfide bonds between cysteine amino-acid residues. Exactly how a protein achieves its 'native' fold is still unknown; it remains a major challenge for biochemists to understand and predict higher-level structural features from knowledge of the amino-acid sequence. For supramolecular chemists, this kind of self-assembly of biopolymers into 'sculpted' molecules raises the question of whether synthetic polymers can be engineered to exhibit analogous behaviour. The creation of synthetic peptide molecules that adopt particular protein-like folded motifs

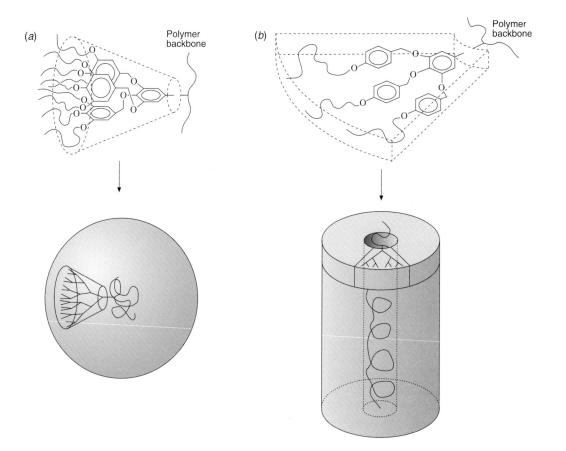

Figure 54
Polymers equipped with bulky, branched side chains will collapse into spherical (*a*) or cylindrical (*b*) conformations, depending on whether the side groups are conical or wedge-shaped, respectively.

– that is, the design of 'synthetic proteins' *de novo* – has been pioneered by Bill DeGrado at DuPont in Delaware, Steve Benner at the ETH in Zürich, Don Hilvert (now also at the ETH), Sam Gellman at Wisconsin and others. DeGrado and co-workers, for example, have investigated the creation of four-helix bundles, a structural motif exhibited by some membrane proteins such as cytochrome *c*. Each helix in the bundle is an alpha-helix secured by hydrogen bonds between amino-acid residues. The assembly of the four helices into a bundle is promoted by the arrangement of water-soluble (hydrophilic) and water-insoluble (hydrophobic) residues around the cylindrical surface of the helix. The helices cluster together in water so as to sequester the hydrophobic regions on the inside.

Other groups have used nature's inspiration but not its materials. Motivated by the cylindrical and spherical shapes of viruses (the coats of which are comprised of protein molecules), Virgil Percec at Case Western Reserve University in Ohio and Dieter Schluter at the Technical University of Berlin have designed polymers with pendant groups along the backbone whose shapes promote spherical and cylindrical conformations of the chain. For example, Percec has shown that polystyrene and polymethacrylate chains with highly branched, cone-shaped side groups will curl up into spheres with the cones radiating outwards, whereas less-branched, wedge-shaped substitutents promote a coiled, cylindrical shape (Figure 54). Here the process of self-assembly is driven largely by simple packing considerations: the polymers condense into shapes that allow the side groups to fit together efficiently.

Figure 55

The cell membrane (*a*) is comprised of a liquid-crystalline arrangement of amphiphilic phospholipid molecules (*b*), within which are embedded membrane proteins.

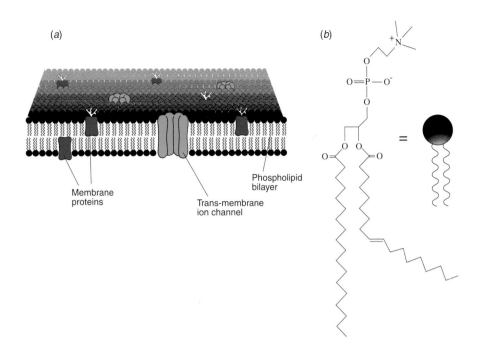

(*a*)

Membrane proteins

Trans-membrane ion channel

Phospholipid bilayer

(*b*)

Self-organizing Surfactant Arrays

The walls of biological cells are liquid crystals of a kind. They consist primarily of linear molecules (phospholipids, or lipids for short) arranged side by side in oriented double layers (bilayers) (Figure 55). A lipid bilayer qualifies as a liquid crystal because all of the molecules have a preferential orientation, even though they remain mobile. (The molecular organization of biological membranes is not, however, merely a consequence of packing effects. The integrity of the structures is maintained by other kinds of interactions too, both covalent and non-covalent. For example, webs of protein filaments called the cytoskeleton lend the membrane robustness.)

Phospholipids belong to a class of molecules known as amphiphiles, which contain both water-soluble (hydrophilic) and water-insoluble (hydrophobic, generally oil-soluble) parts. The basic cellular superstructure – a hollow bilayer sac called a vesicle – will self-assemble from a solution of phospholipids in solution. (Generally some stimulation is required to initiate self-assembly, such as low-frequency sound.) Ordered mesophases whose assembly is driven by the hydrophobic and hydrophilic natures of their molecular components are called lyotropic mesophases.

Soaps are amphiphiles too – they have an ionic head group (typically a carboxylate or sulfonate group) and a fatty hydrocarbon tail (Figure 56(*a*)). Soaps are commonly called surfactants – short for 'surface-active molecules' – because of their tendency to segregate to the interface between water and grease, with the fatty tails embedded in the grease and the head groups in the water. This is how soaps render fatty, greasy particles soluble.

In addition to bilayers and vesicles, amphiphiles will form a variety of other non-covalent molecular structures that shield the hydrophobic tails from water. One such is the micelle, in which the molecules form a loose bundle with the tails pointing inwards and the heads outwards (Figure 56(*b*)). Spherical micelles form in a surfactant solution at the so-called critical micelle concentration; at higher concentrations, they may coalesce into cylindrical micelles and, if the concentration of these aggregates is high enough, they acquire a still higher level of order by packing into columns, like a stack of logs. There is an analogy here with the columnar phases of discotic liquid crystals.

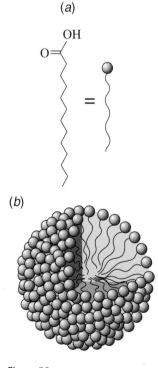

(*a*)

(*b*)

Figure 56

Fatty-acid soap molecules (*a*) will aggregate into organized structures called micelles (*b*).

Surfactants at the surface of water will arrange themselves into a layer one molecule thick, with the tails pointing into the air (Figure 57). The first systematic studies of these layers were conducted at the end of the nineteenth century by Agnes Pockels in Cambridge, but, because of the subsequent pioneering work by the American chemist Irving Langmuir, they have acquired the name Langmuir monolayers. By compressing the layers with a movable barrier in a trough, the degree of ordering of the molecules can be enhanced – at higher density, they pack together more efficiently and can eventually be transformed into a regular, two-dimensional crystalline phase at the water's surface. In 1912 Langmuir and his student Katharine Blodgett showed that these organized films can be transferred onto a solid substrate (a glass slide, say) by a careful dipping procedure. These immobilized layers are called Langmuir–Blodgett films (Figure 57).

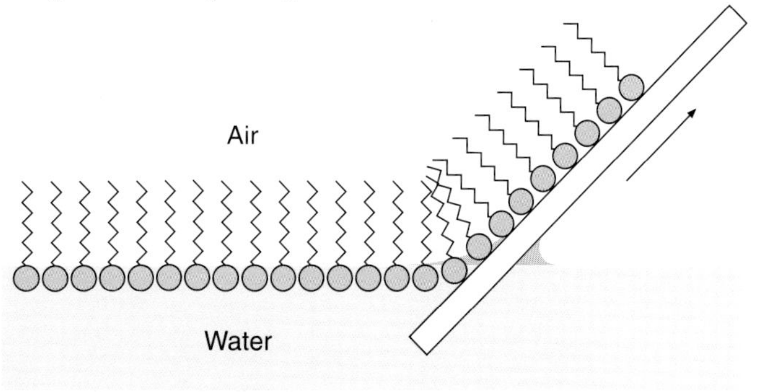

Somewhat similar oriented molecular films can be created from amphiphilic molecules whose head groups form covalent bonds with a solid substrate. For instance, the sulfhydryl group (–SH) of alkanethiols will attach itself to a gold surface via a sulfur–gold bond, while alkylsiloxanes will bind to silica glass. The result is a robust monolayer film with very homogeneous surface properties (Figure 58). These films, called self-assembled monolayers, can substantially modify the surface properties of a material, for example making it highly water-resistant because the alkyl tails are uppermost. The formation of self-assembled monolayers involves only simple 'wet' chemistry and these films can be patterned at the sub-micrometre scale using a kind of printing technique to form structured surfaces with spatially varying wettability characteristics, or masks for etching or metallization of circuit patterns without the need for the repeated photolithographic steps and high-vacuum procedures used conventionally for this sort of surface patterning.

All of these amphiphilic structures provide a rich palette for the supramolecular chemist who wants to extend the scale of molecular organization from discrete supramolecular assemblies to mesoscopic or infinite, self-organizing arrays. The scope of this emerging field, which combines organic chemistry and biophysics with surface and colloid science, polymer science and inorganic materials chemistry, is too vast for us to be able to do more than suggest here a few of its possibilities. At root, much of this work takes its inspiration from the life sciences, specifically from the example provided by the cell membrane (Figure 59). This can be regarded as a self-healing, multifunctional material that employs self-organization, molecular recognition, vectorial transport, molecular switching and many of the other principles that we have discussed so far. In particular, recognition events carried out by proteins and glycoproteins at the cell's surface play a central role in many of the body's processes, such as the immune response, neurotransmission and cell-to-cell trafficking and communication generally. They are also implicated in the way that viruses such as HIV attack cells. The science of supramolecular systems provides the opportunity not just to learn from these natural systems but also to obtain a deeper understanding of them, with all the medical benefits that this could engender.

Even relatively crude mimics of the cell membrane such as phospholipid vesicles have much to offer. These structures (commonly called liposomes in medical research) are being used as capsules for the delivery of drugs. When liposomes are formed in the presence of the drug molecules, some of these molecules become encapsulated inside. The liposomes are injected into the bloodstream and they bear the drugs to their target

Figure 57

A Langmuir monolayer – a single layer of amphiphilic molecules arrayed at the air–water interface – possesses orientational order. This order can be preserved when the film is transferred to a hydrophilic substrate (such as a glass coverslide) by a straightforward dipping procedure. The substrate-bound film is called a Langmuir–Blodgett film.

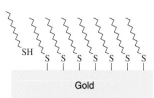

Figure 58

Alkylthiols will form self-assembled monolayers – uniform, oriented molecular films – on the surface of gold.

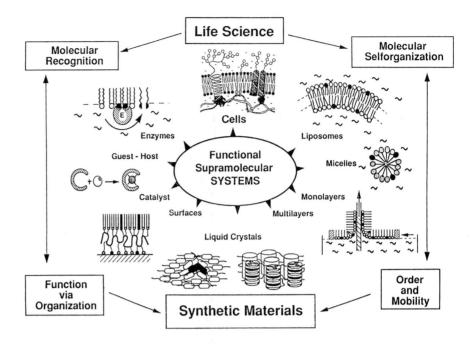

cells in a biocompatible container. In this way, anti-cancer drugs can be delivered to tumours without their toxic effects killing healthy cells too. Liposomes can be fitted with recognition groups on their outer surfaces which control their mutual interactions, leading to multi-compartmental architectures (Figure 60).

Liposomes are loosely bound assemblies. It can be advantageous in some applications to create a more robust structure that is less susceptible to breaking up or being degraded by enzymes. An envelope of cross-linked polysaccharides confers stability on bacterial cell walls. This trick of reinforcing lipid bilayers with polymer chains can be extended to artificial membranes and vesicles by equipping the amphiphile tails with polymerizable groups such as carbon–carbon double bonds, which can be burst open to link the molecules together in just the same way as that in which ethylene (ethene) monomers are linked up in polyelthylene (polyethene). Helmut Ringsdorf's group at the Johannes Gutenberg University in Mainz has pioneered the formation of polymerized vesicles (Figure 61) and their investigation as mimics of the cell's surface. Other researchers have polymerized Langmuir monolayers to form flat sheets that can be peeled off the surface of water. Deborah Charych and co-workers at the Lawrence Berkeley Laboratory in California have made polymerized films and vesicles that can act as sensors for detecting viruses by changing colour when the viruses become bound to receptors at the film's surface. They made cross-linked amphiphilic monolayers from molecules whose tails contain diacetylene (diethyne) units – two carbon–carbon triple bonds side by side. To the ends of some of the amphiphilic chains the researchers attached sialic acid groups, which feature in glycoproteins and to which the influenza virus is known to bind. The polymerized films are a deep blue colour; but, when viruses become bound to the sialic acid groups in the film, the event of binding causes a change in the conformation of the cross-linking polymerized backbone, bringing about a change in the film's colour from blue to red – a kind of 'litmus' indicator for the presence of the virus.

In addition to pre-organizing molecules for covalent synthesis, self-organization in amphiphilic arrays can be used to promote the formation of non-covalent supramolecular arrays. Helmut Ringsdorf has made ordered, two-dimensional crystals of the protein streptavidin by exploiting molecular recognition at the lower surface of a

Figure 60

Liposomes equipped with surface sites that are recognized and bound by multivalent receptor molecules (here the protein streptavidin, which binds biotinylated lipids) can be assembled into multi-compartmental aggregates. This work was conducted by Joe Zasadzinski and his colleagues at the University of California at Santa Barbara.

Lipid
Biotin
Streptavidin

Liposome

(a)

Double bond

(b)

Double bond

Figure 61

Lipid vesicles and other bilayer structures can be cross-linked into robust polymers by incorporating polymerizable groups in the lipids, either in the head groups (a) or in the tails (b).

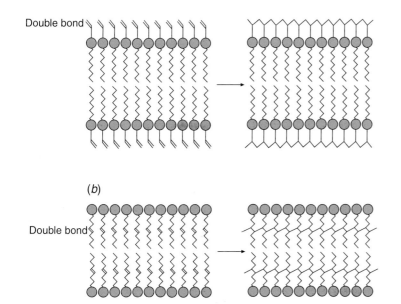

compressed Langmuir film. Streptavidin binds to the small molecule biotin very tightly. So amphiphiles bearing biotin on their head groups, when they are compressed into an ordered monolayer, provide a template for the formation of a close-packed layer of streptavidin (Figure 62). Moreover, because streptavidin has more than one biotin-binding site, multilayer films can be built up in a modular fashion by sequential binding of biotinylated units. Roger Kornberg at Stanford University and co-workers

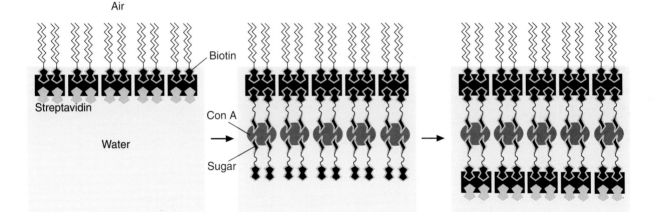

Figure 62

Receptor–substrate interactions can be used to organize proteins such as streptavidin into ordered two-dimensional crystals at the air–water interface. Here a Langmuir monolayer of biotinylated lipids is used to assemble a layer of streptavidin. A second layer of biotin units is then attached to the free binding sites of the streptavidin crystals. These biotin groups are linked to sugar molecules, which are recognized and bound by the protein concanavalin A (Con A). Free biotins attached to Con A then provide the docking sites for a further layer of streptavidin. In this way a sequence of two-dimensional crystals is assembled in a layered fashion.

have used Langmuir films as templates for forming two-dimensional ordered layers of proteins that do not readily form three-dimensional crystals. Because the determination of protein structures by diffraction methods requires well-ordered crystals, this approach makes possible structural investigations that had previously been impossible. In some cases Kornberg's team found that a two-dimensional crystal formed at the interface with a Langmuir film could itself act as a template for the nucleation of three-dimensional protein crystals, even though these would not form spontaneously in solution.

Nucleation of crystals at the surface of organic layers occurs in nature during the formation of biominerals such as nacre (mother-of-pearl). Here the crystals are of an inorganic material, calcium carbonate, and the organic layers are made from proteins containing acidic groups. In this biomineralization process, the protein layer exerts a critical influence on the nature of the mineral phase, determining both the crystal structure (that is, the specific polymorph formed) and the orientation of the crystallites. This influence probably involves molecular recognition between the ions of the crystal and ion-binding groups in the proteins, although the details of this process have yet to be elucidated.

Several groups have used Langmuir films to mimic biomineral templating. Meir Lahav and co-workers at the Weizmann Institute of Science in Rehovot have shown that crystals of the amino acid glycine and of sodium chloride grown from supersaturated solutions at the lower surface of a phospholipid layer have a well-defined orientation, with their crystal planes parallel to the organic layer. By changing the chemical nature of the surfactant they could alter the orientation of the crystals, since this alters the process of molecular recognition between the crystal face and the head groups of the monolayer. By expanding the distance between the head groups (using a Langmuir trough with a movable barrier) they could suppress the nucleation of crystals entirely, perhaps because there was then no longer a near match between the spacing of the ion-binding sites and the spacing of the ions in the crystal lattice.

Templating of inorganic materials with self-organizing organic molecular assemblies can be scaled up to achieve structural control on larger scales. Researchers at the Mobil Research Laboratories in Paulsboro and Princeton have shown that micellar aggregates can impose a regular and uniform pore structure on silica precipitated from a solution of silicate ions. This gives rise to a form of mesoporous silica, in which the inorganic material is perforated by a hexagonal honeycomb of 'mesopores' up to 10 nm across. In the simplest (and probably over-simplified) picture of this process, the mesoporous silica network is cast around a columnar mesophase of cylindrical micelles, with the silica filling the spaces between the stacked cylinders (Figure 63).

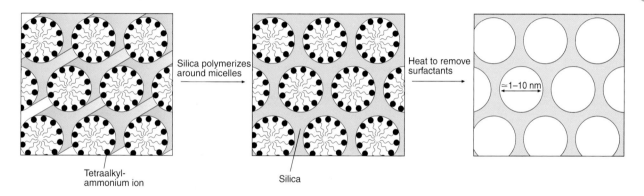

Tetraalkyl-
ammonium ion

Silica polymerizes
around micelles

Silica

Heat to remove
surfactants

≈1–10 nm

These mesoporous materials act as molecular sieves, admitting into their pores only those molecules that are small enough or have the right shape. By inserting or adsorbing catalytic ions or molecular groups into or onto the inorganic pore walls, they can be converted into size-selective catalysts with properties similar to those of the aluminosilicate zeolites used by the chemicals industry, but with pore sizes about ten times larger.

Things to Come

Technology is getting smaller. The average size of microelectronic components has been decreasing steadily for the past few decades, as has the bit size of memory devices. Between 1985 and 1995, for example, the storage density of commercial disc drives increased by almost two orders of magnitude. The rule of thumb coined by Gordon Moore, cofounder of Intel, is that the power of commercial computers doubles every year and a half. Since the earliest days of computing in the 1950s, this power has increased by a factor of around ten billion. The driving force behind these advances has been miniaturization. Today you can hold in your hand a 20-cm silicon wafer containing more electronic components than there are people in the world. So far, the reduction in scale has been made possible by photolithographic procedures that use photochemical patterning followed by selective etching to carve microscopic structures into semiconductor wafers. However, this approach has fundamental limits of resolution, set by the wave optics of the patterning beams: at present, a resolution of 0.1 μm for commercial photolithography remains the Holy Grail. Electron-beam, hard ultraviolet and X-ray lithographies offer still finer resolution, but currently at greater cost; these are not yet routine industrial techniques. Yet, if current trends prevail, the scale of miniaturization will approach the scale of large molecules – a nanometre or so – in just a few decades. Once this happens, completely new technologies will be needed.

As an alternative to the 'top-down' techniques of conventional semiconductor engineering, it now seems expedient to take seriously the option of 'bottom-up' approaches that build nanostructures from molecular components. That is, rather than building devices and patterning materials by 'reduction' – by carving them from larger, monolithic blocks – we can think of achieving this goal by synthesis, putting the structures together molecule by molecule, atom by atom. The methods of standard chemical synthesis provide one option; but, by exploiting self-assembly and self-organization, we might attain the same ends in a spontaneous, pre-programmed and less labour-intensive manner. The creation of organized molecular structures and devices is the objective of the nascent field of nanotechnology, whose adherents propose to conduct engineering on the scale of nanometres.

Some believe that nanotechnology will arise from the manipulation of individual atoms – that molecular-scale devices will be built by putting their constituent atoms into position one by one. That has undoubtedly become a real possibility in the four

Figure 63

Inorganic materials with ordered arrays of pores in the mesopore range (roughly 1–10 nm) can be synthesized from a mixture of silicate ions and tetraalkylammonium amphiphiles. The latter are thought to aggregate into cylindrical micelles that act as a kind of template for the solidification of amorphous silica. The solid is then left with an imprint of the hierarchically self-organized organic material.

decades since the physicist Richard Feynman asked the prophetic question 'What would happen if we could arrange atoms one by one the way we want them?' The development of the scanning tunnelling microscope (STM) in the 1980s provided a tool not just for seeing atoms but also for moving them about. When in 1991 Don Eigler and colleagues at IBM's Almaden Research Center in San Jose wrote their company's name in 35 xenon atoms on nickel by dragging the atoms into position with the tip of the STM, it seemed that nanotechnology had arrived. Maybe, some suggest, we can use these tools to position carbon atoms in such a way as to build diamond-like bearings and joints with the dimensions of protein molecules.

However, the fundamental building block of nanotechnology will be not the atom but the molecule. Except at very low temperatures, atom-pushing is susceptible to the disrupting influence of thermal fluctuations. More fundamentally, atomic building blocks cannot be pre-programmed – each atom must be put in place 'by hand', which is a laborious task. However, molecules can be designed to assemble in highly specific ways, in a gentle manner that uses non-covalent interactions and allows the option of annealing to remove defects. Moreover, the rules of chemistry may work against atomic assembly – there is far more to linking up real atoms than pushing together effigies in a ball-and-stick model. Molecular manipulation, on the other hand, can use chemistry to its benefit.

Moreover, a molecular nanotechnology will be fashioned at least in part from soft materials, not diamond-like components. What we recognize as hardness at the macroscopic scale may be less relevant or less useful at the atomic scale – a rigid protein would be as effective as a concrete pillow. What matters with a molecular-scale device is not that it be robust in the way that an engine part be robust, but rather that it be resilient and unreactive in the environment in which it functions. The body works not by making its molecular machines immune to extreme conditions, but by designing them so that they do not need extreme conditions in order to function.

So what will the components of a molecular nanotechnology be? They will comprise sets of instructed components endowed with photo-active, electro-active, iono-active or switching features, which are capable of spontaneously organizing themselves into functional supramolecular systems and architectures via molecular-recognition processes. We have shown that some of these components exist already and that they can be programmed to organize themselves into specific and complex architectures. We suspect that some will be imported from biology either in their pristine forms or in modified forms.

One of the primary objectives of this technology will be information processing at the molecular and supramolecular scale – a field that, by analogy with the existing technologies of electronic and photonic information processing, we might call chemionics. Molecular circuits will need wires, channels, switches and non-linear signal-processing elements such as diodes and photodetectors, which together will provide functions including storage, processing, amplification and transmission of data. Such systems need not function in the same way that today's electronic devices work – ion channels can be rectifiers of a sort, for instance, and proteins signal to one another by means of covalent modification and diffusion. Moreover, it is not necessary (and probably not wise) to assume that the functions of molecular circuits will be the same as those of electronic or photonic ones. There is no clear need, for instance, for an all-molecular computer, although molecular electronics undoubtedly has other useful things (such as associative memories based on the protein bacteriorhodopsin) to offer information technology. Rather, we can imagine a molecular nanotechnology having a central role to play in fields such as biomedicine, which it might supply with miniaturized, biocompatible biosensors for monitoring the release of glucose or neurotransmitters.

Indeed, a goal-oriented technological perspective should not be allowed to narrow the limits of supramolecular chemistry, which should instead develop into an expansion of the scale and of the creative potential of chemistry on all fronts – into nanochemistry. There are many exciting aims that could be pursued through this expanding window. Materials chemistry is one – the design of materials at the molecular level, through a modular approach to synthesis of materials in which the building blocks are programmed to self-assemble into sophisticated composites. There is a tremendous potential here to learn from the way in which nature uses molecular recognition and templating to make materials with enhanced mechanical characteristics. The inorganic–organic nanocomposite structure of nacre, for instance, gives it a toughness a thousand times greater than that of the pure mineral phase. Chemistry can also offer scope for creating new porous materials, materials with functional gradients and smart materials that respond and adapt to their environment.

As just one example of the directions that this emerging field might take, we can cite the marriage of inorganic materials science, supramolecular chemistry and biotechnology in attempts to control the assembly of nanocrystals of semiconductors and metals by using molecular recognition. Nanometre-sized crystals of semiconductors such as cadmium selenide have optical properties (such as photoluminescence) that can be tuned by altering their size, owing to size-dependent quantum-mechanical effects. Such nanocrystals are already finding use in light-emitting devices and it is hoped that they will also offer new kinds of optical information-storage technologies. However, these applications will demand the ability to control the assembly of nanocrystals into higher-order structures. One approach is to attach to the surfaces of the crystals molecular groups that can recognize one another with high selectivity. Chad Mirkin and co-workers at Northwestern University in Illinois have demonstrated the assembly of nanoparticles of gold via DNA linkers with 'sticky ends' (see page 337) that recognize complementary sequences. Others have controlled the assembly of nanoparticles by using receptor–substrate interactions like those illustrated in Figure 62, or antibody–antigen couplings.

Energy conversion is another attractive objective. Modular molecular assemblies containing chromophores, charge relays and organizing scaffolds can already mimic many of the features of natural photosynthesis for the conversion of light into electrical energy. Whereas most conventional semiconductor solar cells follow a single design principle, the scope for supramolecular photocells seems limitless – designs have included vesicle-based charge separation, zeolite scaffolds with charge relays in the pores, and assemblies of inorganic nanoparticles photosensitized with surface-bound organic dyes. All show promise and many are inspired by, but do not slavishly mimic, nature's own design, the compartmentalized chloroplasts of leaves.

The key to these prospective technologies is information. Supramolecular chemistry is paving the way towards comprehending chemistry as an information science, in which molecules are instructed by judicious structural design to interact in specific ways – with each other, with light, with their environment, with natural systems. Supramolecular chemistry is developing from its origins in recognition processes, through assembly and organization, to a paradigm in which information is stored, retrieved, transferred and processed at the molecular and supramolecular level.

Towards Complexity

There is no question that molecular science can by itself make these things possible: biology shows us that. DNA is a digital molecular data bank, whose genetic information is read out during the transcription of messenger RNA. This information is used to make a programmed molecular or supramolecular material: a protein. Feedback mechanisms act to modulate and regulate this process of information retrieval, so that the expression of certain genes takes place only at the correct stages of the cell cycle, or in the correct tissues. There are editing processes, parallel processing pathways and redundancies to make the system robust against errors or breakdowns. The system is self-regulating, self-regenerating and self-replicating.

These are complex chemical systems! 'It is worth considering to what extent artificial supramolecular systems of this kind can be designed from first principles and to what extent their creation might instead have to involve a process of self-selection in the face of competitive pressures.

'Complex' is not the same as 'complicated'. The beauty of evolution lies not in the very complicated structures of its products but in the simplicity of its principles. Similarly, while the creation of complex, functional supramolecular systems will involve considerable synthetic sophistication, we hope that these systems will be characterized by elegance and economy of design.

In chemical and biological systems, complexity commonly implies a degree of organizational hierarchy, defined by several length scales. Because each level of hierarchy generally contains features that do not and even cannot exist at the level below, supramolecular science cannot be a reductionist discipline but rather must be an integrative one, connecting one level to the others by integrating species and interactions to describe the increasing complexity of behaviour. Much of the appeal lies also in an exploration of the mesoscales, described by Wolfgang Ostwald in 1915 as 'the world of neglected dimensions', where molecular properties have been left behind but bulk properties not yet attained. This is perhaps the scale at which complexity is manifested most profoundly: before the homogeneity of the bulk, after the discreteness of the molecule. We can see the tension between these two extremes emerging in a number of fields whose links with supramolecular chemistry are evident: colloid and cluster science, nanotribology (molecular-scale friction), quantum electronics. Perhaps most enticingly of all, somewhere in this middle ground we cross the no-man's land between inanimate matter and life – and the bridge is provided by chemistry and its ever more complex entities.

Creativity

It might be said that physics is ultimately a quest to understand the universe and biology a quest to understand life. For any scientist engaged in the transformation of matter, meanwhile, there are no pre-ordained goals. Chemistry as a whole is perhaps unique in that it is able, indeed it is compelled, to set its own agenda. Biology is continually an inspiration, a source of ideas, but the challenge for supramolecular science is not merely to mimic nature. Biology is extremely complex (and complicated!), but its scope, in terms of the materials it uses and the classes of problems it solves, is limited. There are good evolutionary reasons for that, but they need not similarly shackle chemistry. Rather, the diversity of materials and of aims is limited only by the imagination of the chemist. What the study of supramolecular science has to offer chemistry is not so much a new set of synthetic challenges but rather a way of seeing new relationships between disciplines, of importing and adapting ideas, of making connections – in several senses! Supramolecular chemical research lies at the intersection of chemistry, biology, physics, materials science and engineering. Because it draws on the physics of organized, generally 'soft', condensed matter and extends into the biology of large molecular assemblies and functional units, while deriving motivation and objectives from the demands of applied sciences and engineering disciplines, the horizons of supramolecular science are wide indeed. They demand an abandonment of compartmentalized specialization and a rich cross-fertilization of diverse disciplines.

As the scope of chemistry becomes broadened by this perspective, it will be important to distinguish the daring and visionary from the utopian and illusory. On the other hand, the chemist can afford broad ambitions – in fact, they rest on the very essence of chemistry: its creativity. As Marcelin Berthelot recognized in 1860, chemistry continually creates itself.

Further Reading

1. J.-M. Lehn, *Supramolecular Chemistry* (VCH, Weinheim, 1995).
2. L. F. Lindoy, *The Chemistry of Macrocyclic Ligand Complexes* (Cambridge University Press, Cambridge 1989).
3. D. Philp and J. F. Stoddart, Self-assembly in natural and unnatural systems, *Angewandte Chem. Int. Edn Engl.* 1996, **35**, 1154–1196.
4. D. J. Cram, Molecular container compounds, *Nature* 1992, **356**, 29–36.
5. H. Ringsdorf, B. Schlarb and J. Venzmer, Molecular architecture and function of polymeric oriented systems: models for the study of organization, surface recognition, and dynamics of biomembranes, *Angewandte Chem. Int. Edn Engl.* 1998, **27**, 113–158.
6. M. Antonietti and C. Goltner, Superstructures of functional colloids: chemistry on the nanometer scale, *Angewandte Chem. Int. Engl.* 1997, **36**, 910–928.
7. G. M. Whitesides, Self-assembling materials, *Sci. Am.* 1995, **273**, 114–117.

13 Advanced Materials

Paul Calvert

Introduction

For the purposes of this chapter we can draw the line between a chemical and a material at the point where a liquid or powder is converted into a functional solid object, a film, a coating or a fibre. The behaviour of a material is directly traceable back to its chemical bonding but a relatively minor change in composition or atomic arrangement can have a huge affect on the useful properties of the material. Thus, marble, chalk and pearl are all calcium carbonate but differ in crystal structure, crystal size and density. As a result the mechanical and optical properties are quite different. Similarly, polyethylene and candle wax differ only in the number of $(-CH_2-)$ links in the molecular chains but are quite different mechanically. One, possibly apocryphal, example of the relationship between structure and properties is that of the soldiers of Napoleon's army, who marched into Russia wearing tin buttons. At low temperatures, metallic β (white) tin recrystallizes to powdery, semiconducting α (grey) tin. As a result the buttons all disintegrated and the army froze.

Materials science as a discipline developed from metallurgy during the 1960s, with the need for high-performance materials for aircraft, electronics and nuclear reactors. Before 1940, engineering was concerned mainly with iron and steel and the practical problems of metallurgy concerned extraction and refining. With the advent of aluminium alloys for aircraft, we needed to understand why the strength of a metal was so dependent on small amounts of alloying elements. This understanding came from the use of the electron microscope by physical metallurgists. A key discovery was that of dislocations, linear defects that control the ability of metals to deform. Having learnt the rules for metals, we then needed the same kind of knowledge for ceramics and glasses, semiconductors, polymers and composites. Thus metallurgy became materials science and materials engineering.

In one view, science is driven by the wish to know. However, there is a parallel track that is more dedicated to exploration than it is to explanation. Materials science originated from a desire to understand why steel was so strong, nylon so tough and glass so transparent. In recent years, the balance has shifted from explanation of materials' properties to the development of new materials. There has also been a change in focus from the mechanical properties of solids to the optical and electrical properties of films, in support of the development of electronics, data storage and communications. With these changes has also come a closer integration of chemistry and materials science. As the structures become more exotic, collaborations with synthetic chemists are needed in order to produce the raw materials before they can be processed into structures and studied. At the USA's National Science Foundation, 'materials research' now includes condensed-matter physics (explanation), solid-state chemistry (exploration) and materials science (exploitation). Nobody has a clear idea of where the boundaries between these disciplines or between materials science and materials engineering lie. The tensions are sometimes quite evident.

It would be nice to predict the properties of a new material directly from its chemical composition. Theory, combined with a heavy dose of extrapolation from existing materials, can give us good predictions of some important properties of materials such as their densities, stiffnesses, electrical conductivities and refractive indices. These properties all depend on forces between fixed atoms. Predictions are relatively poor for properties that depend on atoms moving, such as strength, diffusion, permeability to gases and freezing and melting behaviours. We have also been strikingly unsuccessful at predicting the usefulness of a new material. In the real world new engineered materials often fail because of secondary effects, such as fatigue after many cycles of loading and unloading, slow restructuring at high temperature or corrosion. The result is that an apparently revolutionary solution for problem A often turns out to be unsuccessful for that, but a good solution for problem B. The road from a successful laboratory demonstration of a material with promising properties to an actual application can thus be long; 20 years is typical. Often this involves developing control of impurities and structural defects in order to attain reproducible behaviour and chemical stability. A 'new' cellulose fibre that recently came on the market was based on a solvent-spinning process discovered about 30 years ago and developed sporadically ever since.

What Materials are Useful?

In chemical terms the world of materials is not broad; most of the elements in the Periodic Table are not much use. The mechanically useful metals sit in the middle of the Periodic Table. Most are alloyed with small amounts of other elements to improve their strength or resistance to corrosion. A few metals are blends of two close neighbours, for example brass, which is made from copper and zinc. Mixtures outside these constraints are usually too brittle to be useful in structures. However, brittle intermetallic compounds with interesting magnetic, optical and conducting properties do keep showing up. Niobium–tin compounds, for instance, are weak and brittle but are useful low-temperature superconductors.

Ceramics are compounds of the metals with boron, carbon, oxygen or nitrogen in which the short strong bonds to the small atoms make the material hard. After a great deal of research, the 'fine ceramics' such as aluminium oxide and silicon carbide have not found the widespread use in engines that had been expected. However, following the principle that applications appear where you least expect them, they have found many uses in electronics.

The semiconductors come from group IVA of the Periodic Table (C, Si, Ge) or mix-and-match combinations of elements from groups IIIA and VA, such as gallium arsenide and indium antimonide. Mechanically, they are quite brittle and much effort goes into protecting silicon electronics against stresses generated during vibration or heating and cooling.

Almost all practical polymers, natural and synthetic, are carbon chains with varying amounts of oxygen, nitrogen and hydrogen as side groups. Chlorine, fluorine, silicon, phosphorus and sulfur show up occasionally. Most other elements will result in structures that are unstable against water or oxygen. Despite this, the known polymers are only a small set from a vast array of possible chain molecules. In fact, we are mostly interested in small, simple chain units and don't really expect behaviour outside the envelope we know. Sometimes, such as with Kevlar, we are surprised.

Chemistry, however, is only the start and microstructural organization is as important as composition. Thus, we get stiff, strong fibres from causing polymer chains to be strongly aligned instead of in their normal, randomly coiled, state. Stacks of thin layers of differing semiconductors are the basis for diode lasers and control of the defects at the interfaces is crucial for optimizing their performance. Zirconia alloyed with small amounts of other oxides is a tough ceramic but is too heavy and expensive to be widely used. However, a small fraction of modified zirconia grains entrained in an alumina ceramic toughens and strengthens the brittle matrix.

To illustrate this interplay between chemistry and structure, there follows a number of examples of applications for advanced materials. Since failure can be more instructive than success, the discussion is cast in terms of the gaps in technology that are driving our need for better materials.

Diamond and the Problem of Heat in Electronics

Crystalline silicon is the core of modern electronics, not really because it is the best material for all purposes but because we now have so much skill in its purification and manipulation. The early history of the transistor involves germanium, which lies immediately below silicon in the Periodic Table. Germanium has a lower melting point and so is easier to process, especially in the purification by zone refining that is vital if one wants to make semiconducting material. The softer interatomic bonding that gives germanium its lower melting point also gives germanium a smaller band gap than that of silicon (0.7 versus 1.1 eV) (see Box 1) and so makes its electronic properties more temperature sensitive.

Packaging

The performance of tomato sauce is coupled to the performance of the bottle, which protects the contents yet allows controlled delivery. Semiconductor packaging is as crucial to chip performance in the electronics industry as is integrated circuit processing and requires as intensive an effort. In standard memory chips, the millimetre-square silicon RAM is stuck to a flat alumina sheet (a substrate). Wires connect the chip to metal lines on the substrate. These lines radiate out to the edge, where they couple to leads to the outside world. The whole assembly is then encapsulated in a silica-filled epoxy resin to keep moisture and air out. Alumina is selected for the substrate because it is a very good thermal conductor that is also an electrical insulator. The heat from the chip spreads across the substrate and then through the encapsulant to be radiated from the (black) surface. In microprocessors and CCDs (video chips) the heat-flow problems are much more serious because the power levels are higher.

As a scratching test with a penknife will demonstrate, the earlier Intel 386 and 486 microprocessors were packaged in a sealed ceramic box, rather than being encapsulated in resin. The alumina was black for good thermal radiation and then, with the Pentium, a fan was also mounted on the package. Designers of mainframe computers have gone to even more elaborate solutions with cooling channels for gas to flow through the package or, in some early supercomputers, total immersion in liquid nitrogen. A microprocessor package is normally made by ceramic multilayer processing based on stacking 'green' ceramic sheets (ceramic powder plus a polymer binder), which are then fired to dense ceramic (see Figure 3). The base of the box is a series of ceramic sheets, printed with metal ink lines to bring in power. Punched-through metal-filled vias (holes) connect between the layers. Next come layers with the hundreds of signal lines. The sides of the box are similar layers with a hole cut out for the chip. The base and sides of the box are fired, with the hope that the metal lines will stay intact but that the ceramic will close perfectly around them with no porosity or cracking. The chip is put in the box, attached to the thermally conducting base and connected to the metal lines and then a ceramic top is sealed on with low-melting-point glass. The development of the ceramic package is very much associated with the decision of Kyoto Ceramics (Kyocera) that this would be an interesting challenge for a fine-china company, at a time when integrated circuits were still new.

The materials problem is thus to develop electrical insulators with higher thermal conductivities. They should be compatible with the rest of packaging technology and have low dielectric constants so that the transmission of electrical pulse is not slowed. This will get the heat from the tiny chip to the larger package and the engineer has then to worry about getting the heat out of the package. Thermal conductivity depends on the transmission of vibration (phonons) from atom to atom. For this to work well,

Box 1 Bands, gaps and devices

In isolated atoms, gases and molecular solids and liquids, the electrons are strongly bonded to individual atoms or are shared between pairs of atoms in bonds. In metals, semiconductors and covalent solids, some of the electrons are free to travel throughout the solid. These free electrons, which are only weakly bonded to individual atoms, give rise to the electrical conductivity of metals. In the semiconductors, the outermost electrons need a small kick, an increase in energy, to become free. This energy jump moves them from the valence band, containing electrons that are localized on atoms, to the conduction band, where they have enough energy to roam freely. As they travel, they also leave behind an atom which is one electron short and this 'hole' can travel by being passed from atom to atom, acting is if it were a slowly moving positive charge (see Figure 1).

The band gap is the energy needed to move an electron up to the conduction band. Up to about 1.5 eV (1 eV = 96 kJ mol^{-1}), enough electrons pick up this energy from thermal vibrations that the material conducts fairly well at room temperature. Light will increase the conductivity by promoting more electrons to the conduction band. By adding a few parts per million of impurities from group IIIA (one electron short) or group VA (one too many), extra holes (p-type) or electrons (n-type) can be put into the structure. Combinations of p–n junctions give us diodes, transistors, solar cells, computer gates and the rest of electronics.

The starting point for all semiconductor devices is thus the p–n junction diode (see Figure 2). A piece of silicon is made with p-doping, that is which a very small concentration of impurities from group IIIA in the Periodic Table, B to In. This leaves the crystal one electron short for each impurity atom. Into this are diffused atoms from group VA, N to Sb, to give a zone with an excess of n-type impurity and hence too many electrons for the lattice of silicon to accommodate easily. This junction will readily carry a current of electrons from the electron-rich to the electron-poor side and of 'holes' in the opposite direction. If the polarity is reversed, no current flows and so the device is a diode. Two diodes in the form of a p–n–p or n–p–n arrangement give us the transistor, for which the current between the two outer layers is very sensitive to changes in the voltage applied to the central zone. For all this to work, we need a very pure and perfect single crystal of silicon with precisely controlled levels of dopants. Fifty years of hard work have taught us how to do this very well.

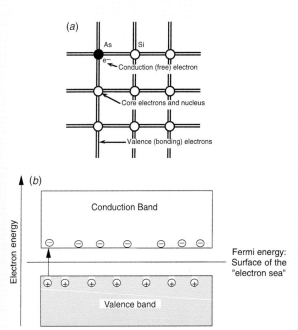

Energy Diagram for an intrinsic (pure) semiconductor. At normal temperatures a few electrons jump from the valence to conduction band leaving positive holes

Figure 1
Diagrams of bands and gaps. (a) A schematic diagram of the silicon lattice with an n-type dopant (an arsenic atom). (b) The band picture of silicon, with a 1.1 eV band gap between the valence and conduction bands.

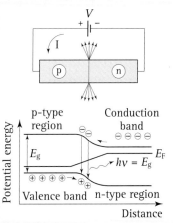

Figure 2
A schematic diagram of a p–n junction as a light-emitting diode. Free electrons driven in from the n-type region meet holes from the p-type region and combine to release energy as light. Redrawn with permission from *Nature* vol. **351**, copyright 1997 Macmillan Magazines Ltd.

Figure 3

A semiconductor package for multi-layer processing. (*a*) A diagram of routes to make alumina packages (courtesy of Kyocera Corp.).

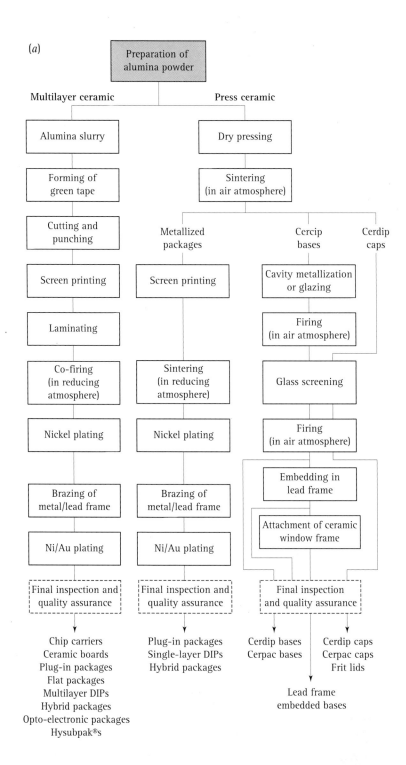

(*a*)

Preparation of alumina powder

Multilayer ceramic

Press ceramic

Alumina slurry → Forming of green tape → Cutting and punching → Screen printing → Laminating → Co-firing (in reducing atmosphere) → Nickel plating → Brazing of metal/lead frame → Ni/Au plating → Final inspection and quality assurance

Dry pressing → Sintering (in air atmosphere)

Metallized packages: Screen printing → Sintering (in reducing atmosphere) → Nickel plating → Brazing of metal/lead frame → Ni/Au plating → Final inspection and quality assurance

Cercip bases / Cerdip caps: Cavity metallization or glazing → Firing (in air atmosphere) → Glass screening → Firing (in air atmosphere) → Embedding in lead frame → Attachment of ceramic window frame → Final inspection and quality assurance

Chip carriers
Ceramic boards
Plug-in packages
Flat packages
Multilayer DIPs
Hybrid packages
Opto-electronic packages
Hysubpak®s

Plug-in packages
Single-layer DIPs
Hybrid packages

Cerdip bases
Cerpac bases

Cerdip caps
Cerpac caps
Frit lids

Lead frame embedded bases

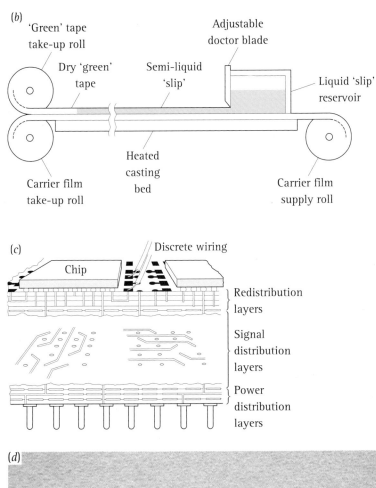

(b) 'Green' tape take-up roll

Adjustable doctor blade

Dry 'green' tape

Semi-liquid 'slip'

Liquid 'slip' reservoir

Carrier film take-up roll

Heated casting bed

Carrier film supply roll

(c)

Discrete wiring

Chip

Redistribution layers

Signal distribution layers

Power distribution layers

(d)

(e)

Figure 3 (*cont.*)
(*b*) A schematic diagram of a tape casting system for multilayer ceramic processing. (*c*) A diagram of a cross-section of a multi-layer, multi-chip ceramic package. (*d*) A photograph of a 'Cerdip' package showing the chip in the central well before sealing. (*e*) A photograph of a pin grid array package showing empty package, left, package containing chip, centre and sealed package, right.

Table 1	Thermal conductivities of electrical insulators in W m^{-1} K^{-1}
Glass	1
Aluminium oxide (85% pure)	14
Aluminium oxide (99.9% pure)	40
Aluminium nitride	150
Beryllium oxide	250
Diamond	2600

atoms should be light, strongly bonded and very regularly arranged. (A harmonic oscillator, of two weights joined by a spring, vibrates rapidly if the spring is strong and the weights are small.) Defects or impurities in the crystal will scatter phonons in the same way as that in which the droplets in clouds scatter light. Table 1 gives the thermal conductivities of some prime candidates for packaging. Alumina is the standard and should be as pure as possible, although purer material is also more difficult to sinter to full density. The obvious step from alumina to boria (B_2O_3) is no good, for the latter is soluble in water. Beryllia is very good but the powder can produce very severe allergic reactions and so it is very expensive to process. We can also go from oxygen down to nitrogen (aluminium nitride, AlN) and much effort has gone into this, but it is a difficult material with which to work. Many other choices have been pursued, particularly at IBM's Fishkill research laboratories, whose 'Next Generation Packaging' project was famous throughout the 1980s for the large amount of money consumed in a heroic but only modestly successful programme.

The obvious answer is diamond. The atoms are light and the bonds are strong, so the thermal conductivity is excellent and it is isotropic, since the crystal is cubic. The dielectric constant is a bit high but acceptable. Unfortunately, unlike alumina, it is not readily available in large, thin sheets. Good big crystals, that could be sliced, are also not cheap. Graphite, of course, is readily available but is electrically conducting and hence no good.

Synthetic Diamond

Large cylindrical crystals of diamond cannot be made in the same way we can make silicon, sapphire (and ruby) and many other materials, mostly because the melting point is inaccessibly high. We can grow diamond crystals from solution, as we can crystallize salt from water, but the solvent is molten iron at a high temperature and a very high pressure, so the process is very inconvenient. Industrial diamonds for cutting tools and polishing paste are made this way.

The existence of the diamond cartel DeBeers' makes the gem-diamond market economically and politically very delicate. Since the 1960s there has been an arrangement whereby small, yellow, synthetic diamonds are widely used in industrial cutting and grinding whereas clear gem-quality diamonds either could not be, or were not, made. Rumours have long circulated that General Electric, the developer of the synthetic-diamond process, could make huge gem-quality diamonds, but chose not to. The need for diamond substrates may change this balance, but the need is really for potato-sized diamonds, not pea-sized diamonds.

In the last few years there has emerged the possiblity of growing diamond sheets or films directly. We all thought that diamond could not be made at normal pressure because graphite is the stable phase of carbon at pressures below 10 000 atm. In 1977 it was reported that diamond films could be formed by heating a methane–hydrogen mixture over a cooled substrate. Unfortunately, this report was not initially believed. It came from the laboratory of B. V. Derjaguin, a respected Russian physical chemist who had previously had the misfortune to publish the discovery of polywater, which later turned out to be quite wrong.

The method for making diamond was repeated about 5 years later in Japan and 5 years after that in the USA. After this, research on diamond films exploded. This process does deposit mostly graphite, as expected, but the presence of hydrogen in the feed gas rapidly etches the graphite away again and leaves the hard diamond. This unexpected development has led to intense research on the production of diamond films. For now the process is still very slow and, at a rate of $1 \ \mu m \ h^{-1}$, it would take over a month to form a diamond substrate.

Replacing Silicon

Instead of better cooling for semiconductors, a preferable solution to handling high power densities is to use a semiconductor with such a large band gap that it is not temperature sensitive. Since silicon is better than germanium, diamond will be better than silicon. Actually, with a band gap of 6 eV, it should be good enough to be stable above 1000 °C and that is probably enough for all time. Diamond is the ultimate answer but we would have to be able to dope it to p-type and n-type. Boron can be used to make p-type diamonds but nitrogen is excluded from the crystal lattice and so will not make n-type diamonds. Hence the diamond diode was not made until 1991, when n-type diamond was made by including some phosphorus during the growth of diamond film. Whether practical devices will be made remains unknown.

Diamond is not the only choice for a large-band-gap semiconductor. Silicon carbide, SiC, is being studied intensively and can be thought of as a compromise between silicon and diamond. Taking groups IIB–VIA in the Periodic Table, we can increase the band gap by going up or by going out to compounds. In a compound semiconductor we keep the same average number of electrons per atom and so the same essential behaviour by combining a group IIIA element and a group VB element, or a group IIB element with a group 16 element. For a long while gallium arsenide, GaAs, was the semiconductor of the future. It has proved to be very expensive to make and the standard joke is that it will always be the semiconductor of the future.

Gallium arsenide has a band gap of 1.5 eV, making it more temperature resistant than silicon. The details of the electronic structure (including a direct band gap) also mean that one should be able to make devices that are smaller, faster and less radiation sensitive than silicon devices from it. This is an important issue for a number of reasons.

In the world of ballistic missiles and anti-missiles, there is a desire to harden electronics against damage by the pulse of radiation from a local nuclear explosion. This drove the development of gallium-arsenide electronics for many years. Silicon also failed quite dramatically at Chernobyl, where robots sent to sweep the roof clean of debris stopped working in a very short time. The job then had to be completed by human volunteers. The presence of traces of radioactive thorium in the silica filler used in epoxy-resin encapsulants is enough to cause occasional random errors in memory chips.

The problem with making gallium arsenide is that it is much more difficult to get an exact 1:1 Ga:As composition than it is to get absolutely pure silicon. Just as impurities in silicon give rise to current carriers, so a small excess of Ga or As will give hole or electron carriers in GaAs. To grow a crystal of GaAs, in exactly 1:1 stoichiometry, you must start not with a 1:1 liquid but with a non-stoichiometric liquid, with a slight excess of arsenic. As a result, the composition of the liquid changes continually as the crystal grows and must be adjusted. Furthermore, arsenic has a high vapour pressure and so is easily lost. All of this means that the better performance of gallium arsenide never quite catches up with the convenience and cheapness of silicon.

Electronics and Light

Consider a small CCD camera sitting on your computer monitor to record your picture. It is puzzling that this tiny camera takes the picture but we need 22 kg of vacuum-tube technology as a display, instead of something with the dimensions of a postcard. Our ability to generate and control light is not impressive in comparison with our ability to generate and control electrical signals. It would be sensible to build computers with an optical bus to communicate between boards and modules by the use of light beams, rather than using wiring. However, optical switches are large and slow in comparison with electronic switches. Optical data storage also continues to disappoint us, promising much, but never getting convincingly ahead of magnetic technology.

Laser Diodes and LEDs

As we saw above, gallium arsenide has applications in computing only when speed is needed at any cost. However, it and other compounds of a group IIIA element with a group VA element do dominate the world of light-emitting diodes and diode lasers. This is because the band gap is direct, a subtlety we will not go into here, but which leads to efficient interconversion of electrical energy and photons. For gallium arsenide, the band gap corresponds nearly to the energy of the infrared light that powers most remote controls. By adding some aluminium, we can increase the band gap to the energy of red light and it can be tuned. As Figure 4, shows, these band-gap games can take us across the whole spectrum from sources and detectors that work deep in the infrared to those that work in the blue.

During the last few years, there has been a competition to produce reliable, long-lived blue diode lasers and light-emitting diodes (LEDs). Zinc selenide and silicon carbide were tried but could not provide the brightness and lifetime needed. In the last few years blue gallium-nitride LEDs and lasers have been made. The winner is Nichia Chemical Industries, a small Japanese company with a business in phosphors (for televisions and fluorescent lighting), which was looking for something else to do and seems to have cracked the problem of controlling defects in gallium nitride. We can expect the blue diode laser to have a big impact on CD technology, both because the shorter wavelength gives a higher information density and because blue light is more chemically active and so the process of writing information can be faster and the chemistry simpler. Blue LEDs can also be used with fluorescers to give white light from LEDs, quite possibly eliminating the familiar flashlight. Of course, now we have blue, we will want UV, then X-rays and, maybe, will get to diamond again.

Displays

Liquid-crystal-display panels have become very familiar and are obviously destined to become ubiquitous. It is high time the cathode-ray tube went the way of the steam engine. A simple liquid-crystal display, such as that in a watch, has the fluid sandwiched between two thin sheets of glass, each of which is coated with a layer of a transparent conductor, indium tin oxide (ITO). (See Box 2.) For a watch we can rely on light reflected from a back mirror but, for a computer, the whole layer must be backlit. This structure is fine for expensive computers but is no good for all the displays we would like to add to cars and appliances. The glass is too brittle, the whole structure is too temperature sensitive and the displays are not yet bright enough. The needs are obvious: displays must be made flexible, strong and insensitive to temperature; better backlighting is needed for illuminated displays and a display that works in reflected light, like a sheet of paper, is needed. In addition, present displays are slow, with response times of about 100 ms, so that a flying baseball looks like a comet on a liquid-crystal display.

One way forwards would be to change to flexible plastic supports with a liquid-crys-

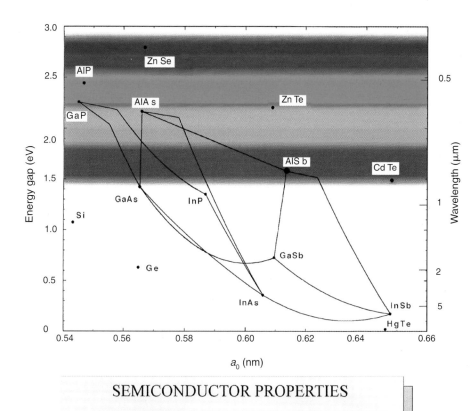

Figure 4
Band gaps. The band gap of semiconductors increases as we go from group IVA to IIIA–VA and IIB–VIA compounds. It decreases as the atomic size increases. The band gap determines the diode's colour. The mobility determines the signal's speed. Thanks to L. R. Dawson, S. T. Picraux and A. J. Hurd, Sandia National Laboratories.

SEMICONDUCTOR PROPERTIES

	a_0(Å)	Eg(eV)	Mobility @ 300K (cm^2/V sec)	
			Electrons	Holes
Si	5.4310	1.11	1400	470
Ge	5.6461	0.67	3900	1900
GaP	5.4506	2.26	110	75
AlP	5.4625	2.45		
GaAs	5.6535	1.42	8500	400
AlAs	5.6605	2.17	280	
InP	5.8688	1.35	5000	150
InAs	6.0584	0.36	33000	460
GaSb	6.0954	0.72	5000	850
AlSb	6.1355	1.58	900	450
InSb	6.4788	0.17	80000	1250
ZnSe	5.6676	2.80	530	
ZnTe	6.0880	2.20	530	130
CdTe	6.4816	1.49	700	60

talline-polymer active layer. A polymeric active layer would not slosh around when the display was flexed and so there would no longer be the need for rigid supports. Liquid-crystalline polymers can be made but respond much too slowly to be useful. Polymer films with embedded liquid-crystalline droplets have also been studied but scatter light and so there is no prospect of a display using them. These have been used to make switchable windows that go from clear to cloudy as the liquid crystal drop-lets re-orient. A second problem is that no one yet seems to know how to put a good, transparent conducting coating onto polymer sheets and large thin sheets of glass are vulnerable. (This is the main reason why you should not drop your lap-top computer.)

Liquid-crystal displays are backlit by a lamp in the edge of the sheet and a white

Box 2 Liquid-crystal displays

Liquid crystals have a structure similar to loose matches in a box. All of the molecules are parallel but, if they are seen from the end, there is no regular arrangement. This state occurs for many stick-like or plate-like molecules and is intermediate between the crystalline and liquid states. Most liquid crystals will become crystalline and rigid if they are cooled and completely fluid if they are heated. Hence a watch display will stop switching if it is frozen and will go black if it is overheated, but will start working again when it is returned to room temperature (don't do this with an expensive watch).

The components of a simple liquid-crystal display can be seen by sacrificing a cheap calculator and pulling the display to pieces. Light from the front travels through a sheet of polarizer, through the liquid-crystalline layer between two sheets of conducting glass and through a rear polarizer which polarizes light at 90° to the front one. It then hits a mirror and makes the return trip. The glass sheets are rubbed with a cloth, apparently to introduce fine channels on the surface, like those on a ploughed field. The surfaces either side of the liquid-crystalline layer are rubbed to form grooves in perpendicular directions, parallel to the directions of polarization of the front and back polarizers. The stick-like liquid-crystal molecules tend to lie parallel to the grooves at each surface and progressively change direction through the film. This causes the light to rotate its direction of polarization as it goes through the film. Thus it enters polarized left–right, rotates to up–down polarization, gets through the back polarizer, hits the mirror and then makes the reverse journey. When the voltage is applied across the film, the molecules line up parallel to the field and do not rotate the light. As a result it does not make it through the back polarizer.

This is only the start. Colours can be introduced by including dyes in the liquid crystal. Normal nematic liquid crystals align with the field quite quickly (within a few milliseconds) but go back to their twisted state rather slowly (taking tens of milliseconds). Ferroelectric liquid crystals switch quickly in both directions but have other disadvantages.

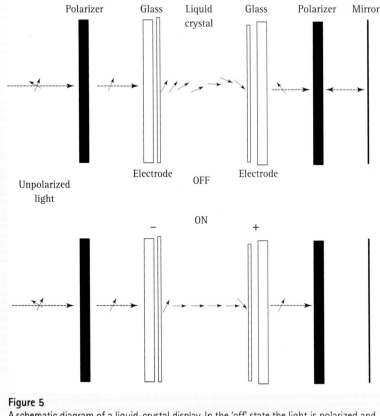

Figure 5
A schematic diagram of a liquid-crystal display. In the 'off' state the light is polarized and then rotated to travel through a second, perpendicular, polarizer. In the 'on' state the polarized light is not rotated and is blocked at the second polarizer; the display looks dark.

backing layer that diffuses this light forwards. The fluorescent lamp requires high voltages and is fragile. A better choice for an illuminated display would be to use a film which is electroluminescent and could provide a uniform sheet of light. Here there has been some notable progress in the last few years. The current generation of digital watches has a blue electroluminescent backlight that is probably a layer of zinc sulfide between two transparent electrodes. The present focus of research is very much on organic and polymeric electroluminescent materials.

It has long been known that aromatic organic compounds, such as anthracene, will glow blue when current is passed through a thin layer beween two different metals. The structure is very similar to a p–n junction in a semiconductor diode. Electrons enter the organic layer from a low-work-function metal (weakly bound electrons) and leave into a high-work-function metal (strongly bound electrons). When the electrons drop in energy inside the organic layer, light is emitted. The structures have become more sophisticated, with two layers of organic compounds so that the light is emitted at this interface and the colour can be tuned by choosing the difference between electron-orbital levels within the compounds. In 1997, these displays were commercialized by Pioneer for use in clocks and automobiles. White light can be produced by putting red, green and blue layers one behind the other. We can then either try to break the display up into small, independently addressable, pixels or control the light with a liquid-crystalline screen.

Aromatic organic compounds are weak powdery materials and so must be embedded or dissolved in some tough polymeric matrix to make a useful device. During the 1980s there were intensive studies of polymers that are electrically conducting, for possible applications in wiring, as semiconductors and in batteries. In 1993 it was discovered that some of these polymers glow when a voltage is applied across the film. Since then, a range of colours has been produced, luminescence efficiencies have increased (to about 18 lumens W^{-1}, 5–10 times those of backlit liquid-crystal displays) and lifetimes of devices have become respectable. Figure 6 shows the band structure and cross section of a polymer LED.

In principle, it should be possible to devise a reflected-light display with electrochromic materials (materials that change colour in response to a voltage), changing

Figure 6
Polymer light-emitting diodes. (a) The band structure of a polymer LED, showing high- and low-work-function metal electrodes and two polymer layers with emission of light from the polymer/polymer interface. (Reproduced with permission from CDT Ltd.) (b) The cross section of a polymer LED. (Reproduced with permission from CDT Ltd.)

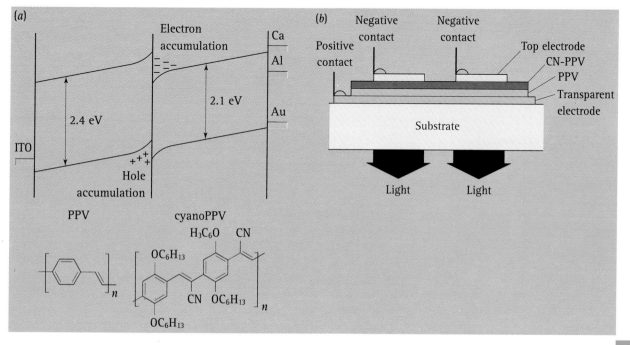

colour pixel by pixel. As far as I know, there is no prospect of such a thing at present. The classical electrochromic system is a tungsten oxide which changes from colourless to dark blue during reduction and oxidation. New self-dipping car mirrors have thin layers of this oxide, an electrolyte and a counter-electrode, like a very thin battery, between two layers of glass. This system is too slow for a display and the contrast is too low. There has been a lot of interest in electrochromic glass for office windows to provide heat control but they are not yet really successful.

Optical Switches and Data Storage

A liquid-crystal pixel is an electro-optical switch: by applying the field we can turn the transmission of polarized light off. However, turning the field on switches the signal in a few milliseconds while the base state returns a few tens of milliseconds after switching off. We would like to switch light much more quickly. In optical-fibre telephone communication, the present method is to pick up an optical signal with a gallium-arsenide photodetector, convert it into an electrical signal, manipulate the signal electronically and then send it into the next optical path from a laser diode. Direct optical switching would be much neater. Likewise, as mentioned above, it would make sense for computers to have an optical bus, with chips communicating via an optical signal that travels rapidly round the whole machine. To switch data in a 200 MHz computer, we need switching times of about 1 ns.

Lithium niobate is a ceramic, closely related to the piezoelectrical perovskites (discussed later), whose refractive index changes in an applied field. Switching is reasonably fast, within microseconds, but the change in refractive index is not large, so the switch must be 1 cm long to transfer an optical signal from one branch to another at a fibre Y-junction. Similar effects can be achieved in polymer films containing polar dye molecules, but the switching is faster, taking nanoseconds, because only electrons move to produce the change in refractive index, whereas ions must move in the perovskites.

The discussion of semiconductor band gaps, see above, established the connection between the wavelength of light absorbed, or emitted, and the band gap. The history of laser diodes has been one of increasing power and decreasing wavelength, from the infrared diodes in remote control to the progressively brighter laser pointers. This progress also greatly influences the ability of CD drives to read smaller spots and to write more quickly. Pitted against this is the ability of the makers of magnetic drives to make the media thinner and smoother and the heads more sensitive, so increasing the density of data and reducing the reading and writing times. There is a continuing battle between optical data storage and magnetic discs. So far, magnetic read/write storage has always managed to stay ahead of optical. Even the CD-ROM with slow but high-capacity storage is vulnerable to newer magnetic systems. Both are driven by continuous improvements in materials and processing.

Optical data processing became important with the advent of optical fibres with low losses and huge carrying capacities. What should be clear from the above is that the availability of materials to control light is lagging behind our needs.

Addressing Displays

Before leaving silicon, let us consider two other areas for which silicon-crystal wafer technology is not suitable. Solar photovoltaic cells can be made from crystalline silicon, but the slicing, mounting and connecting makes it very expensive. Really we would like to be able to print large areas, complete with connectors, onto rolls of plastic. The best current compromise is amorphous silicon, which is chemically deposited by decomposing gaseous silicon compounds over a glass sheet. Being amorphous and irregular, the silicon is full of defects, where one silicon atom has three or two neighbours instead of the normal four. It was found that annealing the film with hydrogen neutralized these defects to the point at which they do not badly degrade the semiconducting properties. The result is the familiar solar cell of calculators and watches. It is not very efficient but it is not too expensive. There are many schemes for cheap, efficient, large-area photovoltaic cells but we really do not have the answer.

In passive matrix displays, each pixel is just a simple intersection between one set of parallel wires above the liquid-crystal layer and another set below. This means that the display can only be addressed one line at a time and we must wait for each line to finish switching before moving on. It also makes it rather easy for a signal to leak into neighbouring pixels. In an active matrix display, each pixel has its own transistor to switch it on and off. Now we have to make about one million transistors in the back of a display. Sticking on crystalline-silicon switches, one at a time, is not a good solution. Amorphous silicon has been used because a film can be deposited onto glass and patterns of dopant added to make the transistors. Polysilicon is our current best bet. This is a thin film of silicon, formed as amorphous silicon on a substrate and then annealed to allow it to crystallize. It is not a single crystal and has grain boundaries that interfere with the motion of carriers. Thus it is worse than a single-crystalline wafer, but it is still better than amorphous silicon. With masks and photoresists, we can pattern the whole substrate at once and so make all the switches needed for a screen in parallel.

Muscles and Actuators

If you reach out and pick up a book, you extend an arm by contracting the triceps on the back of the upper arm, which pulls on the tendon sliding smoothly through your elbow and straightens the arm. Muscles can't push, so to bend the arm back again you contract the biceps and pull on the tendon through the inside of the elbow. Muscle is not very fast, the timescale is 0.1 s or more, and it cannot exert large forces per unit mass or cross section. Compare, for instance, the thin hard tendons on the back of your hand which must carry similar loads to those borne by the plump muscles of the palm. What muscle can do is provide a large, controllable, linear contraction. With small robots we try to achieve the same with motors and screw drives or pulleys but these are slow and hard to control. Large equipment, such as the arm on a mechanical digger, uses hydraulic cylinders, which do work well but cannot be miniaturized. At this point we can build a good dinosaur but a very poor rabbit.

Muscle converts chemical energy to force. Driven by GTP, myosin clicks between bent and straight conformations and so hauls itself along actin threads, to which it is stuck by hydrogen bonding. Unlike a ratchet or a carabiner, the myosin does not lock to the actin. Thus, to exert a steady force, the myosin must keep moving forwards as fast as it slides back, like dogs trying to pull a sledge up an icy slope. This may seem inefficient but it does give good control. Viewed as an engine, muscle gives a peak power of about 200 W kg^{-1}, a maximum stress of about 0.3 MPa and a maximum strain of about 25%. The power is comparable to that from a lead–acid battery; the stress is low compared with what a normal material can stand without breaking but the large strain is important for control.

Electricity is our standard source of chemical energy, so we would like an electrically powered muscle. The closest we come in present use are the piezoelectric ceramics, of which the prime example is lead zirconate titanate (PZT). In this perovskite crystal structure, a Ti^{4+} ion sits in the middle of six oxygens, O^{2-}, with Pb^{2+} at the corners of the cube, Figure 7. The combination of lead and oxygen squeezes in on the titanium so that it does not sit well at the centre of the cell but tends to pop slightly off centre. By replacing the right amount of titanium by its analogue zirconium, it can be arranged that all the central ions in a region of the crystal are displaced in the same

Figure 7
The perovskite structure. (*a*) The crystal structure of perovskite. Note that the small (4+) titanium ion is pushed on by six surrounding (2−) oxygen ions (red). (*b*) Two views showing how titanium ions respond to the crowding by being pushed very slightly off centre, causing the electrical dipole which distorts the crystal slightly. In reality the atoms are nearly touching, not separated as in these sketches.

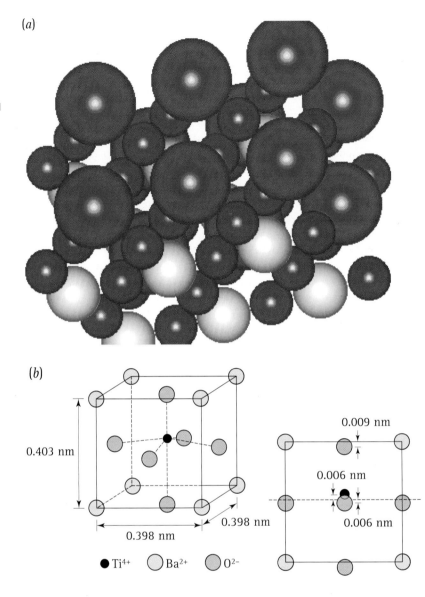

(*a*)

(*b*)

direction. Each unit cell now has an electrical dipole and the crystal is ferroelectric. We next arrange that all domains in the crystal are polarized the same way by applying a large electrical field and warming the crystal. If we now squeeze this crystal, the length of all these dipoles changes very slightly, thus changing the internal electrical field and causing a voltage to appear on the surface. This effect is used in barbecue lighters, in which a small hammer hits a crystal of PZT and causes a spark between wires attached to opposite surfaces. Alternatively, if we apply a field across the crystal, the dipoles lengthen or shorten and the whole crystal gets longer or shorter.

This effect turns an electrical field into linear motion and is used to drive small loudspeakers and ultrasonic generators. The response is fast and large forces can be generated but, unfortunately, the motion is small. Strains are a fraction of a per cent at fields of kV cm^{-1}. There is also not much prospect of increasing this strain. These ceramics are very brittle and start to crack at high strains or after many cycles.

In principle one can always increase the motion via a lever but this is clumsy and precision is lost if there is any looseness in the system. There are schemes for increasing the motion. One is to combine sheets of PZT into a hollow lens shape, such that contraction around the circumference results in the surfaces popping out. Another is to combine three sections as a cylinder sliding on a central metal bar. Alternate clamping at one of the ends and expansion or contraction of the centre results in 'inchworm' motion along the bar. The motion is not very fast and the motor would lose a race to a caterpillar, but it is compact.

There are other choices of material. Magnetostrictive materials contract in response to the application of a magnetic field. Electrostrictive materials contract in response to the square of the applied electrical field. Shape-memory alloys can give reversible changes of shape when they are heated or cooled, but the motion is hard to control.

The materials most similar to muscle are the polymeric gels. Water-soluble polymers can readily be made into cross-linked gels, which swell in water to make a stiff jelly. If the polymer is acidic, for instance cross-linked polyacrylic acid or a copolymer, it will tend to swell very strongly at high pH (base) but contract at low pH (acid). This is because the ionized ($-COO^-$) chains will collapse together and exclude water. It is thus possible to make a muscle from a collection of threads of hydrogel that contract strongly when acid is added and expand again in base. The motion is large and the force is great enough to be useful for a robot arm, but exchanging the acid and base is fairly slow and very messy. (Though it has to be said that muscle is also wet and messy.) It is possible to drive these gels electrically, effectively using the electrodes to shift the pH locally. Most demonstrations of this effect depend on a gel moving in water with at least one electrode external to the gel, a swimming fish is one example, but these systems can run in air if they are not allowed to dry out, Figure 8. This may lend itself to underwater robots but is not yet the versatile artificial muscle that we need.

Where do we go from here? With the hydrogel polymers the forces originate from hydrogen bonding and ionic interactions just like those in muscle, so the forces and extensions are comparable. This suggests that we should focus on the structural organization, rather than on the chemical composition. We want an organization that allows groups of molecules to contract and displace water internally, rather than forcing water out of the system over a long distance. Some kind of a grid might allow such a change in shape without a change in volume. The structure also has to be highly oriented so that the change in shape is a linear expansion and contraction. We also need some way of delivering chemical, electrical or optical energy on a fine scale to drive the system uniformly. Obviously the problem is really one of structure rather than of material.

This takes us into the area of micro-electromechanical systems (MEMSs). There has been rapid development recently in the use of lithography to etch small motors and sensors out of silicon. Much more recently there has been a growth in efforts to do this with ceramics and polymers also. Thus muscle is not really a material; rather, it is a machine, with several components working together on a fine scale. Irrespective of whether we build micro-inchworms or networks of fibre held together by contractile threads, we need to think in terms of blurring the interface between material and machine.

Tough Ceramics

One lesson of the second law of thermodynamics is that engines are more efficient when they run hotter. Thus there is a constant need to raise the operating temperatures both of internal combustion engines and of jet engines. In addition there is a need to lighten engines in order to improve their efficiency at using fuel. The use of ceramic parts is the obvious answer to both problems. A ceramic piston engine would run hotter, should not need a cooling system and might not need lubrication. With no radiator, no water pump and no coil circulation, a car's engine could be much smaller.

These targets have been pursued diligently for the last 40 years and we have

Figure 8

(*a*) A gripper, with fingers made of electro-active polymer, picking up a small stone. (*b*) The accumulation of dust on windows is a problem for the Mars rover. An electro-active polymer wiper is proposed. Pictures courtesy of Y. Bar Cohen, Jet Propulsion Laboratory. See http://eis.jpl.gov/ndeaa/ nasa-nde/lommas/aa-hp.htm.

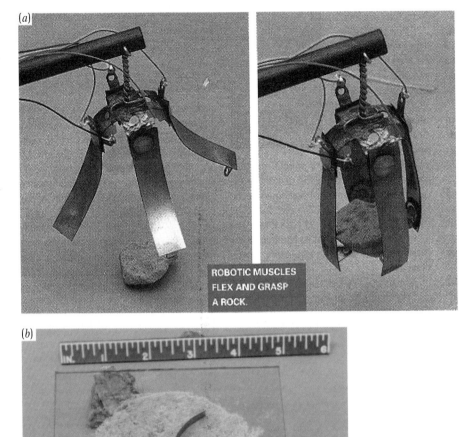

managed to make ceramic turbochargers and valves, but the progress is not as impressive as we had hoped. While so much attention and effort were being focused on structural ceramics, the use of ceramics in electronics grew quite unexpectedly to be a huge business. The problem is that of making strong ceramics reliably. Similar problems have prevented us from using the high-T_c ceramic superconductors that could change the world, if they were stronger and tougher.

Ceramics are the crystalline oxides, nitrides, carbides and borides of the metals and metalloids, especially aluminium and silicon. The small atoms give rise to strong, highly directional covalent bonds. As a result their melting points are high and they are very stiff (the elastic moduli, E, are high). Also, because the atoms are light, the densities are low. The melting points are too high for melting and casting of parts, so they are normally sintered. Powder is pressed into the desired shape and then heated to around 1500 °C, whereupon diffusion slowly causes the piece to shrink by about 30% and become fully dense. The firing of clay is the same process, although the temperature is lower than that needed for fine ceramics. Glasses, which are based on silicon dioxide, are chemically similar but are not crystalline and melt at low enough temperatures for them to be cast into shape.

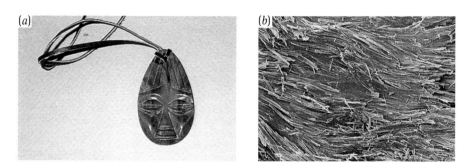

The core problem with ceramics goes back to Griffith's explanation of the strength of brittle materials in the 1930s. Before that, most engineered structures used tough steels. Toughness is the characteristic of exhibiting large local deformation before fracture: a material bends before it breaks. If you happen to be in or on the structure at the time, you get a chance to save yourself. In contrast, in brittle fracture, a small crack can grow catastrophically and result in rapid total failure. A famous example is the World War II Liberty ships, in which a crack would start at a sharp angle, in the hull next to the deck, and suddenly zip around the ship, breaking it in two. Generally, alloying and heat treating a metal to raise the strength also increases the brittleness, so the problem has become worse as the materials got better.

The equation developed by Griffith gave the strength of a brittle material as inversely proportional to the square root of the length of the largest flaw or crack in the sample. This means that, unlike tough materials, a brittle material has no defined strength but rather its strength depends on whether there is already a crack or defect and how big it is. This concept is very unpopular with engineers because it gives no basis upon which to design a part to withstand a certain load. The strength of ceramics is characterized by a Weibull modulus, which describes the breadth of the distribution of strengths in a particular collection of parts. The two choices, both bad, are to design for the lowest expected strength, or to put a part in and see whether it breaks.

According to the Griffith equation, the strength is also proportional to G, the square root of the energy needed to grow a crack by unit area. G is very low for glasses and for aluminium oxide because a crack simply cuts through one layer of chemical bonds. G is very high for jades (Figure 9), a family of minerals with a fibrous crystal structure. Here fibres pull out during fracture and this results in a microscopically rough fracture surface. In consequence, jades can easily be machined into decorative shapes whereas glasses cannot. Silicon nitride is also fibrous, in one of its forms, and hence it was the focus of a great deal of research in the 1960s. However, it was never possible to make parts without big internal cracks and flaws resulting from the pressing and sintering processes.

Attention then switched to decreasing the flaw size, c. In the 1980s we tried to ensure that the starting powder had a diameter of 1 μm or less, with no large grains that might interfere with sintering and leave a big hole. We also tried to pack the powder in the unsintered 'green body' perfectly, with no voids or cracks. As a result we could make excellent thin sheets and rods but it was very hard to make large or complex parts. It also needed very good, fine and uniform, powder to start with and only alumina was available with such high quality. We then went back to trying to raise G again by adding short, crack-deflecting fibres to the ceramic. Carbon fibre and silicon-carbide fibre can work quite well when they are added to alumina, but turn out to be very vulnerable to oxidation in a hot engine. A host of other tricks has been tried along the way, but there are still no good general answers.

Meanwhile, examples of very tough ceramics lay around us. The inner nacre

(*a*)

(*b*)

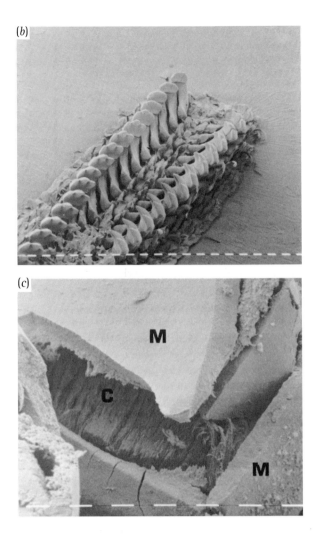

(*d*)

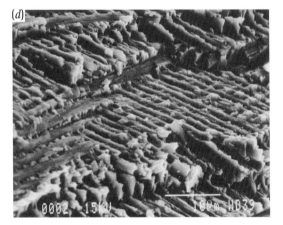

(*c*)

Figure 10

Some tough biological ceramic structures. (*a*) Nacre from a clam shell, showing 0.5 μm-thick plates of calcium carbonate (aragonite) separated by thin layers of protein (seen as small ridges at crystal junctions). (*b*) Chiton teeth of a mollusc are dragged across rock during grazing. The radula of *Plaxiphora albida* has two rows of teeth. Bars are 100 μm. (*c*) Each tooth has an outer layer of hard magnetite (M) and an inner, tough, fibrous core (C). Bars are 10 μm. Thanks to John Webb and Lesley Brooker, Murdoch University. (*d*) A rat's tooth, showing crossed layers of hydroxyapatite fibres.

('mother of pearl') of mollusc shells is a layered structure with plates of calcium carbonate separated by thin tough layers of protein (Figure 10). The protein acts as a crack deflector, though in this case one designed to turn the crack parallel to the surface, and makes the shell far stronger than would be expected for a carbonate ceramic. This would be a good answer for engines, if molluscs would only make shaped silicon carbide and if the protein could withstand 1000 °C. However, not all shells have such high levels of protein. Other shell structures, and that of our tooth enamel, seem to depend on complex arrangements of fibrous or ribbon-shaped crystals for their toughness via crack deflection.

For the moment the effort at developing strong ceramics is at a low ebb, but the need is still there. The problem is one of getting the right microstructure, developing a process for making shapes reliably and efficiently and then persuading designers to try ceramics again.

The Plastic Battery

Batteries and transformers are the curse of electronic equipment. They are heavy, bulky and generate heat that must be dissipated. For portable electronics, the compact mass of the battery dictates much of the design and causes most of the internal damage if the unit is dropped. Simulations of dropping a lap-top computer show that the screen usually breaks when the battery rebounds. We would much prefer a battery in the form of a flexible sheet distributed around the case.

A secondary (rechargeable) battery represents a set of interlocking problems of materials science. The electrodes should be light and be able to store large amounts of charge without swelling, shrinking and breaking up, as the battery is cycled. The electrolyte should be able to withstand the large potential difference (1–4 V) without decomposing and should be able to transport ions rapidly. Small batteries are usually encased in a rigid metal can, because the contents are so reactive, but it would be much preferable to put a battery in a soft, light blister pack.

The ultimate purpose of a battery is to store the maximum energy per unit weight or volume. A glance at the Periodic Table suggests that what we need is lithium and fluorine, two light elements with big differences in electronegativity such that transferring an electron from one to the other will release most energy per unit weight. Because it is a very reactive gas, fluorine is not easy to package or use but we can compromise by using lithium and thionyl chloride, which is a tractable liquid. This is one type of the lithium primary (unrechargeable) batteries used in pacemakers. The chloride has to be combined with some conducting material, such as graphite felt, to get the current out. As the reaction proceeds, lithium chloride will form as a dissolved salt.

For the battery to be rechargeable, we need to be able to reverse the reaction to regenerate the lithium. The best answer at present, the 'lithium-ion' battery in the current generation of lap-top computers, has a metal-oxide positive electrode, usually cobalt oxide.

The 'salt' now forms in the solid oxide anode, not in the electrolyte. For this to work, the oxide has to have a crystal structure with pores large enough to allow the lithium to move in and out rapidly and stable enough that the solid doesn't crumble as it expands and contracts during charging and discharging. Even given all this, it is not a wonderful answer, because the oxide is heavy and bulky, per stored lithium ion. Likewise, lithium metal is too brittle to be a good cathode and is replaced by a lithium–carbon mixture with a further cost in weight, since the carbon is inactive. A core problem is that electrodes must take up and release material and so must expand and contract to some extent. It is much more difficult to engineer with materials that change size, whether as electrodes or due to thermal expansion or changes in humidity. This is why it is so hard to make furniture that does not tear apart when it is moved from a wet climate to a dry one.

The electrolyte in this cell must transport the lithium ions from the cathode to the anode. Water would do this well, but would react violently with the lithium metal. Highly polar organic solvents, which will dissolve salts of lithium with large anions, are used. Typically lithium hexafluorophosphate ($LiPF_6$) in solution in ethylenecarbonate is used. The enormous size of the PF_6^- ion means that there is not much attractive energy between it and Li^+ in the solid salt and, hence, it dissolves easily. However, the heavy ion again adds more weight. It would be very attractive to replace the liquid organic electrolyte by a polymer gel. This would add mechanical integrity, would stabilize the spacing between the electrodes, would remove worries about boiling or leakage of electrolyte and would lend itself to rolling up batteries from thin sheets of electrode and electrolyte. There has been much research on polyethyleneoxide, a waxy polymer that melts at 60 °C, as an electrolyte. A plasticizing organic liquid can be added to lower the melting point, but the diffusion of ions is much slower than that in the liquids and this really limits the current which a battery can supply. New designs of very thin batteries can use a polymer electrolyte because the ions do not have far to move.

The answer to slow diffusion and to many of the problems with batteries may be thinner layers. The distances for diffusion then become shorter. The differential strains developed as the electrodes expand and contract are reduced and the current densities are much lower because we have a much larger surface area of electrode. Several organizations are working on 'Swiss-roll' configurations in which thin layers are rolled up into a sausage. Capacitors have long been made like this, but they are built from stable, strong and unreactive materials. Batteries are much more difficult to make.

The case of a lithium battery has to prevent the permeation of water or oxygen for a 5-year lifetime of the battery. So far, it has not proved possible to do this with anything other than a steel can at a great cost in weight and loss in flexibility of design. One would think that some combination of metal foil and polymer could do the trick but, so far, it hasn't happened. Likewise, tough and light polymer-based electrodes would be ideal but none has been made to work. Electrical vehicles are now powered by lead–acid batteries. Lithium batteries are too expensive and there are concerns about their behaviour in crashes and fires. However, the poor energy/weight ratio of the lead–acid battery severely limits the range of electrical cars. Much cheaper and more efficient batteries are needed.

Another choice would be to look back at 100 years of battery technology, admit that we really have not made much progress and concentrate on other small power sources. Fuel cells are one answer, though current designs run at high temperatures in very corrosive environments and so have severe problems with materials. They would be unsuitable for powering portable electronics. It may be that we can learn from biological systems here. A small mammal is essentially a chemically powered device with computing and mobility functions. The glucose–oxygen power system certainly has a competitive power/weight ratio. This does suggest that small, low-temperature, fuel cells should be possible.

Rapid Prototyping

Making a material with outstanding properties is only part of the solution to any problem. It must also be processed into the desired form and must function without degradation for some years. Most processing involves melting the material and shaping it in a mould or through an extruder. Free-form fabrication methods may change the relationship among materials producer, designer and manufacturer by allowing the direct production of parts from a computer design without the need for a mould.

The origin of this approach was the stereolithographic method for rapid prototyping, Figure 11. A part is designed in three dimensions using a computer-aided-design package and the solid representation is broken down into a series of thin slices, like bacon. The information describing the slices is fed sequentially to a laser system that writes onto the surface of a bath of photosensitive resin. As each slice is converted to solidified resin, the supporting platform drops to allow a new layer of liquid to cover the top surface so the next layer can be made. After some hours, the platform is raised and the part can be picked up. This method has been adopted widely by manufacturers to provide a model part that can be used in demonstrations and to check its fit in an assembly. Once stereolithography became widely known, it was obvious that any other method for converting a liquid to a solid at a point could be used to build three-dimensional objects.

From prototyping in fragile cross-linked polymer the emphasis of development has shifted to making functional parts in thermoplastics, ceramics, composites and metals and to the direct making of moulds for casting metals or injection of polymers. These methods are shortening the design cycle for structural components. Parallel developments on a finer scale are allowing rapid prototyping of electrical and electronic assemblies.

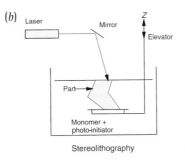

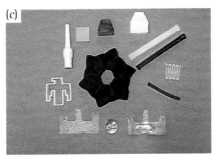

Figure 11
Stereolithography. (*a*) One example of a commercial free-form fabrication system (fused deposition modelling). (Photograph kindly provided by Stratasys Inc.) (*b*) A schematic diagram of another free-form fabrication method (stereolithography). (*c*) Parts made in a range of materials by free-form fabrication. Clockwise from the coin: polyvinylpyrrolidone, alumina, zirconia, aluminium, silicon carbide, polyoxymethylene, epoxy resin, carbon-fibre-filled epoxy resin, acrylic/silica hybrid material, polyacylamide gel, agarose gel and (centre) carbon-fibre-reinforced polycarbonate. (Thanks to Advanced Ceramics Research, Tucson, Arizona and Sandia National Laboratories.)

The Future

The development of integrated circuits has had a great effect on the direction of materials research. There still is interest in stronger fibres, tougher ceramics and high temperature polymers but more effort is going into complex combinations of materials for optical and electronic devices, for medical implants and for 'smart' devices with built-in sensors and actuators. These structures cannot be simply moulded but must be grown via many sequential steps, such as lithographic and free-forming methods.

There is also a great deal of effort going into the interface between biology and materials. As well as biomedical materials, there is work on biosynthetic routes to plastics and inorganic particles. Examples are the cloning into bacteria of silk and related proteins for production by fermentation and the production of nanometre-sized magnetic particles by bacteria. The field of biomimetic materials concerns the development of composites with microstructures similar to those in structural biological materials, especially the hierarchy of scales that seems to lead to impressive combinations of toughness and stiffness. This argument can also be taken to the level of design. Natural objects tend to be built of soft materials with curved surfaces whereas manufactured objects are made from hard materials with right angles.

There is also considerable new interest in nanotechnology. The novelty of nanostructured materials arises from the much greater importance of surface forces at these small scales. Another major driving force is the power of the atomic-force microscope for viewing these structures under ambient conditions. With the electron microscope sample-preparation methods and beam damage limit our ability to study living systems.

A good example of these new directions is the 'gene chip'; a pattern of DNA probes on a glass slide allows the sequence of a cluster of genes to be read out. Soon it should be possible for the whole of the human genome to be read out. The device arises from a marriage of integrated-circuit processing and biotechnology. The next stage is to include the chemical processing steps in the form of small channels, valves and reservoirs cut into the glass slide itself. This also brings us into new regimes of behaviour for soft materials. Thus, in transferring liquids on the micrometre scale, gravity is no longer important and surface tensions control everything.

For the next ten years, do not expect so much progress in stronger, harder materials but rather developments in the integration of devices into structures, active materials with embedded sensors and actuators, more soft, flexible structures and hybrids of biopolymers or cells with synthetic materials. In making these, new processes for building such complex combinations by lithographic and printing methods will be developed.

Further Reading

1. D. A. Davies, *Waves, Atoms and Solids* (Longman, London, 1978).
2. R. R. Tummala and E. J. Rymanszewski, *Microelectronics Packaging Handbook* (van Nostrand Reinhold, New York, 1989).
3. L. M. Brown, Dawn of the diamond chip, *Nature* **350**, 561 (1991).
4. V. M. Chernousenko, *Chernobyl, Insight from the Inside* (Springer Verlag, Berlin, 1991).
5. S. Nakamura, Blue–green light-emitting diodes and violet laser diodes, *Mater. Res. Bull.* 29–35 (1997).
6. N. Greenham, S. Moratti, D. Bradley, R. Friend and A. Holmes, Efficient light-emitting-diodes based on polymers with high electron-affinities, *Nature* **365**, 628–630 (1993).
7. Y. Xia, J. McClelland, R. Gupta, D. Qin, X. Zhao, L. Sohn, R. Celotta and G. Whitesides, Replica molding using polymeric materials: a practical step toward nanomanufacturing *Adv. Mater.* **9**, 147–149 (1997).
8. L. Kuhn-Spearing, H. Kessler, E. Chateau, R. Ballarini, A. Heuer and S. Spearing, Fracture mechanisms of the *Strombus-gigas* conch shell: implications for the design of brittle laminates, *J. Mater. Sci.* **31**, 6583–6594 (1996).
9. P. F. Jacobs, *Rapid Prototyping and Manufacturing* (SME, Dearborn, MI, 1992).
10. www. Many sites can be found by searching for rapid prototyping. See also The Rapid Prototyping Resource Center: http://cadserv.cadlab.vt.edu/bohn/RP.html.
11. P. Calvert, Biomimetic ceramics and hard composites in *Biomimetics*, edited by M. Sarikaya and I. A. Aksay (AIP Press, Woodbury, NY, 1995), pp. 145–161.

Molecular Electronics

Bob Munn

Introduction

When I was a boy, over 40 years ago, 'electronics' meant vacuum electronics. Radios used thermionic valves, which depended on heating a filament to boil off electrons into an evacuated glass container where the current was controlled by wire grids at varying electrical potentials. Valves were bulky (up to 10 cm long) and fragile because of the glass; they took a long time to warm up and a lot of electrical power. By the time I became a teenager, 'electronics' had begun to mean solid-state electronics. Radios now used transistors, which depended on electrons already present in pieces of semiconductor deliberately doped with impurity atoms in various ways. Transistors were quite small (up to 1 cm long) and robust; they started work on demand and used little power. Thus the transistor radio ran on batteries and was portable, giving people music wherever they went – or wherever teenagers went.

Transistors were wired into circuits much as valves had been, but people then realized that this wasn't essential. You could treat a piece of semiconductor to create a number of transistors on it and you could also evaporate metal tracks to connect them. Electronics then came to use these integrated circuits more and more while increasing the scale of integration, i.e. the number of components in a given area. This gave us the ubiquitous microchip with very-large-scale integration (VLSI) corresponding to components with dimensions of the order of 1 μm (a millionth of a metre). We rely on these chips to tell us the time, run our washing machines, control the ignition of our cars, give us silly but enthralling games to play and, in general, perform scarcely believable feats of rapid calculation, control and display.

Progress has certainly been spectacular. If we draw a graph of the size of a component against time (Figure 1) we see a steady decrease. Since the size axis is logarithmic – each division corresponds to a change in size by a factor of ten – the decrease is exponential in time, by a factor of ten every 10 years. This 'law', which was originally pointed out by Gordon Moore, the founder of Intel, still seems to be holding. I first drew my own version of it some 15 years ago and there has been no major deviation yet.

So where is it all leading? If we assume that the same rate of progress is to be maintained, then in about 20 years' time we shall need components with sizes of only 1 nm (a billionth or a thousand millionth of a metre). Since individual atoms are only two or three tenths of a nanometre in diameter, such components will have to be made of rather few atoms. In that case, the argument runs, why not use molecules as the components? We could design them to perform the necessary functions and chemists will surely be able to synthesize what we have designed. In this way we reach a new phase of electronics – and of chemistry – namely *molecular electronics*.

Why should we bother? Well, for one thing, as the arguments already given suggest, molecular electronics will give us even larger scales of integration, up to a million times larger in a conventional planar array. This means smaller devices, perhaps ones that surgeons could safely implant in the body (like a heart pacemaker) to replace or supplement impaired bodily functions. Smaller devices in turn mean shorter distances for signals to travel and hence faster devices. Quite another argument appeals to biology. Nature achieves its purposes through molecular systems such as the brain. Our brains cannot compete with electronics for rapid and accurate calculations, but they

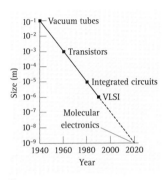

Figure 1

The decrease in feature size with time.

are extraordinarily good at tasks such as pattern recognition – for example, recognizing a man you last saw 20 years ago who has subsequently gone grey and acquired a bushy beard. Nature also grows these systems from simple ingredients at a low cost in energy and can even repair damage to some extent. So what is commonplace in biology proceeds by molecular-electronic mechanisms that we should do well to emulate. This is a challenging scientific task with major practical implications.

What I have described is *electronics at the molecular level*, employing individual molecular electronic functions, and this is the future of molecular electronics. However, molecular electronics has a present, which is *molecular materials for electronics*, employing assemblies of molecules to make new materials that perform electronic functions. This aspect of molecular electronics is analogous to the partial replacement of steel by plastics in car-making when there are advantages of light weight and low cost. Besides generating useful devices, developing molecular materials for electronics also serves as a proving ground for concepts and techniques required for electronics at the molecular level. In molecular materials, the molecules retain their separate identities. The material's properties then arise from a combination of the molecular properties, the molecular arrangement and the intermolecular interactions. Success in molecular electronics depends on understanding and controlling all these aspects, as we shall see in the examples later, which also show the interplay among materials, phenomena and applications.

Finally, in considering the scope of molecular electronics, one should note a common generalization. Electronics concerns information processing, for example, combining two inputs, storing the output and displaying it. So the storage and display of information properly come under the aegis of electronics, even though they may use techniques other than electronics, such as optics – after all, the venerable technique of photography undoubtedly stores information and so does the even more venerable abacus. So for present purposes molecular electronics encompasses all forms of manipulating, storing, displaying and transmitting information using molecular materials. This definition includes not only electronics properly so called, in which information is processed by controlling the flow of electrons, but also what is sometimes called photonics, in which information is processed by controlling the flow of photons, the basic constituents of a beam of light.

Conduction

Copper conducts electricity – we use it to make wires. Materials like rubber and poly(vinylchloride) (PVC) do not conduct electricity – we use them to insulate copper wires. Silicon does not conduct electricity at all well, but it gets less bad as we raise the temperature and much better as we 'dope' it by putting in other kinds of atoms as impurities – it is a semiconductor. What accounts for these differences in behaviour? Why is it that the familiar metals and semiconductors are not molecular materials whereas the familiar insulators are?

Conduction requires a supply of mobile negatively charged electrons. Electrons are always present in matter, so the question is that of how they can be made available to move and so conduct electricity. In an isolated atom, ion or molecule the electrons occupy specific energy levels. When atoms, ions or molecules come together in a solid, new sets of energy levels develop. An electron is influenced not only by the charge on the nucleus of its original atom, ion or molecule but also by the charges on the nuclei of the surrounding atoms, ions or molecules. This influence of neighbouring species provides a mechanism for electrons to move. It also changes the energy levels of the isolated species into bands of allowed energies for electrons in the solid. Strong interactions between neighbouring species lead to wide bands that allow electrons to move rapidly. How fast they move depends on how strongly they are scattered; and how strongly they are scattered depends on how perfectly the solid is ordered. All solids vibrate, so even a perfect crystal is perfectly ordered only on average: at any given instant the vibrations cause deviations from the average ordered structure that scatter electrons. A crystal may be ordered, on average, but contain imperfections such as a vacancy (a site from which an atom is missing), an atom in the wrong place or an atom of the wrong kind. These too constitute deviations from perfect ordering that scatter electrons. Finally, the solid need not be ordered even on average (for example, glass or a polymer) and these deviations from order scatter electrons very strongly.

All electrons can move in this way and so contribute to conduction in principle. However, in practice every full band gives zero net contribution: for each electron moving in one direction there is another moving equivalently in the opposite direction. Hence only electrons in partly filled bands are mobile and able to contribute to electrical conduction. Each energy level in an isolated species or in a band can accommodate up to two electrons, the lowest levels filling first. The electron has a spin of $\frac{1}{2}$, the spin being a quantity that accounts for the magnetic moment of a particle (like a very small bar magnet). A spin s can adopt $2s + 1$ different orientations. An electron's spin can thus adopt two orientations, usually described as 'up' and 'down', and these account for the occupancy of an energy level. Most molecules contain an even number of electrons that occupy the available energy levels in pairs. If N molecules come together to form a solid, any given energy level gives rise to a band containing N energy levels. The isolated molecules contain N pairs of electrons and these are just enough to fill completely the N energy levels in the resulting band. Hence a molecular solid chosen at random is likely to be an insulator, as in the cases of rubber and PVC already mentioned. Similar considerations apply to many ionic solids, such as sodium chloride (common salt), in which the sodium and chloride ions contain even numbers of electrons.

However, the copper atom contains an odd number of electrons. One of its atomic levels contains a single unpaired electron. In the solid, N such levels combine to form a band containing N energy levels. These can accommodate N pairs of electrons, but only N single electrons are available. These then occupy the lowest $\frac{1}{2}N$ energy levels in pairs, the band is only half full and conduction characteristic of a metal results. Electrons are always available for conduction, but raising the temperature causes the solid to vibrate more. This scatters the electrons more, so the metal conducts less well: a decreasing conductivity with increasing temperature is characteristic of a metal.

We might, emboldened by this successful interpretation, then note that the iodine atom also contains an odd number of electrons and so by the same argument should be a metal. This is not so: the iodine atoms prefer to combine in pairs to form the familiar I_2 molecules, which have an even number of electrons that yield filled bands. Thus instead of the odd electron of N atoms yielding a half-filled band of N energy levels we find the paired electrons of $\frac{1}{2}N$ molecules yielding a filled band of $\frac{1}{2}N$ energy levels. This illustrates the important interplay between electronic band structure and geometrical structure (since the formation of I_2 molecules results in close pairs of bonded iodine atoms and remoter pairs of non-bonded atoms).

Semiconductors, not surprisingly, fall in between insulators and conductors. An intrinsic semiconductor like silicon has paired electrons that form a network of bonds that hold the solid together. Because the electrons are paired, they form filled bands. However, the energy gap between the top of the filled band formed from these valence electrons and the bottom of the next empty band formed from the next unoccupied energy levels of the atoms is fairly small. It is small enough for available thermal energy to excite a small fraction of the electrons from the filled valence band to the empty band. In the empty band, they are mobile and can conduct, so it is called the conduction band. As the temperature increases, more electrons are excited to the conduction band. Although the scattering increases as the temperature increases, as we saw for a metal, this is more than offset by the rise in the number of mobile electrons; hence the conductivity of the semiconductor increases with increasing temperatures, in contrast to the behaviour of a metal. Excitation of electrons to the conduction band also leaves the valence band less than completely full. Since a full band yields no conduction, it proves convenient to ascribe the conduction of the partly emptied bands to positively charged 'holes' corresponding to the absence of an electron. Each electron excited to the conduction band leaves a hole in the valence band.

In conventional semiconductors we assist the process of thermal excitation by doping. Replacing a silicon atom by a phosphorus atom has two effects. It donates an extra electron and it creates a filled donor energy level for that electron in the original band gap, a little way below the bottom of the conduction band. Hence doping provides a controllable source of extra electrons that are more readily available for conduction because they require less thermal energy to reach the conduction band. Similarly, a boron atom replacing a silicon atom accepts an extra electron and creates a vacant acceptor energy level a little way above the top of the valence band. This in turn yields a controllable source of extra holes created in the valence band by thermal excitation of electrons to the new vacant energy levels.

These considerations are conventionally illustrated by the energy-band diagrams shown in Figure 2. This shows schematically the energies of the bands plotted vertically, with occupied energy levels shaded. Such diagrams appear in elementary descriptions of solid-state physics. Molecular electronics builds chemical understanding on to this structure.

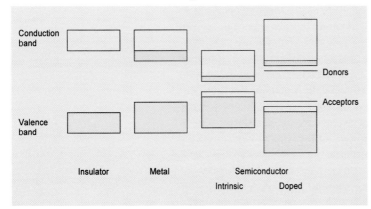

Figure 2
Energy-band diagrams illustrating the characteristic features of insulators, metals and semiconductors.

Left to their own devices, then, molecular materials tend to be insulators. To make them conduct we must excite or unpair electrons. If thermal energy is insufficient to excite them, we can instead use optical energy. Light provides energy in amounts or quanta $E = hf$, where h is the Planck constant (6.626×10^{-34} J s) and f is the frequency of the light. For green light, f is about 6×10^{14} s^{-1} and E is then about 4×10^{-19} J, or 2.5 eV, where the electron volt (eV) is the energy involved in moving the electronic charge through an electrical potential difference of 1 V. Since electronic energy levels are typically spaced by $1 - 10$ eV, visible or near-ultraviolet light has the requisite energy to excite an electron from the valence band to the conduction band where it can carry a current. This is the process of *photoconduction*.

Many aromatic and hetero-aromatic molecules yield photoconducting crystals or thin films. One example is carbazole (1). Under illumination it gives quite a good yield of photo-excited charge carriers, which remain mobile for some time before becoming trapped at imperfections and ceasing to carry current. These properties can be used in photocopying. An image projected on to a photoconductor carrying a layer of surface charge renders it conducting in the illuminated areas, from which the charge can then drain away. The remaining charge in the dark areas can attract pigment that is transferred to paper to produce a copy of the image. Most photocopiers and laser printers now use organic photoconductors, rather than the original amorphous selenium, but we shall see later that technical factors require that carbazole be modified for this purpose.

Aromatic hydrocarbons composed of fused planar hexagons – 'chicken-wire molecules' – played an important early role in developing conducting properties of molec-

1

ular materials. They are thermally stable, crystallize readily and have rather simple energy-level structures. Some examples are anthracene (**2**), pyrene (**3**) and perylene (**4**). The ultimate aromatic hydrocarbon, consisting of an infinite sheet of hexagons, is graphite, which is known to be an electrical conductor. As the number of fused rings gets larger, the gap between the valence and conduction bands falls, so an obvious target is to make a very large aromatic hydrocarbon. In practice, trying to make molecules more than about twice the size of perylene usually ends in making graphite unless extreme care is taken. Hence an alternative is to make a large hydrocarbon and use chemical energy to produce mobile charge carriers.

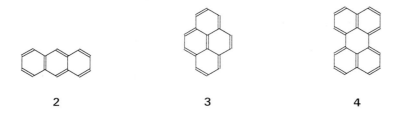

2   **3**   **4**

A suitably mild and convenient way of doing this is to expose perylene to iodine vapour. Iodine is a mild oxidizing agent, that is, it tends to attract electrons to form I^-, I_3^- or higher ions. The electrons it attracts from perylene leave holes in the valence band and hence the material conducts, its conductivity being about 0.5 S cm^{-1} (S denotes the siemens, the unit of electrical conductance, formerly called the mho or reciprocal ohm, Ω^{-1}). This process resembles doping silicon with boron and yields an equivalent material, an *organic semiconductor*. Aromatic hydrocarbons are obliging but not especially good electron donors, so related species containing atoms such as nitrogen and sulfur, for example phenazine (**5**), can replace them. Equally, better electron acceptors than iodine can be devised, containing nitro groups (−NO$_2$) or cyano groups (−CN) that are able to accept electrons and spread them over the molecule. Of these, the classic example is TCNQ (tetracyanoquinodimethane, **6**), with its four cyano groups connected by conjugated pathways providing extensive delocalization of electrons and hence powerful electron-accepting ability. Pairs of donors and acceptors like this form donor–acceptor crystals that are typically semiconductors.

5   **6**

Thus doping or going further to produce a donor–acceptor crystal changes the energy-level structure to yield a semiconductor. Once we have two components with paired electrons and a tendency to donate and accept electrons, we can try to persuade electrons to transfer permanently, yielding two components with unpaired electrons that could yield partly filled bands and hence a conductor. This argument is illustrated in the energy-band diagrams in Figure 3. The donors and acceptors are assumed to form separate bands, which is a good assumption for those crystals in which the charge transfer occurs readily. Such crystals typically comprise face-to-face stacks of donor (DDD...) and acceptor (AAA...) molecules, like decks of cards; on the other hand, most

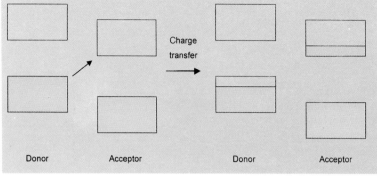

Figure 3
An energy-band diagram illustrating how partly filled bands result from the transfer of electrons between a donor and an acceptor.

donor–acceptor crystals form mixed face-to-face stacks of alternating donor and acceptor molecules (DADADA...) like multidecker sandwiches. As the diagram suggests, the charge transfer costs energy. However, some energy is gained because of the favourable interaction between positively charged donor stacks and negatively charged acceptor stacks, so overall the energetic considerations could favour charge transfer. (In fact, similar considerations apply to the formation of sodium chloride from sodium and chlorine: formation of the gas-phase ions costs energy but the crystal is stabilized by the favourable lattice energy.)

So much for the argument. Actually achieving the charge transfer requires a strong acceptor such as TCNQ (**6**) and also a very strong donor, the archetype of which is TTF (tetrathiafulvalene, **7**). The material TTF–TCNQ has a high conductivity at room temperature, of up to 1000 S cm^{-1}, approaching that of an indifferent metal such as mercury, which exhibits the characteristic metallic behaviour of increasing as the temperature decreases and leads to a shiny metallic lustre. Thus TTF–TCNQ is an *organic metal*, containing no metal atoms but behaving in many ways as conventional metals do.

7

If lowering the temperature raises the conductivity of TTF–TCNQ, how high will it go? There turns out to be a limit. By about 60 K the conductivity has increased roughly a hundredfold, but at lower temperatures it starts to decrease again, behaving not like a metal but like a semiconductor. Such transitions are called *Peierls transitions*, after Rudolph Peierls. He showed that a strictly one-dimensional metal (having interactions along only one direction) must undergo a transition at low enough temperatures. In the transition, the lattice distorts, which costs elastic energy, but the energy bands distort, which gains even more electronic energy. The argument resembles that already given to explain why iodine is an insulator. If we consider a half-filled band for simplicity, then 'dimerization' or alternation of bond lengths in the chain of molecules splits the band into two. As illustrated in Figure 4, the electrons can now fill the lower band, lowering the energy and destroying the metallic conduction.

Of course, TTF–TCNQ isn't really one-dimensional, but the donor and acceptor stacks are well separated so that their interactions are weak. If the interactions were stronger, the Peierls transition would not occur. Armed with this knowledge, one can embark on the procedure characteristic of molecular electronics: modifying molecules to optimize the properties of a material. One can change the size, shape and chemical constitution of molecules. Thus changing TTF to the analogous compound HMTSF (hexamethylenetetraselenafulvalene, **8**) elongates the molecule by virtue of the

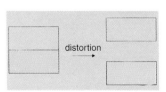

Figure 4
An energy-band diagram illustrating how the Peierls distortion lowers the electronic energy for a half-filled band.

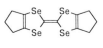

8

additional rings and strengthens the interactions by virtue of the more polarizable sele-niums' replacing sulfurs. Then the complex HMTSF–TCNQ remains metallic down to the lowest temperatures, as a respectable metal should.

However, some metals go further: at very low temperatures (below 25 K and often much lower) they lose all trace of electrical resistance and become *superconducting*. This property occurs because electrons that repel because of their charges can never-theless form pairs if they are strongly attracted to the oppositely charged nuclei. As John Ziman vividly put it, in *Principles of the Theory of Solids* (Cambridge University Press, 1964), they can gain by 'sitting close together in the same depression of the mat-tress'. Stronger interactions turn out to require one of the components to be a small ion rather than a large molecule. Eventually, in 1979 Denis Jérome and colleagues in Paris prepared the first *organic superconductor*, $(TMTSF)_2PF_6$, where TMTSF is tetra-methyltetraselenafulvalene (**9**). This compound became superconducting below a crit-ical temperature T_c of 1.3 K, but only under a pressure of 6500 atm. Pressure squeezes the TMTSF molecules together, so the argument that using a smaller ion than PF_6 should have the same effect as increasing the pressure led to the first ambient-pressure organic superconductor, $(TMTSF)_2ClO_4$, which becomes superconducting below 1.2 K. Numerous organic superconductors have now been prepared by ringing the changes on the TMTSF and on the small ion. Another series is based on BEDT-TTF (*bis*-ethenedithiolato-TTF, **10**). This makes up for having only sulfur atoms by having eight of them, two in a ring attached to each side of TTF. There are also superconductors that are certainly molecular but contain the species $M(dmit)_2$, where M is a metal atom com-plexed with the sulfur-containing donor dmit (1,3-dithiol-2-thione-4,5-dithiolate, see structure **11**). One can now vary the metal atom, even to the extent of making the molecular ion negatively charged.

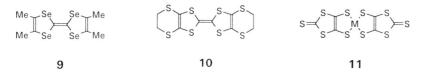

9 **10** **11**

Over 10 years, the highest known critical temperature T_c for molecular supercon-ductors was increased to over 10 K. This increase was much faster than that in the early days of superconductivity, arousing hopes of realizing the room-temperature superconductor predicted some 35 years ago. Then came the oxide high-T_c or warm superconductors, which soon increased the highest known T_c for inorganics by a factor of about five, to over 120 K (though this is still well short of room temperature). The goal of beating T_c for inorganics receded, still leaving much science to be done on molecular superconductors, and then came another twist in the tale. Buckminsterfullerene, C_{60}, and the other fullerenes were characterized as new forms of carbon and as new unsaturated molecules. Their chemistry was naturally investi-gated and compounds with the alkali metals were prepared. These have the formula A_3C_{60}, where A_3 represents three alkali atoms, not necessarily all the same. Some of these proved to be superconductors, the highest T_c at present being 33 K for Rb_2CsC_{60}. Were it not for the oxide superconductors, this would be the highest T_c for any material; and it suggests that there is still major progress to be made in molecular superconductors.

One other property we associate with metals is magnetism, particularly the perma-nent magnetism exhibited by iron and some other metals and alloys. Can molecular materials also be permanent magnets? Answering this question requires a look at the origins of magnetism. Ordinary magnetism occurs because electrons behave as tiny magnets; they do so because they are charged particles with spins, as mentioned earlier.

Left to itself, each electron's spin adopts an arbitrary direction – not only up or down as already discussed, but up or down relative to an arbitrary direction. As a result, there is no permanent magnetism, which requires spins, and hence their associated magnets, to adopt the same direction. So the spins must interact in a way that favours alignment.

Two interactions are important. Correlation favours parallel alignment of spins, whereas exchange favours antiparallel alignment. Exchange usually dominates, which is why permanent magnetic materials are rare. Overcoming this problem by design (instead of by the good fortune of finding iron) starts at the level of atoms. The exchange interaction depends on the overlap between the orbitals the electrons occupy. On a given atom, there is zero overlap and hence zero exchange interaction between different orbitals, which therefore tend to accommodate electrons with spins kept parallel by correlation. This conclusion may be modified by the effect of surrounding atoms, but it proves straightforward to obtain metal ions with spins s corresponding to several parallel electron spins, e.g. Mn^{2+} $\left(s=\frac{5}{2}\right)$ and Cr^{3+} $\left(s=\frac{3}{2}\right)$. Transition metals are favoured because they have five d orbitals available. Making molecular magnets applies these ideas to the molecule and to the molecular material.

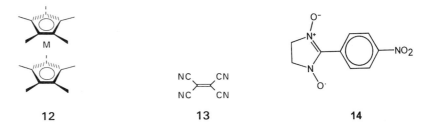

12   **13**   **14**

One kind of building block for molecular magnets is the decamethylmetallocene $[MCp_2^*]$, **12**, where Cp* denotes pentamethylcyclopentadienyl. Varying the metal M varies the spin. This building block is positively charged and so forms complexes with electron acceptors such as TCNQ or its simpler analogue TCNE (tetracyanoethylene, **13**). These become magnetic, but at very low temperatures. Ferromagnetism, the kind of magnetism exhibited by iron, involves spins all aligned in the same direction. An alternative form of permanent magnetism, ferrimagnetism, involves spins aligned in different directions such that there is a net magnetism, as illustrated in Figure 5. This proves easier to achieve, because the exchange energy tends to make adjacent spins line up in opposite directions, so ferromagnetism is favoured if alternate spins are of different sizes, as sketched in Figure 5. This state is achieved by the material $V(TCNE)_x \cdot y(CH_2Cl_2)$, where x is about 2 and y is about $\frac{1}{2}$. Here CH_2Cl_2 (dichloromethane) is the solvent; other solvents can be used. This material is magnetic at room temperature and can be compared in magnetic behaviour to a material containing 1–2% iron. Although metal ions facilitate such work, they are not essential; the purely organic compound p-NPNN (p-nitrophenyl nitronyl oxide, **14**) is ferromagnetic, but only below 1 K.

Figure 5
(a) Ferromagnetism refers to all spins being aligned in the same direction. (b) Ferrimagnetism refers to different sets of spins being aligned in different directions, leaving a net spin.

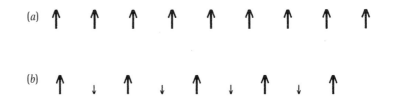

Polymers

I have already described how electronics has changed since my childhood. Over the same period of time, another major technological change has been the growth in the use of synthetic polymeric materials. Polymers are ubiquitous as textiles in clothing and soft furnishing, as structural materials in household goods and vehicles and as financial instruments in credit cards to buy them! They can be made from petroleum products, modified for diverse uses and processed at relatively low cost in energy into fibres, thin films and mouldings. Hence a natural goal has been to develop polymers that combine electronic functions with these mechanical advantages.

An obvious approach is to attach functional pendant groups to a polymer backbone. Carbazole (1) has already been mentioned as a photoconductor, but is not directly suitable for photocopying, for which one requires a large-area photoconductor (the size of the copy) thin enough for photogenerated charge to move through the film before it is trapped. However, poly(vinyl carbazole) (PVK, structure 15 – so abbreviated to avoid confusion with PVC) can be spread as a large-area thin film that retains adequate photoconductivity to be suitable for photocopying. Photocopying technology also uses films of inert polymers incorporating photo-active species not chemically bonded to the polymer chains. Technical developments like this have meant that polymeric photoconductors now dominate photocopying and laser printing.

Nevertheless, more attention has been given to incorporating conducting behaviour into the polymer's backbone. Our previous look at band theory indicates that this might be achieved with a polymer composed of odd-electron units that give a half-filled conduction band. Indeed, this proves to be the case for the polymer $(SN)_x$ (structure 16): it was the first metallic conducting polymer and also the first molecular superconductor, with a T_c of 0.3 K. However, $(SN)_x$ is rather a one-off, affording little scope for the chemists' favourite game of preparing a range of derivatives with diverse properties. In any case, can't we get anywhere with a carbon backbone?

| 15 | 16 |

The simplest sensible building block is perhaps the CH unit. It yields the polymer $(CH)_x$ and, since C_2H_2 is acetylene (its systematic name is ethyne), the polymer is called polyacetylene (PA). Hideki Shirakawa of Tokyo prepared shiny black films of PA by admitting acetylene (ethyne) gas into a flask that was being swirled around to give a film of catalyst solution on its surface. The material is fibrous, air-sensitive and intractable: it may be synthesized in the all-*cis* form 17a but soon reverts to the stable all-*trans* form 17b, especially at high temperatures. Polyacetylene prepared like this isn't much of a conductor, more a semiconductor, with a conductivity of perhaps 10^{-5} S cm^{-1}. With the structure represented as 17b, poor conductivity is not surprising. We don't have a uniform chain of equally spaced CH units but rather alternating bond distances between adjacent CH units. This can be regarded as another instance of the Peierls distortion and it leads to a full valence band and an empty conduction band. In the circumstances, the conductivity isn't too bad, apparently because defects and left-over catalyst act as dopants.

| 17a | 17b |

Very well, let us dope PA deliberately. Alan Heeger and Alan MacDiarmid in Philadelphia showed that doping with AsF_5 increased the conductivity to some 200 S cm^{-1} at room temperature. This doping produces holes because AsF_5 is an avid collector of electrons. Doping with species such as alkali metals that are keen to dispose of electrons reduces the conductivity, apparently by cancelling out the doping effect of the residual impurities. As a result, the conductivity of PA varies over an unprecedentedly wide range – at least 11 orders of magnitude. The fibrous nature of the material makes it quite suitable as an electrode and PA electrodes doped with various organic counter-ions yield a metal-free battery.

Impressive though all this may be, we have not yet secured all the advantages sought from a conducting polymer. In fact, the conjugation that helps to give us conductivity also gives rigidity and reactivity. If we want to process polymers, they are better not conjugated. So instead of making PA at once, we make a 'precursor polymer' that is processable but can be converted into PA once it has been processed. This concept was realized by Jim Feast at Durham using the precursor polymer **18**. The polymer chain is not conjugated, so it is flexible and leaves the polymer colourless. The ring system attached to the backbone is designed to detach itself readily at about 60 °C, leaving PA behind in whatever form the precursor polymer adopted. This has advantages in preparing device configurations, as we shall see. It also allows the precursor to be stretched to several times its original length. This aligns the chains very effectively, leading to aligned PA that can be doped to a conductivity of some 10^5 S cm^{-1}, higher than that of lead and approaching that of copper.

Semiconducting PA prepared via the precursor route can be deposited onto electrode arrays to make polymer diodes and transistors. Richard Friend at Cambridge has demonstrated that these have excellent characteristics; for example, they allow currents 10^5 times larger for a potential difference of 1.5 V in one direction than in the opposite direction (rectification). Such devices are not about to supplant those based on silicon, but they do provide proof of the validity of the concept and give insights into the physical processes occurring. They also lead to further types of device.

Many conjugated polymers in which the conjugation pathway runs through a ring system have been prepared. One such is poly(*para*-phenylene vinylene) (PPV, **19**). This can be prepared via a different sort of precursor, **20**, that is processable because the side group makes it soluble. With Donal Bradley and Andrew Holmes, Richard Friend

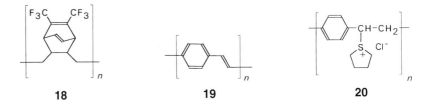

18	**19**	**20**

has shown that PPV affords a light-emitting diode or LED. Electrodes that inject electrons on one side of the film and inject holes on the other are chosen. The electrons and holes combine and annihilate, producing yellowish-green light – a process that is essentially the inverse of photoconductivity. The colour is determined by the polymer's energy levels and can be altered by chemical modification of the polymer such as attaching side groups (often alkoxy groups, RO–) to the phenyl ring or to the vinyl carbons. Moreover, alternating sections of two polymers in a copolymer allows one to tune the colour, which can also be confined to patterned regions by using a mask to control where the conjugated polymer is produced. Hence one produces a light-emitting display.

Rings containing hetero-atoms often feature in conducting polymers, for example poly(pyrrole) **21** and poly(thiophene) **22**. These are conveniently prepared electro-chemically, because the material is deposited onto the electrodes in a conducting form that allows further growth. More complicated, but perhaps inevitably more interesting, is poly(aniline). This involves the nitrogen in the conjugation path and, moreover, exhibits several forms. That shown as structure **23** (where A^- is a suitable anion) is

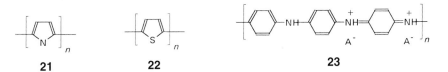

21　　　　**22**　　　　　　**23**

called emeraldine salt, indicating its colourful green appearance. Such materials are finding uses in electromagnetic shielding, to protect sensitive electronic apparatus from stray electrical and magnetic fields, and in batteries, where large electrode areas increase the energy stored.

Langmuir–Blodgett Films

Stretching the polyacetylene precursor polymer has the beneficial effect of increasing the conductivity of the ultimate product, as we have seen, which serves to remind us of the role of molecular arrangement in determining material properties. In crystals, the molecular arrangement is fixed; it may change to another one as the temperature changes and, sometimes, if a different solvent is used to grow the crystal, but in either case there are few variants and we must accept what nature provides. Since we can influence the molecular arrangement in polymers, can we go further in organizing the molecules to improve the electronic properties of the material? Indeed we can, as will appear in scattered examples in later sections of this chapter, but Langmuir–Blodgett films constitute a class of materials of controllable structure and so deserve the present section to themselves.

Oil and water proverbially do not mix. Moreover, substances that dissolve readily in water, termed hydrophilic (water-loving), do not dissolve in oil, whereas those that do not dissolve readily in water, termed hydrophobic (water-fearing) or sometimes lipophilic (fat-loving), do dissolve in oil. Now consider a molecule such as octadecanoic acid (**24**). The acid group $-CO_2H$ is hydrophilic; if we left out the chain of 16 $-CH_2-$ groups we would have ethanoic acid, which happily dissolves in water in the familiar form of vinegar. The long hydrocarbon chain $CH_3(CH_2)_{16}-$, on the other hand, is hydrophobic; if we left out the acid group we would have a greasy oil. Hence octadecanoic acid is termed amphiphilic (both-loving).

$$CH_3(CH_2)_{16}CO_2H$$
24

Now dissolve octadecanoic acid in a volatile solvent like dichloromethane (which both groups *quite* like) and drop a small quantity on a trough of water. The solvent evaporates and the octadecanoic acid molecules have a problem: the hydrophilic heads want to dissolve in the water and the hydrophobic tails definitely do not. Fortunately, both can be reasonably happy if the molecules stay on the water's surface with their heads sticking down into the water and their tails sticking up away from the water, like ducks dabbling. In this way, a surface layer is formed. (Similar molecules and principles produce the action of detergents, whereby hydrophobic tails are attracted to grease particles, leaving hydrophobic heads outside so that the composite particles can be dispersed in the washing water.)

Irving Langmuir, while working in the USA in the 1920s, showed that one can use a moving barrier to compress this floating monolayer quite readily, up to a point. Beyond this point, further compression requires a much higher pressure. This corresponds to a situation in which the hydrophilic heads have been forced adjacent to one another with the tails well out of the water and quite snugly packed together. Later, Langmuir and his associate Katharine Blodgett showed that the compressed floating film can be transferred to a substrate such as a glass slide dipped into the trough, provided that the barrier is simultaneously moved to keep the film compressed. This process yields a *Langmuir–Blodgett film* and can be repeated to yield a multilayer structure. Much more recently, various ingenious arrangements to allow different layers to be deposited alternately or even in more complicated sequences have been devised. This can now be done under automatic computer control, whereby one dials up the number and sequence of layers – a far cry from the manual procedure used by Michio Sugi in Japan when first depositing hundreds of layers: he even needed a bed in the laboratory so that he could rest after hours of turning handles.

Langmuir-Blodgett films are typically dense and highly ordered, though not crystalline. The thickness of each layer is uniform, being determined by the length of the molecule and any tilt in the packing of the molecules in a layer. Multilayers of specified thicknesses can therefore be deposited. Since multilayers of specified sequences can also be deposited, we no longer have to accept the structure nature provides, at least in the direction of the layer stacking.

To use this ability to control structure, we have to build functionality into the film-forming molecules – or, as most people would see it, we have to build amphiphilic behaviour into functional molecules. Clearly, this is the job of the chemist; and a huge job it is. Crystals of anthracene (2) emit blue light after absorbing white light (fluorescence) or when currents of electrons and holes recombine (electroluminescence, as in a light-emitting diode). If this property could be incorporated into a Langmuir– Blodgett film, it would constitute progress towards an electroluminescent display. Then let us furnish anthracene with a hydrocarbon tail and an acid group and let us do so at the middle carbon atoms opposite one another in the central ring, since these are the most reactive. Anthracene itself is hydrophobic, so we should not expect to need as long a chain as in octadecanoic acid, and anthracene is also rather bulky, so we should not expect to obtain the right amphiphilic balance if we attach the acid group directly to the ring. This still leaves plenty of scope, but the right sort of balance is obtained with a hydrophobic chain four carbons long and a chain two carbons long joining the ring to the acid group (structure 25), which yields Langmuir-Blodgett films that exhibit essentially the luminescence of anthracene. Similar sorts of procedures lead to Langmuir-Blodgett films with unusual optical properties, as we shall see on pp. 396–8.

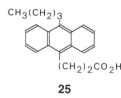

$CH_3(CH_2)_3$

$(CH_2)_2CO_2H$

25

In practice, Langmuir–Blodgett films are often not very stable, especially when they are functionalized as just described. This is not really surprising: by depositing amphiphilic molecules at the air–water interface, we force them to adopt the least bad arrangement; by compressing the film, we squash them together; and by depositing successive layers, we give them little choice of arrangement. If the resulting structure is not too different from the crystal structure that the molecules choose when left to their own devices, then it might not be worth the molecules' while trying to rearrange. However, if the film and crystal structures are greatly different (and if they are not much different, why bother with the Langmuir–Blodgett film?), then the film will wriggle into a more comfortable arrangement when it is left around. A layer you think has been deposited safely onto the substrate may even sneakily rearrange while you are depositing the next layer.

If we want to control the structure, such contumacious behaviour cannot be allowed. It can be stopped by incorporating groups that allow them to be polymerized into the amphiphiles, say by using light or irradiation. Double or triple bonds in the hydrocarbon chain provide simple polymerizable groups, although these reduce the flexibility of the chain that is useful in deposition. Films formed from ω-tricosenoic acid (26) deposit very well and are readily polymerized. The ω denotes that the double bond is at the far end of the chain from the acid group, where it interferes minimally with the packing in the layer. This material can be polymerized by an electron beam, which can be rastered over the surface like that inside a television tube. This process can be controlled to produce patterning of the film consisting of polymerized areas in a monomer film. Since the monomer but not the polymer can be dissolved away, this process is of potential use in making increasingly small-scale conventional electronic circuits, aided by the uniformity and thinness of the Langmuir–Blodgett film, which acts as a *resist*.

A more complicated polymerizable group is diacetylene (diethyne) (27). Single crystals of dialkynes with suitably bulky side groups R and R′ can sometimes be

$$H_2C=CH(CH_2)_{20}CO_2H$$

26

$$R-C\equiv C-C\equiv C-R'$$

27

polymerized to give single crystals of polydiacetylene directly. This indicates that polymerization can occur with rather little disruption of the lattice. If we are trying to stabilize a Langmuir– Blodgett film, we certainly want to avoid disrupting it in the process, so incorporating a diacetylene group looks like a promising strategy, which indeed proves successful in yielding easily polymerizable films. In principle, there is a choice between polymerizing each layer on the trough before depositing it or on the substrate after having deposited it. In practice, however, polymerized layers are usually not flexible enough to deposit well.

However, Gerhard Wegner in Mainz has obtained a very versatile class of polymer films. The molecules have no hydrophilic groups, but consist of a rigid rod-like polymer core to which are attached short flexible hydrocarbon chains, yielding what are referred to as 'hairy-rod' polymers. The side chains are liquid-like and it is rather as though each rod carries with it its own shell of solvent, which helps to overcome the rigidity problems encountered previously. Once they have been deposited onto a water surface in the usual way, the hydrophobic hairy rods lie parallel to the surface, without the need for a hydrophilic group. On compression, the surface film forms a liquid monolayer with no preferred overall orientation of the rods within it. Transfer to a substrate is achieved in the usual way by dipping and, because the layer is fluid, it is able to flow and become oriented by the dipping process itself. Thus an ordered molecular film is produced, but not in the conventional way; and the orientation is within the layers rather than across them.

Several types of hairy-rod polymer have been designed. One is based on phthalocyaninato(polysiloxane), the repeating unit of which is shown as structure **28**. The flat phthalocyanine ring has a silicon at its centre joined to silicons in adjacent rings through Si–O bonds to give a 'shish–kebab' polymer, with the siloxane chain as the skewer and the parallel rings as the chunks of meat. The side chains R and R′ that make the rod hairy can be the same, but the necessary liquid-like disorder is enhanced if they are of different lengths, for example CH_3 and C_8H_{17}.

These materials form films of excellent quality. The disordered side chains screen the

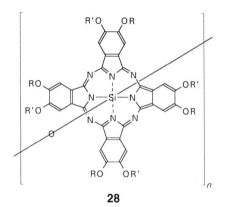

28

rods from direct contact and so obviate the problem of competing structures with different orderings. Multilayers are readily formed and may as usual contain sequences of different molecules. The side chains may contain polymerizable groups that can serve to stabilize the film or allow it to be used as a resist. The solvent-like side-chain environment is also well suited for dispersing active molecules that are not readily derivatized to form Langmuir–Blodgett films of their own. If the active molecules complex selectively with species such as sodium and hydrogen ions, such materials can act as ion-selective sensors.

Langmuir–Blodgett films, then, constitute a broad class of materials that can have a wide range of properties – another example is the incorporation of TTF and TCNQ groups to give charge transfer and conductivity. They also represent a step towards self-assembly of materials in ways determined by how we design the molecules. Making the molecules amphiphilic makes them assemble spontaneously at the air–water interface. After that, however, the assembly is done mechanically, by the compression and dipping.

Liquid Crystals

Calculators and tiny televisions usually contain them, watches and portable computers often do. They occur in telephones, in car instruments and in fridges. What are they? They are liquid crystals, which are now finding widespread use in displays of all sorts. The materials themselves have been known for a century or more, but it is only for the last 20 years that their properties have been exploited in electronics. In no small measure has this development relied on the skills of chemists.

Liquid crystals result from molecules that have conflicting desires about whether to melt, much as Langmuir–Blodgett films result from molecules that have conflicting desires about whether to dissolve. As the name suggests, liquid crystals are materials that have some properties characteristic of liquids – they flow – and some characteristics of crystals – they have different properties in different directions. This behaviour arises because of the way the molecules arrange themselves; and they arrange themselves in special ways because of their special shapes.

Liquid-crystalline behaviour can be produced in one of two ways. One is to change the concentration of the active molecule in a solution. Such liquid crystals are called *lyotropic* (changing with solvent). They are important practically, for example as structured liquids in household cleaning products, but not for their electronic behaviour and so we concentrate on the other class of liquid crystals. These are produced by a change of temperature and so are called *thermotropic* (changing with heat).

Thermotropic liquid crystals become ordinary liquids at high enough temperatures and ordinary crystals at low enough temperatures. They therefore constitute a phase between the liquid and crystalline phases, or a *mesophase*. Consequently the molecules that can generate such a phase are called *mesogens*. What is special about the shape of a molecule that makes it a mesogen is that it is very elongated (like a ruler) or very squashed (like a coin). In a crystal, molecules are fixed in position and in direction, whereas in a liquid they are fixed in neither. Because of their exaggerated shapes, mesogens are much easier to fix in direction than to fix in position. So a liquid crystal flows because the molecules are not fixed in position, but it has different properties in different directions because the average direction of the molecules remains fixed. Individual molecules are not fixed in direction – after all, they can flow past one another – but a series of snapshots would show that the molecules tend to point in the same specific direction. This can occur in several characteristic ways.

Ruler-shaped molecules produce *calamitic* liquid-crystal phases, named after the horse-tail plant so persistent in gardens. These come in three groups. The *nematics* are the simplest: they just tend on average to point along what is known as the director, or equally, in the opposite direction, but not at right angles to this direction. The *cholesterics* are a variant of nematics that occur when the mesogens are chiral, that is, they differ from their mirror images like a right hand from a left hand. As a result, they cannot quite pack to form a nematic but instead form a structure that looks like a nematic in a small region but in which the director follows a helix. The molecules are thus arranged rather like the stairs on a spiral staircase. Finally, the *smectics* have some degree of order of position as well as of direction. Thus the molecules not only tend to point along a particular direction but also tend to lie within layers. This is rather like a Langmuir–Blodgett film, except that the molecules are still in a fluid wherein they can move around, albeit sometimes with difficulty.

Coin-shaped molecules produce *discotic* liquid-crystal phases and these too come in several varieties. There are *nematic discotics*, in which the planes of the molecules tend to lie across the director, or, equivalently, the short axes of the molecules tend to lie along the director. This is like a heap of coins. There are also *columnar discotics*, which resemble smectics in having some positional order, but now this consists of a tendency for the molecules to stack in columns that then pack together, like piles of poker chips.

An example of a calamitic mesogen is 4-*n*-pentyl-4′-cyanobiphenyl (5-CB, **29**). This is fairly typical in having a flexible fairly short tail, a reasonably rigid elongated core and a polar group at one end that helps the ordering. Obviously there is great scope for changing these building blocks to design new materials. Discotic mesogens, on the other hand, need an extended flat core surrounded by flexible fairly long tails. An example here is provided by the hexa-esters of 2,3,6,7,10,11-hexahydroxyltriphenylene (HHTP, **30**), where R may be alkyl, alkylphenyl or alkylcyclohexyl, the alkyl chain being often about eight carbon atoms long.

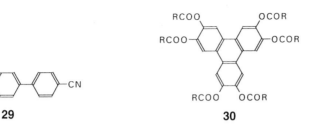

29   **30**

Various properties of liquid crystals depend on direction, but perhaps the most easily studied and useful property of a liquid crystal is its refractive index. This determines by how much light is slowed down while passing through the crystal. A light wave consists of an oscillating electrical field and an oscillating magnetic field that are at right angles to one another and to the direction in which the light wave is travelling. The direction of the electrical field is called the polarization of the light wave. Usually light consists of all possible polarizations – this is true of the natural sunlight or artificial lamp by which you are reading this book, for instance. However, it is possible to use a polarizer to select just one direction. If you imagine holding one end of a long skipping rope fixed at the other end, you can send a wave along by shaking the end you are holding. The direction in which you shake corresponds to the polarization. Now suppose that the rope passes through a set of railings on its way from the fixed end to your hand. You can send a wave along the rope only by shaking up and down; if you try any other direction, the rope hits the railings and the wave gets broken up. This is like a polarizer, letting through only a set polarization.

If we take the simplest case of an ordinary nematic liquid crystal, light polarized along the director has a different refractive index from that of light polarized across the director. In fact, the light polarized along the director has the higher refractive index. As we know, an electrical field tries to move charges. In a light wave detectable by our eyes, the oscillation of the electrical field occurs perhaps a hundred million million times a second. This is so fast that only the lightest charges can move at all, which means only the electrons. Now recall that the electrons are confined in molecules shaped like rulers that tend to lie along the director. This means that the electrons can be moved much further by an electrical field along the director than they can by one across it and this interaction slows down light polarized along the director much more than it does that polarized across it.

In these cases in which the light is polarized exactly along or across the director, nothing happens to the polarization. That means that, if we observe the light coming out of the liquid crystal through a second polarizer set to pass light polarized at right angles to that passed by the first one (an arrangement called crossed polarizers), we shall see nothing, just darkness. The light reaching the second polarizer (or analyser) is polarized at right angles to that which can pass through. What if the light is polarized in some intermediate direction? It is subjected to slowing at different rates along and across the director, which causes the polarization to be rotated. The situation is rather like that when the nearside wheels of a car suddenly run into a deep puddle and are slowed relative to the offside wheels, which tends to make the car swerve towards the kerb while you struggle to steer straight. In this case, then, the initial polarization is modified to contain a contribution from the opposite polarization. This does pass through the analyser, so we now see a bright field.

This phenomenon can be used to study liquid crystals. If a crystal melts but the liquid still gives a bright field through crossed polarizers (typically in a polarizing microscope), then you have a liquid crystal. If you keep heating, the temperature at which the bright field goes dark is that at which the liquid crystal has changed into the normal liquid. In between these temperatures, the bright field is not uniform but often displays beautiful patterns known as textures. These arise from regions where the liquid crystal's structure points in different directions and they can be used to identify different types of liquid crystal. For example, nematic means thread-like, after the characteristic texture seen for nematics. (Smectic, on the other hand, just means soapy, from the feel of the liquid.)

Most of the uses of liquid crystals in displays that I listed at the beginning of this section rely on this phenomenon, plus the fact that liquid crystals can be aligned in a controllable way. So once again we encounter the idea of controlling structure to control properties. For liquid-crystal displays, the alignment has first to be made uniform and then to be changed in a defined manner to convey information.

Uniform alignment is achieved by treating the surfaces in contact with the liquid crystal. For a display that relies on light passing through the liquid crystals, a thin layer saves too much light being absorbed. Since a display has to be of a reasonable area to be seen easily, using a thin layer also keeps the volume of expensive material required down. So displays usually consist of a layer of liquid crystal as little as 10 μm thick (ten millionths of a metre – the diameter of a fine hair) sandwiched between two glass plates. At the plates the alignment of the liquid crystal is fixed by polishing them, to produce parallel fine grooves, or by attaching a surface layer to them, to produce a specific chemical environment. These procedures can serve to align the molecules either flat along the plate like a raft of logs on a river – one can imagine the elongated molecules preferring to lie along the grooves in a polished surface, for instance – or else sticking up perpendicular to the plate like a forest of fir trees. Once the molecules next to the surface have been lined up, the others through the bulk of the liquid crystal follow by the usual processes of responding to each other's shape.

Changing this alignment is achieved by applying an electrical field (a magnetic field will also do, but is much less convenient to arrange). For definiteness, consider ruler-shaped molecules. These are helped to form a liquid-crystal phase by having an electrical dipole moment. This occurs when the centres of positive and negative charge do not coincide. They do not coincide because some atoms are successful at attracting more than their fair share of electrons: a fair share would leave them electrically neutral, but they end up negatively charged owing to the extra electrons, leaving other atoms short of electrons and hence positively charged because they can't compete successfully for them. The dipole moment has a direction, along the line between the centres of charge, and this direction tends to lie along the ruler shape or across it – seldom in between. For example, in 5-CB (**29**) the nitrogen atom gets the lion's share of the electrons and the resulting dipole moment lies along the axis of the molecule. On the other hand, in *para*-azoxyanisole (PAA, **31**) the formal charges shown on nitrogen and oxygen indicate a dipole moment across the axis. By incorporating various structural motifs like this, chemists can synthesize mesogens with either direction of dipole moment.

$$CH_3O-\!\!\!\bigcirc\!\!\!-N\!\!=\!\!\overset{O^-}{\underset{+}{N}}\!\!-\!\!\!\bigcirc\!\!\!-OCH_3$$

31

The significance of the dipole moment is that it provides a 'handle' by which an electrical field can turn a molecule. Consider a dipole moment placed between two oppositely charged electrodes, as in Figure 6. The end of the molecule with extra electrons is attracted towards the positively charged electrode, while the other end, being short of electrons, is attracted towards the negatively charged electrode. Overall, the result is a tendency to rotate the dipole moment to lie along the electrical field, pointing directly from one electrode to the other. For a sufficiently strong electrical field, the molecules all snap into alignment like soldiers coming to attention. The electrical field increases with increasing voltage across the electrodes and with decreasing separation between them. This is another reason for having a very thin layer of liquid crystal: it brings the threshold voltage required to produce the alignment into the range supplied by the familiar 'button' batteries.

Like other properties of liquid crystals, the threshold voltage depends on the direction in which it is applied. If it is applied along the director of a material in which the dipole moments lie along the axes of the molecules, alignment is relatively easy – after all, the molecules already tend to lie along the director. If the voltage is applied across the director, alignment is much harder – the molecules have to be moved in such a way as to make them point at right angles to their original preferred direction and, being elongated, they tend to get in one another's way while turning round. If the dipole moments lie across the axes of the molecules, the easy and hard directions swap round:

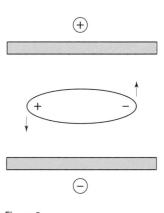

Figure 6
The electrical field produced by two electrodes (top and bottom) tends to rotate a dipolar molecule to align it along the field.

now a voltage across the director reinforces the existing alignment (easy, low threshold) whereas one along the director opposes it (hard, high threshold).

These properties are exploited in displays in various ingenious ways. The most common is the twisted nematic display. As the name suggests, it uses a nematic liquid crystal. This is enclosed between glass electrodes treated to make the director align along the surface, but the direction of alignment at the top electrode is twisted through a right angle from that at the bottom electrode. So the molecules tend to point east–west, say, at the bottom, but north–south at the top. In between, the director gradually twists round from one direction to another, following a sort of spiral staircase arrangement. What is produced in this way is in fact an artificial cholesteric structure whereby we create the twist instead of the molecular shape doing so.

Now shine light on to the assembly through a polarizer arranged to transmit light polarized along the director at the bottom electrode. As the light travels through the liquid crystal, its polarization rotates with the director. By the time it reaches the top electrode, it is polarized at right angles to its original direction and so passes through a top polarizer crossed relative to the bottom one. This gives a bright field.

The bright field can be switched off by applying a voltage across the electrodes. This requires a conducting material transparent to light, typically indium tin oxide. The molecules are designed with dipole moments along the axis, which we remember tends to lie along the electrodes (though of course in directions at right angles to one another at the top and the bottom). Applying a large enough voltage produces a field between the electrodes that realigns the molecules to point from one electrode to the other (except right at the electrodes, where the surface treatment keeps them lying along the surface). This effectively produces an ordinary nematic, which affects the light as already discussed, i.e. not at all. Since the light is polarized across the director, it stays polarized that way. When it reaches the top polarizer, it is polarized the wrong way to pass through, so a dark field is produced.

This is the effect you can probably see in your watch or calculator. The background is bright and the numbers and symbols are dark. The display consists of a number of regions (typically seven for a numeral, for example) that can be subjected to separate voltages to control their appearance. The voltages in turn are controlled by the microelectronic circuits inside the device. Liquid-crystal televisions and computer displays use similar principles, with colour filters and special methods of getting information to each element of the picture. They also use 'supertwist' technology, in which the top and bottom electrodes are twisted by more than one right angle to improve visual contrast.

Reaching practical devices like this does require developments additional to those already outlined. For example, it is convenient to look *at* a watch display rather than *through* it. Putting a mirror at the back of the assembly to reflect the light to the observer readily achieves this. One advantage of such displays is that they use the light that is already available. The battery then provides power only to rearrange the liquid crystal; it does not have to generate light. Light-emitting displays are very power-hungry, as owners of the earliest digital watches discovered from the rate at which batteries had to be changed.

Most of the developments required for practical devices like this fall outside the realm of chemistry. However, one development essential for the initial commercial success of liquid-crystal displays does not. Though liquid crystals had been known for many decades, they formed at temperatures well above 100 °C. This is not convenient for the owner of a wristwatch. . . . One would like a watch to work in the depths of winter in Edmonton, Alberta, at perhaps −40 °C, and at the height of summer in Luxor, Egypt, at +40 °C. To lower and broaden the range of temperatures at which liquid crystals exist has required several approaches to be pursued. One is the preparation of new families of mesogen, such as 5-CB (28). Another is the systematic study of the various members of the family, for example the various cyanobiphenyls like 5-CB but with different chain lengths. These approaches indicated that the target was achievable. Actually achieving it required the judicious use of mixtures of mesogens.

Mixtures tend to be more stable than their separate components. For example, salty water freezes at a lower temperature and boils at a higher temperature than does fresh water. This is why ponds freeze before the sea does and why we sprinkle salt on icy paths. Similarly, additives are put into the cooling systems of cars to act as anti-freeze. Much the same happens when mesogens are mixed: the mixture yields a liquid crystal over a wider temperature range than would either mesogen separately. Detailed quantitative study was required to elucidate the physical principles and devise mixtures suitable for displays; modern systems use complicated mixtures of as many as four components.

Liquid-crystal displays are ubiquitous, as already noted. Current research could serve to make them even more so, by developing materials that switch in an electrical field and stay switched when it is turned off, hence saving power. Two kinds of liquid crystal offer particular promise in this area. One is the ferroelectrical liquid crystal. This occurs in some smectic phases made of chiral molecules. An electrical field causes the molecular dipole moments to swing round from one direction to another. They then stay there when the field is switched off until a field is applied to reverse the process. The two sets of directions for the dipole moments change the liquid crystal's optical behaviour so that it can be used for a display in much the sort of way already described.

The other kind of liquid crystal is one that is also a polymer. One type has mesogenic groups in the main chain separated by flexible chain segments that allow the groups to align with neighbours in the same chain (like a concertina) or in adjacent chains. The other more common type has the mesogenic pendant groups hanging from the main chain, which again needs enough flexibility to allow the groups to align with their neighbours along the backbone, this time like a litter of puppies being nursed. Because the chain's flexibility is important, these materials form liquid-crystal phases only above the glass-transition temperature at which the polymer softens. This allows us to align a polymer liquid crystal at a high temperature and then to fix the alignment by cooling the polymer below the glass-transition temperature, so storing the information contained in the alignment. One way of achieving this is to hold the glassy polymer in an electrical field strong enough to align the liquid crystal and then to write information onto the polymer with a laser beam. Where the laser beam touches the polymer, it heats it above the glass-transition temperature and the electrical field aligns the liquid crystal. When the polymer cools after the laser beam has passed on, the alignment is retained in a glassy disordered matrix and can be viewed through crossed polarizers. In this way, information can be stored, much as on a microfilm, for years. It can be erased completely by heating or partly by a laser beam without an electrical field.

Liquid crystals illustrate very well the processes of molecular recognition and self-assembly. Mesogenic behaviour is well understood and new families of materials can fairly readily be devised. Building in suitable optical and electrical behaviour is also feasible and practical problems of producing devices can also be overcome. Thus liquid crystals offer a perspective of the possible scope for other molecular electronic materials.

Photonics

Electronics has proved enormously powerful and successful, although, as we have seen, that fact doesn't hinder attempts to make it even more so. However, it has known limitations: for example, stray electrical or magnetic fields can interfere with electrical currents carrying information. The information that can be carried in a current is also limited by how fast the current can be switched on and off. Electrons don't weigh much, but even the little they do weigh makes them resist switching. Both these problems are avoided, or at least much reduced, if we use photons, the elementary quantities of light, instead of electrons. Having no charge, they are not directly affected by fields and, having no rest mass, they can be switched very rapidly. The analogue of electronics is then sometimes termed *photonics*.

The advantages of photonics can be realized without molecular materials. For example, we already find optical glass fibres transmitting information around telephone and computer networks. Nevertheless, molecular materials are promising for at least some uses in photonics (though the optical fibre looks safe from competition for some time). Liquid-crystal devices exemplify this promise. We saw that they can interact with light to display or store information. Molecular materials have also been developed to interact with light in the ancient area of dyes and pigments, both natural and, more recently, synthetic. The skills used to determine hue and interaction with fabric ought to be capable of being adapted for other kinds of interaction with light.

Basically, there are two classes of behaviour that can be used in photonics: using materials to control light in some way and using light to control materials in some way. The former class would include the twisted nematic display, for which we use an electrical field to control the material. The latter would include writing information by laser in a polymeric liquid crystal, although the effect of the laser is primarily thermal rather than optical. We shall look at the first class here and the second in the next section.

The first method of controlling light is simply guiding it. This can be achieved by surrounding a material of a high refractive index by one of a lower refractive index. When light in the high-index material strikes the boundary with the low-index material, it is bent towards the high-index material. This is rather like the effect in a liquid crystal to which we have already referred, whereby the plane of polarization is affected by different refractive indices; again, the difference of speeds causes a slewing round. If the light strikes the boundary nearly straight on, it escapes, but if it does so at an oblique angle (which depends on the difference between the refractive indices), it is bent so far that it is bent back into the high-index material. It is then not just bent (refracted) but reflected, by what is called total internal reflection. This phenomenon is used in optical fibres and in the prisms found in cameras and binoculars. Molecular materials are not especially predisposed to guiding light in this way, but they are certainly capable of it. This is important for the possibility of developing integrated optics (like integrated electronics) in molecular materials, such that the difference of refractive index between a monomer and its polymer, say, could be used to guide light between active molecular elements of the sort we shall discuss shortly. It should also be noted that there are polymers of excellent optical quality used in light-weight spectacle lenses.

More interesting and useful things can be done with *non-linear* optical materials. Normally, if you shine light of a single colour onto a material it will be transmitted to a greater or lesser extent and if you make the light twice as bright then what is transmitted will be twice as bright. However, if the material is non-linear, some of the light that is transmitted will be a different colour and this will happen increasingly as you make the light brighter. Non-linear materials also have a refractive index that depends on the intensity of the light and, sometimes, on an applied electrical field. As we saw, the refractive index is used to guide light, so, if we can change the refractive index, we can switch the light from one direction to another.

Molecular non-linear optical materials offer a number of advantages. One is that we can modify what colours of light they absorb, by using the knowledge developed for dyestuffs. This ability is important because it is not much use changing the colour of light in a non-linear material, perhaps in order to transmit information faster or with less absorption in an optical fibre, only to have the material promptly absorb the light it has obligingly produced. Molecular materials also have highly non-linear responses in many cases, with good resistance to damage from laser light. This allows reasonably efficient photonic devices to be foreseen. Finally, for molecular materials the non-linear response comes mainly from that of the molecules, which in turn comes from that of the electrons. Since electrons are very light, this means that the response is very fast compared with that of inorganic materials (of which one of the best is lithium niobate, $LiNbO_3$), for which the response comes mainly from the much heavier and hence more sluggish nuclei.

Designing molecular materials for non-linear optics thus means designing molecules with highly non-linear responses. (It also turns out to mean designing molecules that can be arranged in suitable ways, as we shall see later.) Since it is the electrons that are going to make the response, there have to be electrons readily available, somewhere for them to move to and some way of getting there. This can be achieved in molecules that contain a donor group that likes to give up electrons, an acceptor group that likes to collect them and a conjugated pathway linking the two. One could regard this design as a one-molecule version of the organic metals. A molecule embodying these features in a simple way is *para*-nitroaniline (*p*NA, **32a**). The $-NH_2$ group is the donor, the $-NO_2$ group is the acceptor and the benzene ring can transmit the charge between them. One guide to the ease of this charge transfer is that complete charge transfer gives a structure obeying all the usual rules of bonding (**32b**). Measurements in solution confirm that *p*NA is highly non-linear, whereas its isomer *meta*-nitroaniline (*m*NA, **33**) is less so, because the charge cannot move so far and in this arrangement the benzene ring is less obliging about transmitting it.

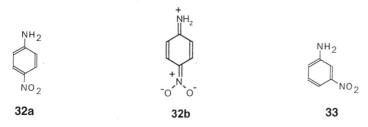

32a   32b   33

One of the non-linear optical properties already mentioned is that some of the light produced may be of a different colour from that of the incident light. The simplest case is when two identical photons pool their energy to give one of double the energy. This means that the new output light has twice the frequency of the input light (which corresponds to having half the wavelength). By analogy with music, this is called the second harmonic and the phenomenon is called *second-harmonic generation* (SHG). For example, two infrared photons produced by a laser at a wavelength of 1 μm can produce a green photon with a wavelength of 0.5 μm. However, this can happen only if the material lacks a centre of symmetry. Molecules of the sort we have been discussing already lack a centre of symmetry – they have a clear built-in direction corresponding to the direction of charge transfer. Materials for SHG must contain these molecules arranged so that molecules pointing one way are not balanced by an equal number of molecules pointing exactly the opposite way, which would in effect cause the second-harmonic light produced by one set to be cancelled out by the second set. Materials must also lack a centre of symmetry in order to exhibit the *linear electro-optical effect*, in which the refractive indices change in proportion to an applied electrical field.

Unfortunately, *p*NA forms crystals with a centre of symmetry and hence is no use for SHG. On the other hand, *m*NA crystals lack a centre of symmetry and are found to produce quite strong SHG. So *m*NA adopts the right arrangement, but *p*NA would be better in the same arrangement because it has a higher molecular response. How can we get the best of both worlds? We need the donor and the acceptor opposite one another as in *p*NA, but we also need to disrupt the fairly 'streamlined' structure to something more awkward such that as in *m*NA. These arguments led to the synthesis of the molecule 2-methyl-4-nitroaniline (MNA, **34**). This duly lacks a centre of symmetry and exhibits strong SHG, assisted by an increase in molecular response caused by the methyl group originally introduced merely for its shape or steric effect.

Another strategy adopted to avoid having a centre of symmetry is to make the whole molecule chiral with all molecules having the same handedness. There is then no way of packing them with a centre of symmetry, as can be verified by experimenting with a pair of right-foot shoes, say. Start them alongside one another on their soles and then turn one to point in the opposite direction. They are still both on their soles, so turn one over – but now the curves of the insteps go in the same direction. . . . One successful product of this design strategy is the molecule *N*-(4-nitrophenyl)-prolinol (NPP, **35**; the star marks the chiral centre), which can be synthesized from the amino acid

34   **35**

proline, which is available naturally in pure chiral form. Joseph Zyss and his colleagues at the CNET, the French telecommunications laboratory, have extensively studied NPP.

Crystals are fine for studying non-linear optics and sizable crystals are essential for some applications. Growing such crystals requires a considerable investment of effort in optimizing the conditions and of time waiting while they grow (since rapid growth gives crystals of poor optical quality). So, when crystals are not required, other materials are of potential interest. Since the molecular arrangement is the problem for obtaining SHG, the Langmuir–Blodgett technique suggests itself as a method of controlling structure. Now the commonest and usually the most stable way of depositing Langmuir–Blodgett films is for alternate layers to be deposited pointing up and down. If the molecules are perpendicular to the substrate, each pair of layers has a centre of symmetry and SHG is not possible; if they tilt, the centre is lost but SHG is possible only along the film, which is not the most useful direction.

This problem can be overcome by using a double trough able to deposit alternate layers of different molecules. The two sorts of molecule are then designed so that one sort has the donor at the hydrophilic head and the other sort has the acceptor at that end. This makes the directions of charge transfer in the two molecules opposite, so that, when the molecules are deposited in opposite directions, the directions of charge transfer are the same. In this way their effects are made to reinforce rather than cancel out and useful SHG can result.

A pair of molecules designed for this purpose consists of the hemicyanine **36** and the stilbene **37**. Their structures are both drawn with the electron-accepting group at the lower end; for the hemicyanine this is the hydrophobic end, but for the stilbene it is the hydrophilic end, so that alternating layers deposit with the molecules as they are drawn, whereupon their responses reinforce each other.

Making both molecules in an alternating layer contribute significantly to the SHG is ideal. Still quite satisfactory is making one molecule contribute while the other is essentially inactive. An example of this approach is provided by the hemicyanine **38**

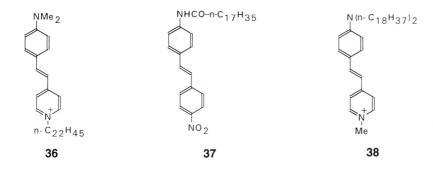

36   **37**   **38**

and the dimeric species **39**. These molecules are thought to form good films because the long chains pack neatly in between one another instead of just end to end. As always, polymeric films are attractive for their extra mechanical strength and another example of an alternating-layer film uses the hemicyanine polymer **40** with the polymeric spacer **41**.

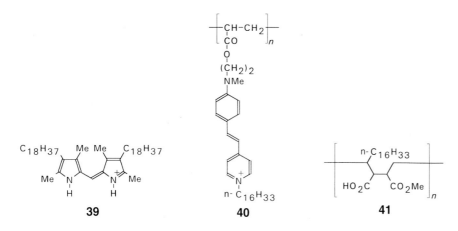

39   **40**   **41**

In special cases, molecules of a single species can be persuaded to deposit pointing in the same direction in every layer. If the molecules have a permanent separation of charge, the positive end of one molecule will be strongly encouraged to line up with the negative end of the next molecule, so that successive layers all line up nose to tail. A molecule of this sort is shown as structure **42**.

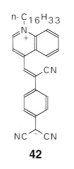

42

Polymeric Langmuir–Blodgett films have the advantage of high stability. However, designing a non-linear-optics (NLO)-active polymeric Langmuir–Blodgett film poses three chemical challenges: making a NLO-active group, making it amphiphilic so that it can deposit as a Langmuir–Blodgett film and making it into a polymer sufficiently flexible that it *will* deposit as a Langmuir–Blodgett film. This prompts the thought that perhaps the advantages of using a polymer might be acquired more directly. A glassy polymer can have excellent optical properties, but also has a centre of symmetry, so once again the problem is that of how to remove it. A process known as poling achieves this.

A polymer is made active by attaching NLO-active side groups to it or else by mixing NLO-active molecules with it, say by melting the two components together. As we have seen, the NLO-active groups or molecules must have strong charge transfer. This in turn means that they have large dipole moments, so that they tend to turn to point along an electrical field (like the molecules in a liquid crystal). Hence the polymer is heated to soften it enough for the NLO-active groups to move fairly freely. While it is soft, a high electrical field is applied to line up the NLO-active groups via their dipole moments. With the field still being applied, the polymer is cooled, so freezing in place the order produced by the field, which can then be removed. This process readily lends itself to large-scale production and so is attracting commercial interest. It remains a challenge to incorporate enough NLO-active groups and to achieve a high enough degree of order to make the polymer sufficiently active. It is also important that the order produced by poling not be lost when the polymer is left at room temperature.

Processing Information by Changing Colour

Having seen how materials can control light, we now turn to how light can control materials by changing their colour and so process information. In fact, light is a potent reagent. A photon can contain just the right amount of energy to make a chemical change proceed, for example in polymerization. To process optical information, we need light to change materials in a detectable way. While we are using light anyway, an obviously useful change is in the colour of a material. Changing the colour of a material by light is called *photo-chromism*. This can be observed in the bleaching of clothes exposed to the sun for long periods. Jeans fade, in what may be a fashionably desirable way, but not underneath the belt used to hold them up. Thus information on the presence of the belt is stored in the pattern of fading. Of course, this is the same idea as a contact print in photography and photography is an excellent way of storing information. The detail in a photograph is limited by the size of the silver grains, but, if we use a molecular material, individual molecules are the limit, at least in principle.

A molecular photo-chromic material should have the following properties. It must absorb light at one wavelength that changes its structure in such a way that it will absorb light at another wavelength. In this way information is written at the first wavelength and read at the second. The two structures will each absorb over a range of wavelengths and these wavelength ranges must not overlap. Otherwise, when we read a region in the original 'colourless' state we run the risk of changing it to the 'coloured' state and so destroying the stored information. This requirement is illustrated in Figure 7. The coloured structure must also stay coloured for a long time without spontaneously reverting to the colourless form, which would again destroy information. It is useful, however, if we can make the coloured structure revert to colourless (perhaps also optically) so that we can erase, correct or update the information. In particular, this allows prompt correction of errors because photo-chromic materials do not require a lengthy development process like that used in photography. Finally, this cycle from colourless to coloured and back again must be free of fatigue: it must really regenerate the original material each time, just as responsive as before.

These are demanding criteria, but they translate into challenges for chemical design and synthesis. One class of materials that shows how to meet these challenges is the *fulgides* (from the Latin verb 'to shine'). Their colourless and coloured forms are related by making and breaking a six-membered ring, as shown in structures **43a** and **43b**. This class of reaction is governed by rules deduced by R. B. Woodward and R. Hoffman,

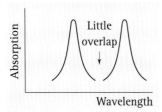

Figure 7

For a photo-chromic material to be useful, the absorption of the colourless form must not overlap with that of the coloured form.

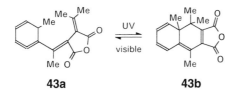

43a   **43b**

who won Nobel Prizes on separate occasions. These rules prescribe the way groups of atoms twist during the reaction, depending on whether it is produced by light or by heat. In the fulgides, the three methyl groups at the top of the molecule physically obstruct the twist required in the thermal reaction, but not that required in the photo-chromic reaction. As a result, these compounds are thermally very stable in either form unless changed deliberately by light. Their colours can be controlled in the now-familiar way by introducing suitable substituents at other positions and by replacing the oxygen atom in the five-membered rings by other atoms or groups such as S and NH. The same concepts can also be used to develop similar molecules whose thermal reaction is *not* hindered, so that they stay coloured under illumination but promptly revert to colourless when the light level falls – just what is required for photo-chromic sunglasses.

Electro-chromism, in which the colour of a material changes with an applied electrical potential able to add or remove electrons (reduce or oxidize the material), is also met. Molecular electro-chromic materials are not common, but the concept has been combined ingeniously with photo-chromism. The azobenzenes, the top structures shown as **44a** and **44b**, readily interconvert optically between the *trans* (Latin 'across') or *E* (German *entgegen*, 'contrary to') form with the phenyl groups on opposite sides of the N=N bond and the *cis* (Latin 'on this side') or *Z* (German *zusammen*, 'together') form with the phenyl groups on the same side. Unfortunately for information storage, azobenzenes also readily interconvert thermally. However, this can be prevented by applying an electrical potential that reduces the N=N bond to a single N−N bond with the addition of two hydrogen atoms. The resulting derivative **44c** has a different colour

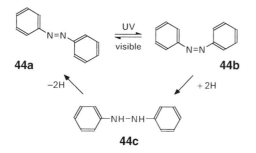

and is stable with respect to heat, thus storing the information. Applying the reverse potential removes the additional hydrogen atoms, erasing the information and directly regenerating the starting material **44a**.

Optical information storage like this is convenient in many ways: it is rapid, it can be used to store whole pictures at once and it is limited by the molecular size rather than some grain size. In fact, though, something else intervenes to prevent information storage at the level of individual molecules. Light exhibits wave properties and its most useful wavelengths for information storage lie in or near the visible region around 0.5 μm. This makes a beam of light fuzzy on the same scale, so that two spots of light blur into one another unless they are at least one wavelength apart (the diffraction limit). As a result, regions storing different pieces of information need to be

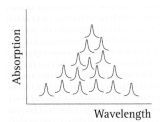

Figure 8

Narrow optical absorption spectra of individual molecules dispersed in a glass give a broad-band total absorption spectrum.

at least this far apart, equivalent to hundreds of molecular diameters. Hence the potential advantage of molecular resolution is lost.

Rather surprisingly, dispersing the photo-chromic molecules in a disordered medium, i.e. a glass, can circumvent this problem. Each molecule now has its own absorption spectrum, like that shown in Figure 7 but modified by the glassy environment. Because the glass is disordered, it provides a wide range of different environments, so the individual absorption spectra are distributed over a range of wavelengths. If we study the absorption with light covering a broad range of wavelengths, we produce a broad absorption made up of all the individual ones (Figure 8). However, if we use the very narrow range of wavelengths that a laser can produce, we produce a narrower absorption made up of just those individual ones that correspond to molecules in just the right environment to absorb.

Now let there be photo-chromism. The molecules may change, their interaction with the glass may change or the glass itself may change and, in favourable cases, this change (which need not be drastic) may persist after the light has been switched off. With a high enough light intensity and a narrow wavelength range, essentially all molecules may absorb and be converted to a form that no longer absorbs at the same wavelength. As a result, the previous broad absorption has a dip where these absorptions used to be and the spectrum is said to have a hole burned in it. If the laser is tuned in tiny steps of wavelength across the whole broad band, then at each step it targets a small group of molecules that we may choose to burn out of the spectrum or not. By this process we can encode digital information as a pattern of holes burned or material not burned at each point in the sequence of wavelength steps. There can be as many as a thousand such bits of information in one band – all within one spatial region 1 μm or so across. In this way, the wavelength dimension is used to increase the density of information stored within given spatial dimensions. We cannot pick our groups of molecules by their spatial location, so we do so by their frequency location, analogously to how a radio signal is broadcast to all sets but received only by correctly tuned sets.

As always, there is a snag. This time, the materials must be cooled to liquid-helium temperature to make the individual absorption lines narrow enough. Systems that exhibit persistent hole burning include free-base phthalocyanine **45** in a polymer glass and pentacene **46** in an alcohol glass. In the phthalocyanine system, the

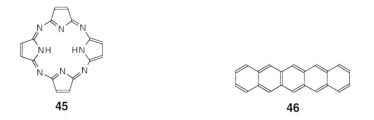

45

46

photochromism arises from transferring the hydrogen atoms on either side of the central ring to the nitrogen atoms at the top and bottom of the ring; since bonds in the guest are broken and made, this is referred to as photo-chemical hole burning. In the pentacene system, the photo-chromism arises from rearrangements of the hydrogen bonds in the alcohol glass; since no bonds in the guest change, this is referred to as photo-physical hole burning.

Colour changes are familiar in nature, where they convey information on such matters as the ripeness of strawberries, which is indicated by their redness. Biological phenomena provide much inspiration for molecular electronics, as noted in the introduction, and some people believe that 'real' molecular electronics is necessarily bioelectronics. More to the point is *how* molecular electronics uses biology: does it simply copy the phenomena or does it rather understand the principles and adapt them to its purposes? One example that certainly starts at the former extreme is the use of bacteriorhodopsin for optical storage of information.

Bacteriorhodopsin (BR) is a protein (similar to the visual pigment in our eyes) that grows in a salt-loving bacterium when it is deprived of oxygen. BR is contained in a structure called the purple membrane, which causes the purple colour that can be seen in the basins of salt works such as those in San Francisco Bay. Lack of oxygen stops the bacterium obtaining energy in the usual way by respiration, but then the BR takes over and provides energy via photosynthesis. It does so by a multi-step process which is very efficient and the BR is also very stable against light and heat, reflecting the role of natural selection in evolved systems. Optical detection media can be produced by embedding the BR in an inert polymer such as poly(vinyl alcohol). Owing to the various steps in the photosynthetic cycle, information can be written, read and erased with various combinations of wavelengths that convert from one chemical species to another. In the natural state, the BR uses the light energy to separate charges and, once it has done this, it returns to the beginning of the cycle in a few hundredths of a second. The charge separation produces a photo-voltage that allows the BR to be used as a photo-detector.

However, by site-directed mutagenesis (genetic engineering), variants of BR for which the decay to the initial state is so slow that it may take months have been produced. Hence the information in a light pattern can be stored in them for a correspondingly long time. Such media can be used effectively in areas such as pattern recognition, for example tracking the motion of an object by comparing the moving image with a static one. In nature the bacterium needs the BR to complete the cycle reasonably promptly in order to process another photon and generate more useful energy. By understanding how the BR is regenerated and by blocking this process through the design of the artificial variants of BR, we have moved away from merely using biology towards the latter extreme of adapting its principles to our own ends.

Using optical means to change optical properties is elegant, provided that we can avoid thermally induced reverse changes. Sometimes, though, we can use thermal changes themselves, i.e. *thermo-chromism*. Useful thermo-chromic materials must switch between states of different colours as the temperature changes. To store information, they must then remain in the coloured state when the temperature changes back. This phenomenon, known as hysteresis (Greek 'coming late') is illustrated in Figure 9. The wavelength of maximum absorption changes little with increasing temperature until it rapidly changes to a different value at temperature T_h, after which further heating again produces little change. However, cooling from above T_h produces little change until the lower temperature T_c is reached, whereupon the previous low-temperature value is regained. Between T_c and T_h the system is *bistable*: it exists in one of two forms that depend on how it has been treated. (The same sort of behaviour is manifested in the magnetization of soft iron, which is produced by a magnetic field but remains after the field has been removed.) By developing a material with T_c well below room temperature and T_h well above it, we can write information by localized heating, store it at room temperature and erase it by localized cooling.

Promising materials for thermo-chromic storage of information rely on changes in the arrangement of electrons' spins in transition metal ions. This arrangement governs the characteristic colours of transition metal compounds. The electrons in question occupy the five d orbitals, each of which can accommodate two electrons with opposed or paired spins, giving between them zero total spin. Other things being equal, the electrons prefer have the same spins (with the result that the total spin of the system is high), which means that they have to occupy different orbitals; they can then avoid

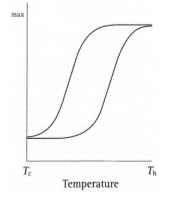

Figure 9

Hysteresis in thermochromism. The wavelength of the maximum absorption λ_{max} changes abruptly at one temperature, T_h, on heating and at a lower one, T_c, on cooling.

one another and reduce the energy of repulsion between their charges. However, the groups or ligands around the transition metal ion stop things being equal. Because of the way the different orbitals are directed in space, their energies are affected differently by the surrounding ligands. Typically two orbitals have higher energies than the other three. Then systems of four, five, six or seven d electrons are presented with a dilemma. They can keep down the repulsive energy by remaining unpaired as far as possible, but then they also increase the energy by having to use the higher-energy pair of orbitals. What actually happens depends on the relative importance of these two energies: if repulsion predominates, the electrons' spins remain parallel and the system adopts a high-spin state, whereas if orbital energies predominate, the electrons' spins pair up and the system adopts a low-spin state. The balance between the two energies depends on the nature of the ligands and their geometry around the metal ion. If it is sufficiently delicate, a change of temperature may tip it the other way, so that a transition between low- and high-spin states results. When the molecules interact to reinforce this effect, it tends to exhibit the desired sharp transition with hysteresis.

Olivier Kahn in Paris developed spin-transition compounds like this to a very promising stage. The iron(II) triazole complex $[Fe(Trz)(HTrz)_2]BF_4$ changes from a low-spin purple state to a high-spin colourless state on heating, with T_c around 85 °C and T_h around 125 °C. Between these two temperatures it stays purple on heating or colourless on cooling and the effect persists for weeks. A display using resistive heating and thermoelectrical cooling to turn each pixel on or off, giving a bright display on a dark background, has been constructed.

At the beginning of this section, I observed that light is a potent reagent. This potency may not only encourage reactions but also inhibit them. This phenomenon has been used to good effect in a light-sensitive oscillating reaction that can process images. Some complicated sequences of reactions can, under suitable conditions, oscillate in time and space, as can readily be seen when species of different colours are involved. In a shallow layer of solution containing the reagents, the result is a series of chemical waves of different compositions and hence different colours as the reaction propagates through the solution.

Such reactions involve a catalyst and, by making this sensitive to light, we can affect the reaction. Now consider projecting a black-and-white half-tone image onto the reacting solution. The chemical waves develop at different rates, depending on the level of illumination, so that first the image is generated and then it is processed. Under different conditions, processing may consist of alternating production of the positive and negative versions of the image; producing an outline of the image; or smoothing the image, for example by joining dots as in children's puzzles. Because the reacting medium is continuous down to the molecular level, these processes are capable of very high resolution. They also exemplify new approaches to computation using molecules, to which we now turn.

Molecular Computation

Perhaps the oldest device that could be described as a computer is the abacus. Beads that can slide on wires store a number by virtue of their position. Moving the beads along the wires allows this number to be combined with another. Experienced operators can reckon very rapidly with an abacus, but the inertia of the beads provides a limit to the speed attainable. Smaller beads could be faster and, if we were using mechanical rather than electronic computing, the smallest bead would be a molecule. Then, pursuing the abacus analogy further, the molecule needs to enclose a rod along which it can move without slipping off.

That this turns out to be entirely feasible was shown by Fraser Stoddart in

Birmingham, England. He synthesized a class of molecules called *rotaxanes*, shown schematically in structure **47**. The 'bead' consists of two doubly charged bipyridinium units linked by two benzene rings. The 'rod' consists of two benzene rings connected by chains of ethoxy groups to one another and also to bulky Si(CHMe$_2$)$_3$ (tri-*iso*propylsilyl) groups to act as stoppers. The charges on the bead are balanced by those

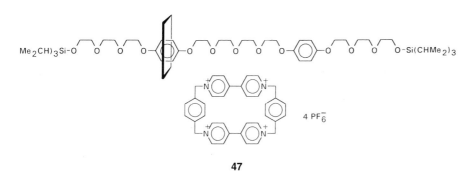

47

on four PF$_6^-$ ions. Synthesizing rotaxanes is less chancy than might perhaps have been expected, because the bead is stabilized when it wraps around either of the benzene rings. This allows it to form quite readily round the rod, because the open-chain molecule is already attracted to the benzene ring and, once there, is happy to cyclize. This means that it is also possible to make the molecule by threading the uncapped rod through the bead and then putting on the caps.

Thus we have a bead confined to a rod. Because there are two benzene rings on the rod, the bead has two preferred positions. Experiments show that it moves quite rapidly between them. Hence we have several of the ingredients of the molecular abacus. It is also possible to make the rod into a ring, so that no stopper groups are required, and to provide more than two preferred positions for the bead, so that it becomes like a train stopping at stations on a circular track or even like a bead necklace. What we do not have is the ability to move the bead at will in individual molecules.

This is where the electron scores, its charge providing a 'handle' by means of which an electrical field can move it. Another molecular analogue of an existing device, the shift register, has been proposed to use this fact. A shift register is a form of memory. It consists of a line of cells each storing one bit of information. At regular intervals, one bit is input to the line of cells, each cell transfers its contents to the next one along and the end one gives up its bit as output.

The *molecular shift register* uses ideas taken from photosynthesis, in which the photon generates charges that move rapidly and efficiently apart, as we saw for bacteriorhodopsin. The register consists of a polymer with repeating units A, B and C. Flashes of light can excite an electron from the ground state of A to an excited state. Judicious choice of B and C gives them energy levels such that the excited electron can transfer downhill in energy to B, to C and then to the next A unit. Even more judicious choice of A, B and C ensures that the net transfer from one A to the next occurs much faster than does any transfer backwards. Then any electron in a given A is transferred at each flash; if there is no electron in a given A, there is nothing to transfer, so shift-register action is produced.

One proposed but not yet synthesized polymer is shown as structure **48**. The components A, B and C are all analogues of those found in the photosynthetic apparatus. An alternative molecular shift register uses a Langmuir–Blodgett film of alternating layers of species X and Y. These have different affinities for electrons and so present energy wells of different depths. Applying an electrical field lowers the energy of the electron progressively across the film. At a high enough field, the depths of the wells

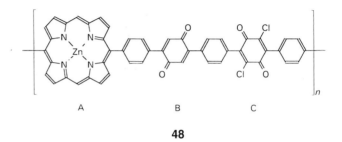

48

become equal and electrons can move along the register, just like water cascading from one basin to another in a fountain.

Clearly we can move electrons and we can use that ability, at least in principle, in these ingenious devices. However, for electronics at the molecular level we need to move individual electrons in individual molecules. This *addressing* problem has long been recognized as an obstacle to molecular electronics, but now progress towards overcoming it is being made.

One approach comes from the drive towards ever smaller conventional devices. This has led to the production of slender metal tracks with a gap of only 5–10 nm between them. This gap is the size of a large molecule, which could therefore provide communication between the metal tracks like a cantilever bridge across a river. Ari Aviram at IBM in New York has proposed connecting two such pairs of tracks by a molecule consisting of two segments of poly(thiophene) joined at right angles through a special twisted 'spiro' group. Either segment can conduct if it loses an electron and, under suitable conditions, an electrical field along the axis of the *spiro* group can cause an electron to switch from one segment to the other. This modifies the conducting pathway so that electrons pass from one pair of tracks to the other and such switching can be used for computation. The necessary molecules are not quite ready, but the structure **49** has been prepared and is some 3 nm across. It is also important to note that the

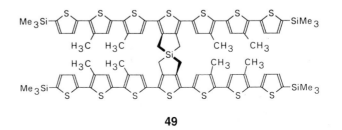

49

synthesis of this molecule is designed to produce it as a single product rather than one among several species obtained by different combinations of the same reactants (as happens in polymerization, for example).

Optical methods are now also addressing single molecules. In photo-chemical hole burning we saw that the different molecular environments in a glass give different spectra spread out over a range of wavelengths. Out in the wings of the total spectrum, the molecular environment is very different from the average and, far enough out, only one molecule may be responding in a diffraction-limited spot. This has been verified by the hole-burning experiments in which the hole comes back in the same place after a while – the hole is persistent, but evidently not permanent. Emission from an excited molecule gives similar results, for which statistical analysis confirms that the response is due to a single molecule. As before, we cannot choose which individual molecule we address; it is the one with the right environment; the situation is now more like possessing a winning lottery ticket than a correctly tuned radio. So far these experiments show that single molecules can be addressed without yet yielding single-molecule devices

A final technique allows individual molecules to be touched and manipulated. *Scanning-probe microscopy* is the name for a family of techniques that use a very fine probe that is scanned across the surface of a sample, like the electron beam on a television tube. At the same time its height above the sample is precisely controlled, typically to within 10 pm (1 pm is a millionth of a micrometre). This precise control is possible because the effective tip of the probe consists of a single atom, all others being too far away to interact with the surface through forces that fall off very rapidly with distance. The original technique of the *scanning tunnelling microscope* or STM used an electrical current between the probe and a metal substrate. To keep the current constant as the probe is scanned across the sample, the tip has to be moved up and down, which movement models the topography of the surface. If currents are undesirable or impossible, for example for thick insulating layers, the probe can be mounted on a tiny lever and the force kept constant during a scan, very much like the stylus in a record player. Under carefully controlled conditions, the scans can be built up into a picture of the surface in which individual molecules or even atoms are resolved. This has given much valuable information about the structure and perfection of materials such as Langmuir–Blodgett films.

If you can detect individual molecules, you can interact with them. Certainly isolated atoms can now be moved. Xenon atoms on a nickel surface accumulate at low-energy sites and, at very low (liquid-helium) temperatures, they stay there. A STM probe can locate a xenon atom, lift it from the surface, move it elsewhere and then release it by turning off the current. In this way structures such as letters to write messages and enclosures to modify the surface energy states can be made. An inverse technique makes structures by using pulses of current that remove atoms from a crystal surface.

Possibly the ultimate refinement of this technique has been the construction of an atomic switch. Don Eigler and collaborators at IBM San Jose have moved a xenon atom reversibly between two positions by applying suitable voltages. The conductance between the two positions reads whether the switch is 'on' or 'off'. This demonstration is still far from a usable molecular switch for computing, but it shows that atoms are addressable, especially ones as large as xenon. One remaining problem is speed. Switching one atom is feasible but takes a human sort of time scale of seconds. This will not enhance our computing ability until we can do it much faster and for many atoms at once.

Molecular electronics has come a long way in the 20 years or so since it was first systematically studied. Addressing individual molecules is just one of the challenges successfully met along the way. Inevitably the subject strays away from 'pure' chemistry by drawing upon physics, electronics and biology. Chemistry remains at the core, though, since molecules are central to molecular electronics. So far chemists have met the many synthetic challenges with great success and often the problem has been more to design the right molecule than to make it once it has been designed. It therefore seems likely that new chemistry will continue to involve making molecules to help us process information.

Further Reading

1. R. W. Munn, Molecular electronics, *Chem. in Britain*, **20**, 518 (1984).
2. R. W. Munn, Molecular electronics, *Phys. Bull. 39*, 202 (1988).
3. P. Day, Future molecular electronics, *Chem. in Britain*, **26**, 52 (1990).
4. P. Day, D. C. Bradley and D. Bloor (editors), Molecular chemistry for electronics, *Phil. Trans. Royal Society of London*, **330**, 61 (1990).

5. M. R. Bryce, Molecular electronics, *Chem. in Britain*, **27**, 707 (1991).

6. D. Bloor, Breathing new life into electronics, *Phys. World*, **4** 36 (1991).

7. M. R. Bryce, Organic conductors, *Chem. in Britain*, **24**, 781 (1988).

8. D. D. C. Bradley, Molecular electronics – aspects of the physics, *Chem. in Britain*, **27**, 719 (1991).

9. A. E. Underhill, Molecular systems, *Chem. in Britain*, **27**, 708 (1991).

10. K. C. Fox, The electric plastics show, *New Scientist*, 34 (5 March 1994).

11. G. Cooke, Laying it on thin, *Chem. in Britain*, **33** 54 (1997).

12. D. Pugh and J. N. Sherwood, Organic crystals for non-linear optics, *Chem. in Britain*, **24**, 544 (1988)

13. S. Allen, Materials with a bent for light, *New Scientist*, 59 (1 July 1989).

14. J. B. C. Findlay, Across the biological membrane, *Chem. in Britain*, **27**, 724 (1991).

15. J. F. Stoddart, Molecular Lego, *Chem. in Britain*, **24**, 1203 (1988).

16. J. F. Stoddart, Making molecules to order, *Chem. in Britain*, **27**, 714 (1991).

17. J. Leckenby, Probing surfaces, *Chem. in Britain*, **31**, 212 (1995).

18. J. Gimzewski, Molecules, nanophysics and nanoelectronics, *Phys. World*, **11** 29 (1998).

19. J. R. Barker, Building molecular electronic systems, *Chem. in Britain*, **27**, 728 (1991).

Electrochemical and Photoelectrochemical Energy Conversion

Andrew Hamnett and Paul Christensen

Introduction

Successive generations have sought efficient *portable* electrical power for many uses: for traction, for light, for personal electrical appliances such as tape recorders and video cameras, for transistor radios, for remote-area supply systems such as households, navigation devices, remotely operated and exploration vehicles, among innumerable other articles that add quality to life but whose operation depends on power sources that are portable or self-standing. There are several types of power source that fall into this category: the internal combustion engine in its various forms, the gas turbine, two electrochemical systems, the battery and the fuel cell, and self-standing solar-powered devices, based on photovoltaic or photoelectrochemical principles. Batteries and fuel cells have considerable advantages: they possess high thermodynamic efficiencies, particularly in the case of fuel cells when they are used with hydrogen as a fuel, excellent characteristics when they are used under partial as well as full-power output, short response times, low pollution-emission behaviour, simplicity of mechanical engineering, good power/weight ratio and their modularity of construction makes them ideally flexible in the provision of power. Such advantages should have led to rapid and extensive exploitation both of batteries and of fuel cells in a wide variety of environments, but, although batteries have become very widespread in their uses, as we shall see below, fuel cells remain under-utilized, for reasons ultimately of cost. Similarly, solar-powered systems potentially offer remarkable advantages in terms of the transformation of solar power into electrical energy, which can, in turn, be stored in batteries or in the form of a chemical fuel, or used directly; again, however, problems of conversion efficiency and weak solar irradiance have impeded commercial exploitation.

The fundamental principle of operation of batteries and fuel cells is the same: under appropriate conditions, the chemical free energy associated with a particular reaction can be converted into electrical energy, often with extremely high efficiency. The conditions for successful conversion are the following. First of all, it must be possible to break the overall chemical reaction down into component *oxidation* and *reduction* reactions, the first involving the transfer of electrons from one set of reactants to the *anode* and the second involving the transfer of electrons from the *cathode* to a second set of reactants. Secondly, it must be possible to arrange the electrochemical configuration such that both electron-transfer reactions are rapid and, ideally, completely reversible in chemical terms. During operation, electrons flow from the anode to the cathode, constituting an electrical current that can be used to drive a device. The main difference between batteries and fuel cells is that, in the former, the chemical reactants are an inherent part of the device; a battery, in other words, carries its fuel around with it, whereas fuel must be supplied to a fuel cell from an *external* source. Unlike a battery, a fuel cell cannot 'go flat'; so long as fuel and combustant gases are supplied, the fuel cell will generate electricity. The problem of re-charging is, therefore, peculiar to batteries, though the fuel-management systems characteristic of most advanced fuel cells are not normally needed in the operation of batteries.

The question of utilization is ultimately economic; unless there are over-riding reasons for ignoring cost, such as the compelling advantages of the low thermal signature of fuel cells in military applications and the very high power/weight ratio in space probes, then at the end of the day the additional expense often associated with electrochemical conversion devices must be justified. As an example, electricity from the National Grid in the UK is sold at about £0.07 per kW h whereas the cost of electricity from a lead–acid battery is about £0.5–1.5 per kW h and for small lithium batteries the cost can reach £300 per kW h. The latter represents the extraordinary premium we are prepared to pay for convenience. We can also represent costs in terms of installed power (i.e. the capital costs of the battery and its installation): for a lead–acid battery, the cost is about £200–300 per kW, whereas costs for NiCad batteries are closer to £1000 per kW, with a slightly cheaper price for the experimental ZEBRA batteries described below. Fuel cells, in contrast, particularly with their associated fuel-processing systems, cost usually in excess of £1000 per kW, far greater than current competing technologies, though the latter have benefited enormously from economies of scale.

Aqueous Batteries

To illustrate the chemical conversion process, consider the operation of the most familiar of all batteries, the lead–acid battery. This battery is present under every car bonnet and its net chemical reaction is

$$Pb + PbO_2 + 2H_2SO_4 \rightarrow 2PbSO_4 + 2H_2O \tag{1}$$

The free energy, $\Delta G°$ of this reaction under standard conditions of pressure and solution concentration (or more accurately solution *activity*) and at room temperature is -393.9 kJ mol^{-1}. The voltage obtainable from a cell whose net reaction is (1) under these same standard conditions can be shown to be

$$E° = -\Delta G°/n\,F \tag{2}$$

where n is the number of electrons involved in the process and F is a fundamental constant, the Faraday, with a value 96485 C mol^{-1}. As we will see in a moment, the value of n for reaction (1) is 2, so that E has the value 2.041 V under standard conditions. To realize a battery based on reaction (1), we must divide the reaction into two parts, corresponding to the anodic and cathodic processes, as shown in Figure 1. For the anodic process, we have

$$Pb + H_2SO_4 \rightarrow PbSO_4 + 2H^+ + 2e^- \qquad E° = -0.356 \text{ V} \tag{3}$$

and for the cathodic part

$$PbO_2 + H_2SO_4 + 2H^+ + 2e^- \rightarrow PbSO_4 + 2H_2O \qquad E° = +1.685 \text{ V} \tag{4}$$

where the $E°$ values are the standard electrode potentials of the two half-cell reactions, whose difference is 2.041 V as expected.

In practice, the cell delivers a rather lower voltage than the predicted 2.041 V when current is drawn, since there is a number of sources of irreversibility associated with the working cell. Three concern us directly: the first two can be traced to limitations in the rates of the partial reactions (3) and (4). No electrochemical process takes pace at infinite speed, but the rates do increase rapidly (in fact exponentially) as the potential is raised (for an anodic reaction) or lowered (for a cathodic reaction) with respect to the $E°$ values quoted above. A third source of irreversibility is the internal resistance of the cell; as this increases, an increasing fraction of the cell voltage is actually dropped within the cell and is not available for external work. The effects of these sources of loss are shown graphically in Figure 2. The net cell voltage is the difference between the partial current–voltage curves for the anode and cathode; these curves are initially exponential in form at low current densities, but the curves become essentially linear at higher current densities as the main additional source of loss becomes the internal resistive loss of the cell. To minimize this effect, concentrations of the electrolyte, sulfuric acid, of up to 36% by weight are used in practical lead–acid batteries.

Although the lead–acid battery was developed in the middle of the last century, it remains highly successful in certain applications, particularly those for which large currents are needed for short periods. Its incorporation as an essential accessory into the starting circuit for internal combustion engines ensured the latter's triumph over battery-driven vehicles in the early part of this century and it has also found application in back-up power units for telecommunications to ensure that, for example, the telephone network will continue to operate even in the event of a mains power failure. The theoretical energy density of the lead–acid battery is quite high, 167 W h kg^{-1} (corresponding to ≈ 0.6 MJ kg^{-1}), but practical densities of about 40 W h kg^{-1} and power densities of 250 W kg^{-1} (at 300 mA cm^{-2}) are attainable. The number of re-charges possible is about 300–1500, depending on the battery's design and conditions.

There have, of course, been significant improvements to the lead–acid battery since 1859: these improvements include not only such practical engineering changes as the incorporation of high-impact polypropylene cases, heat-sealed plastic covers, safety vents for sealed cells, through-the-wall cell interconnects and automatic watering systems for traction purposes but also scientific improvements aimed particularly at the cathode, or positive grid for which the castability and adherence of the electro-active PbO$_2$/PbSO$_4$ paste proved problematic in the past. The problem of the evolution of gas at the electrodes during re-charging has also led to serious problems: as the battery becomes nearly fully charged, the charging voltage can rise to the point at which some electrolysis of water to hydrogen and oxygen can take place. In extreme cases, this can lead to a serious risk of explosion; in more normal circumstances the battery steadily loses water and needs to be 'topped up' from time to time. Recent years have, however, seen the introduction of 'maintenance-free' batteries, in which gassing is controlled by careful choice of the composition of the lead alloys used and in particular by the replacement of the antimony (3%) present in older batteries by calcium (0.1%). In addition, finite-element modelling has been used to improve grid designs and separators using organic polymers with inorganic fillers are now employed to control the pore-size distribution and pore volume.

The lead–acid battery is very widely manufactured: some 300 million units are made

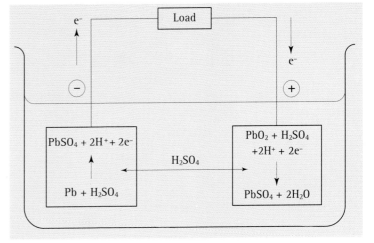

Figure 1
A schematic diagram of the processes taking place in the lead–acid battery.

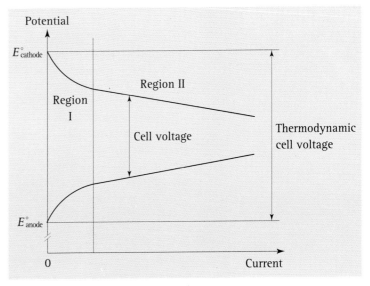

Figure 2
A schematic diagram showing the variation of the cell's voltage with the current for a typical battery.

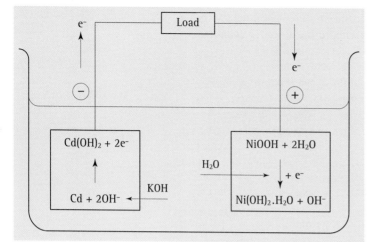

Figure 3

A schematic diagram of the processes taking place in a nickel–cadmium battery.

each year, ranging in size from 2 W h cells to 100 W h starting, lighting and ignition (SLI) systems and up to 40 MW h load-levelling modules. For SLI units, short bursts of very high current are needed, so the construction is designed to minimize the internal resistance. Units normally have nominal voltages of 12 V and 30–100 A h capacity for cars and 24 V and up to 600 A h for heavy goods vehicles. For batteries used in traction, the units must be able to sustain prolonged and deep discharging followed by long re-charging. Voltages for such batteries normally lie in the range 12–240 V with specific energies in the range 20–30 W h kg^{-1} (cf. the theoretical value above). Similar considerations apply to batteries used in stationary back-up power applications, such as telecommunication systems, railway signalling and point de-frosting systems, for which reliability, long life and low rates of self-discharging are primary requirements. Similar considerations apply to load-levelling, for which the cycling efficiency is critical.

The lead–acid battery has, in fact, been so successful that many of its properties have become paradigmatic for all potential batteries: these include high power, energy and efficiency; good safety characteristics, low self-discharging, fast re-charging, low sensitivity to mishandling, low environmental impact of constituent materials and ease of servicing. The main problems of the lead–acid battery are, however, important to recognize. The most acute is the low specific energy associated with the very high density of lead and the low utilization of active material. Much of the lead in lead–acid batteries in fact simply acts as a conductive and mechanical support for the electro-active paste and as a current collector, so active investigation of lighter alloys is in progress. The utilization of the lead is also inhibited by the nature of the reaction product, $PbSO_4$, which is non-conducting and tends to 'choke off' the supply of reactants to the underlying grid. Ways of controlling the porosity of the $PbSO_4$ by the addition of inert polymer and by other methods are now being investigated, but the effects of $PbSO_4$ have been found not only to be inimical to the usage of lead but also to lead to the formation of PbO, both on the cathode and on the anode, as a result of swings in pH caused by the $PbSO_4$ inhibiting the diffusion of the H^+ and SO_4^{2-} ions. This leads to a slow and irreversible decay of electrodes. These types of problem also complicate the search for ways to achieve faster re-charging of the lead–acid battery, which will be essential if the system is to have any application to traction.

Although the lead–acid battery is one of the most commonly encountered aqueous acid systems, it is not the only successful aqueous battery. Other systems, particularly ones based on nickel, have been developed. The nickel–cadmium (NiCad) battery dates from the end of the last century and relies on the following reactions, as shown in Figure 3. At the anode

$$Cd + 2OH^- \rightarrow Cd(OH)_2 + 2e^- \qquad E^\circ = -0.809 \text{ V} \qquad (5)$$

and at the cathode

$$NiOOH + 2H_2O + e^- \rightarrow Ni(OH)_2 \cdot H_2O + OH^- \qquad E^\circ = +0.450 \text{ V} \qquad (6)$$

This is usually carried out in 20–28% KOH, under which condition the overall cell reaction gives an open-circuit voltage of about 1.30 V. The cathodic reaction in this cell is of great intrinsic interest, since the structure of the NiOOH form is layered and the process involves diffusion of protons into the layers and concomitant addition of electrons to the low-spin Ni(III) to give octahedral Ni(II). The environment of the low-spin Ni(III) centres is expected to be strongly distorted owing to the so-called Jahn–Teller effect, which is particularly prominent for ions with odd numbers of electrons in the upper d orbital energy level. Low-spin Ni(III) has the electronic configuration $t_{2g}^6 e_g^1$ and such ions invariably distort to raise the degeneracy of the E_g manifold. As a result, the solid is expected to exhibit considerable distortion during oxidation, and adhesion and resistivity are both problems that need to be surmounted; in addition, the resistivity of the $Cd(OH)_2$ also presents problems. Technical solutions involve mixing the hydroxides with graphite or other high-conductivity additives, or using porous sintered nickel frames into the pores of which are added the hydroxides. The energy density is comparable to that of the lead–acid battery, but the power density, up to 700 W kg^{-1}, is much higher and the battery can withstand up to 3000 re-chargings. The battery finds a wide range of uses, for example as storage systems in aviation and in many types of electronic and electrical equipment, including power tools and garden and kitchen appliances. Sealed cells also have important military and aerospace applications: the battery for the Viking Mars spacecraft consisted of 26 sealed 30 A h NiCad cells. Larger versions are under development as potential traction units for Renault and Peugeot cars.

Related to this cell is the nickel–iron cell, whose constitutive reactions are

$$Fe + 2OH^- \rightarrow Fe(OH)_2 + 2e^- \qquad E° = -0.877 \text{ V} \qquad (7)$$

at the anode and

$$NiOOH + 2H_2O + e^- \rightarrow Ni(OH)_2 \cdot H_2O + OH^- \qquad E° = +0.450 \text{ V} \qquad (6)$$

at the cathode. This cell was developed by Eddison in 1900 since Fe is cheaper than Cd and not so toxic. However the Ni/Fe cell is not so good in low-temperature operation and extensive gassing takes place during re-charging, owing to the competing low-temperature electrochemical reactions evolving hydrogen (at the iron electrode) and oxygen (at the nickel electrode). It also has a rather low efficiency. The main applications are in railway lighting systems and as traction units in mine locomotives, tractors and trucks.

The above batteries have the immense advantage that they can be re-charged when they go 'flat'. Many other re-chargeable, or secondary, aqueous-based batteries are known, including the silver–zinc battery, based on the cell reaction $Zn + AgO + H_2O \rightarrow Zn(OH)_2 + Ag$, with a voltage of 3.1 V at open circuit and an extremely high power density, giving it considerable attraction in military, aerospace and other specialist applications, including as the power source for the Ranger lunar-photography spacecraft. A battery based on Zn/NiOOH has also been developed and has attracted some attention as a possible battery for traction applications, since the very negative $Zn/Zn(OH)_2$ couple (-1.25 V) ensures that it has a high energy density; however, zinc is amphoteric and hence tends to dissolve in the KOH solution, leading to migration, densification and formation of dendrites on re-charging. To overcome these difficulties, zinc electrode systems in neutral solution have been developed, including the zinc/chlorine battery, which is based on an aqueous $ZnCl_2$ electrolyte and the reactions

$$Zn \rightarrow Zn^{2+} + 2e^- \qquad E° = -0.76 \text{ V} \qquad (8)$$

at the anode and

$$Cl_2 + 2e^- \rightarrow 2Cl^- \qquad E° = +1.36 \text{ V} \qquad (9)$$

at the cathode, where the chlorine itself is stored as the clathrate $Cl_2 \cdot 7.3H_2O$, from which it is released by mixing with warm water, or by heat derived from the battery itself during discharging. On re-charging, the chlorine gas is simply added to cold water. These batteries have been investigated as very-high-current storage devices and for vehicle traction. Cells based on bromine, in which the Br_2 is stored as the Br_3^- ion, have also been studied.

Metal–gas cells have become the objects of a renewal of attention in recent years, though mainly as primary batteries, i.e. batteries that cannot be re-chargeable *in situ*. These batteries are in some ways analogous to fuel cells (which are discussed in detail below), in that the cathode reaction uses a combustant supplied externally, such as air. The most familiar are the Zn/air and Al/air batteries, both of which are based on the reactions

$$M + yOH^- \rightarrow M(OH)_y + ye^- \qquad (M = Zn, y = 2; M = Al, y = 3) \qquad (10)$$

at the anode and

$$O_2 + 2H_2O + 4e^- \rightarrow 4OH^- \qquad E° = 0.401 \text{ V} \qquad (11)$$

at the cathode. Both of these have been examined as possible power sources for traction, but their essential irreversibility means that some infrastructure capable of recovering large amounts of metal hydroxide and re-converting this into the metal would be needed. In addition, there are substantial losses at the cathode arising from the very low rate of reaction (11) save at high overpotentials (i.e. at potentials well below the thermodynamic value of 0.401 V), reducing the attainable cell voltage substantially.

In addition to metal–air electrodes, metal–hydrogen cells, in which the hydrogen replaces the metal at the anode side, have also been proposed. An obvious and interesting example is the nickel/hydrogen cell to replace the Ni/Cd cell. The reaction at the cathode is now

$$H_2 + 2OH^- \rightarrow 2H_2O + 2e^- \qquad E° = -0.828 \text{ V} \qquad (12)$$

and the hydrogen can be supplied as a gas under high pressure or by warming certain alloys, such as Li–Ni, that absorb considerable quantities of H_2 at low temperatures. Such batteries can be re-charged provided that the hydrogen can be re-stored, but this can present real problems in practice and their relatively poor self-discharging characteristics have been the subject of much optimization in recent years. However, the attraction of eliminating Cd, with its severe environmental impact, has led to a huge programme of investment in which several leading battery producers, including Varta and Duracell in Europe, have begun producing nickel–hydride batteries in considerable quantities for the '3C' market (camcorders, cellular telephones and computers). Some development of these batteries for traction has also been carried out and they are increasingly replacing NiCad batteries.

Perhaps the best known simple primary battery is the Leclanché cell, which is still the basis of many 'throw-away' batteries. The basic cell reactions are shown in Figure 4:

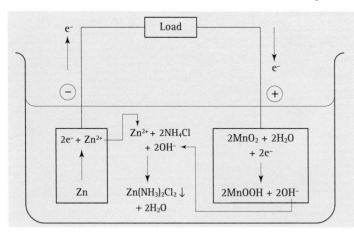

Figure 4
A schematic diagram of the processes taking place in a Leclanché cell.

$$Zn + 2NH_4Cl(aq) + 2OH^- \rightarrow Zn(NH_3)_2Cl_2 + 2H_2O + 2e^- \qquad (13)$$

at the anode and

$$2MnO_2 + 2H_2O + 2e^- \rightarrow 2MnOOH + 2OH^- \qquad (14)$$

at the cathode, with a cell voltage of 1.5–1.6 V. The zinc is normally present as the cylindrical container (which may be surrounded by steel to prevent loss of electrolyte);

but the MnO_2 is not a good conductor of electricity and hence is normally mixed with graphite to form a paste, with a central graphite rod acting as the current collector. The aqueous NH_4Cl electrolyte is usually thickened with polymeric material such as starch or methyl cellulose and the resultant cells are reasonably stable against self-discharging. The main route to self-discharging is via the evolution of hydrogen:

$$Zn + 2H_2O \rightarrow Zn(OH)_2 + H_2 \tag{15}$$

which clearly does not involve a net transfer of electrons. This process can, in turn, be inhibited by the addition of small amounts of mercury to the electrolyte paste, but this has serious environmental consequences. Indeed, it is environmental concerns that may well spell the end of the primary Leclanché cell in the near future, for there is a concerted move towards re-cycling in most advanced industrial nations and a strong interest in research into improving re-chargeable batteries as a result. That said, however, re-chargeable cells based on the Leclanché principle have proved extremely difficult to develop: a concerted research effort by Kordesch at the University of Graz has led to the development of so-called RAM cells, which are now being marketed by Raynovac and are based on the structure

$$Zn(s)|KOH(aq.)|MnO_2, C(s)$$

Provided that the discharge is not permitted to go beyond $MnO(OH)$ at the cathode, this cell can be re-charged, though it is found that the capacity of the cell falls steadily with the number of cycles, particularly if relatively deep discharging is permitted. However, for applications such as mobile phones, personal audio equipment, electronic organizers, cameras, toys and games they are highly suited and have already significantly penetrated the market, helped by their excellent shelf-life compared with other secondary systems.

Non-aqueous Battery Systems

We saw above that energy densities of up to 700 W kg^{-1} are possible in aqueous systems, but recently non-aqueous systems of far higher potential power densities have been developed on the basis of the very light metal lithium. Almost all of these systems have lithium or lithium-alloy cathodes and a wide variety of anode systems, depending on the application. For portable power systems, for watches, calculators, cameras, memory-back-up systems, pacemakers and the like, for which lightness, a long shelf-life and mechanical reliability are essential, primary cells based on lithium, in which the cathode normally acts to intercalate lithium, are employed. Such electrodes can also be used as the basis of re-chargeable cells. A simple example would have the form

$$Li(s)|LiPF_6 \text{ in an inert organic solvent}|MO_2(s)$$

with the reactions

$$xLi(s) \rightarrow xLi^+(solvent) + xe^- \tag{16}$$

at the anode and

$$xLi^+(solvent) + xe^- + MO_2(s) \rightarrow Li_xMO_2 \tag{17}$$

at the cathode. Clearly, to operate effectively, Li_xMO_2 should be electronically conducting or at least miscible with an appropriate inert conducting adduct and its formation, through the diffusion of Li ions, must be reasonably facile and, ideally, highly reversible if the battery is to be re-chargeable. The latter requirement suggests that layered structures, allowing the Li^+ ions to diffuse easily throughout the bulk of the crystal, are likely to be the most effective. Suitable metal oxides include MnO_2 and V_6O_{13}: higher voltages may be obtained by using $Li_{1-y}CoO_2$ and $Li_{1-y}NiO_2$, since these oxides possess the transition metal in a high oxidation state, which allows a high cell voltage to develop (given that the intercalation process is, in essence, reduction of the metal). These oxides are also capable of accommodating large quantities of lithium per formula unit and have low relative molecular masses, giving rise to high power and energy densities.

Importantly, while they can intercalate lithium relatively easily, they do not co-intercalate solvent, since the interlayer separation in these structures is too small. They are also stable in contact with the solvent, are of low cost, are easily fabricated into electrodes and are neither of them environmentally problematic. Some, such as MnO_2, exhibit spin-state changes that lead to Jahn–Teller type distortions: the spinel $LiMn_2O_4$ is cubic, for example, but $Li_2Mn_2O_4$ is tetragonal. Such changes do lead to excellent voltage–discharge characteristics, but militate against reversibility, so considerable effort is now being directed towards optimizing the two features together. Cells based on oxide cathodes are now manufactured widely and have found much use in low-drain long-term applications such as computer-memory retention. Higher-current devices can also be fabricated for use in electrical motors in automatic cameras and toys.

In addition to oxides, polycarbon fluorides, of general formula $(CF)_x$, mixed with carbon black to increase the conductivity can be used in primary cells. This type of cell has an overall cell reaction of the form

$$nxLi(s) + (CF)_x \rightarrow nC(s) + nxLiF(s)$$

where the detailed mechanism involves an intermediate intercalate. Cells of this type have found application in radio transceivers, surveying equipment, computer-memory back-up etc.

The solvent used may be an organic ether, which must, of course, be thoroughly dried. However, polymeric ethers, such as Li-salt-impregnated polyethene oxide, have particularly attractive properties for certain applications and, especially when they are combined with electronically conducting intercalation polymers on the cathode side, have the immense advantage of being amenable to continuous fabrication using fast polymer-film processing techniques. Such thin-film batteries can be fabricated into essentially arbitrary shapes, giving considerable design flexibility, and they have large surface areas, which makes thermal management more straightforward. However, there are disadvantages: the initial set-up costs for manufacturing them are high and the need for critical gas-tightness means that manufacturing tolerances are very tight.

The performance of lithium batteries can be extended: medium-energy/high-power systems are of particular military importance, since they allow very large current pulses to be drawn and, at the same time, are very light. Such batteries are based on Li anodes and inert cathodes capable of electrochemically reducing such species as $SOCl_2$ and SO_2 and experimental Li/BrF_3 batteries capable of up to 1000 W h kg^{-1} are now being developed. These batteries are usually supplied in 'reserve' form; that is, the battery is supplied without its electrolyte and will operate once the electrolyte has been added. For the $Li/SOCl_2$ battery, an electrolyte of the form $SOCl_2$–$LiAlCl_4$ must be added, whereupon the battery generates electricity from the cell reaction

$$4Li + 2SOCl_2 \rightarrow 4LiCl + SO_2 + S$$

High-energy/high-power batteries based on lithium have also been developed. The simplest of these is the Li/Cl_2 battery, which uses a LiCl–KCl–LiF eutectic electrolyte melting at 450 °C. The overall cell reaction was similar to that of the Zn/Cl_2 cell described above, but the cell voltage is very high (3.46 V) and the theoretical energy density enormous (2200 W h kg^{-1}). However, the actual operating temperature has to be rather high (600 °C) and there are severe corrosion problems, caused in part by the solubility of the lithium in the electrolyte. This type of cell has now been refined and developed into the Sohio battery, which uses an electrolyte with a rather lower melting point (LiCl–KCl at an operating temperature of 400 °C) and a Li–Al alloy as the anode, to reduce solubility problems. The overall cell reaction for this system is now

$$4LiAl + 2Cl_2 \rightarrow 4LiCl + 4Al \qquad (18)$$

with a voltage of 3.2 V and a high peak current density of 2 A cm^{-2}. The main disadvantages of this cell are that the discharge voltage is dependent on the state of charging and the energy density is considerably lower (62 W h kg^{-1}).

A second type of system, developed from the analogous Na/S battery described below, is based on the cell reaction

$$2Li + S \rightarrow Li_2S \qquad (19)$$

with a LiCl–KCl eutectic electrolyte and a cell voltage of 2.25 V. This cell has an extremely high theoretical energy density (2624 W h kg^{-1}) and a high power density, but suffers from the solubility of Li in the electrolyte and from the high vapour pressure of sulfur at the operating temperature. Both of these problems have been addressed in the LiAl/FeS system, which is based on Li alloy anodes and FeS_2 cathodes with a lower-temperature molten-salt eutectic. This battery has a high specific power, high rate capability and a long shelf-life, and has exciting possible applications to traction. The cell reactions are

$$LiAl \rightarrow Li^+ + Al + e^- \qquad (20)$$

at the anode and

$$2e^- + 2Li^+ + FeS \rightarrow Li_2S + Fe \qquad (21)$$

or

$$4e^- + 4Li^+ + FeS_2 \rightarrow 2Li_2S + Fe \qquad (22)$$

at the cathode, with an electrolyte of LiCl–LiBr–KBr with a (relatively) low melting point (400–450 °C). However, the cell voltage is only 1.3 V for (19) and 1.6 V for (20), so the energy density is rather low, particularly for (19). In addition, the cells are not tolerant of overcharging and there are some rather difficult materials problems to overcome. Nevertheless, the development of this battery has proceeded apace and demonstrator units are now being contemplated, specifically with traction in mind.

There are many areas for which re-chargeable lithium batteries are desirable, particularly for portable equipment with relatively high demands on power, such as cellular telephones, camcorders, portable CD players and televisions and implantable medical devices. The re-chargeable devices are based on similar types of cell to those of (16) and (17), the basic difficulty being that the replating of lithium leads to a loss of electrical contact between particles due to the formation of insulating layers that are assumed to arise from chemical reaction with the solvent or adventitious water. Multiple cycling also leads to expansion of the lithium anode's surface with associated non-uniformities (such as the formation of dendrites). Such non-uniformities can lead to hot spots and in turn to disastrous failure of the battery by explosion or fire. This type of problem has been tackled by replacing the lithium anode by a second intercalate, such as Li_xC_6, giving a battery with an anode reaction

$$Li_xC_6 \rightarrow Li_{x-y}C_6 + yLi^+ + ye^- \qquad (23)$$

Carbons that have been studied include natural and synthetic graphites, petroleum coke, carbon fibres and mesocarbons, all of which differ in degree of crystallization and stacking order, but all of which have the characteristic structural feature of graphite, namely planar layers of carbon atoms forming fused six-membered rings and separated by the intercalate.

The cathodes can be oxides, of the type described in detail above, layered sulfides, such as TiS_2, or electronically conducting polymers such as polyacetylene, polypyrrole and polyaniline. The latter are fascinating modern materials that can be oxidized by removing electrons from the polymer's backbone and simultaneously inserting anions to maintain charge neutrality:

$$P(s) + X^- \rightarrow (P^{y+})(yX^-) + ye^-$$

Discharge is the reverse of this process, in which the LiX salt in the electrolyte is reformed.

Rechargeable lithium batteries were first marketed by Sony Energytech in 1990 and have rapidly become established in the area of electronic consumer goods. Scaling them up to traction units may, however, also be feasible and research programmes with this aim are going on in Japan, Europe and the USA.

Although the $LiAl/FeS_2$ system has considerable potential as a traction-power battery, there are inherent difficulties of cost that need to be addressed. Lithium is not a rare metal, but it is not inexpensive; and considerable engineering and scientific problems need to be overcome before a reliable re-chargeable traction system can be developed. The Ford Motor Company, seeking an inherently cheaper solution, announced in 1967 the development of a wholly new battery, the sodium–sulfur (Na/S) system, a far less expensive alternative. The battery was innovative in a number of other ways, mainly in the inversion of the usual engineering configuration in which two solid electrodes are separated by a liquid electrolyte. In the Na/S battery, liquid sodium is used as the anode, liquid sulfur as the cathode, carbon felt as the conductor/current collector and the electrolyte is a solid ceramic, which has a high sodium-ion conductivity. We have already seen that oxides can intercalate lithium ions, but very high ionic conductivities are needed if the ceramic is to serve as an electrolyte. Ceramics capable of sustaining such conductivities are termed superionic conductors. For sodium, the best known such superionic conductor is sodium β-alumina, with the formula $Na_2O \cdot 11Al_2O_3$. The structure of this exceptionally interesting material consists of slabs of alumina spinel separated by layers of sodium ions, the latter being highly mobile at temperatures of 300 °C or higher. At 350 °C, for example, the specific conductivity of the ceramic is a remarkable 0.2 Ω cm^{-1}, comparable to that of 2 M H_2SO_4 at room temperature.

The basic cell reaction involves the formation of catenated sulfide species such as S_3^{2-}:

$$2Na + 3S \rightarrow Na_2S_3 \tag{24}$$

with a cell voltage of 2.067 V when the battery is fully charged and 1.76 V at discharge. This battery gives very high power densities, but overcharging can lead to an almost insulating phase and, given that the cell operates at temperatures near 350 °C and that the external aluminium can would melt at 650 °C, there is not much room for error in the control of temperature. The biggest problems have been with sealants, but, in spite of this, a pilot production plant has been set up in the UK and field trials have been carried out.

A promising descendent of the Na/S battery is the so-called ZEBRA cell, based on the configuration

$$\text{Na(l)} | \text{ Na–β-alumina } | \text{NaAlCl}_4\text{(l)} | \text{MCl}_2$$

in which M can be Cu, Ni, Co, Fe or Cr, with open-circuit voltages of 2.35 V for Fe and a higher 2.58 V for Ni. Just as for the Na/S battery, the Na–β-alumina acts both as solid electrolyte and as a separator, but, unlike the Na/S battery, there is a second liquid electrolyte, a molten salt, that greatly facilitates the kinetics and reversibility of the battery. The basic cell reaction has the form

$$2Na + NiCl_2 \rightarrow Ni + 2NaCl \qquad E° = 2.59 \text{ V} \qquad (25)$$

where the NaCl dissolves in the molten salt. The cell is stable against both overcharging and over-discharging, the relevant reactions being

$$Ni + 2NaAlCl_4 \rightarrow 2Na + 2AlCl_3 + NiCl_2 \qquad E° = 3.05 \text{ V} \qquad (26)$$

$$3Na + NaAlCl_4 \rightarrow 4NaCl + Al \qquad E° = 1.59 \text{ V} \qquad (27)$$

and the cell is normally assembled in the over-discharged state, without liquid sodium. The advantages of this cell are that there is no effect of high rates of discharging on the cell's capacity, the cell is also very stable with respect to cycling, there is almost no corrosion and it has an inherently high reliability.

The ZEBRA cell has been developed for traction purposes and is perhaps the most advanced of the 'new' batteries in this regard. The requirements from car companies for batteries suitable for traction are severe: the battery should be light-weight, inexpensive, safe in the event of an accident, robust, capable of providing very high power for short periods, re-chargeable at least 1000 times and have a 5-year lifetime. In testing by Mercedes, the ZEBRA batteries manufactured by the Beta Battery Company in the UK appear at the moment to come closest to these ideals, the main drawback being the cost, which is mainly ascribable to the nickel content. The safety of these batteries was initially thought to be a major drawback, especially in the event of a serious accident in heavy rain, but tests have shown that even accidents that result in puncturing of the outer steel cladding of the battery case do not lead to catastrophic chemical reactions. However, much more testing will need to be done in this regard before the battery can be declared truly safe.

Fuel Cells

A very simple realization of a fuel cell is shown in Figure 5, which demonstrates the mode of operation of an alkaline fuel cell, so called because the electrolyte is aqueous KOH. The cell consists of two electrodes: at the anode, hydrogen is oxidized to water:

$$H_2 + 2OH^- = 2H_2O + 2e^- \qquad E° = -0.828 \text{ V} \qquad (28)$$

and at the cathode, oxygen (usually supplied as air) is reduced to hydroxide ions:

$$\tfrac{1}{2}O_2 + H_2O + 2e^- = 2OH^- \qquad E° = +0.401 \text{ V} \qquad (29)$$

The net chemical reaction, the sum of the reactions at the anode and cathode, is then

$$H_2 + \tfrac{1}{2}O_2 = H_2O \qquad (30)$$

which can be seen essentially to be the combustion of hydrogen in air. Such a reaction can, of course, be used directly to drive an internal combustion engine, but, if electrical power is required, the direct conversion of the free energy associated with reaction (30) in a fuel cell is an exceedingly attractive alternative, at least in principle.

However, it will be noted that the fuel used is hydrogen and therein lies one of the major problems for fuel cells. All attempts to drive fuel cells directly using primary fuels, such as coal, oil and natural gas, have failed, at least with low-temperature devices, such as that shown in Figure 5, and either high temperatures or reforming of fuel or both have proved necessary, leading to high costs associated with the use of expensive catalysts or complex engineering. These interconnected problems have bedevilled efforts to build commercially attractive fuel-cell systems. Since the work of

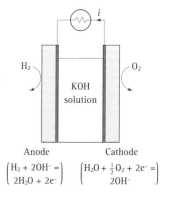

Figure 5
A schematic diagram of an alkaline fuel cell, with the flow of current shown conventionally.

Grove, more than 150 years ago, which led to the realization of the first simple fuel cell (it was based on reactions (28) and (29)), hydrogen–oxygen fuel cells of steadily increasing efficiencies have been fabricated, but this enhancement of performance has not been sufficient to justify the costs of isolating H_2 from the primary fuels available. The dominance of coal in the nineteenth century led to innumerable attempts to employ the latter's electrochemistry, which were abandoned only when it was realized not only that the ash would severely contaminate the molten KOH favoured as the electrolyte but also that this electrolyte actually reacted directly with carbon when moisture was present.

The end of the nineteenth century saw, initially in the USA, the rise of oil as a primary fuel, which led to a renewal of efforts to realize fuel costs based on direct oxidation of hydrocarbons at the anode. Unfortunately, facile splitting of the non-polar C–C bond has never been achieved at low temperatures by electrochemical means and the activation of the C–H bond to allow controlled electrochemical oxidation also remains a very difficult reaction. As we shall see below, internal reforming of natural gas (i.e. its conversion with steam to CO_2 and H_2) remains a real possibility for at least one type of modern fuel cell, which is based on molten carbonate, but both for coal and for heavy hydrocarbon fractions, reforming externally with steam remains the only option, resulting in fuel-cell systems that are both technically complex and prone to the production of sulfur-containing poisons.

The emergence of natural gas as a major energy vector has altered the equation in a number of ways: methane can be reformed with high efficiency and considerable ingenuity has been expended on the design of reformers that are significantly less complex than those required for heavier fuels. The innate purity of natural gas has also led to considerable advantages in terms of poisoning, allowing the possibility of internal reforming, as well as the coupling of external reformers to low-temperature fuel cells that are particularly prone to poisoning. Methane can even be used directly in very-high-temperature fuel cells of the solid-oxide type discussed below. Even more attractive is the possibility of using methanol as a major fuel, since this is a liquid that can be transported easily and would be ideal for traction. Methanol is appreciably more reactive than methane, it can be reformed at temperatures of about 300 °C and is even sufficiently electro-active for direct oxidation at the anode at temperatures in excess of about 60 °C. A direct methanol fuel cell, which is implied by the latter possibility, would indeed be a serious competitor for traction applications, since the costs of reforming need no longer be considered. The downside of natural-gas-derived fuels is, inevitably, the limited lifetime of the current supplies, though new discoveries are likely to prolong its availability well into the twenty-first century.

In rehearsing the problems of the anode, it would not be appropriate to lose sight of the fact that, even for a hydrogen-based fuel cell, there are other problems that reduce both efficiency and practicability. Perhaps the most serious is the relatively poor performance of electrocatalysts on the cathode side. We have already seen the effects of electrode irreversibility in our discussion of the inefficiencies found with the metal–air battery and there are similar problems for fuel cells, particularly in the case of the reduction of oxygen. This does not take place at an appreciable rate until relatively high overpotentials (≈ 0.3 V) and a typical current-potential curve for a fuel cell is shown in Figure 6. This illustrates the very large losses at reasonable current densities (≈ 0.5 A cm^{-2}) associated both with the necessity of providing high overpotentials and, at higher current densities, with ohmic losses associated with the finite ionic resistivity of the internal electrolyte. At very high power loads, an additional loss associated with the transport of fuel is encountered: the finite solubilities of hydrogen and oxygen in the ionic electrolyte inevitably lead to difficulties at high enough current densities, at which the cell's performance declines catastrophically.

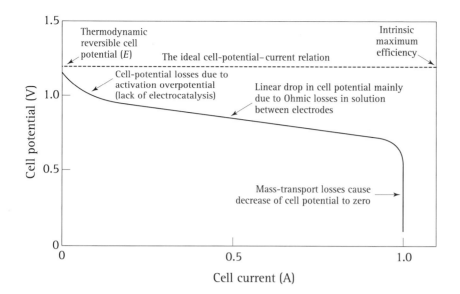

Figure 6
A typical plot of the cell's potential versus the current for fuel cells, illustrating regions of control by various types of overpotential.

The alkaline fuel cell using H_2 as a fuel is the simplest low-temperature fuel cell in concept and operation. It is shown schematically in Figure 5. It operates at temperatures of about 70 °C and, even at room temperature, has power levels of about 50% of those at the nominal operating temperature. Within Europe, stacks have been constructed and tested, by Elenco and Siemens particularly, and considerable experience in the operation of fuel cells in general has been gained.

The electrode's construction is critically important; as in all low-temperature systems, the primary requirements are (a) good electronic conductivity to reduce ohmic losses, (b) adequate mechanical stability and suitable porosity, (c) chemical stability in the rather aggressive alkaline electrolyte and (d) stable electro-activity of the catalyst with time. These are quite demanding considerations, especially insofar as the cell must operate for a number of years. Normally, the electrodes are fabricated from a mixture of carbon and PTFE, the latter controlling both the macro-porosity of the electrode and its hydrophobicity. By using a multi-layer structure, with an outer uncatalysed but highly hydrophobic layer to allow ingress of gas without egress of electrolyte, free-standing electrodes can be made. Furthermore, the cost of the current collector in these systems is relatively low, since nickel mesh can be used.

The primary cost of these electrodes lies in the catalyst, most particularly for the cathode. The Elenco electrodes used platinum, albeit with a relatively low loading (0.6 mg cm^{-2}), but an immense advantage of alkaline electrolytes is that the oxygen-reduction reaction is relatively facile and non-noble-metal catalysts can be used: the Siemens electrodes, for example, use Ti-doped Raney nickel at the anode and silver at the cathode, the latter with a rather high loading of 60 mg cm^{-2}.

The primary problems associated with the alkaline fuel cell are: (a) operation in air is problematic owing to the presence of CO_2, which is absorbed into the alkaline electrolyte, generating the relatively insoluble K_2CO_3, which, in turn, can deposit on and foul the cathode, so that CO_2 must, therefore, be scrubbed from the air if it is used as the oxidant; (b) the poisoning of the electrodes, which is particularly severe for platinum, since the anode can be poisoned by traces of CO in the hydrogen arising from sub-optimal efficiency of the reformer, or by sulfur-containing compounds derived from the primary fuel; and (c) removal of the main combustion product, water, which would otherwise dilute the KOH and reduce performance. This is achieved in the Siemens cell by a remarkable feat of engineering, whereby the electrolyte is pumped through the cell, carrying away waste heat, and is then passed through an evaporator, in which a hydrophobic diffusion membrane eliminates excess water from the electrolyte, passing it back into the stack at its normal strength. It should, however, be emphasized that the evaporator is comparable in size to the stack, giving a rather large overall size for the system.

The alkaline fuel cell is perhaps best known for its role in the Apollo programme, for which its compact structure, high power/weight ratio and reliability made it ideal. The hydrogen and oxygen could be carried in cryogenic form and the product, water, used in the capsule. By using high catalyst loading, the performance of the cell was made to exceed 70% in terms of efficiency. There has recently been a revival of interest in such cells as the basis of a regenerative system in which a store of water can be electrolysed to hydrogen and oxygen with photovoltaics. These gases are then stored for use during the satellite's dark period, when they can be recycled through the fuel cell.

The other main area of interest has been in military applications, for which the low thermal signature, silent operation and pollution-free exhaust of the fuel cell are extremely attractive in terms of hindering detection by common acoustic and infrared equipment. Operation of submarines under these conditions is especially important, but in principle any military vehicle could be adapted. Progress in the adaptation of alkaline fuel cells to traction more generally has been much slower, partly because the economics of using fuel cells in vehicles must ultimately address the problem of cost. The costs of internal combustion engine technology are now very low indeed, with figures of £30–60 kW^{-1} quoted for a normal family car; current fuel-cell technology is far more expensive, with the result that exploration of niche markets is essential. One such market is the powering of city buses, for which environmental considerations are playing an increasingly important role. The overall costs of operation for buses are quite different from those associated with private vehicles and the depreciation of the cells plays a much smaller role in the economic assessments that have been carried out. However, the difficulties of operation with air as an oxidant in these applications, which would be essential if the overall weight of the system is to be minimized, have militated in favour of the solid-polymer-electrolyte systems described in more detail below.

The difficulties associated with the use of air as an oxidant in alkaline fuel cells sparked considerable interest in acidic media, with which CO_2 would present fewer problems. The common mineral acids, however, present quite serious problems in this regard: adequate conductivities can be attained only at temperatures close to boiling, at which the thermal instability of the commoner strong oxy-acids, particularly in contact with powerful noble-metal catalysts, can lead to the formation of decomposition products that poison the electrodes. Perchloric acid was found to be explosively unstable in contact with the fuel, the hydrohalic acids are extremely corrosive and perfluorosulfonic acids, though possessing considerable advantages such as high ionic conductivity, thermal stability, high solubility of oxygen and low adsorption on the platinum catalysts, have led to flooding of electrodes through their high wettabilities on PTFE and have presented considerable concentration-management problems as well as being much more expensive. The restrictions on temperature associated with these acids were particularly troublesome because the kinetics of the oxygen-reduction reaction in acid at low temperature are poor, a drawback that can most easily be overcome by increasing the temperature to take advantage of the positive enthalpy of activation.

Phosphoric acid at room temperature is only slightly dissociated, exhibiting a very low conductivity even in concentrated solution and the reduction of oxygen at low temperatures on platinum has very poor kinetics owing to the strong competitive adsorption of phosphate ions. However, at temperatures above about 150 °C, the pure acid is found predominantly in the polymeric state, as pyrophosphoric acid, which is a strong acid of high conductivity and the large size and low charge density of the polymeric anions leads to low chemisorption of them on platinum, facilitating the oxygen-reduction reaction. Other advantages of phosphoric acid include good fluidity at high temperatures, tolerance of CO_2, a low vapour pressure, a high solubility of oxygen, a low rate of corrosion and a large contact angle. The result is that, since its introduction in 1967, the phosphoric-acid fuel cell has undoubtedly come to dominate the low-temperature fuel-cell market and is the only commercially available fuel cell.

The basic principles of the cell are shown in Figure 7 and the realization of the stack is shown in Figure 8. The reactions taking place are

$$H_2(g) = 2H^+ + 2e^- \tag{31}$$

at the anode and

$$2H^+ + \tfrac{1}{2}O_2 + 2e^- = H_2O(g) \tag{32}$$

at the cathode. The basic design of the stack is bipolar, with separator plates used to delineate the individual cells. Within each cell, impervious graphitic carbon sheets are used for the bipolar plates and each electrode is formed of a ribbed porous carbon-paper substrate onto which is affixed a catalyst layer composed of a high-surface-area carbon powder compacted with PTFE, which acts both as a binder and as a means of controlling the hydrophobicity of the electrode. Between the anode and the cathode there is a porous matrix formed from PTFE-bonded silicon carbide impregnated with phosphoric acid that acts as the electrolyte and also as a separator to prevent the oxidant and fuel gases mixing; since there is some loss of electrolyte, means of replenishment are incorporated into the cell. The operating temperature lies in the range 190–210 °C and active cooling systems, using air, water or a dielectric liquid, are essential.

The performance of the phosphoric-acid fuel cell (PAFC) has been studied carefully. The greatest loss of efficiency is in the stack itself: the efficiency of the stack is given by

$$\eta_{FC} = \frac{\text{electrical energy generated in the cell}}{\text{hydrogen energy consumed in the cell}} \times 100\%$$

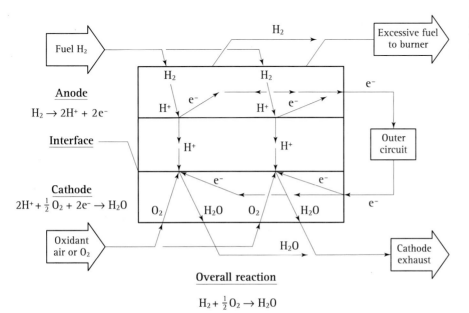

Overall reaction

$$H_2 + \tfrac{1}{2}O_2 \rightarrow H_2O$$

Figure 7
The principles of operation of a phosphoric-acid fuel cell.

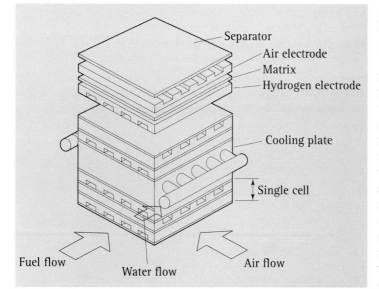

Separator
Air electrode
Matrix
Hydrogen electrode

Cooling plate

Single cell

Fuel flow
Water flow
Air flow

Figure 8
A model of a phosphoric-acid fuel-cell stack system showing the principal components and the water-cooling system.

and for the PAFC, operating under standard conditions of 150–350 mA cm^{-2}, at atmospheric pressure and with a cell voltage of 700–600 mV, $\eta_{FC} \simeq 40$–50%. Other losses in the overall system include: (a) the fuel-processing efficiency (the ratio of the heating value of hydrogen supplied to the fuel cell to the heating value of natural gas supplied to the reformer), which is normally in the range 80–85%; (b) the efficiency of the inverter (the ratio of the AC output power from the inverter to the DC input power from the stack), which normally is very high, about 96%; and (c) the auxiliary power factor, η_m (defined as the ratio of the AC output to load to the AC output power from the inverter) which is usually also very high, about 97%. Combining these together yields a value of about 35–40% for the overall efficiency of a PAFC.

The primary difficulties encountered in the phosphoric-acid system have been the poisoning of the platinum electrocatalyst by H_2S and CO, both of which are present in the reformed fuel. It is vital to incorporate a desulfurizer in the fuel stream to reduce the former to ppm levels, but, by operating above 180 °C, the sensitivity of the platinum to CO has been much reduced, so that it is possible to operate with levels of CO as high as 1–2%. Even so, the demands on purity of the fuel do lead to considerable complexity of engineering in the design of the overall system, with heat management a very significant problem. Even a simplified flow system for the fuel supply consists of a desulfurization stage, a steam reformer, a two-stage shift converter and a separator. Incorporation of these stages into a properly engineered system substantially reduces the flexibility of operation of the overall stack, so the PAFC has been developed mainly for stationary power applications, for which operation under steady-state conditions can take place.

The commercial importance of PAFCs has led to considerable research on the lifetimes of cells, which in turn are known to depend on operating conditions such as working temperatures, voltages, pressures and modes of operation, such as start-up and shut-down conditions. All PAFCs are observed, however, to deteriorate with time, which has been traced to agglomeration of the platinum catalyst particles, corrosion of the supporting carbon, particularly at the cathode, progressive flooding of electrodes and loss of acid. The oxidation of the carbon support has important implications for operation of the cell; oxidation of carbon to CO_2 is known to take place at the cathode at potentials above about 0.8 V and experimental data do show that there is a much higher rate of deterioration at cell voltages in excess of this value. In addition, rapid changes in demand are found to be highly detrimental to performance, since uncontrolled pressure differences that can lead to flooding or starvation of the electrode can build up. In this regard, proper control of the reactant gases supplied to the cell is essential, particularly under conditions of varying demand.

The development of the PAFC has taken two directions. The more important is the construction of relatively large stationary power units, such as the 11 MW system at Goi, near Tokyo in Japan, with smaller systems in the USA and Europe, which have allowed technologists to study the operation of these systems under real conditions. Smaller-scale local combined heat and power (CHP) systems are also being actively developed, their key advantage being the lack of any efficiency penalty for partial-load operation, even under 50% load; this balance is particularly suitable for applications such as in small hospitals. There has also been considerable investigation of vehicular applications, using reformed methanol as the fuel, which has been directed towards use for buses as the primary commercial goal. Again, it would seem that niche markets may dictate the success of such systems in the next few years.

The necessity of employing noble-metal catalysts and the sensitivity of the phosphoric-acid fuel cell to significant quantities of CO are serious drawbacks; only by working at higher temperatures can the kinetics of the oxygen-reduction reaction be sufficiently accelerated for the use of cheaper catalysts and, as indicated above, high temperatures also favour the desorption of adsorbed CO. In addition, heat from higher-temperature fuel cells can be used in combined-cycle operation, allowing us to increase the system's efficiency quite substantially, albeit at the cost of flexibility of operation. An additional advantage of high-temperature operation is an improvement in compatibility with the operating temperature of the reformer, allowing us, at least in principle, the attractive prospect of combining the reformer with the fuel-cell stack.

The earliest high-temperature fuel cell to be developed was the molten-carbonate fuel cell (MCFC), whose basic principle of operation is shown in Figure 9. At the cathode, made from porous lithiated NiO ($Li_xNi_{1-x}O$, $0.022 \leq x \leq 0.04$), the reaction is

$$\tfrac{1}{2}O_2 + CO_2 + 2e^- = CO_3^{2-} \tag{33}$$

The electrolyte is a eutectic mixture of 68% Li_2CO_3 and 32% K_2CO_3, at a temperature of 650 °C, retained in a porous γ-$LiAlO_2$ tile. At the anode, made from a porous Ni–10 wt% Cr alloy (the Cr preventing sintering), the reaction is

$$H_2 + CO_3^{2-} = H_2O + CO_2 + 2e^- \tag{34}$$

In a practical cell, the CO_2 produced at the anode must be transferred to the cathode, which can be carried out either by burning the spent anode stream with excess air (to remove any unspent H_2) and mixing the result with the cathode's inlet stream after removal of water vapour or by directly separating CO_2 at the anode's exhaust point. A major advantage of this cell is that CO, far from being a problem, can actually serve as the fuel, probably through the water-gas-shift reaction.

$$CO + H_2O = CO_2 + H_2 \tag{35}$$

which at 650 °C equilibrates very rapidly on Ni.

We have seen that, at least in principle, the temperatures of the reformer and stack are sufficiently comparable for the two to be combined. Three configurations have been suggested: (a) indirect reforming, in which a conventional reformer, operating separately from the stack, is provided; (b) direct internal reforming, in which the reformer catalyst pellets are incorporated into each of the anode's gas-inlet channels and therefore operate at 650 °C, which temperature is maintained by direct heating of the anode compartment; and (c) indirect internal reforming, in which the reformation takes place in separate catalyst chambers installed in a stack between a set of cells so that waste heat from the fuel-cell stack is supplied to help maintain the reformer's temperature during the endothermic process of reformation. The most elegant solution is (b), which is also the most cost-effective solution, but contamination from electrolyte vapour in the anode space leads to rapid degradation of the catalyst's performance.

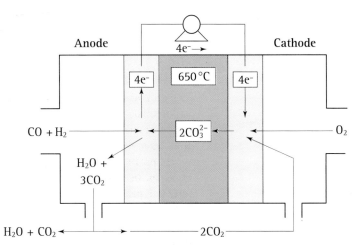

Figure 9

A schematic diagram of a molten-carbonate fuel cell.

The overall performance figures for the MCFC reflect the fact that the $E°$ value for reaction (30) above is reduced from the room temperature value of 1.23 V to a value of 1.02 V at 650 °C. The cell's voltage drops almost linearly with increasing current; at ambient pressures, state-of-the-art cells have voltages of 0.70–0.75 V at 150–160 mA cm^{-2}, though this can be appreciably enhanced by increasing the pressure.

Problems with the MCFC are (a) the oxygen-reduction reaction remains relatively slow and its mechanism depends critically on the electrolyte, the initial reduction product being either peroxide in Li-rich mixtures or superoxide in K-rich electrolytes, with both present at the eutectic composition; and (b) dissolution of NiO leads to serious problems since the Ni(II) can migrate to the anode where, at the low potentials, it is deposited as the metal. The deposition of metal grains in the pores of the tile can lead eventually to the formation of electronically conducting pathways through the electrolyte, effectively short-circuiting the cell. The dissolution of Ni is related to the partial pressure of CO_2 and may involve the reaction

$$NiO + CO_2 = Ni^{2+} + CO_3^{2-} \tag{36}$$

Furthermore, (c) water must be present in the feed gas in order to avoid the deposition of carbon through the Boudouard reaction:

$$2CO = CO_2 + C \tag{37}$$

and (d) management of the electrolyte is a key problem in the MCFC; the control of the three-phase boundary relies on controlling the extent to which the molten carbonate is drawn out of the tile by capillary action at the cell's operating temperature. By careful choice of the pore size in the electrode and tile and by partially filling the porous electrodes themselves in advance, an initially optimum distribution of the electrolyte can be ensured, but, throughout the lifetime of the cell, a slow but steady loss of electrolyte has been found to occur, leading to a gradual decay in performance. This decay may, however, become catastrophic if loss of electrolyte from the tile leads to the tile becoming permeable to the fuel gases; the crossover of gas is usually accompanied by the formation of intense local hot spots and a rapid loss of performance, so it is essential that a liquid layer be maintained at all times.

At a sufficiently high temperature, all kinetic limitations at the cathode will disappear, and it also becomes possible to utilize solid ceramic-oxide ion conductors that have very high conductivities above about 900 °C. The principle of operation of this type of solid-oxide fuel cell (SOFC) is shown in Figure 10. The electrolyte is typically ZrO_2 with 8–10 mol% Y_2O_3; the latter not only stabilizes the fluorite structure, preventing transition at lower temperatures to the baddeleyite phase with resultant shattering of the ceramic, but also confers a substantial ionic conductivity through the presence of mobile O^{2-} ions. There have been attempts to produce improved electrolytes, which must: (a) have high oxygen-ion conductivities and minimum electronic conductivities; (b) exhibit good chemical stability with respect to the electrodes and inlet gases; (c) have a high density to inhibit the crossover of fuel; and (d) undergo thermal expansion compatible with other components. Materials such as Bi_2O_3 and doped CeO_2 do have higher oxide-ion conductivities at lower temperatures than that

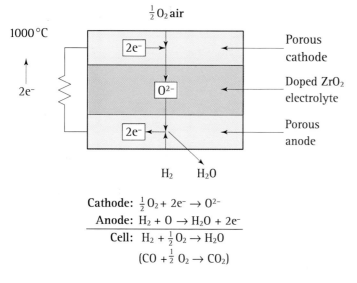

Cathode: $\frac{1}{2}O_2 + 2e^- \rightarrow O^{2-}$

Anode: $H_2 + O \rightarrow H_2O + 2e^-$

Cell: $H_2 + \frac{1}{2}O_2 \rightarrow H_2O$

$(CO + \frac{1}{2}O_2 \rightarrow CO_2)$

Figure 10

The principles of operation of a high-temperature solid-oxide-electrolyte fuel cell.

of stabilized zirconia, but are more easily reduced to electronic conductors at the anode.

The requirements on the anode are that it be an effective oxidation catalyst; have a high electronic conductivity; be stable in the reducing environment; undergo thermal expansion compatible with the electrolyte and other fuel cell components, have a physical structure offering low resistance to the transport of fuel, exhibit chemical and mechanical stability and be tolerant of sulfur-containing contaminants. The anode most closely satisfying these requirements is a porous about 35% $Ni-ZrO_2/Y_2O_3$ cermet (i.e. an intimate mixture of ceramic and metal) with good electronic conductivity. Unlike conventional electrodes, there is no necessity for a well-defined three-phase boundary, since oxidation of the fuel gas can take place over the entirety of the electrode's surface in this mixed electronic/ionic conductor. The anode reaction is

$$H_2 + O^{2-} = H_2O + 2e^- \qquad (38)$$

The cathode must (i) exhibit a good electrocatalytic activity for the reduction of O_2 and (ii) have a good electronic conductivity since it must serve as the current collector. The combination of a high temperature and an oxidizing atmosphere leads to severe materials problems at the cathode, which must be stable over very wide ranges of partial pressure of oxygen and involatile, must not undergo any destructive phase changes, should adhere strongly to the electrolyte over a very wide range of temperature, possessing, therefore, a coefficient of expansion essentially identical to that of ZrO_2, and should form a junction of very low resistance with the zirconia. The cathode material most closely satisfying these requirements is porous perovskite manganite of the form $La_{1-x}Sr_xMnO_3$ ($0.10 < x < 0.15$), which undergoes a transition from small-polaron to metallic conduction near 1000 °C. It also exhibits mixed ionic/electronic conduction, again allowing reduction of the oxygen to take place over the entire surface. The cathode reaction is

$$\tfrac{1}{2}O_2 + 2e^- = O^{2-} \qquad (39)$$

Realization of the cell in practical terms is through three competing designs, a planar geometry, similar to conventional designs, a tubular design, shown in Figure 11, and a monolithic design. All these designs, however, still need to overcome quite serious problems with interconnection and fabrication technology.

Although the solid-oxide fuel cell (SOFC) clearly possesses a number of advantages, there remain some severe problems.

(a) The thermodynamic cell voltage is only 0.9 V, at 1000 °C. However, because there are essentially no kinetic limitations on the reactions at the cathode and anode, reasonable current densities can be obtained at voltages of 0.75 V.

(b) The materials problems, particularly those related to thermal expansion and stability against mechanical damage, are proving extremely difficult to solve.

The intolerance of SOFCs towards repeated shut-down–start-up cycles, which is related to the problem of compatibility of thermal expansion, suggests that the most likely type of application to be envisaged for SOFCs is that of stationary power generation under steady-state

Figure 11
A schematic diagram showing the Westinghouse tubular solid-oxide fuel-cell system in its 'bundle' configuration.

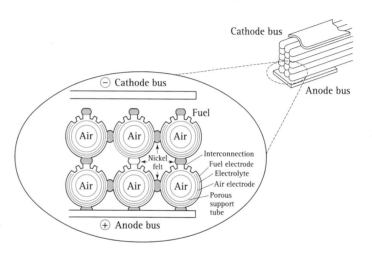

Figure 12

A schematic diagram of the operation of a solid-polymer-electrolyte H_2/O_2 fuel cell.

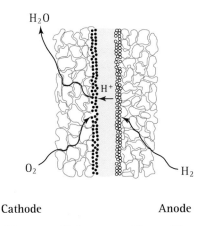

H_2O

H^+

O_2

H_2

Cathode	Anode
$\frac{1}{2}O_2 + 2H^+ + 2e^- \rightarrow H_2O$	$H_2 \rightarrow 2H^+ + 2e^-$

conditions. Westinghouse, in particular, has constructed 25 and 100 kW systems and there is strong interest both in the USA and in Australia in developing this technology to commercialization.

By analogy with the solid-oxide-based fuel cells above, which are dependent on oxide-ion-conducting electrolytes, it should be possible to find fuel cells based on solid proton conductors. In fact, *high-temperature* solid proton conductors appear not to exist; at these elevated temperatures, all hydrated oxides tend to lose water and, if conduction takes place at all, it will be through the metal or the oxide ion. However, at low temperatures (below about 150 °C), there are several types of solid proton conductor, both inorganic and organic, that can be used in so-called solid-polymer-electrolyte (SPE) fuel cells that operate in this low-temperature regime.

A schematic diagram of this type of cell is shown in Figure 12: the electrolyte itself is a polymeric-membrane proton conductor and the anode and cathode are, in the simplest designs, formed either directly from metal particles or from catalysed carbon particles bound to the membrane. The current collectors are porous plates of carbon or graphite and the cell reactions are

$$H_2 = 2H^+ + 2e^- \tag{40}$$

at the anode and

$$\frac{1}{2}O_2 + 2H^+ + 2e^- = H_2O \tag{41}$$

at the cathode. The low operating temperatures dictate the use of noble-metal catalysts and particular problems are experienced at the cathode.

Critical to the success of the solid-polymer-electrolyte fuel cell (SPEFC) is the electrolyte and the revival of interest in this configuration is due to the success of Du Pont and latterly Dow Chemical Co. in developing polymers containing perfluorinated sulfonic acid moities. These consist of a PTFE backbone to which pendant sulfonic-acid groups are bonded as shown in Figure 13. These materials have the secondary advantage of being dispersible in certain organic solvents, such as ethanol, allowing fabrication of membranes with a variety of thicknesses and also allowing the use of this sol as a binder for the catalysed carbon particles, thereby greatly increasing the utilization of the electrode by its conversion to a mixed conductor.

The performance of such cells, particularly under a few bars pressure, is quite remarkable, as is shown in Figure 14, in which power densities in excess of 1 W cm^{-2} can be seen. These high performances should not, however, disguise the fact that, if application of the SPEFC to transport is considered, particularly if methanol is used as the fuel and a reformer employed, there are several additional sources of loss. Operation of the fuel cell at 5 bar leads to an air-compressor loss of about 12% of the stack's output and reformer losses of 10–15% of the heating value of the input methanol fuel are expected. Further losses to ancillary units and in power converters etc. can lead to overall power efficiencies on going from the chemical energy of methanol to power used to drive wheels of about 15%.

In contrast, furthermore, to the high-temperature cells above, the SPEFC is very sensitive to contaminants in the fuel gas; in particular, the level of CO must be reduced to well below 10 ppm to avoid deterioration of the anode's performance. This type of purity can be achieved only by using multi-stage reformers and considerable effort has recently been expended on developing promoted-platinum catalysts that are less sensitive to residual CO.

The high power density of SPEFCs makes them extremely attractive for traction applications, for which the size and weight of the fuel cell should be as small as possible. Ballard has recently designed and constructed a bus that uses compressed hydrogen in cylinders as the fuel and has been testing this in Vancouver.

All the cells previously described have been based on hydrogen as a fuel, but, as the Ballard bus shows, transport of this fuel clearly imposes a considerable weight penalty. The alternative, which is to use methanol as a fuel and use a reformer, is both costly and fraught with engineering difficulties. Highly desirable would be a fuel cell that could directly oxidize methanol at the anode, but retained the high power/weight ratio of the SPEFC described above. Such a fuel cell is now under active development in Europe and the USA and the principles are shown in Figure 15. The anode reaction is

$$CH_3OH + H_2O = CO_2 + 6H^+ + 6e^- \qquad (42)$$

and the cathode reaction is

$$\tfrac{3}{2}O_2 + 6H^+ + 6e^- = 3H_2O \qquad (43)$$

Methanol possesses a number of advantages as a fuel: it is a liquid and therefore easily transported and stored and dispensed within the current fuel network; it is cheap and plentiful; and the only products of combustion are CO_2 and H_2O. The advantages of a direct methanol fuel cell are that changes in demand for power can be accommodated simply by alteration of the supply of the methanol feed; that the fuel cell operates at temperatures below about 150 °C, so there is no production of NO_x; and that methanol is stable in contact with the acidic membrane and easy to manufacture.

The basic problems currently faced by the direct methanol fuel cell (DMFC) are that (a) the reaction at the anode has poor electrode kinetics, particularly at lower temperatures, making it highly desirable to identify better catalysts and to work at as high a temperature as possible; (b) the reaction at the cathode, the reduction of oxygen, is also slow, though the problems are not so serious as are those with aqueous mineral-acid electrolytes (nevertheless, the overall power density of the direct methanol fuel cells is much lower than the 600 mW cm^{-2} or more envisaged for the hydrogen-fuelled SPEFC); and (c) perhaps of greatest concern at the moment is the permeability of the current perfluorosulfonic-acid membranes to methanol, allowing considerable

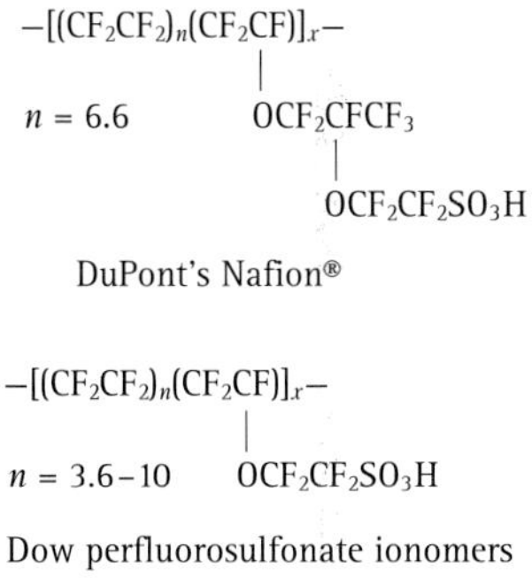

$$-[(CF_2CF_2)_n(CF_2CF)]_x-$$

$n = 6.6 \qquad \underset{|}{}OCF_2CFCF_3$

$$OCF_2CF_2SO_3H$$

DuPont's Nafion®

$$-[(CF_2CF_2)_n(CF_2CF)]_x-$$

$n = 3.6 - 10 \qquad OCF_2CF_2SO_3H$

Dow perfluorosulfonate ionomers

Figure 13
A comparison of Du Pont's Nafion and Dow's perfluorosulfonate ionomer membrane.

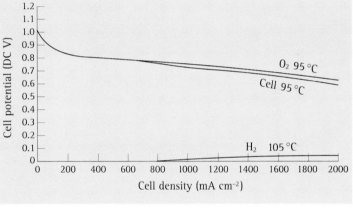

Figure 14
Cell and half-cell potentials versus the current density for a single cell with Dow membrane (thickness 125 μm) and Pt-sputtered Prototech electrodes (Pt loading 0.45 mg cm^{-2}) operating at 95 °C with H_2/O_2 at 4–5 atm.

Figure 15

The solid-polymer-electrolyte fuel-cell configuration for the direct methanol fuel cell.

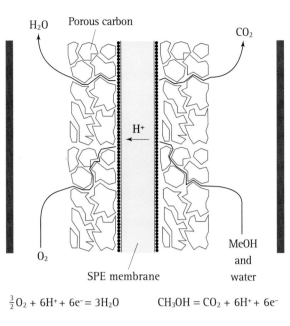

$$\tfrac{3}{2}O_2 + 6H^+ + 6e^- = 3H_2O \qquad CH_3OH = CO_2 + 6H^+ + 6e^-$$

Figure 16

(a) The effect of the partial pressure of oxygen on the cell's performance for a DMFC operating at 97 °C with a vapour feed of 2 M methanol.
(b) The effect of the partial pressure of oxygen on the power output of the cell in (a).

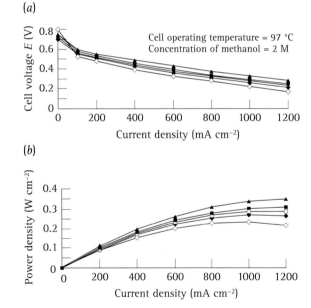

crossover of fuel. This leads both to degradation of performance, since a mixed potential develops at the cathode, and to a deterioration of the utilization of fuel. Methanol vapour also appears in the cathode's exhaust, from which it would have to be removed.

In spite of these difficulties, the DMFC does have the capability of being very cheap and potentially very competitive with the internal combustion engine, particularly in niche city-driving applications, for which the low pollution and relatively high efficiency at low load are attractive features. Performances from modern single cells are highly encouraging: an example from our own work is shown in Figure 16. It can be seen that, in oxygen, power densities of up to 0.35 W cm^{-2} are possible; in air a power density of 0.2 W cm^{-2} has been attained with a pressure of 5 bar.

Photoelectrochemical Cells

We have seen how both batteries and fuel cells will form essential components of sustainable energy systems, but that both are dependent on prior conversion of energy in the form of hydrocarbons, nuclear sources, solar/wind/wave or other renewable sources into electrical energy or into a secondary energy vector such as hydrogen or methanol. Ideally, we would like to complete a sustainable cycle by finding ways of directly converting solar energy into chemical or electrical energy, thereby breaking free of dependence on fossil fuels. Considerable success has been achieved in this area, with hydro-electrical power a long-standing example of the indirect harnessing of solar power. More recently, the harnessing of solar power through wind and wave conversion has reached the point at which attractive niche markets can be identified, but there would be huge attractions in identifying systems that could convert solar energy directly into chemical energy, essentially mimicking the process of photoelectrolysis. This review concludes with a brief description of the development of such processes, using electrochemical principles.

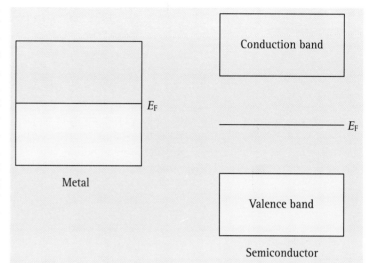

Figure 17
Electronic structures of metals and semiconductors showing occupied (hatched) and unoccupied levels.

With the development of the transistor during the late 1940s, the attention of the Bell group turned to other applications of semiconductor technology, including increasing our understanding of the electrochemical behaviour of semiconductors. These studies were driven, in the first instance, by the need to improve our understanding of the semiconductor-etching processes, which were recognized to be electrochemical in origin, but very different from the etching processes familiar from the classical electrochemistry of metals. In order to understand these differences, we should first consider the electronic structures of metals and semiconductors, shown in Figure 17. In metals, overlap of the valence orbitals leads to a band of allowed and partially occupied energy levels, in which, broadly speaking, the kinetic energy of the electrons increases the higher the energy level in the band. In a semiconductor, in contrast, the occupied valence orbitals overlap to form a 'valence band' that is filled with electrons; separated in energy from this band is a 'conduction band' of low or zero occupancy formed by the overlap of unoccupied orbitals. For metallic behaviour, two conditions must be satisfied : first, the band of energy levels must be only partially occupied; secondly, the extent of the overlap of orbitals and the resultant width of the band of levels must be such that sufficient kinetic energy to overcome the changes in Coulombic repulsion encountered on moving an electron from one metal atom/ion to the next is available to the electrons. This second condition leads, for example, to the fact that CoO is not a metal, in spite of the fact that it possesses a partially filled d orbital band: this band is extremely narrow, since d–d overlap between Co^{2+} centres is very low, and considerable energy is required to effect the process $Co^{2+} + Co^{2+} \rightarrow Co^{3+} + Co^{+}$.

The first condition is strictly necessary for metallic behaviour; only if the allowed band of levels is partially occupied can there be a net drift of electrons on application of an external electrical field. This is because, in the absence of an external electrical field, there are as many electrons with momentum in any direction in the crystal as there are in the opposite direction. To obtain a net drift, more electrons must be excited into levels with the momentum in the direction of the field than there are in levels with the momentum opposed to the applied field. For the filled valence band in a semi-conductor such excitation is not possible and, given the low number of electrons likely to be thermally excited across the bandgap into the otherwise unoccupied conduction band, the conductivity of the material will be very low, as shown in Figure 17. Indeed, at temperatures close to 0 K, the conductivity of any semiconductor will tend to zero, but there are two mechanisms that allow some conductivity at higher temperatures. The first is thermal excitation of electrons from the valence band to the conduction band: this begins to populate the latter and, provided that the band is broad enough to satisfy the second condition above, the electrons, once they are in the conduction band, are free to move on application of an external field. Interestingly, the excitation of electrons from the valence band leads to the presence of vacancies in the latter; these vacancies are termed 'holes' and these holes behave essentially as positively charged carriers, of mass usually similar to that of the electrons in the conduction band. On application of an electrical field, these holes migrate in the opposite direction to that of the electrons, constituting a second contribution to the conductivity. Semiconductors that derive their conductivity from this mechanism are termed intrinsic.

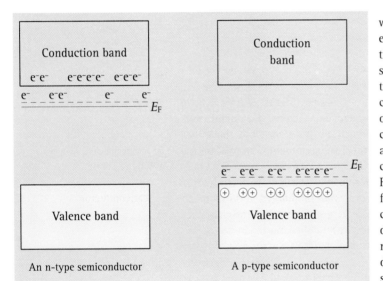

Figure 18
The origin of conductivity in *extrinsic* semiconductors by excitation from donor levels near the conduction band (n-type conduction) or to acceptor levels near the valence band (p-type conduction).

For many semiconductors, particularly those with a bandgap in excess of about 1 eV, thermal excitation of electrons from valence to conduction bands becomes very improbable and so a second mechanism for enhancing the conductivity becomes important. The presence of certain types of impurity, possessing either occupied electronic energy levels just below the conduction band edge or unoccupied levels just above the valence band edge, can give rise to conductivity by the mechanism illustrated in Figure 18. In the former case, thermal excitation from the donor level to the conduction band is clearly easy, though the holes left behind in the donor levels are not themselves normally mobile, so the only mobile carriers are the conduction-band electrons. For this reason, such semiconductors are termed extrinsic n-type. Analogously, in the second case, excitation of electrons from the valence band to the unoccupied levels of the acceptor states can give rise to mobile holes in extrinsic p-type semiconductors.

Thermodynamically, the free energy required to add an electron to the semiconductor or take an electron away from the semiconductor is the chemical potential, $\mu_{e^-}^0$, of the electron, which is termed the electronic Fermi level. In the absence of any external electrical field, the Fermi level must be the same throughout the semiconductor: if this were not so, the electrons would diffuse until equilibrium were reached. In the presence of an electrical field, the thermodynamic quantity of interest is the electrochemical potential, given by $\bar{\mu}_{e^-}^0 = \mu_{e^-}^0 - e_0\varphi$, where φ is the local electrical potential and e_0 is the charge of an electron (1.602×10^{-19} C), and this effectively defines the Fermi level. It is this quantity that remains constant, even in the presence of an electrical field.

For both types of extrinsic semiconductor, the number of carriers per unit volume is determined by the density of impurity states; commonly, such densities lie in the range 10^{13}–10^{19} cm^{-3}, numbers that may seem very high until we realize that the number density of atoms or ions in the crystal will be about 10^{23} cm^{-3} and that this will also be the number density of free electrons in a normal metal. Hence, the density of electrons in a semiconductor will normally lie far below the corresponding density in a metal, with the result that the conductivity will be much lower. This lower carrier density has a second important consequence arising from a fundamental law of physics. It is well known that, when an electrical field is applied to a metal, the potential drop associated with that field must take place outside the metal, usually at the metal's contacts: any change in potential inside the metal is prevented by the mobility and number density of the carriers. This can be understood from the Poisson equation

$$\partial^2 V/\partial x^2 = -\rho/(\epsilon\epsilon_0) \tag{44}$$

which relates the second derivative of the potential in the x direction to the density of charge carriers, ρ, and the dielectric function of the semiconductor, ϵ. The symbol ϵ_0 is given to the dielectric constant of the vacuum and is a fundamental physical quantity, with the value 8.854×10^{-12} F m^{-1}. Equation (44) tells us that the greater the value of ρ the steeper the gradient of potential that can be accommodated. In the extreme, for metals, equation (44) shows that changes in potential of the order of a few volts can be accommodated in that atomic layer right at the surface. However, for carrier densities normally encountered in extrinsic semiconductors, the distance over which a potential drop of 1 V can be accommodated is far greater, often of the order of 1 μm. This situation is illustrated in Figure 19(a), in which a semiconductor/metal junction is shown. The semiconductor is n-type and a positive potential is applied, moving the Fermi level down by an amount $e_0\varphi$. It is apparent that, for the equilibrium situation shown in Figure 19(a), for which the Fermi levels of the metal and the semiconductor are equal, the energy levels of the edges of the valence and conduction bands vary approximately parabolically with the distance from the interface with the metal, becoming constant only at distances more than a few micrometres away from the interface.

A similar situation arises at the junction of a semiconductor and an electrolyte, as shown in Figure 19(b). The reason is that the concentration of charged ions in a molar electrolyte is again far larger than that expected for a normal extrinsic semiconductor. Once again, the change in potential at the interface must be accommodated within the semiconductor and, if there is a fast redox couple in solution from and to which electrons can easily be transferred from the semiconductor, then, at equilibrium, the Fermi level in the semiconductor must be equal to the Fermi level of the redox couple. This concept, it should be said, has had its critics; however, provided that the concentrations of oxidized and reduced forms of the redox couple are equal, the Fermi level can, with little loss of accuracy, be put equal to the redox potential of the couple, $E°$. It is worth pointing out that the value of $E°$ is invariably given with respect to the standard hydrogen potential, which is arbitrarily given the value 0 V. However, the energy levels of the semiconductor and the associated Fermi energy are given with respect to an electron remote from the semiconductor in vacuum. Ideally, we would like to connect these two energy scales, which is far from being an easy thing to do, since we require the electronic energy levels of single hydrated ions. Such information is not available, even in principle, from experiment, but calculations of increasing sophistication have been performed in recent years, suggesting that the standard hydrogen electrode is positioned about 4.7 eV below the vacuum level. Hence, to obtain the position of a redox couple in aqueous solution, we should add 4.7 eV to the value of $E°$.

The situation in Figure 19(b) is appropriate for an n-type semiconductor at positive (sometimes referred to as reverse) *bias*; there is clearly a particular potential at which the energy bands will be flat, i.e. there will be no field inside the semiconductor. This potential is termed the *flat-band potential* and it is clearly of considerable fundamental importance, since, at that potential, we can determine the internal Fermi level and

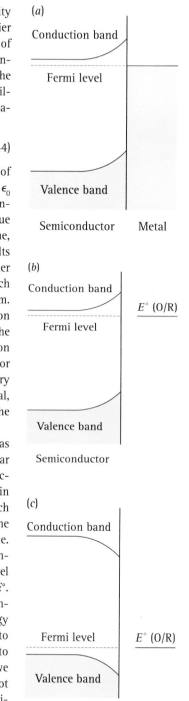

Figure 19
(a) An ideal metal–semiconductor junction; (b) an ideal n-type semiconductor–electrolyte junction; (c) an ideal p-type semiconductor–electrolyte junction.

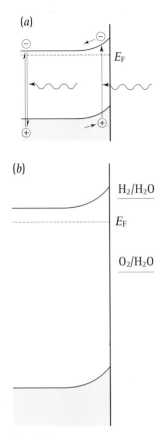

Figure 20

(a) The effect of illumination at an ideal n-type semiconductor-electrolyte interface: photogenerated electrons are shown as ⊖ and the corresponding holes as ⊕. (b) The energetics of the n-TiO$_2$–water interface, showing the relative energies of the band edges and the redox couples associated with the electrolysis of water.

hence make a reasonable estimate of the energy of the conduction band, assuming that the separation of the Fermi level from the latter is small (this is a reasonable assumption for most n-type extrinsic materials). If the semiconductor is biased negatively relative to the flat band, the Fermi level will clearly intersect the conduction band; electrons will flow from the interior of the semiconductor to the surface, increasing ρ and allowing the change in potential to be accommodated over a much smaller region. This situation, termed *forwards bias*, often leads to chemical reaction at the semiconductor's surface; from our perspective this is undesirable, as will be seen below, so we will not consider it further. However, an important variant is the situation with regard to p-type semiconductors, as shown in Figure 19(c). It is clear from Figure 19(c) that, to move the Fermi level in a p-type semiconductor away from the valence-band edge, we must bias the semiconductor *negatively* relative to the flat band (which is sometimes referred to in this context as *reverse* biased). The result, as shown, is that the energies of the valence and conduction bands now curve in the opposite direction from those in a reverse-biased n-type semiconductor, but that the basic principles otherwise remain unaltered.

The fact that, in a reverse-biased semiconductor, the region within which the potential drop is accommodated extends a significant way into the surface has opened the way towards devices that can harness solar energy. If we consider a simple reverse-biased n-type semiconductor in contact with a strong electrolyte, as shown in Figure 20(a), then, on illumination with light of sufficient energy to excite an electron from the valence to the conduction band, two processes may take place. If absorption takes place in a region remote from the interface, then the electron and hole formed by absorption of the light will eventually recombine, usually around an impurity centre. However, if the light is absorbed close to the interface, then the hole and electron are formed in a region where there is an electrical field; this field will tend to drive the electrons towards the interior of the semiconductor, whilst driving the holes to the surface. Once the holes reach the surface, they may either become trapped at the surface site, eventually leading to oxidation of the semiconductor itself, or they may oxidize a suitable redox couple in the solution, essentially by inducing the transfer of electrons from the couple to the hole vacancy in the valence band. For most classical semiconductors, such as n-Si and n-GaAs, the first process is overwhelmingly favoured unless a fast redox couple such as $[Fe(CN)_6]^{3-/4-}$ is present in solution, but, in 1973, Fujushima and Honda in Japan showed that, for n-TiO$_2$, even the highly irreversible O$_2$/H$_2$O couple could be driven. That is, upon illumination, n-TiO$_2$ under positive bias could evolve oxygen without undergoing decomposition. The position of the energy bands of TiO$_2$ in an acid electrolyte with respect to the thermodynamic positions of the H$_2$/H$_2$O and O$_2$/H$_2$O couples is shown in Figure 20(b). It is evident that, although the holes certainly have enough energy to oxidize water, the electrons, by the time they have reached the bulk of the semiconductor, cannot spontaneously reduce water to hydrogen. If we are to do this, we must obviously raise the energy of the conduction-band edge. Goodenough and co-workers showed that this could be done by utilizing the cubic perovskite phase, SrTiO$_3$, in which the Ti–Ti interactions being weaker than those in rutile (TiO$_2$) leads to a decrease in the 3d bandwidth of Ti. This raises the energy of the conduction-band edge sufficiently for spontaneous photoelectrolysis of water to take place on illumination, as shown in Figure 21.

The photoelectrolysis of water is a stunning achievement in electrochemistry, opening the door to a completely sustainable hydrogen economy. Unfortunately, it has proved impossible to make the device shown in Figure 21 sufficiently efficient to make its commercial development worthwhile. The reason is that the bandgap of SrTiO$_3$ is 3.4 eV, which is higher in energy than any visible light. In other words, the cell shown in Figure 21 will work only in the ultraviolet: the fraction of solar irradiation absorbed is a tiny fraction, less than 1% of the sun's rays. Attempts to increase the energy of the valence-band edge have led to unstable materials, poor kinetics for the evolution of oxygen or simply very low quantum efficiencies for electron–hole separation. Attempts to turn the cell round have also been made, by using a p-type semiconductor. A cell based on p-GaP is shown in Figure 22. Such a cell will certainly evolve hydrogen, but a rather high overpotential is needed and the energy of the holes is quite insufficient to drive the oxidation of water at the counterelectrode.

The direct photoelectrolysis of water has, therefore, proved extremely difficult, though we return to some recent and very promising results below. The failure of photoelectrochemical experiments to lead to efficient solar-power conversion did not inhibit the invention of other potential types of solar cell. In principle, it is possible to deconstruct the photoelectrochemical process by first converting the solar irradiation into electrical energy and then using this to drive a conventional electrolysis cell, in which water is directly split into hydrogen and oxygen. The inspiration for this approach came from one of the earliest semiconductor devices, the p–n junction, shown in Figure 23. In this device, contact between a p-type and an n-type semiconductor is made. The Fermi levels in the two materials must be equal at equilibrium and this leads to the formation of an internal electrical field at the junction. Illumination of this junction then generates holes and electrons that are separated by the field, giving rise to an electrical current. A close analogy of this is shown in Figure 24. Illumination of an n-type semiconductor/electrolyte junction in the presence of a redox couple of appropriate reversibility and energy allows current to flow round the circuit from the semiconductor to the metal counterelectrode, again generating an electrical current. This type of device has proved far easier to fabricate and efficiencies in excess of 10% have been reported, which compare not unfavourably with those of solid-state p–n junctions. However, long-term stability has proved a big headache, particularly with the more classical semiconductors such as GaAs and InP, whereas the oxide semiconductors, although they are more stable, have unacceptably high bandgaps.

This problem has been addressed in recent years by combining oxide semiconductors with materials termed sensitizers. The latter are dyes, many of which were developed initially for the photographic industry. The mode of action of a semiconductor sensitizer is illustrated in Figure 25: on illumination, the dye (D) absorbs a photon, with the concomitant excitation of an electron from the highest occupied molecular orbital (HOMO) to the lowest unoccupied molecular orbital (LUMO). Provided that the LUMO lies just above the conduction-band edge in energy, injection of electrons into the conduction band of the semiconductor can take place, leaving an oxidized form of the dye on the surface. Provided that a kinetically facile redox couple is present in solution, this oxidized form of the dye can be re-reduced, allowing the cycle to begin again. The importance of this type of system is that very stable dyes based on ruthenium are known and these dyes can be adsorbed onto colloidal TiO_2 films to form exceptionally efficient solar-conversion devices. Such devices were first fabricated by Grätzel and co-workers in 1991 and have since shown substantial promise, particularly in view of the fact that they are inexpensive and easy to make.

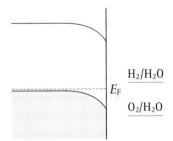

Figure 21
The sustained photoelectrolysis of water with n-SrTiO$_3$ as the anode and a noble-metal cathode.

Figure 22
The energetics of the p-GaP/water inferface, showing the relative energies of the valence and conduction bands in this material and the redox couples associated with the electrolysis of water.

Figure 23
An idealized p–n junction, showing how the absorption of light leads to a photovoltage and a photocurrent.

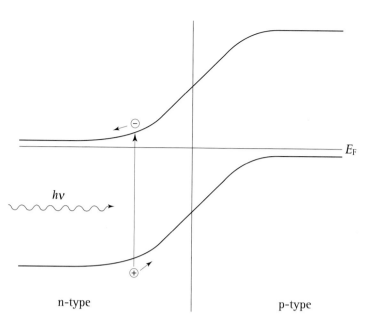

Figure 24
A schematic diagram of a photoelectrochemical cell for the conversion of light into electrical energy.

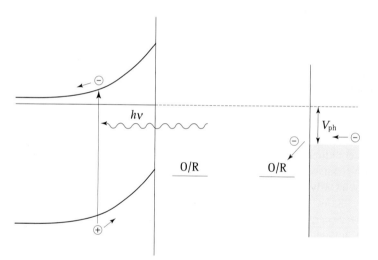

Figure 25
The mode of action of a dye sensitizer attached to the surface of a wide-bandgap n-type semiconductor such as TiO_2.

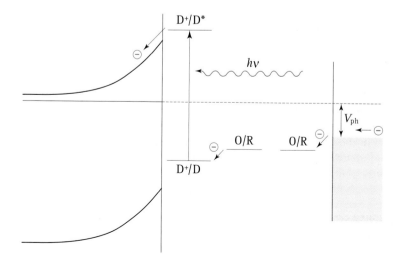

(a)

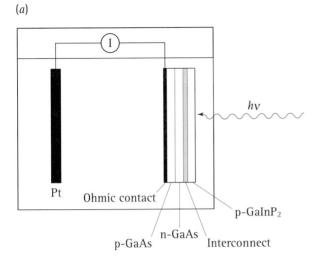

(b)

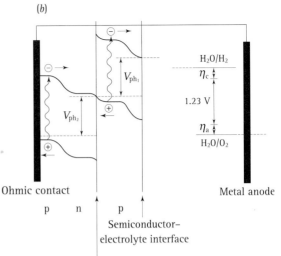

Figure 26
(a) The physical construction of a two-photon photoelectrochemical cell for the electrolysis of water. (b) The principles of operation of a two-photon photoelectrolysis cell.

The comparative success of these 'photovoltaic' cells has not prevented electrochemists from dreaming of creating a single cell in which photoelectrolysis could be achieved. As our understanding of the process of photosynthesis in nature has improved, it has become apparent that our objective of achieving both solar conversion and electrolysis using a single photon energy has been abandoned by nature as simply too difficult. Instead, nature uses a two-photon process, one to drive the generation of hydrogen (or, more accurately, to reduce carbon dioxide to carbohydrate) and one to drive the evolution of oxygen. Several possible devices have been suggested, including some in which two different semiconductor or dye-sensitized electrodes are separately illuminated, but efficiencies remained disappointingly low until very recently, when Khasalev and Turner reported having built a simple monolithic device comprising a fairly high bandgap semiconductor, p-GaInP$_2$, attached to a p–n junction based on GaAs. Hydrogen is evolved at the p-GaInP$_2$/electrolyte interface, whilst light of lower energy than the bandgap of this material then passes through to the p–n junction, allowing the formation of holes sufficiently energetic to oxidize water at a metal/solution contact, as shown in Figure 26. This device is the first photoelectrolysis unit to exceed 12% efficiency directly and, although it is too costly to develop commercially at the moment, it is still a very attractive unit.

As the focus of attention broadened away from the application of semiconductors as light-harvesting catalysts for the conversion and storage of solar energy, interest in plausible alternative photoelectrochemical processes naturally grew. Given the very high oxidation potential of the photogenerated holes in TiO$_2$ and the increasing environmental problems relating to toxic chemicals faced by the world, it was soon realized that TiO$_2$ could have an application as a catalyst for the photoelectrochemical oxidation of organic impurities in water. The 'active agent' responsible for the oxidation of organics at the surface of irradiated TiO$_2$ is now believed to be (but has yet to be conclusively proved to be) the $^\bullet$OH radical, formed via the oxidation of adsorbed hydroxide ions by photogenerated holes, according to

$$OH^-(ads) + h^+ \rightarrow {}^\bullet OH$$

The photogenerated electron can be captured by dissolved oxygen either at a counterelectrode or, indeed, at the TiO$_2$ surface itself to form, initially, the superoxide ion:

$$O_2 + e^- \rightarrow O_2^-$$

Attention initially focused on non-electrochemical systems, in which both anodic and cathodic reactions took place on the surface of the TiO_2, which was present in the form of a slurry of small particles. Since the early 1980s, when the potential of semiconductor photocatalysis using TiO_2 as a method of purifying water first became generally recognized, a wide range of organic impurities, including phenols, chlorobiphenyls and halogenated hydrocarbons, has been shown to be susceptible to decomposition in this way and interest worldwide is now intense.

Most of the studies on photocatalysis using TiO_2 have focused on the activation of the photocatalyst with ultraviolet lamps, principally UVA ones (i.e. those having peak emission at about 365 nm). However, more recently, attention has increasingly turned to *solar*-driven photocatalysis using TiO_2. Irrespective of whether solar irradiation or UV light sources are employed, the slurry approach simply involves adding particulate TiO_2 directly to the contaminated water under illumination. The concentrations of TiO_2 commonly employed are in the range 0.5–10 g dm^{-3}; however, it is generally found that the rate of photomineralization increases with increasing concentration, tending towards a limiting rate at higher concentrations, typically 0.1 wt%, i.e. 1 g dm^{-3}. In the case of more dilute slurries, the average interparticle distance is so low that the rate of degradation of organics is much less than the rate of mass transport of the organic species to the TiO_2 surface.

Two limiting approaches to the design of photochemical slurry reactors have routinely been employed: batch reactors and continuous-flow reactors. In the former, the solution to be treated is placed in the reactor, the TiO_2 powder added and the solution stirred and irradiated. When the detoxification has gone to completion, the solution is removed. In the latter, the slurry solution is pumped through the reactor, either in a single pass or by recirculating it continuously between the reactor and a suitable reservoir. Since the need for the presence of oxygen in the slurry to facilitate the degradation of organics is now well documented, the slurry solutions are generally aerated or oxygenated. This is particularly important in view of the fact that the reduction of oxygen by the capture of electrons from the TiO_2 particles is thought to be the rate-limiting factor in most reactor configurations.

However, there is a significant drawback associated with the application of the photocatalyst in the form of a powder, concerning the small sizes of particles required. As was discussed above, irradiation of a TiO_2 particle with supra-bandgap light results in the formation of an electron–hole pair. There is then a finite possibility that the electron and hole will overcome their mutual electrostatic attraction and become spatially separated; they can then diffuse to the surface where the electron will be captured by O_2 and the hole by adsorbed hydroxide to form $^{\bullet}OH$ radicals. However, before the electrons and holes reach the surface, there is a significant chance of recombination, which is an important source of inefficiency in systems employing semiconductor catalysts for the photochemical conversion of light. In the absence of an electrical field, recombination of charges is the dominant process for the photogenerated electrons and holes and the quantum yield for the generation of $^{\bullet}OH$ radicals at Anatase TiO_2 ((number of $^{\bullet}OH$ radicals generated/number of incident photons) $\times$ 100%) has been calculated to be <5%, in agreement with the low rates of photodegradation that are generally observed. The minority-carrier length, L_p, in TiO_2, i.e. the distance the holes diffuse in a field-free region before recombination with an electron, is about 0.1 μm. Consequently, L_p must be comparable to, or greater than, the particle size if the recombination of charges is not to be a limiting factor and hence there is a severe restriction on the size of the semiconductor particles. Unfortunately, if the slurry contains such very fine particles, then this, in turn, implies that long settlement times are required for removal of the catalyst from the purified water; alternatively, fine filters must be employed. In either case, a severe restriction is placed on the throughput of the treatment process, significantly increasing the cost of the system; this is a very serious drawback of the slurry approach and a significant barrier to its commercialization.

One method of improving the rate of the photodegradation of organics by TiO_2 slurries that was investigated relatively early in the development of the field was the use of solar concentrators, such as those at the PSA facility in Almeria, Spain and the Sandia National Laboratory/Lawrence Livermore National Laboratory in the USA. However, this approach was soon compromised by the fact that the rate of degradation was found to increase linearly with the intensity of the light, I, only up to about one sun equivalent; at higher intensities, the rate increases only as the square root of the intensity and eventually becomes independent of the intensity for very intense irradiation, owing to an increase in the rate of electron-hole recombination.

The disadvantage of particulate methods is now apparent and this has focused attention on ways of immobilizing TiO_2 onto a suitable support material; such methods fall into two broad categories:

(i) immobilization onto a particulate support such as glass microbeads, which makes separation easier, and
(ii) immobilization onto a fixed support such as glass, metal plate or fibre-glass mesh, which renders separation unnecessary.

If an immobilized catalyst, particularly an immobilized film, is employed, it is not difficult to see that mass transport would replace the surface reaction as the rate-limiting step, especially at the low concentrations of organics that are being targetted ($mg\ dm^{-3}$ of water, or parts per million, ppm). This problem can, to some extent, be offset by reverse biasing the TiO_2 as described above; this substantially increases the quantum efficiency for the generation of holes at the surface and is referred to in more recent studies as the electrical-field-enhancement (EFE) effect. It is depicted in Figure 27.

The maximum depth of penetration of the incident light into TiO_2 is $1/\alpha$, where α is the absorption coefficient of the TiO_2 at the wavelength of the incident light. Any holes generated in the depletion layer at depth W will be efficiently transported to the surface. Holes generated still deeper, at a depth of between W and $W + L_p$, may diffuse to the boundary of the depletion layer and will then also be efficiently transported to the surface. Absorption of radiation at depths $> (W + L_p)$ will result in recombination of charges; hence thin films are best suited to the exploitation of the EFE effect. The conduction-band electrons are drawn away from the semiconductor film and migrate to the counterelectrode, allowing the reaction at the counterelectrode (usually the reduction of O_2) to be driven separately and producing a measurable *photocurrent* exactly analogous to the photocurrent produced in solar-energy-conversion devices. Clearly, in contrast to the slurry case, with careful design of the reactor, the reduction of oxygen is no longer rate limiting for the photo*electro*chemical detoxification. For reasonable levels of a dopant within the TiO_2 layer and in *reasonably conducting electrolytes*, the magnitude of the internal electrical field, even at a bias of about 1 V, is sufficient to separate photogenerated electrons and holes efficiently, so enhancing the quantum yield.

The simplest method of generating 'doped' n-TiO_2 films is to heat titanium metal

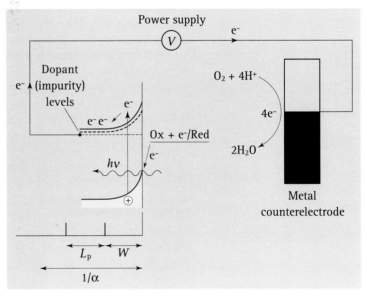

Figure 27

A working photoelectrical system for the remediation of aqueous effluent, showing the effect of the external bias on the semiconducting electrode and the counterelectrode reaction.

in air for a few minutes at about 600 °C; alternatively, anodic films of TiO_2 can be grown on the metal electrochemically. In our own studies in Newcastle, we have employed anodic, thermal and sol–gel TiO_2 films in prototype photo*electro*chemical reactors (with areas of TiO_2 film of 700–1600 cm²), using ultraviolet light; these reactors have shown considerable promise, but need more design work if they are to overcome the mass-transport problems associated with low concentrations of organics. Moreover, although the application of thin films of TiO_2 may be advantageous in terms of the exploitation of the EFE effect, there is, at least at first sight, an additional serious drawback (in addition to the mass-transport problem): the light-harvesting efficiency of thin TiO_2 films is very poor at energies near the bandgap, owing to the low value of the absorption coefficient in this spectral region. This is not a problem encountered with TiO_2 slurries, for it is possible to ensure that all the incident light is absorbed simply by having a large enough number of TiO_2 particles present. Surprisingly, it does not appear to be a problem with immobilized films either, since it has generally been found that immobilized films of TiO_2 have photo-activities only two or three times lower than those observed with the slurries. It does appear, on this basis, that sufficiently innovative engineering approaches are capable of yielding photoelectrochemical reactors of potentially high overall efficiency which will be capable not only of oxidatively removing organic impurities from aqueous solution but also, perhaps in association with intense light sources such as focused solar reactors, of destroying extremely unpleasant waste products, such as those associated with (1) the effluent from the processes employed in the extraction of radionuclides from the waste streams of nuclear reactors, containing primarily citric acid and EDTA, and (2) highly toxic and dangerous organic materials such as the active components of chemical weapons.

Conclusions

Batteries remain an enormous industry, turning over billions of pounds worldwide. The main problems faced by the industry at the end of this millennium are associated with bringing some of the promising new batteries fully to commercialization. Undoubtedly, vehicular traction remains the greatest challenge and the rewards of success in this area would be enormous. However, the industry still uses lead–acid or NiCad batteries. ZEBRA batteries have been developed, but their power/weight ratio, cost and safety all remain to be optimized. Other batteries, such as the Sohio battery, have had substantial development costs expended on them but remain problematic.

A key feature of modern battery manufacture is the close partnership between the design of equipment and the fabrication of batteries. This partnership has been particularly productive in Japan, where it has led to significant gains in efficiency both for established and for novel battery technologies. The future for the battery industry looks set fair: the steady increase in the number of portable electronic appliances in use will, by itself, continue to generate an increase in the market for batteries and continuing technical breakthroughs in the Li-battery arena will guarantee further progress in the complexity and power capacity of such appliances. This fertile interdependence of innovation in the two industries will probably be the most important single driver of the expansion of the market for batteries in the next few years.

Fuel cells remain at the threshold of significant application. Severe reductions in costs are essential if stationary power applications are to be realized and a considerable increase in lifetime would also be highly desirable. For traction applications, which are likely to be driven as much by environmental considerations as by considerations of cost, the outlook is brighter, but there remains much development work to do. Nonetheless, the overall picture is a positive one: the extraordinary progress over the last few years has placed fuel cells firmly on the agenda of energy-foresight panels the world over and there is little doubt that we shall start to see significant encroachment, at least in niche markets, during the next decade.

Photoelectrochemical devices have also made considerable strides in recent years, both in terms of direct-energy-conversion applications and in driving desirable but difficult Faradaic reactions such as the selective decomposition of organic impurities in drinking water. Commercial applications in the latter area are likely to be with us very soon; applications to solar-energy conversion remain some way off, however, because the costs are still too high.

Further Reading

1. C. H. Hamann, A. Hamnett and W. Vielstich, *Electrochemistry* (Wiley–VCH, Weinheim, 1998).
2. C. A. Vincent and B. Scrosati, *Modern Batteries* (Arnold, London, 1997).
3. A. R. Landgrebe and Z.-I. Takehara (editors), *Batteries and Fuel Cells for Stationary and Electric Vehicle Applications, Electrochemical Society Symposium* (Electrochemical Society, Pennington, NJ, 1993).
4. D. A. Corrigan and S. Srinivasan (editors), *Hydrogen Storage Materials, Batteries, and Electrochemistry, Electrochemical Society Symposium* (Electrochemical Society, Pennington, NJ, 1992).
5. L. J. M. J. Blomen and M. N. Mugerwa, *Fuel Cell Systems* (Plenum Press, New York, 1993).
6. A. J. Appleby and F. R. Foulkes, *Fuel Cell Handbook* (Van Nostrand Reinhold, New York, 1989).
7. C. M. A. Brett and A. M. O. Brett, *Electrochemistry: Principles, Methods and Applications* (Oxford University Press, Oxford, 1993).
8. H. Lund and M. Baizer (editors), *Organic Electrochemistry*, third edition (Marcel Dekker, New York, 1991).
9. *Berichte der Bunsenges, Phys. Chem*, **94**, September issue (1990).
10. A. Hamnett and G. L. Troughton, *Chem. Ind.*, 480 (1992).
11. K. Kinoshita, *Electrochemical Oxygen Technology* (John Wiley, New York, 1992).
12. P. G. Bruce (editor), *Solid-State Electrochemistry* (Cambridge University Press, Cambridge, 1995).
13. P. A. Christensen and G. M. Walker, *Opportunities for the UK in Solar Detoxification* (HMSO, London, 1996).
14. K. S. V. Santhanam and M. Sharon (editors), *Photoelectrochemical Solar Cells* (Elsevier, Amsterdam, 1988).

16 Chemistry Far from Equilibrium: Thermodynamics, Order and Chaos

Guy Dewel, Dilip Kondepudi and Ilya Prigogine

Introduction

Classical science emphasized equilibrium and stability. It was admitted that a small change in causes would produce, likewise, a small change in effects. Also, classical science was based on a reductionist point of view: once the basic units (molecules, atoms, . . .) and their interactions were known, all properties of systems formed by these units could be easily predicted.

However, since the late 1950s, a different image of nature has been emerging. Once we are far from equilibrium, new properties associated with 'self-organization' may emerge. Disequilibrium leads to new structures on the macroscopic scale, separated by bifurcation points. Still, all these structures are realized with the same units and the same interactions. These findings are of essential importance for our description of our natural environment and have found applications in a number of fields, from biology to economics and the social sciences.

Chemistry has played a central role in this development. The reasons are partly experimental, partly theoretical. It is easy to build devices such as the continuous-flow-stirred-tank reactor (CSTR), in which we can, through the modulation of the flows of the reactants, show how they go from near-equilibrium to far-from-equilibrium conditions. The results obtained in non-equilibrium physics and chemistry shed new light on properties of matter and on basic concepts, such as time. Far from equilibrium matter acquires new properties, leading to new forms of coherence exemplified by chemical waves and chemical clocks. Basic spatial symmetries may also be broken, which leads to a remarkable analogy with phase transitions.

In many monographs and popular texts, it is claimed that irreversibility, as those factors associated with entropy increase, leads only to disorder and to a forgetting of the initial conditions. Self-organization shows that it is not so. It is through irreversible processes that the most delicate structures around us, such as those associated with living systems, came into existence.

This chapter has been subdivided into two parts. The first deals with thermodynamics and emphasizes the differences among the behaviours of systems at equilibrium, close to equilibrium and far from equilibrium. In the second part, we discuss specific examples related mainly to chemical clocks, chaos and spatial structures (Turing structures). We could present only a small part of the large amount of material on this subject. The further reading cited will contain a wealth of topics for the interested reader.

Irreversible Processes and Order

Thermodynamics is based on two fundamental laws that were formulated during the nineteenth century. The first law is the law of conservation of energy: when a system undergoes transformations from one state to another, the energy is conserved. The second law is about irreversibility of changes we observe in nature; it states that entropy inexorably increases due to natural processes. It associates an arrow of time with all natural processes. Irreversible processes dissipate mechanical and chemical energy into heat; hence they are also called dissipative processes.

Since the formulation of the second law, it has been known that entropy producing irreversible or dissipative processes may lead to disorder. During the past five decades, however, we have realized how *dissipative processes can also create ordered states of matter:* states of matter with coherence and structure. Such structures, whose existence depends on dissipative processes, are called *dissipative structures.* The spontaneous generation of dissipative structures by irreversible processes is a general phenomenon of importance to physics, chemistry and biology.

Although dissipative structures arise in many different ways, there are some general features that can be identified. (i) They appear only when the system is far from thermodynamic equilibrium. (ii) Non-linearity of the laws that govern the behaviour of the system is essential. (iii) The possibility of dissipative structures often depends on the operation of autocatalytic processes. The thermodynamic theory of stability that we shall present in the following sections shows that, as the result of the above three features, a far-from-equilibrium system can become unstable and spontaneously evolve to new organized states.

The formation of dissipative structures is associated with instability and 'bifurcation' to new states of coherence. The system arrives at a juncture at which it becomes unstable and makes a transition to one of several possible new states. Which of the possible states will be realized could be a matter of chance and hence unpredictable. This brings us to an important feature of non-equilibrium chemical systems, viz. the role of fluctuations. At or close to equilibrium, intrinsic fluctuations in concentration, temperature and other quantities, arising as results of random molecular motion, are hardly noticeable and are not of much significance. When a system is far from equilibrium, its evolution may be dominated by fluctuations. This may happen in the following way. From a given initial non-equilibrium state, there may be many possible states to which the system can evolve. Which of these possible states will result depends on the fluctuations. The conventional determinism of chemical kinetics is lost in these situations.

Through bifurcation to more and more complex states, a system can reach a mode of behaviour that is totally chaotic. The state of a system then changes in an unpredictable way. Examples of chaotic systems will be given later on in this chapter. We know many simple mathematical mappings that mimic such behaviour. The mathematical study of such mappings has given us some insight into how such unpredictability can arise even in seemingly simple systems. The study of such 'deterministic chaos' has grown to be a vast subject extending from physics to biology and beyond. It is not our intention to give a detailed account of this vast field. We shall deal only with some examples of chaos in chemical systems as one of the many aspects of chemistry far from thermodynamic equilibrium. The interested reader can find many detailed expositions of this subject. We refer also to some recent discussions of the relation between deterministic chaos and thermodynamics (see also the concluding remarks).

The Thermodynamics of Irreversible Processes

The basic aspects of thermodynamic description are shown in Figure 1. Every system is associated with an energy U and an entropy S. It can exchange energy and entropy with its exterior. In addition, owing to irreversible processes, entropy may be produced within the system. According to the first law or the law of conservation of energy, in any transformation of the state of a system that occurs in a small time dt, the energy gained by the system, dU, is equal to the energy lost by its exterior and vice versa. The change of energy, dU, depends only on the initial and the final state, not on the manner in which the change occurs.

Not all processes that conserve energy are realizable in nature. Processes that are

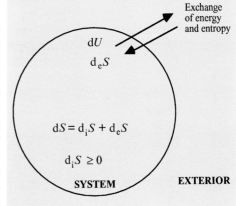

Figure 1

A system can change its state by exchanging matter and energy with its exterior. In any change of state, the energy gained by the system dU is equal to the energy lost by the exterior. In contrast, the change in entropy in the system consists of two parts: the production of entropy due to irreversible processes in the system d$_i S$ and the exchange of entropy with the exterior d$_e S$. According to the second law d$_i S \geq 0$. However, the entropy exchanged, d$_e S$, can be positive or negative.

not realizable, even if they conserve energy, are specified by the second law: the law of entropy. The change in entropy, dS, associated with a change of state that occurs in a small time, dt, is a sum of two parts:

$$dS = d_iS + d_eS \qquad (1)$$

the change due to irreversible or dissipative processes within the system d$_iS$, and the change, d$_eS$, due to exchange of entropy with the exterior (see Figure 1). According to the *second law of thermodynamics*

$$d_iS \geq 0 \qquad (2)$$

Physical processes that result in a negative d$_iS$ are not realizable in nature, even if they conserve energy. d$_eS$, however, can be positive or negative. Like the changes in energy, the changes in entropy, dS, also depend only on the initial and the final states, not on the manner in which this change has occurred.

The fundamental quantity that distinguishes a system in equilibrium from a system not in equilibrium is the rate of production of entropy:

$$P = d_iS/dt > 0 \qquad (3)$$

When a system is in equilibrium, all irreversible processes vanish, resulting in a state in which $P = 0$.

The development of the second law consists of relating the d$_iS$ and d$_eS$ to real physical processes and measurable physical quantities. This can be done for systems in *local equilibrium*. At equilibrium, the entropy, S can be expressed as a function of the energy U, volume V and composition N_k (the number of moles of the constituent 'k'). A system is in local equilibrium if the entropy density, $s(x)$, (the entropy per unit volume at the location x), can be expressed in terms of the local energy density $u(x)$ and concentrations $n_k(x)$ (moles per unit volume of the constituent 'k'). There are formulations of 'extended thermodynamics' in which $s(x)$ depends on supplementary variables, such as the gradient of temperature, but they have until now not led to any significant modification for the situations we shall consider.

In the framework of local-equilibrium thermodynamics, the change in entropy in an open system due to exchange of heat and matter is

$$d_eS = d\Phi/T - \mu_x d_eN_x/T \qquad (4)$$

Here dΦ is the energy exchanged, d$_eN_x$ is the amount of matter exchanged in the form of the chemical species x, μ_x is the chemical potential of the species x in the system and T is the temperature.

For local equilibrium, we can also evaluate the production of entropy, d$_iS$. This is done as follows. Each irreversible process is associated with a 'force' F and a 'flow' J. For example, when diffusion occurs due to a gradient in concentration, the flow of particles corresponds to the flow J and the concentration gradient that causes the flow is related to the force F. In a small time dt, the entropy produced is given by d$_iS = FJ$ dt. Similarly, the entropy produced due to a chemical reaction is expressed in terms of a thermodynamic force A called the *affinity* and a thermodynamic flow v called the velocity of reaction and d$_iS = (Av/T)$ dt. The concept of affinity was introduced by Théophile de Donder (1872–1957) in order to formulate a theory for the production of entropy in chemical systems. When several processes are to be considered simultaneously, the force and flow of each irreversible process are identified with a subscript a. The *rate* of the total production of entropy is the sum of the rates of production of entropy due to each process:

$$P \equiv \frac{d_iS}{dt} = \sum_a J_a F_a \geq 0 \qquad (5)$$

Expression (5) can be extended to inhomogeneous systems that are in local equilibrium by defining the production of entropy per unit volume, $\sigma(x)$, in terms of the flow densities $J_a(x)$ and forces $F_a(x)$ which are functions of the position x. The total production of entropy is then given by

$$P = \int_V \sigma(x)\, d^3x = \int_V \sum_a J_a(x)\, F_a(x)\, d^3x$$

These expressions for the production of entropy are valid for almost all systems that we normally encounter.

As a simple illustration of the computation of the production of entropy P, let us consider the reaction

$$2A + B \rightleftharpoons 2C \tag{6}$$

As noted above, the production of entropy due to a chemical reaction is given by

$$\frac{d_i S_{\text{chem}}}{dt} = \frac{Av}{T} \geq 0 \tag{7}$$

The affinity A can be written in terms of the chemical potentials of the reactants and the products. For the reaction (6), the affinity is

$$A = 2\mu_A + \mu_B - 2\mu_C \tag{8}$$

The velocity of reaction v is the net rate of conversion of the reactants A and B to the product C. If the forwards rate is ρ_f and the reverse rate is ρ_r, the velocity is the difference between the two rates:

$$v = \rho_f - \rho_r \tag{9}$$

As is the case for elementary steps of a chemical reaction, let us assume that for the reaction (6) the forwards and reverse rates may be written

$$\rho_f = k_f[A]^2[B] \qquad \rho_r = k_r[C]^2 \tag{10}$$

in which k_f and k_r are the kinetic rate constants and [A], [B] and [C] are the concentrations of the reacting compounds.

In a closed system, the chemical reaction will eventually drive the system to the state of equilibrium in which the affinity given by (8) and the velocity given by (9) both vanish. This leads to the equations

$$2\mu_A + \mu_B = 2\mu_C \qquad \rho_r = \rho_f \tag{11}$$

which characterize the state of equilibrium. For simplicity, we shall assume that we have an ideal system for which the chemical potentials of a species X may be written in the following form:

$$\mu_X = \mu_{X0}(T) + RT \ln([X]/[X]_0) \tag{12}$$

in which [X] is the concentration and $[X]_0$ is the concentration of the reference state. Usually the reference state is chosen such that $[X]_0 = 1$, hence we will not explicitly include it in the expressions that follow. $\mu_{X0}(T)$ is a function of the temperature T. By combining (11) with (12), the condition for equilibrium can also be written in the following form, known as *the law of mass action*.

$$\frac{[C]^2_{\text{eq}}}{[A]^2_{\text{eq}}\,[B]_{\text{eq}}} = \frac{k_f}{k_r} = e^{(2\mu_{A0} + \mu_{B0} - 2\mu_{C0})/(RT)} = K(T) \tag{13}$$

in which the subscript 'eq' denotes the equilibrium concentrations. The quantity $K(T)$, the equilibrium constant, is a function of the temperature alone.

For a system out of thermodynamic equilibrium, using expressions (10), (12) and (13), it is easy to see that the affinity A given by (8) can itself be written in terms of the forwards and reverse reaction rates:

$$A = RT \ln\left(\frac{\rho_f}{\rho_r}\right) \tag{14}$$

Using (9) and (14), we can now write the rate of production of entropy (7) in terms of the forwards and reverse rates of the reaction:

$$\frac{d_i S_{chem}}{dt} = \frac{Av}{T} = R(\rho_f - \rho_r) \ln\left(\frac{\rho_f}{\rho_r}\right) \geq 0 \tag{15}$$

Thus, for elementary steps of a reaction, if the rates of the forwards and reverse reactions, ρ_f and ρ_r, are known, the rate of production of entropy can be calculated using the above expression. When several reactions occur simultaneously, the total production of entropy is the sum of the productions due to each of the reactions. In a similar manner, the production of entropy for diffusion, heat conduction and other irreversible processes can be written explicitly.

Three Thermodynamic Regimes

Thermodynamic systems can be broadly classified into three regimes: *equilibrium*, *linear near-equilibrium* and *non-linear far-from-equilibrium* regimes. Each regime has certain general characteristic features, which we shall discuss below.

Isolated systems (systems that exchange neither energy nor matter) will eventually reach a state of equilibrium in which all dissipative processes have died out and the production of entropy has ceased, i.e. $P = 0$. Open systems (systems that exchange energy and matter) can also reach a state of equilibrium; in this state the temperature, pressure and chemical potential of the system may become equal to those of the exterior. In special cases, this evolution can be associated with thermodynamic potentials. For example, when a system is under conditions of constant temperature, T, and volume, the relevant thermodynamic potential is the *Helmholtz free energy F* defined by $F = U - TS$. The evolution of the system to a state in which the production entropy of P vanishes can also be described as the evolution of the system to a state of minimum Helmholtz free energy F. Similarly, for systems under conditions of constant temperature T and pressure p, the thermodynamic potential is the *Gibbs free energy* $G = U - TS + pV$, in which V is the system's volume; as the production of entropy P approaches zero, G evolves to its minimum value.

At equilibrium, quantities such as temperature and pressure are uniform and constant throughout the system. The dissipative processes eliminate all inhomogeneities and drive the system to the equilibrium state that is unchanging in time. Generally speaking, the uniformity of the system in space and its unchanging nature in time mean that the system has no macroscopic 'structure' either in space or in time.

When the system is only slightly perturbed from the state of equilibrium, the forces and flows are small. For small perturbations, there is a simple *linear* relationship, $J = LF$, between a force F and the corresponding flow J. When several thermodynamic forces F_α are present simultaneously, the corresponding flows are given by the general linear relation

$$J_\alpha = \sum_\beta L_{\alpha\beta} F_\beta \tag{16}$$

where the coefficients $L_{\alpha\beta}$ are called the Onsager coefficients, after Lars Onsager (1903–1976) who, in 1931, established the fundamental relation

$$L_{\alpha\beta} = L_{\beta\alpha} \tag{17}$$

Equation (16) has an important physical meaning. It means that the various thermodynamic forces and flows are coupled, i.e. a particular thermodynamic force such as a heat gradient causes not only a flow of heat but also a flow of matter or electrical current. The well-known effect of thermal diffusion, in which a difference in temperature causes a flow of matter, is an example. Equation (17) expresses a fundamental relation between the constants that couple different thermodynamic forces and flows. It is called the *Onsager reciprocal relation* and is one of the important general relations valid in the linear non-equilibrium regime.

With regard to entropy, it was shown in 1945 that another general result can be obtained for systems in the linear non-equilibrium regime. Although the production of entropy P vanishes at equilibrium, it takes the minimum possible value consistent with the constraints when the system is maintained in a state slightly away from equilibrium. This result, called the *theorem of minimum production of entropy*, is derived using the Onsager reciprocal relations.

In any particular system, the constraints that prevent the system from evolving to the equilibrium state can be parametrized in terms of a set of variables δ_k. The constraints are a measure of the flow of energy and/or matter through the system. Consider, as an example, the following chemical reaction:

$$A \underset{k_{-1}}{\overset{k_1}{\rightleftharpoons}} X \underset{k_{-2}}{\overset{k_2}{\rightleftharpoons}} B \qquad (18)$$

If there is no flow of reactants, the system will evolve to its state of equilibrium in which the concentrations are $[X]_{eq} = (k_1/k_{-1})[A]_{eq} = (k_{-2}/k_2)[B]_{eq}$. Now, this system can be maintained away from equilibrium through an inflow of A and an outflow of B maintaining the concentrations of A and B at some non-equilibrium values, $[A] = [A]_{eq} + \delta_A$ and $[B] = [B]_{eq} + \delta_B$. In response to these constraints, the concentration of X will assume some value $[X] = [X]_{eq} + \delta_X$. If δ_A and δ_B are small compared with the equilibrium concentrations of A and B, then, according to the theorem of minimum production of entropy, δ_X will evolve to a value that minimizes the production of entropy P (Figure 2).

The total production of entropy for the two reactions in (18) can be calculated using (15), in which the rates of the forwards and reverse reactions are $k_1[A]$ and $k_{-1}[X]$ etc. For a fixed δ_A and δ_B, we can obtain the production of entropy as a function of δ_X:

$$P(\delta_X) = \frac{d_i S}{dt} = R\left(\frac{(k_1\delta_A - k_{-1}\delta_X)^2}{k_1[A]_{eq}} + \frac{(k_2\delta_X - k_{-2}\delta_B)^2}{k_2[X]_{eq}} \right) \qquad (19)$$

This quadratic function attains a minimum when

$$\delta_X = \frac{(k_1\delta_A + k_{-2}\delta_B)}{k_2 + k_{-1}}$$

Under the above-mentioned constraints on [A] and [B], this value of δ_X is identical to the steady-state solution of the kinetic equation

$$\frac{d[X]}{dt} = k_1[A] - k_{-1}[X] - k_2[X] + k_{-2}[B] = 0$$

Thus we see that the actual steady-state value of [X] in the near-equilibrium regime is also the value that minimizes the production of entropy.

More generally speaking, when the constraints $\delta_k = 0$, the system can reach the state of equilibrium in which $P = 0$. When δ_k are small but not equal to zero, the system will evolve to a steady state in which the production of entropy P takes the minimum possible (non-zero) value compatible with the constraints. This result can be represented

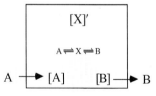

Figure 2
An example of a chemical system in the linear non-equilibrium regime. The inflow of A and the outflow of B maintain the system in a non-equilibrium state. The concentration of X in this case will assume a value that minimizes the production of entropy.

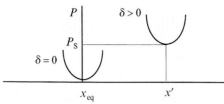

Figure 3
The production of entropy at and near the state of equilibrium. A non-zero value of the constraint parameter δ keeps the system away from equilibrium. The variable takes a particular value depending on the constraint. When δ = 0, the system can reach the state of equilibrium for which $P = 0$; the corresponding value of x is x_{eq}. For a small δ > 0, the system evolves to a state that corresponds to the minimum possible value of P; the corresponding value of x is x'.

graphically as shown in Figure 3. It could be said that, since the system is prevented from reaching the equilibrium state in which $P = 0$, it settles for the next best situation and makes P as small as possible.

The above picture also allows us to understand the *stability* of the equilibrium and near-equilibrium states. In Figure 3, the states x_{eq} and x' are stable in the sense that any small perturbation δx from this state 'decays' or decreases with time and approaches one of the time-independent states, x_{eq} or x' as the case may be. Mathematically, the notion of stability is made precise through the idea of a Lyapunov function $f(x)$. A state x is stable with respect to small perturbations δx if the corresponding change δf associated with the perturbation δx has the following properties:

$$\delta f > 0 \qquad \mathrm{d}\delta f/\mathrm{d}t < 0 \tag{20}$$

For example, for fixed values of the temperature T and the volume V, the Helmholtz free energy $F = U - TS$ takes its minimum value at equilibrium. Any perturbation can only increase the value of F, therefore $\delta F > 0$. When F deviates from its minimum, irreversible processes drive the system back to the state of minimum F: therefore $\mathrm{d}\delta F/\mathrm{d}t < 0$. Hence, F is a Lyapunov function. The production of entropy P is also a Lyapunov function valid both for equilibrium and for near-equilibrium states in the linear regime.

The existence of Lyapunov functions has a fundamental consequence; the system is then *stable* with respect to fluctuations. The fluctuations will die out. One of the main differences between equilibrium or near-equilibrium states and far-from-equilibrium states is that the latter, in general, admit no Lyapunov function. In preparation for the study of far-from-equilibrium situations, it is interesting to consider $\delta^2 S$. Let us express the perturbation of the entropy from the equilibrium state as a Taylor series:

$$S = S_{eq} + \delta S + \tfrac{1}{2}\delta^2 S + \cdots \tag{21}$$

Since the entropy of a system in equilibrium is a maximum, it follows that the first variation $\delta S = 0$ and $\delta^2 S < 0$. The second variation $\delta^2 S$ has a simple physical meaning. Consider the perturbation of energy U. Then according to the classical relation

$$\delta S = \frac{\delta U}{T} \qquad \text{i.e.} \qquad \frac{\partial S}{\partial U} = \frac{1}{T} \tag{22}$$

by using the relations $\partial S/\partial U = 1/T$ and $\delta U = C_v \mathrm{d}T$, in which C_v is the heat capacity at constant volume, we have

$$\delta^2 S = \frac{\partial^2 S}{\partial U^2}(\delta U)^2 = \frac{\partial \frac{1}{T}}{\partial U}(\delta U)^2 = -C_v \frac{(\delta T)^2}{T^2} < 0$$

The inequality $\delta^2 S < 0$ expresses that the heat capacity at constant volume C_v is positive. This is a so-called stability condition. More generally

$$\delta^2 S = -\frac{C_v(\delta T)^2}{T^2} - \frac{1}{T\kappa_T}\frac{(\delta V)^2}{V} - \sum_{i,j}\left(\frac{\partial}{\partial N_j}\frac{\mu_i}{T}\right)\delta N_i\,\delta N_j < 0 \tag{23}$$

in which κ_T is the isothermal compressibility and N_i are the numbers of moles of the species i. The condition $C_v > 0$ corresponds to thermal stability, $\kappa_T > 0$ corresponds to mechanical stability and

$$\sum_{i,j}\left(\frac{\partial}{\partial N_j}\frac{\mu_i}{T}\right)\delta N_i\,\delta N_j > 0$$

corresponds to stability with respect to diffusion.

Now, elementary transformations show that near equilibrium

$$\frac{1}{2}\frac{\partial}{\partial t}\delta^2 S = P = \sum_\alpha J_\alpha F_\alpha > 0 \tag{24}$$

Therefore, $-\delta^2 S$ is indeed a Lyapunov function. Let us emphasize that the situations in equilibrium and in linear near-equilibrium systems are, to a large degree, parallel. However, the condition of minimum production of entropy already implies in general the appearance of some non-equilibrium 'order'.

A simple example is thermal diffusion involving a mixture of two gases, say N_2 and H_2, which are placed in a container consisting of two chambers. If one of the chambers is at a higher temperature than the other, there will be a flow not only of heat but also of matter from one chamber to the other. In the stationary state, one compartment will be richer in H_2 while the other will be richer in N_2. Therefore, the flow of heat leads both to *disorder* (the increase of thermal motion) and to *order* (the separating of the two components). This is typical for non-equilibrium situations. We meet this dual structure in the world around us which contains both disorder (such as that associated with the residual black-body radiation) and order (exemplified by living systems).

A system can be driven further and further from the state of equilibrium. When this is done, something quite dramatic may happen. The dissipative processes that keep the system in a stable state now begin to interact in a complex way and they may destabilize this state. At that point, the system makes a transition to a new state that is organized. As mentioned earlier, such structured states, created and maintained by dissipative processes, are called *dissipative structures*.

From the thermodynamic point of view, the main difference is that, for a system far from thermodynamic equilibrium there is, in general, no Lyapunov function. Fluctuations are no longer doomed to regress.

Though the inequality (23) is still valid far from equilibrium, the production of entropy P becomes more complicated. We have to evaluate the effects of the fluctuation, say δX, both on the flow J and on the force F. We then have

$$\frac{d\delta^2 S}{dt} = \sum_\alpha \delta J_\alpha \delta F_\alpha \tag{25}$$

in which δJ_α and δF_α correspond to the changes in the thermodynamic forces and flows due to the perturbation δX. This quantity is referred to as the *excess production of entropy*. If the condition

$$\frac{d\delta^2 S}{dt} = \sum_\alpha \delta J_\alpha \delta F_\alpha > 0 \tag{26}$$

is satisfied, then (23) and (26) make $-\delta^2 S$ a Lyapunov function; the corresponding non-equilibrium steady state is then stable. The positivity of $d\delta^2 S/dt$ expressed in (26) is *a sufficient but not a necessary condition for stability*. When (26) is not satisfied, i.e.

$$\frac{d\delta^2 S}{dt} = \sum_\alpha \delta J_\alpha \delta F_\alpha \leq 0$$

the non-equilibrium steady state may become unstable, causing a transition to a new state.

Inequality (26) shows the general importance of autocatalysis. This point can be illustrated through the following examples. First, let us consider the following non-autocatalytic reaction:

$$A + B \underset{k_r}{\overset{k_f}{\rightleftharpoons}} C + D \tag{27}$$

For this reaction let us assume that the rates of the forwards and reverse reactions are

$$\rho_f = k_f[A][B] \qquad \rho_r = k_r[C][D] \qquad (28)$$

We assume that this system is maintained out of equilibrium by suitable flows similar to those shown in Figure 2. The production of entropy for this reaction can be written using expression (15). Let us assume that, for a non-equilibrium steady state, the production of entropy is P_s. Relations (9) and (28) specify the velocity v and (14) specifies the affinity. Using these expressions, one can obtain the production of entropy in the steady state. Then it is easy to see that, for a small perturbation δB from the steady state, the *excess production of entropy* (26) can be written in terms of $\delta F = (\delta A)/T$ and $\delta J = \delta v$, the perturbations from the steady states:

$$\frac{d\delta^2 S}{dt} = \sum_\alpha \delta J_\alpha \delta F_\alpha = \frac{\delta A \, \delta v}{T} = Rk_f \frac{[A]_s}{[B]_s}(\delta B)^2 > 0 \qquad (29)$$

in which the subscript 's' indicates the non-equilibrium steady-state values of the concentrations. As discussed above, since $d\delta^2 S/dt$ is positive, the steady state is stable.

The situation is different, however, for an autocatalytic reaction such as

$$2X + Y \underset{k_r}{\overset{k_f}{\rightleftharpoons}} 3X \qquad (30)$$

which appears in a reaction scheme called the 'Brusselator' that we will consider below. For this reaction, we can consider a perturbation δX from a non-equilibrium steady state in which the concentrations are $[X]_s$ and $[Y]_s$. Using the forwards and reverse rates $\rho_f = k_f[X]^2[Y]$ and $\rho_r = k_r[X]^3$ in the expressions (9) and (14) for A and v, we can once again calculate the expression for the excess production of entropy and obtain

$$\frac{d\delta^2 S}{dt} = \sum_\alpha \delta J_\alpha \delta F_\alpha = -R(2k_f[X]_s[Y]_s - 3k_r[X]_s^2)\frac{(\delta X)^2}{[X]_s} \qquad (31)$$

The excess production of entropy can now become negative, particularly if $k_r \ll k_f$. Hence the stability is no longer ensured and the steady state *may* become unstable.

The approach discussed above is summarized in Figure 4. In it the value of the parameter Δ is a measure of the distance from equilibrium. For each value of Δ, the system will relax to a steady state, denoted by X_s. The equilibrium state corresponds to $\Delta = 0$. X_s is a continuous extension of the equilibrium state and it is called *the thermodynamic branch*. As long as the condition (26) is satisfied, the thermodynamic branch is stable. If (26) is violated, then the thermodynamic branch may become unstable.

An important means of deciding whether instability will occur is linear-stability theory. This theory allows us to determine whether a steady state is stable with respect to small perturbations of the variables. In general, the rate equations of a chemical system take the form

$$\frac{dX_i}{dt} = F_i(X_1 \dots X_n, \lambda) \qquad (32)$$

where the X_i correspond to concentrations (such as [X] in (18)) and λ corresponds to fixed concentrations (such as those of A and B in (18)). Suppose that we have determined a stationary solution X_i^0 of (32). This means that

$$F_k(X_1^0 \dots X_n^0, \lambda) = 0 \qquad (33)$$

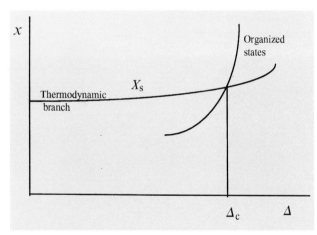

Figure 4

In this figure, each value of x represents a state of the system. The distance from equilibrium is represented by the parameter Δ. When $\Delta = 0$, the system is in a state of thermodynamic equilibrium. When Δ is small, the system is in a near-equilibrium state, which is an extrapolation of the equilibrium state; this family of states is called the thermodynamic branch. In some systems, such as those that have autocatalysts, when Δ reaches a critical value, Δ_c, the states belonging to the thermodynamic branch become unstable. When this happens, the system makes a transition to a new branch of organized states.

Will this solution be stable when we *perturb* it by adding the small quantity $x_i(t)$? Linear-stability analysis provides the answer in the following way. To begin with we have

$$X_i = X_i^0 + x_i(t) \qquad (34)$$

A Taylor expansion of $F_k(X_i)$ gives

$$F_k(X_i^0 + x_i) = F_k(X_i^0) + \sum_j \frac{\partial F_k}{\partial X_j} x_j + \dots$$

In linear-stability analysis the higher-order terms are neglected. Then, since X_i^0 is a steady state, we obtain for $x_i(t)$ the linear equation

$$\frac{dx_i}{dt} = \sum_j M_{ij}(\lambda) x_j \qquad (35)$$

in which $M_{ij} = \partial F_i/\partial X_j$. We then look for normal modes, that is for solutions of the form

$$x_i(t) = A_i e^{\omega_0 t} \qquad (36)$$

If the real parts of the frequencies ω_0, usually written as $\mathrm{Re}(\omega_0)$, are negative, the perturbation will regress; contrariwise, the system will become unstable if $\mathrm{Re}(\omega_0)$ is positive. The linear theory indicates that the small perturbation in x will grow exponentially. *From the thermodynamic arguments we have given, we know that this can happen only far from equilibrium and when autocatalytic effects are involved.* This growth, however, 'saturates' due to the non-linear terms and the system reaches a new stable value. The new values of X_i often represent a state that is more organized than the initial unstable state, i.e. the system makes a transition to a dissipative structure.

Dissipative Structures

Let us consider a few examples of non-equilibrium transitions to a dissipative structure. The first is a simple example related to a fundamental aspect of biochemistry.

It is well known that the building blocks of life, viz. amino acids, DNA and RNA, have a fundamental asymmetry: of the two possible mirror-image molecular structures, only one appears in the chemistry of life. Amino acids are almost all of the L form and the ribose in DNA and RNA is of the D form (Figure 5). As Francis Crick noted 'The first great unifying principle of biochemistry is that the key molecules have the same hand in all organisms'. This is all the more remarkable because chemical reactions exhibit equal preferences for the two mirror-image forms (except for very small differences due to non-parity-conserving electroweak interactions). We do not yet understand the true origin of this fundamental biomolecular asymmetry but we can see how such a state might be realized in the framework of dissipative structures. Through a variation of a model reaction devised by F. C. Frank (see Figure 6) we can see how a set of chemical reactions that produce with equal preference the mirror-image forms X_L and X_D, of a molecule X can nevertheless make a transition to a state dominated by

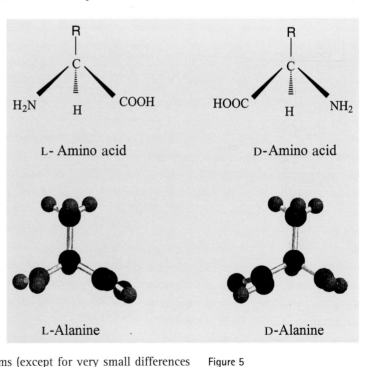

Figure 5

L and D forms of amino acids. With only very rare exceptions, the amino acids in living cells are of the L form.

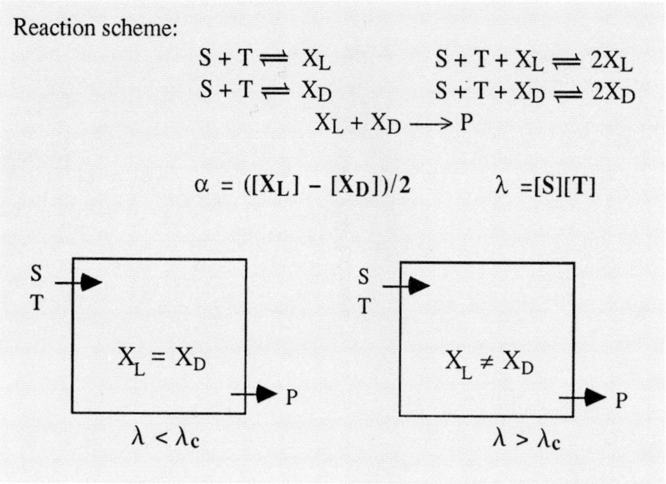

Reaction scheme:

$$S + T \rightleftharpoons X_L \qquad S + T + X_L \rightleftharpoons 2X_L$$
$$S + T \rightleftharpoons X_D \qquad S + T + X_D \rightleftharpoons 2X_D$$
$$X_L + X_D \longrightarrow P$$

$$\alpha = ([X_L] - [X_D])/2 \qquad \lambda = [S][T]$$

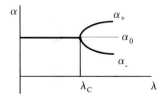

Figure 6

The autocatalytic reaction scheme in which X_L and X_D are produced with equal preference in an open system leads to a dissipative structure in which $[X_L] \neq [X_D]$.

production either of X_L or of X_D. The reaction scheme we consider, which includes two autocatalytic reactions, is shown in the upper part of Figure 6. In a closed or an isolated system, the set of reactants S, T, X_L, X_D and P will reach the state of equilibrium which is a *symmetrical state*, i.e. $[X_L] = [X_D]$. The symmetrical state corresponds to the thermodynamic branch. However, if we consider an open system into which S and T are pumped and from which P is removed, due to the instability of the thermodynamic branch created by the autocatalysis, the system can *break the chiral symmetry* and make a transition to an *asymmetrical state* in which $[X_L] \neq [X_D]$. This can happen if the input of S and T is such that the product of their concentrations exceeds a critical value.

The above behaviour can more conveniently be analyzed by defining a parameter $\lambda = [S][T]$. When λ is greater than λ_c the symmetrical state (the thermodynamic branch) becomes unstable. The symmetrical and asymmetrical states can be associated with a variable $\alpha = ([X_L] - [X_D]/2)$. The transition from a symmetrical state to an asymmetrical state can now be described using the variables λ and α. When the transition to an asymmetrical state occurs, α can be positive or negative. Thus there are two possible asymmetrical states to which the system can make a transition. The variables α and λ obey a general equation of the form

$$\frac{d\alpha}{dt} = -A\alpha^3 + B(\lambda - \lambda_c)\alpha \tag{37}$$

in which A and B are constants. The form of this equation is independent of the details of the particular chemical scheme (such as the one in Figure 6, or any other) that breaks the chiral symmetry; only the numerical values of A and B depend on the details of the kinetics. This is a so-called *normal form*. The steady states of α, obtained from (37) by setting $d\alpha/dt = 0$, are shown in Figure 7. Other examples will be given in the second half of this chapter. The amplitude of α is a measure of the asymmetry. For steady states, we obtain the amplitude equation

$$\alpha = \pm \left(\frac{B}{A} (\lambda - \lambda_c) \right)^{1/2} \tag{38}$$

Even in this simple case (called the pitchfork bifurcation) α varies with λ in a non-analytical way. In this case, the transition takes the system to a more ordered state in which one of the enantiomers is in excess.

The instability described by equation (37) and the associated steady-state diagram are generally valid for any non-equilibrium system that breaks a two-fold symmetry. For each of these systems, parameters α and λ can be defined, which, in the vicinity of the critical point λ_c, obey an equation of the form (37). The 'branching' of α (shown in Figure 7) as λ varies is called the *bifurcation* of states and is a general feature of all non-equilibrium transitions to dissipative structures.

The next example we consider illustrates oscillations and spatial structures in chemical systems. For this purpose, we use a model reaction scheme, which has acquired the name 'Brusselator'. This model proposed by Prigogine and Lefever in 1968 has been used extensively as a prototype to gain an understanding of the various aspects of chemical dissipative structures. Since then, several other models have been studied in great detail both experimentally and theoretically. Many of the general features discovered theoretically have been realized experimentally in the most exten-

Figure 7

A bifurcation diagram showing a mirror-symmetry-breaking transition to a dissipative structure that contains an excess of one enantiomer. This excess is maintained through dissipative chemical reactions.

sively studied Belousov–Zhabotinsky (B–Z) reaction. An extensive study of the B–Z reaction was conducted by Field, Körös and Noyes, who identified the most important steps of this reaction. Today the study of dissipative structures is an active field of chemistry.

The reaction scheme of the Brusselator is shown in Figure 8. We assume that the concentrations of the reactants A and B are maintained at a desired non-equilibrium value through appropriate flows. The products D and E are removed as they are formed. All the reverse reactions are assumed to be slow enough that they can be neglected. The rates of reaction are assumed to obey mass-action kinetics. For example, the rate of the first reaction in the scheme is $k_1[A]$, that of the second reaction is $k_2[B][X]$ and so on. Since all other concentrations are fixed, the state of the system is determined by the concentrations [X] and [Y]. Transport of X and Y occurs through diffusion. With these assumptions, it is straightforward to write the kinetic equations for the concentrations of X and Y. For mathematical convenience, one can redefine 'normalized' concentrations in such a way that all the rate constants are equal to unity. These normalized concentrations, which are denoted by X, Y etc., obey the following equations, in which D_x and D_y are the diffusion coefficients:

$$\frac{\partial X}{\partial t} = A - (B+1)X + X^2Y + D_x \nabla^2 X \tag{39}$$

$$\frac{\partial Y}{\partial t} = BX - X^2Y + D_y \nabla^2 Y \tag{40}$$

Here we shall state some of the main features. For appropriate values of the diffusion coefficients and of the concentrations of A and B, the concentrations of X and Y may exhibit homogeneous oscillations, form stationary spatial patterns or form time-dependent patterns such as propagating waves. Almost every feature observed experimentally in systems such as the B–Z reaction appears in this model system under appropriate conditions.

When the two diffusion coefficients are nearly equal, for a fixed value of A, as the value of B increases, the thermodynamic branch becomes unstable. The reason for this instability is the autocatalytic reaction $2X + Y \rightarrow 3X$. When D_x is nearly equal to D_y, the condition for the instability is

$$B > 1 + A^2 + \frac{\pi^2 m^2}{l^2}(D_x + D_y) \tag{41}$$

This condition can be easily derived by the linear-stability analysis outlined in the previous section. In this equation, l is the size of the system and m is an integer that corresponds to the wave number of the spatial mode that becomes unstable. Clearly, the smallest value of B that will satisfy this inequality corresponds to $m = 0$, which is the homogeneous mode. The value of B at which the mode m becomes unstable is shown in Figure 8(c). Here we see that, when B reaches the critical value $B_c^H = 1 + A^2$, the homogeneous mode becomes unstable. This instability leads to the growth of homogeneous oscillations of X and Y. The amplitude of these oscillations depends on the value of $B - B_c^H$.

(a)

(b)

(c)

(d)

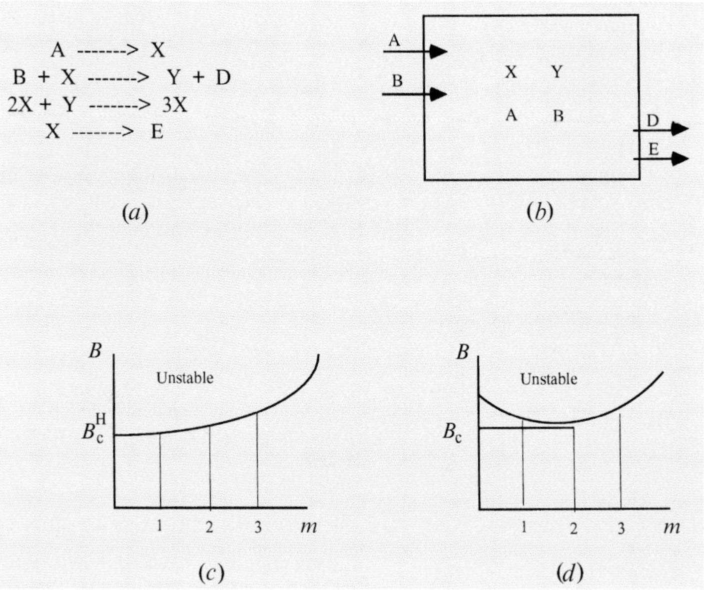

Figure 8

(a) The reaction scheme of the Brusselator. (b) The conditions under which dissipative structures are formed are indicated. The concentrations of reactants A and B are maintained at constant values through an inflow and the products D and E are removed. Under these conditions, the concentrations of X and Y may oscillate in time or form spatial patterns, depending on the values of the diffusion coefficients D_x and D_y of X and Y. (c) A stability diagram showing the value of B for which modes of wave number m become unstable when D_x and D_y are nearly equal. The instability occurs when $B = B_c^H$, at which $m = 0$, leading to homogeneous oscillations. (d) A stability diagram showing the value of B for which modes of wave number m become unstable when the difference between D_x and D_y is large. In this case, when $B = B_{c1}$ the corresponding mode $m = 2$ becomes unstable, leading to the formation of a spatial pattern.

Spatial patterns can arise when the difference between D_x and D_y is large. The precise condition for this instability of the thermodynamic branch is given by the following expression:

$$B > 1 + \frac{D_x}{D_y}A^2 + \frac{A^2}{D_y\pi^2m^2}l^2 + \frac{D_x\pi^2m^2}{l^2} \tag{42}$$

In this case the minimum value of B that will satisfy this condition corresponds not to $m = 0$, but to a non-zero value of m. If an equality is used in (42), thus defining B as a function of m, B attains its minimum value:

$$B_{min} = \left[1 + A\left(\frac{D_x}{D_y}\right)^{1/2}\right]^2 \quad \text{when} \quad m^2 = m^2_{min} = \frac{Al^2}{\pi^2(D_1D_2)^{1/2}} \tag{43}$$

If $B > B_{min}$, all the integer values of m that satisfy (42) represent unstable modes. This is shown graphically in Figure 8(d). Each of the unstable modes can grow. The details of the spatial patterns that form when several modes become unstable are discussed in the second half of this chapter.

Similarly, under appropriate conditions, the Brusselator also shows that propagating waves may arise. Thus this simple model shows how, under non-equilibrium conditions, oscillations, stationary spatial patterns and propagating waves can all arise in chemical systems.

Self-Organization in Chemistry

In the preceding part, we have stressed that non-linear autocatalytic chemical systems can become subject to symmetry-breaking instabilities when they are driven beyond a critical distance from equilibrium. In this part we describe the nature of the solutions that branch off at these bifurcation points.

We have shown that, for the Brusselator model, when $B > B_c^H = 1 + A^2$ (obtained from equation (41) for $m = 0$), the *homogeneous steady state* becomes unstable with respect to homogeneous perturbations that drive the system away from this reference state; the corresponding rate of growth, $\text{Re}(\omega_0)$ (equation (36)), which is derived from the linear-stability theory, takes the form

$$\text{Re}(\omega_0) = \frac{B - B_c^H}{B_c^H} \equiv \mu \tag{44}$$

This exponential growth cannot, however, last forever. After a certain time, the non-linear interactions generate harmonics that moderate the growth, leading eventually to the saturation of the instability. By building on models of phase transitions developed by Landau, bifurcation techniques to describe the new solutions near threshold have been developed.

First one defines a *complex order parameter*, W, the amplitude of the critical mode that is marginally stable at threshold. In the case of the Brusselator model one gets for instance

$$\binom{X}{Y} = \binom{A}{B/A} + \binom{1}{(i - A)/A}(We^{i\omega_0 t} + \text{c.c.}) \tag{45}$$

in which c.c. stands for the complex conjugate. For the Brusselator, calculation shows that $\omega_0 = A$. In the absence of spatial modulations, W obeys the following evolution equation (called the complex Ginzburg–Landau equation), which can be derived exactly from the original reaction–diffusion equations:

$$\frac{\partial W}{\partial t} = \mu W - (\beta_r + i\beta_i)|W|^2 W \tag{46}$$

Such amplitude equations present generic features. Symmetry-breaking instabilities occurring in physically diverse systems are described by equations of the same type that depend only on the nature of the bifurcation. The particularities of each system appear only in the expressions of the coefficients (e.g. β_r and β_i) in terms of the parameters of the corresponding model. This property justifies the use of toy models to describe realistic situations.

Upon writing $W = Re^{i\phi}$, equation (46) yields

$$\frac{dR}{dt} = \mu R - \beta_r R^3$$

$$R\frac{d\phi}{dt} = -\beta_i R^3 \tag{47}$$

These equations describe a *supercritical Hopf bifurcation* leading to homogeneous oscillations. The stationary amplitude R_0 of this oscillation is obtained by setting $dR/dt = 0$ in (47):

$$R_0 = \left(\frac{\mu}{\beta_r}\right)^{1/2} \tag{48}$$

The frequency of the limit cycle is $\Omega = \omega_0 + d\phi/dt$. Using the fact that $\omega_0 = A$ and the value of $d\phi/dt$ from (47), we obtain

$$\Omega = A - \beta_i R_0^2 = A - \frac{\beta_i}{\beta_r}\mu \tag{49}$$

In most chemical systems, the coefficient β_i (non-linear dispersion) is positive; the frequency of the oscillations then decreases when the corresponding amplitude increases.

The onset of such a limit cycle clearly breaks the continuous group of time translation down to the discrete one given by $t \rightarrow t + n\tau$, where τ is the period of the oscillations and n is an integer. When $\beta_r < 0$ in equation (47) higher-order terms must be included, yielding a subcritical bifurcation (see Figure 9). The oscillations then appear suddenly with a finite amplitude. This instability is the analogue of a first-order transition and the system also exhibits bistability between the homogeneous steady state and the limit cycle.

Numerical simulations of various models show that, far from threshold, the oscillations generally lose their harmonic character and exhibit a strong relaxational behaviour. Farther changes of the constraints may lead, through a cascade of secondary bifurcations, to a regime of chemical chaos.

From the experimental point of view, oscillating chemical systems were observed early this century. Most of these studies were, however, carried out on closed configurations (batch reactors). The dissipative structures then appear only transiently since the system inexorably relaxes towards equilibrium. The key experimental development in the study of dynamical phenomena in homogeneous chemical systems was the adoption of the continuous-flow-stirred-tank reactor (CSTR) commonly used in chemical engineering. The reactants are continuously pumped into a tank provided with a suitable overflow exit. The temporal evolutions of concentrations of some major species can be followed using potentiometric or optical probes (Figure 10).

Since homogeneity inside the reactor is ensured by vigorous stirring, such systems can be modelled by a set of ordinary differential equations similar to (32):

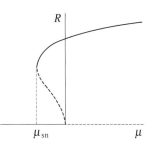

Figure 9
A schematic bifurcation diagram of a subcritical Hopf instability at $\mu = 0$. The amplitude R of the limit cycle is plotted as a function of the bifurcation parameter μ. Thick lines indicate the stable branch while dotted lines indicate the unstable branch of the solution. μ_{sn} corresponds to the lower stability limit of the limit cycle.

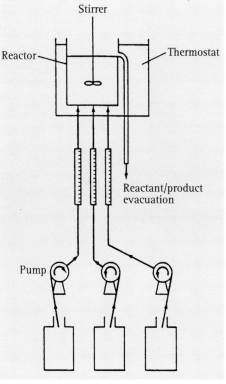

Figure 10
A schematic drawing of a CSTR. In the configuration shown, up to three different solutions can be pumped into the reactor. (Reprinted with permission from *Exploring complexity* G. Nicolis & I. Prigogine, 1989, W. H. Freeman NY.)

$$\frac{\mathrm{d}X_i}{\mathrm{d}t} = F_i([X_i]) + k_0 (X_i^0 - X_i) \tag{50}$$

The first term on the right-hand side describes the chemical reactions involving the reactants X_i. X_i^0 is the concentration of the species X_i in the inlet flow and k_0 is the rate of flow (i.e. the ratio of the volumic flow to the reactor's volume).

When the residence time $1/k_0$ is very large (low k_0), the system approaches chemical equilibrium (that is, the thermodynamic branch in Figure 4). On the other hand, when the flow is high, the chemicals are unable to react significantly and they leave the tank with concentrations that are close to the input values (the flow branch). The rate of flow is therefore a measure of the distance away from thermodynamic equilibrium. Non-trivial behaviours can occur only when k_0 is varied between these two extreme steady states. For certain values of the feed concentrations X_i^0, there is indeed a critical rate of flow, k_0^c above which the thermodynamic branch becomes unstable with respect to the formation of a limit cycle. For $k_0 > k_0^c$, the system oscillates with a clock-like precision. The most thoroughly studied oscillating system is the Belousov–Zhabotinsky reaction, which involves the cerium–catalysed oxidation of an organic substrate (e.g. malonic acid) by a sulfuric-acid solution of bromate.

Aside from chemical oscillations, one behaviour that can be shown in a CSTR is *bi-stability*: the existence of two different stable steady states under the same operating conditions. What state is actually achieved depends upon the past history of the system. J. Boissonade and P. De Kepper have shown that a bistable chemical system can be converted into an oscillating one by adding an appropriate feedback species that reduces the width of the region of bistability. The application of this algorithm has allowed the discovery of dozens of new isothermal chemical oscillators besides the B–Z reaction. Similarly, thermokinetic oscillations have been studied in gas-phase combustion reactions, for which thermal feedback replaces autocatalysis. Chemical oscillations have also been observed in heterogeneous catalysis, for instance, in the case of the oxidation of CO by platinum surfaces.

A very convenient means of characterizing the various behaviours of a dynamical system is provided by the power spectrum $P(\omega)$, which is the modulus squared of the Fourier transform of the time-series data of a relevant concentration $X(t_k)$, $k = 1,2, \ldots, N$:

Figure 11
Time series and power spectra observed in the Belousov–Zhabotinsky reaction: (*a*) the periodic regime and (*b*) the chaotic regime. (Reprinted with permission from C. Vidal *et al.*, *Journal de Physique* **43** (1982) 7.)

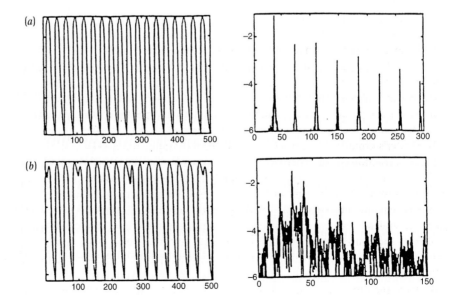

$$P(\omega_j) = |\overline{X}(\omega_j)|^2 = \left| \frac{1}{\sqrt{N}} \sum_{k=1}^{N} \exp\left(-1\frac{2\pi k}{N}j\right) X(t_k) \right|^2 \qquad (51)$$

where $\omega_j = j(N\Delta t)$, $j = 0, \ldots, N-1$ and $\Delta t = t_k - t_{k-1}$. In the oscillating regime, $P(\omega)$ presents isolated lines corresponding to the sharp fundamental frequency and its harmonics. After a further increase of k_0, $P(\omega)$ sometimes exhibits a broad-band continuous spectrum containing an important low-frequency part indicating that the system is aperiodic (see Figure 11).

More elaborate techniques based on an analysis of the phase portraits constructed from measurements of a single concentration have allowed one to determine that this noisy spectrum is characteristic of deterministic chaos. This apparently stochastic behaviour is a manifestation of a complicated trajectory of the orbit of a solution on a *chaotic attractor* (also called a 'strange attractor'), a region of phase space that attracts all nearby trajectories and presents an *extremely sensitive dependence on the initial conditions*. Neighbouring trajectories, stemming from initial conditions that differ but slightly, depart very quickly from each other (Figure 12).

The strangeness of this attractor consists of the fact that all trajectories are diverging and yet occupy a finite volume and the corresponding state-space path never retraces itself. This sensitivity to initial conditions is the root of the unpredictability of chaos known as the 'butterfly effect': one can never predict exactly how a chaotic system will behave over long periods. Various techniques for getting the characteristics of the flow on such a strange attractor, which is the very essence of a chaotic system, have been developed. It is indeed entirely determined by the basic equations of the model and the values of the parameters. The main sequences of the transition to chaos have been observed in the experiments on the B–Z reaction in a CSTR: they are the period-doubling cascade, intermittency and the periodic-quasi-periodic-chaotic sequence. As an example, the period-doubling route to chaos is illustrated in Figure 13.

In this scenario, if the rate of flow is increased somewhat in the oscillatory regime, the limit cycle becomes unstable and the system produces a period-two attractor, that is, it takes two broad cycles before the path retraces itself. If k_0 is increased again, the period-two orbit is also destabilized and a period-four orbit will emerge. If k_0 is sufficiently large, all periodic orbits become unstable and a chaotic attractor will appear. The rapid convergence of the doubling sequence has prevented the observation of more than three period doublings in experiments.

Because chaos is considered uncontrollable and unreliable, it has been regarded as a troublesome property by most engineers who have tried merely to avoid it. This point of view has just begun to change. Scientists have indeed recently designed devices that exploit the advantages of chaos, namely a fast response and flexibility. A chaotic attractor can be viewed as a reservoir of unstable periodic orbits. Ott, Grebogi and York have shown that a chaotic system can be encouraged to follow one of the regular behaviours by applying small perturbations to the system's constraints. In this framework, K. Showalter's group has succeeded in stabilizing a periodic orbit in the chaotic regime of the B–Z reactions (Figure 14).

They could also switch that behaviour rapidly from one period to another by making modest adjustments to the rates of flow. The control of chaos in living systems holds out the promise of new therapeutic and diagnostic tools for diseases ranging from heart disease to epilepsy. A variant of this method has already been applied with success to stabilizing the fast erratic contractions of a small part of a rabbit's heart tissue. The real challenge in this field remains that of controlling spatio-temporal chaos in spatially extended systems.

We have seen that the homogeneous steady state of the Brusselator model can be

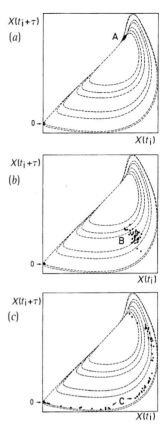

Figure 12
An experimental example illustrating the sensitivity to initial conditions of the strange attractor of the B–Z reaction. The signal at time $(t + \tau)$, which is the potential of a specific electrode, is plotted as a function of its value at time t. The position of the set of points at time $t = 0$ is 0; (a), (b) and (c) correspond to the positions of the points at $t = 80\Delta t$, $t = 105\Delta t$ and $t = 210\Delta t$, respectively.

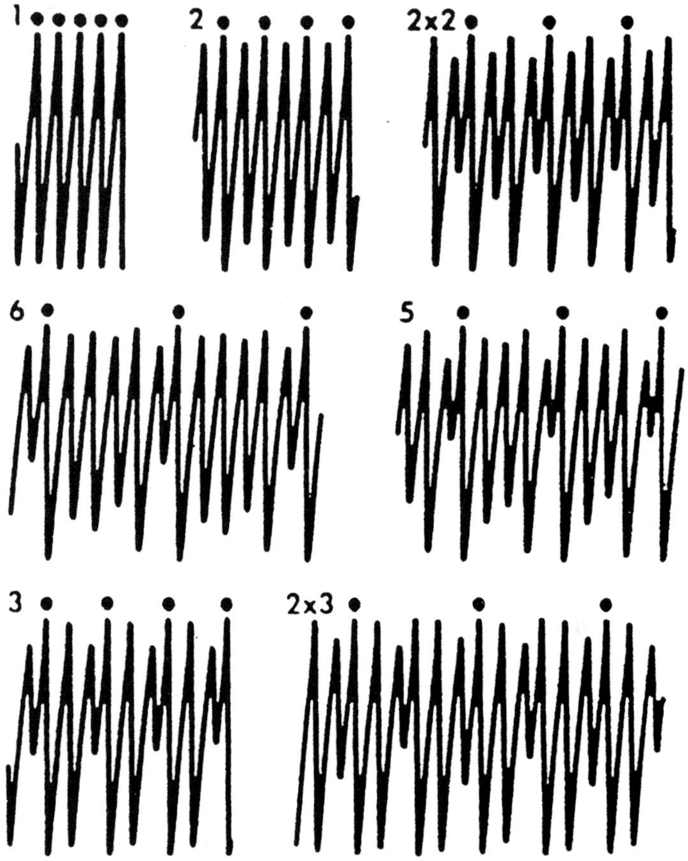

Figure 13
Observed bromide-ion-potential time series with periods τ (115 s), 2τ, $2 \times 2\tau$, $6t$, 5τ, 3τ and $2 \times 3\tau$. The dots above the time series are separated by one period. (Reprinted with permission from Simoyi *et al.*, *Phys. Rev. Lett.* **49** (1982) p. 245.)

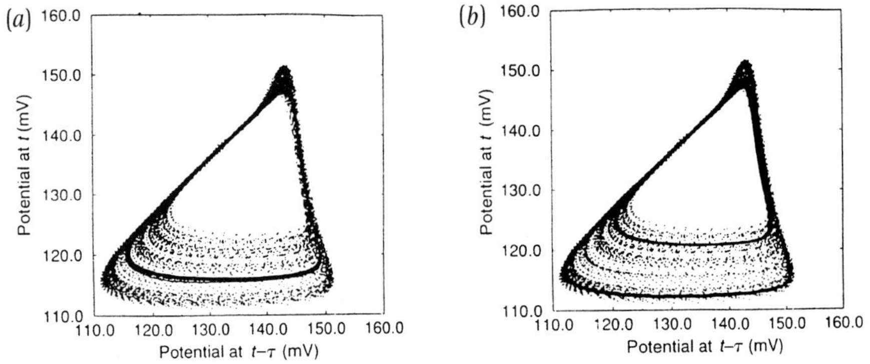

Figure 14
Stabilized period 1- and period 2-limit cycles embedded in the strange attractor of the B–Z reaction. Scattered points exhibit a chaotic trajectory in time-delay phase space. (Reprinted with permission from *Nature* **361**, pp. 240–243, V. Petrov *et al.* Copyright 1993 Macmillan Magazines Ltd.)

destabilized by perturbations with non-zero wave number. In sufficiently large systems this occurs when the control parameter exceeds a critical value $B_c^T = (1 + A\eta)^2$ with $\eta = (D_x/D_y)^{1/2}$. If the ratio D_x/D_y is sufficiently small, i.e.

$$\frac{D_x}{D_y} < \left(\frac{(1 + A^2)^{1/2} - 1}{A}\right)^2 \tag{52}$$

this instability appears before the Hopf bifurcation discussed in the preceding section and steady concentration patterns can then be obtained.

In contrast to most convective cellular structures, the wavelength of this 'chemical crystal', depends on intrinsic parameters: rate constants and diffusion coefficients.

Useful analogies with symmetry-breaking phase transitions in condensed-matter physics may then be drawn.

This instability results solely from the interplay between chemical reactions and diffusion. The beauty of Turing's idea lies in this counterintuitive organizing role of diffusion that usually smears out any inhomogeneity in concentration. The term *diffusion-driven instability* has been coined to characterize this mechanism that has been shown to be pertinent to the formation of spatially dissipative structures not only in chemistry and biology but also in many other fields.

From the thermodynamic analysis of the preceding section, it follows that only those systems that present feedback loops of some type can exhibit this instability. In the two-variable models one can generally identify an activator that stimulates its own production and is controlled by an inhibitor. A pattern can appear only when the local balance between these antagonistic species is broken, for instance when the diffusion coefficient of the inhibitor is larger than that of the activator, i.e. $D_x/D_y \ll 1$ in equation (52). Most of the models that have been introduced in this field are indeed based on some disguised form of the principle of *local activation and long-range lateral inhibition*.

More recently an alternative way of spatially decoupling these counteracting species was proposed. In this case the homogeneous steady state (HSS) is destabilized by flows of the activator and inhibitor at different rates (e.g. by selectively binding one component to a support). Because the rates of flow are easily tunable parameters, this new mechanism for the formation of spatio-temporal patterns could be operative in a large class of chemical, physical and biological systems.

The first experimental evidence of sustained steady chemical patterns was obtained only recently, nearly four decades after their prediction by Alan Turing. The experimental difficulties that had prevented their observation have finally been overcome by the design of open spatial gel reactors (Figure 15).

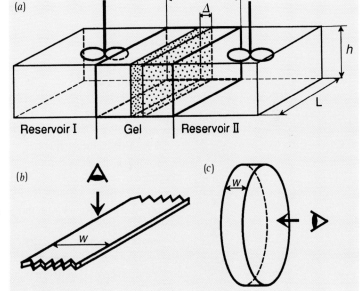

Figure 15

Sketches of the open spatial reactor. (*a*) Basic principles. A block of hydrogel ($L \times h \times w$) is in contact with the contents of reservoirs I and II. L and h are the dimensions of the feed surfaces and w is the width of the gel. Δ is the characteristic width over which the chemical pattern develops. (*b*) A thin-strip reactor. (*c*) A standard disc reactor. (Reprinted with permission from *Physica A* 213, copyright 1995 Elsevier Science.)

The core of these reactors is a block of gel with two opposite sides in contact with the contents of stirred tanks containing non-reacting sub-sets of reagents. The reaction system is maintained at a controlled distance from equilibrium by feeding the tank continuously with fresh reactants. The gel damps out convective fluid motions so that the only processes within the gel are indeed reaction and diffusion.

Up to now, Turing instabilities have been demonstrated experimentally only in one family of liquid-phase chemical reactions, the chlorite–iodide–malonic-acid (CIMA) reaction and its variants. They are redox reactions characterized by positive feedbacks due both to autocatalysis by iodine and the kinetics of inhibition of the substrate by iodide.

The linear-stability analysis allows one to determine the critical value B_c^T of the control parameter above which various modes with wave numbers near k_c begin to grow exponentially, leading eventually to a new patterned solution. The nature of these solutions strongly depends on the aspect ratio, namely the ratio of the characteristic size of the reactor to the critical wave number λ_c.

In small-aspect-ratio systems, the geometry of the reactor plays a decisive role in isolating specific patterns. This effect has been illustrated nicely in the simulations that

J. D. Murray performed to explain the variety of coat markings of animals found in nature. In this model, chemical species called morphogens, which react and diffuse through the cells, are subject to a diffusive instability. The corresponding Turing structure then acts as a 'pre-pattern' laying down a kind of scaffolding for subsequent development. Each cell then reacts by producing a pigment according to the concentration of the morphogen at its location.

In contrast, in large systems such as those characterizing the recent experiments on the CIMA reaction, we face a large degeneracy of destabilizing modes as soon as the stress parameter exceeds its critical value. In such systems, for which the boundaries are far away, all the wavevectors belonging to the critical manifold $|k| = k_c$ are equally amplified. The rates of growth of these modes take the following form

$$\text{Re}(\omega_k) = \mu - \frac{\xi_0^2}{4k_c^2}(k^2 - k_c^2)^2 \tag{53}$$

in which $\mu = (B - B_c^T)/B_c^T$ measures the distance from the Turing instability point and ξ_0 is the coherence length. In this case all values of k for which $\text{Re}(\omega_k) > 0$ become unstable. This infinite degeneracy reflects the rotational symmetry of large isotropic systems.

On the other hand, as soon as $B > B_c^T$, a continuous band of unstable wave numbers appears for each orientation. The width of this band $|k| = k_c + Q$ is given by

$$\frac{-\sqrt{\mu}}{\xi_0} < Q < \frac{\sqrt{\mu}}{\xi_0} \tag{54}$$

These *active modes* lie within a critical annulus in two dimensions and within a spherical shell in three dimensions. The presence of this bandwidth is also at the origin of the wavelength-selection problem.

Because of these degeneracies, the pattern that finally emerges is not uniquely determined by the linear stability analysis of the reference state but rather the non-linear couplings between the active modes play a crucial role in the mechanism of pattern selection.

Standard bifurcation techniques have been developed to tackle this problem near the threshold where the amplitudes of the branching solutions are still small. They are based on the separation of time and length scales between the active and the passive modes. The latter are linearly damped above B_c^T but they are continuously regenerated through non-linear interactions by the active modes.

The analysis of the patterns is done as follows. First, one approximates the concentration field by a linear superposition of the critical modes. For the Brusselator model, one writes for instance

$$\begin{pmatrix} X \\ Y \end{pmatrix} = \begin{pmatrix} A \\ B/A \end{pmatrix} + \begin{pmatrix} 1 \\ -\frac{\eta}{A}(1 + A\eta) \end{pmatrix} \left(\sum_{i=1}^{m} A_i e^{ik_i \cdot r} + \text{c.c.} \right) \tag{55}$$

where $|k_i| = k_c$. Because they determine the degree of order in the system, the complex amplitudes A_i play here also the role of order parameters corresponding to this symmetry-breaking instability.

The competition among the critical modes can be described by writing a set of equations for these amplitudes. If the saturation instability occurs for the cubic order, they take the following form in the absence of spatial modulation:

$$\frac{dA_i}{d_t} = \mu A_i + v \sum_j \sum_k A_j^* A_k^* \delta(k_j + k_k + k_i)$$

$$-g_D\,|A_i|^2-\sum_j g_{ND}(i,j)\,|A_j|^2 A_i$$

$$-\sum_j\sum_k\sum_m \gamma(j,k,m)\,A_j^* A_k^* A_m^* \delta(k_j+k_k+k_m+k_i) \qquad (56)$$

The coupling terms $g_{ND}(i,j)$ and $\gamma(j,k,m)$ are functions of the angles between the inter-acting wavevectors. Here also these equations have a universal character and all of the information concerning a particular system is contained in the structure of the coefficients.

These amplitude equations admit only fixed points as the attractors which correspond to steady patterns. Each set of wavevectors indeed defines a structure. In two dimensions, we have $m=1$ stripes, $m=2$ squares or rhombs and $m=3$, with $k_1+k_2+k_3=0$, structures of hexagonal symmetry. Higher-order superpositions of the active modes correspond to quasicrystalline order.

The investigation of the stability of the bifur-cated solutions goes through the redeployment of the linear-stability analysis of these states to determine secondary bifurcations, which generally correspond to structural transitions between patterns. The difficulty arises from the fact that, for each state, one has to discuss the stability with respect to all possible perturbations: amplitude, orientation of the wavevectors and also resonant perturbations with different symmetries.

We now present briefly typical bifurcation diagrams that have been obtained in two and three dimensions for various reaction–diffusion systems. In two dimensions, a hexagonal pattern first appears subcritically where it may co-exist with the homogeneous steady state for $\mu_H^- < \mu < 0$ (see Figure 16).

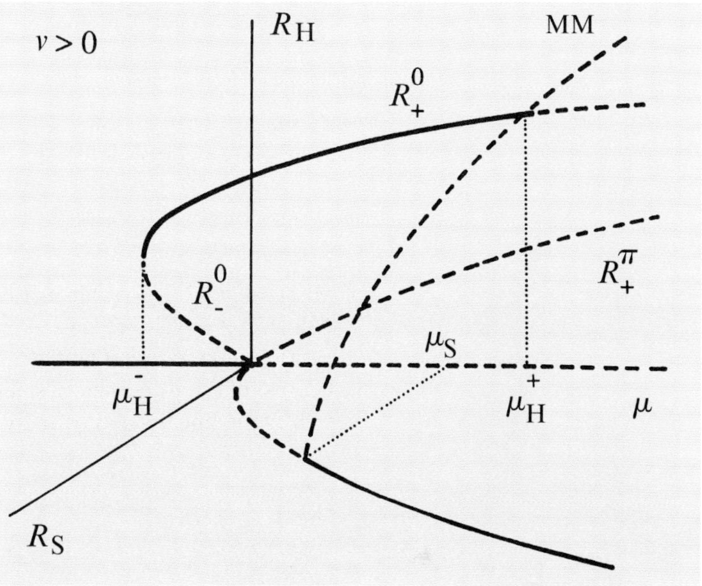

Depending on the sign of the quadratic term v in equation (56), the extrema of concentration form either a triangular lattice (H_0) or a honeycomb lattice (H_π); they lose stability for $\mu>\mu_H^+$. Beyond the critical point, stripe structures may also appear supercritically. The three types of patterns obtained by solving the Brusselator model numerically are displayed in Figure 17.

These stripes are at first unstable with respect to resonant perturbations but they become stable for $\mu>\mu_s$. Since $\mu_s<\mu_H^+$ there is a domain of bistability between the two types of structures. In this range one recovers thus the standard 'hex–stripes' competition that comes up in various fields such as hydrodynamics and optics. They have also been obtained in thin-disc reactors, which allow one to observe quasi-two-dimensional structures subjected to uniform constraints (Figure 18).

Under the experimental conditions, the concentration profiles created by the process of feeding influence the selection and orientation of the patterns. The simplest principle that comes to mind is that a structure will develop in the region of space where the local values of the parameters allow it to be stable in the corresponding uniform system. As a result, these profiles generate the unfolding in space of the bifurcation diagram obtained under uniform conditions. This is shown in Figure 19(a) for the hex–stripe competition. Numerical simulations, however, reveal that the stability limits are shifted by the ramps. This competition can be experimentally observed in various

Figure 16
A schematic bifurcation diagram, showing the amplitude as a function of the bifurcation parameter for the case in which the quadratic coupling v is positive. A pattern of hexagonal symmetry (H_0) first bifurcates subcritically. Supercritically one has the bifurcation of another branch of hexagons (H_π) and a branch of stripes. A branch of mixed modes (MM) is also represented. Thick (broken) lines indicate stable (unstable) branches of solutions. (Reprinted with permission from *Physica A* **213**, copyright 1995 Elsevier Science.)

(a)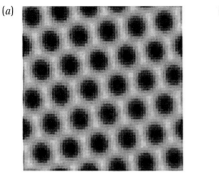

(b)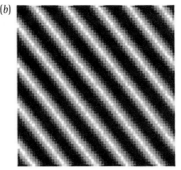

(c)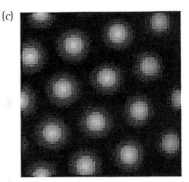

Figure 17

The three basic two-dimensional patterns for the activator (X) of the Brusselator model. The grey scale interpolates between the maximum (white) and minimum (black) concentration. The maxima respectively form (a) a honeycomb, (b) a striped lattice and (c) a triangular lattice. (Reprinted with permission from *Physica A* **213**, copyright 1995 Elsevier Science.)

Figure 18

Stationary chemical patterns formed in open disc reactor: (a) and (c) show hexagonal patterns and (b) shows stripes. (Reprinted with permission from Onyang & Swinney, 'Transition to Chemical turbulence', *Chaos* 1, pp. 411–420. 1991, American Institute of Physics.)

reactor geometries. The strong localization of chemical structures by the feed-concentration profiles is illustrated for thin-strip reactors in Figure 19(b).

Three-dimensional patterns can develop in reactors whose geometrical dimensions are larger than λ_c. The pattern selection in three-dimensional systems proceeds along the same lines as before. On increasing the bifurcation parameter the following sequence emerges. The first structure to appear subcritically is characterized by six pairs of wavevectors. In real space, the maxima of concentrations form a body-centred cubic lattice. It is followed also by hexagonally packed cylinders ($m = 3$), which are the natural extensions in three dimensions of the two-dimensional hexagonal patterns. Finally, a smectic-like order of lamellae arises supercritically. These structures have been observed in that order in numerical simulations of the Brusselator. For a range of values of the parameters these structures may co-exist and the system exhibits spatial tristability. The actual experimental determination of the three-dimensional structures inside gel reactors is a technically difficult problem. However, observations made under various angles have revealed the existence of hexagonal prisms and body-centred-cubic lattice structures.

The structures described above are generated by a diffusional instability of the steady state of a reaction–diffusion system. They appear *spontaneously* when some parameter exceeds a critical value. A new type of concentration pattern has recently been discovered experimentally in a bistable chemical system, the iodate-ferrocyanide–sulfite reaction. For some values of the concentrations, the system traces a hysteresis loop between the two homogeneous steady states as the flow rate is varied. A labyrinthine pattern can be initiated only by a *finite-amplitude* perturbation with ultraviolet light. The system also exhibits an unusual mechanism of growth that has also been observed experimentally in the CIMA reaction, involving self-replicating chemical spots. After the first spots have emerged from the uniform state, they grow

(a)

(b)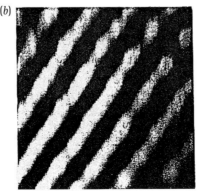

(c)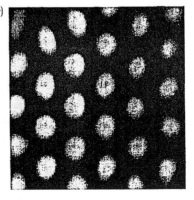

(a) 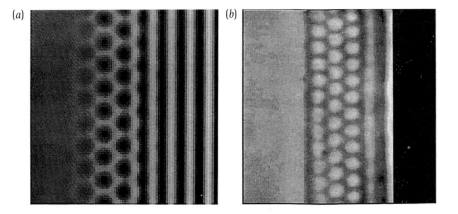 (b)

Figure 19
(a) A numerical simulation of the Brusselator model showing the spatial co-existence of hexagons and stripes resulting from the existence of a linear ramp in the bifurcation parameter B. The value of B_c is imposed for the third of the reactor from the left-hand boundary. (b) Stationary concentration patterns in a thin-strip gel reactor, showing obtained localized staggered rows of spots and stripes. (Reprinted with permission from *Physica A* 213, copyright 1995 Elsevier Science.)

in size and they elongate when their edges reach a minimum length. Thereafter, the elongated spots eventually divide into two spots of nearly identical sizes. The process repeats itself for the new daughter spots, leading finally either to a hexagonal pattern or to a chaotic state in which the spots compete continuously for territory. This dynamics is illustrated by the series of snapshots shown in Figure 20. Such a mechanism has also been obtained in a numerical simulation of the model of Gray and Scott. This mechanism of growth is reminiscent of that of some biological systems. Here, however, it takes place in a monophasic system.

Unpredictability in Non-equilibrium Systems

In the previous sections we have seen how autocatalytic systems under far-from-equilibrium conditions can generate dissipative structures. These systems undergo a transition to an organized state that is driven by fluctuations. What happens when, as a result of instability, the system can make a transition to more than one possible state? To which of these possible states the system will evolve may depend on the fluctuations. Some transitions require a particular type of fluctuation that is somewhat rare; in these cases the time at which the transition will occur becomes unpredictable. These situations bring to light a new domain of chemistry, one in which knowledge of the kinetic rate laws does not give us the ability to predict the exact state to which the system will evolve or the precise times at which major changes will occur. Let us look at some simple examples.

A reaction known in the literature as the 'clock reaction' involves chlorite and thiosulfate ions. It is called a clock reaction because, depending on the reaction conditions, it can undergo a sharp change in colour at a particular time, in a clock-like manner. The change in colour is due to a rapid decrease in pH after an initial induction period. This behaviour is due to the autocatalytic production of H^+ with a rate proportional to $[H^+]^2[Cl^-]$. A small increase in $[H^+]$ grows with great rapidity and causes a sharp change in colour. In a typical clock reaction, the change in colour may occur after about 7 min to within a few seconds in repeated trials. Epstein and Nagypál have shown that, under a different set of reaction conditions, this clock becomes 'crazy' and unpredictable: the sharp change in colour occurs randomly at unpredictable times, however carefully the initial conditions and all conditions of the reaction (temperature, light intensity etc.) are kept the same for repeated trials (Figure 20). Theoretical modelling of such systems was done by Peters, Baras and Nicolis.

The unpredictable behaviour of the clock reaction arises when the rapid decrease in pH is a consequence of a rare event, such as a build up of H^+ in a small region which then quickly spreads to the entire system. Similar behaviour can also be seen in the stochastic nature of ignition times in the kinetics of cool flames.

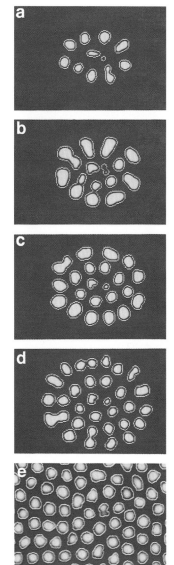

Figure 20
Pattern-growth dynamics obtained by spot-splitting. This series of snapshots was taken at 4-min intervals. (Reprinted with permission from P. De Kepper *et al.*, *Int. J. Bifurcation and Chaos* 4 *(1994) 1228.)*

The unpredictability becomes more striking if there is an inherent symmetry in the system such as mirror reflection symmetry. Even though it is not a chiral molecule, $NaClO_3$ crystallizes in a chiral form. The unit cell consists of four molecules of $NaClO_3$ arranged with a twist that gives it a sense of handedness. The crystals of $NaClO_3$ exhibit optical activity, which can be easily detected with a pair of polarizers. When a supersaturated solution of $NaClO_3$ is placed in a Petri dish and allowed to crystallize through evaporation of the solvent, statistically equal numbers of *laevo-* and *dextro*-rotatory crystals are generated. If the crystallization is done in a beaker while the solution is constantly stirred, then, remarkably, almost all (more than 98%) of the crystals in each crystallization batch are *laevo-* or *dextro*-rotatory (Figure 22). It is not possible to predict whether the crystals in a particular crystallization will all be *laevo-* or *dextro*-rotatory. There is a 50% chance that they will be *laevo-* and a 50% chance that they will be *dextro*-rotatory. Even though the kinetics of crystallization for the two types of crystals are identical, one of them will dominate in stirred crystallization, i.e. the mirror-image symmetry in the kinetics is broken by this system.

The unpredictability is a consequence of autocatalytic secondary nucleation. In a stirred crystallization, a *laevo-* or *dextro*-rotatory crystal can generate crystals with the same enantiomeric form autocatalytically through a process called 'secondary nucleation'. Thus, the 'first crystal' that quickly begins to generate secondary crystals autocatalytically (within about 20 min) produces a large number of daughter crystals with the same handedness. Owing to the rapid generation of the crystals, the concentration quickly drops and reaches a value at which the rate of nucleation of new crystals is virtually zero. Thus, one crystal, through autocatalysis, is able to multiply rapidly and cause all the crystals produced do be either *laevo-* or *dextro*-rotatory. Since nucleation of a crystal is a rare event that is amplified through secondary nucleation, the distribution of products (i.e. the percentages of *laevo-* and *dextro*-rotatory crystals) is random, varying dramatically between nearly zero and nearly 100% from trial to trial. The process is akin to a successful mutation in biological evolution, but here we have only two species, '*laevo*' and '*dextro*'. We see that the dominance of one 'species' through random fluctuation and autocatalysis (or self-replication) can be realized even in simple non-equilibrium chemical systems.

Concluding Remarks

To conclude, we would like to quote form a recent report by C. K. Biebricher, G. Nicolis and P. Schuster. They wrote that

The maintenance of the organization in nature is not – and cannot be – achieved by central management; order can be maintained only by self-organization. Self-organizing systems allow adaptation to the prevailing environment, i.e., they react to changes in the environment with a thermodynamic response which makes the system extraordinarily flexible and robust against perturbations of the external conditions. We want to point out the superiority of self-organizing systems over conventional human technology which carefully avoids complexity and hierarchically manages nearly all technical processes. For instance, in synthetic chemistry, different reaction steps are usually carefully separated from each other and contributions from the diffusion of the reactants are avoided by stirring. An entirely new technology will have to be developed to tap the high potential of self-organizing sytems for guidance and regulation of technical processes. The superiority of self-organizing systems is illustrated by biological systems, in which complex products can be formed with unsurpassed accuracy, efficiency and speed.

In spite of the progress summarized in this chapter, we are only at the beginning. We may expect that the gap between self-organizing systems in chemistry and in biology will be narrowed during the coming decades.

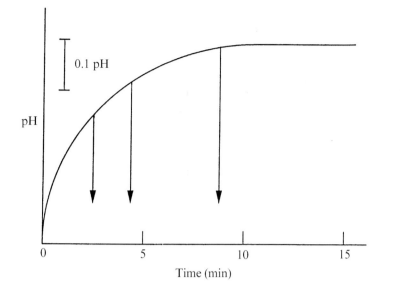

Figure 21
Stochastic behaviour in the chlorite–thiosulfate 'clock reaction'. The traces show the variation of pH for different runs. Although each trial starts with identical initial conditions, the time at which a rapid drop in pH occurs is random, varying considerably from trial to trial.

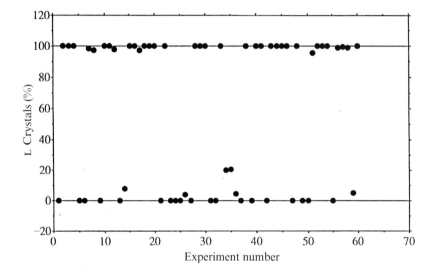

Figure 22
Stochastic behaviour in the symmetry-breaking crystallization of $NaClO_3$. The percentage of *laevo* crystals in stirred crystallization is unpredictable. In each case it is either close to zero or close to 100%.

Further Reading

1. G. Nicolis and I. Prigogine, *Exploring Complexity* (Freeman, New York, 1989).

2. R. Kapral and K. Showalter (editors), *Chemical Waves and Patterns* (Kluwer Academic Publishers, Dordrecht, 1995).

3. D. Kondepudi and I. Prigogine, *Modern Thermodynamics* (John Wiley and Sons, New York, 1998).

4. I. R. Epstein and J. A. Pojman, *An Introduction to Nonlinear Chemical Dynamics* (Oxford University Press, Oxford, 1998).

5. A. Goldbeter, *Biochemical Oscillations and Cellular Rhythms: The Molecular Bases of Periodic and Chaotic Behaviour*, (Cambridge University Press, Cambridge, 1996).

6. F. Baras and D. Walgraef (editors), *Nonequilibrium Chemical Dynamics: From Experiment to Microscopic Simulation* (Special Issue of *Physica*) (North-Holland, Amsterdam, 1992).

7. *Biological Asymmetry and Handedness* (John Wiley, London, 1991).

8. P. Gray and K. S. Scott, *Chemical Oscillations and Instabilitites* (Clarendon Press, Oxford, 1990).

9. R. J. Field and M. Burger (editors), *Oscillations and Traveling Waves in Chemical Systems* (John Wiley, New York, 1985).

10. I. Epstein, K. Kustin, P. De Kepper and M. Orbán, *Sci. Am.*, 1983. **248**, 112.

11. C. Vidal and A. Pacault (editors), *Non-Linear Phenomenon in Chemical Dynamics* (Springer, Berlin, 1981).

12. G. Nicolis and I. Prigogine, *Self-Organization in Nonequilibrium Systems* (Wiley-Interscience, New York, 1977).

Chemistry in Society

Colin Russell

Introduction

The notion that chemistry had anything to do with society would have seemed distinctly novel – even laughable – until relatively recent times. That, no doubt, reflects at least as much changing views of society as the changing nature of chemical science. Until about 250 years ago some of its early practitioners, the alchemists, were scolded for their duplicity in pretending to make gold, for inflicting intolerable smells on their neighbours or for both offences at once. This is not to say that they did not sometimes have an uncomfortably high social profile; some monarchs (such as Rudolf II of Prague) regarded alchemy as a hobby or had an alchemist as part of their courtly entourage, much as they would employ a court jester. When these favoured individuals failed to deliver the expected goods, however, their fall from grace was spectacular. Alchemists in general were banned by various religious orders and even by the famous Decretal of the Pope in 1317. Dante charitably assigns them, together with forgers and other scoundrels, to the tenth gulf of the Inferno. It appears that society did not altogether cherish these predecessors of the modern chemist, as may be seen in the alchemist described by Chaucer whose coat 'is not worth a myte . . . it is all filthy and to-tore'.

Today it might be argued that public perception of chemistry is almost as ill-informed as it was in the Middle Ages, though the reasons are a little different. There is a widespread misunderstanding of those blights that might be fairly laid at the doorstep of chemistry and a breath-taking unawareness of the benefits the science has conferred on mankind. That is not pro-science propaganda but a simple, if deplorable, fact. This final chapter will, therefore, concentrate on two main issues. First it will attempt to achieve some kind of objective survey of what chemistry actually has done, and is doing, for society. Then it will examine the crucial contemporary problem of the image of chemistry widely held throughout society, seeking to understand and assess the reasons for the present situation. A brief concluding section will offer some tentative suggestions for the future.

The Social Impact of Chemistry

It may seem strange in a book about *New Chemistry* even to glance at developments that took place 200 years ago or more. The reason is twofold. First, these developments changed the face of civilization and thus indicate the crucial role chemistry has already played in society and (by implication) could play again. Second, they have left their mark on the way the general public perceives the place of chemistry today. It also happens to be a fact that most students of social history are completely oblivious to the part that the science of chemistry played in those turbulent times. So we begin with the Industrial Revolution.

The time when chemistry started to make a really large-scale impact on society may be identified as the middle of the eighteenth century. Until then there had been a continuous history of the extraction of metals, purification of alum and a few other common minerals, the manufacture of pigments and paints and the production of vinegar and of a handful of natural drugs purified by the art of the apothecary or druggist. How far they can readily be called 'chemicals' and how far the rudimentary theory behind their manufacture was properly called 'chemistry' must be a

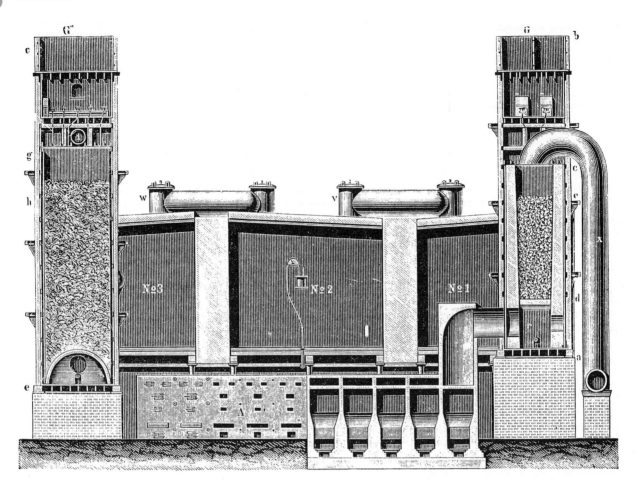

Figure 1
A sulfuric acid plant.

matter of debate. Just about the time when Europe was priding itself on its 'Enlightenment' and the French Revolution was imminent, some strange chemical production processes were being scaled up to such an extent that people could hardly avoid noticing them.

In the Midlands of England and in the Clyde valley in Scotland several entrepreneurs were beginning to make sulfuric acid on a considerable scale, by atmospheric oxidation of moist sulfur dioxide in large though fragile glass globes and later in vast air-cooled chambers made of the acid-resistant lead (Figure 1). At about this time Liebig made his famous remark that the consumption of sulfuric acid is a barometer of a nation's commercial prosperity. Much was needed for a new process for making soda, for which the vital raw material was sodium sulfate, made by the time-honoured reaction

$$H_2SO_4 + 2NaCl = Na_2SO_4 + 2HCl\uparrow$$

The soda was made by a process invented by Nicholas Leblanc and summarized by the equation

$$Na_2SO_4 + CaCO_3 + 4C = Na_2CO_3 + CaS + 4CO$$

Meanwhile something else was happening. Towards the end of the eighteenth century British society, though exempt from the political upheavals of revolutionary France, was to experience a convulsion no less dramatic and far-reaching. It was, in fact, shaken to its very foundations by what we now call the Industrial Revolution. Most obviously this was an explosion in the production of textiles made possible by the mechanization of such processes as spinning, carding and weaving. The new technolology was powered at first by water-wheels and later by steam. There were unprecedented changes in demography with a rural society accepting urbanization on a huge scale. The movement of people, and still more of raw materials like coal and iron, called for vastly improved means of transport and the canals of the eighteenth century were followed by the railways of the nineteenth. These increased still further the pace of social transformation of early Victorian Britain. Less than 20 years elapsed between the opening of the Stockton and Darlington Railway in 1825 and the rise of the Railway Mania, in which speculation went wild in an attempt to cover the country in a network of railways, a pattern that was closely followed overseas. In addition to the rise of great industrial towns, especially in the north, there came about an increase in the mobility of individuals and families, the rise of the seaside holiday and the emergence of a new and militant working class.

In all these revolutionary changes to society it has not been customary to regard chemistry as having made a significant contribution. Indeed, most elementary textbooks of modern British history do not even mention it and it is usually ignored in more advanced treatments of economic and social history. Yet the fact is that recent scholarship has demonstrated beyond doubt that chemistry was not merely an important, but even a critical, determinant in the rise of an industrial society. The eighteenth century saw the demise of the old phlogiston theory and its replacement by the modern (oxygen) theory of combustion. Paradoxically, one of the most famous casualties of the French Revolution, Antoine Lavoisier (Figure 2; who was guillotined in 1789), played a principal part in establishing what is now often called 'the chemical revolution'. However, certain other developments took place at this time, just as important and having a more obvious impact on society. These have also been termed a 'chemical revolution' and amply chronicled by Archie and Nan Clow in their classic book of that name. They involved the rise and rapid growth of the heavy chemical industry. At its heart lay the manufacture of soda, sulfuric acid and chlorine.

As the rate of production of textiles soared, it became woefully evident that problems that had so far been minor irritants were now to threaten the success of the whole enterprise. These were the processes of washing and bleaching the fabrics. Hitherto they had been conducted in a manner so extravagant in its use of labour, land and time that the inefficiencies could be tolerated only on a very small scale.

Washing is needed at many stages of the production of textiles, particularly that from animal raw materials like wool. For this purpose soft soap (from potash) is less suitable than its sodium counterpart. This had been made from soda, which until the eighteenth century was usually produced by ignition of barilla, a form of sea-weed whose harvesting from the coasts of northern Britain was a long and labour-intensive operation. The discovery of an alternative source, by Nicholas Leblanc, came in the nick of time at the end of the century. It became possible to use this 'synthetic' soda to satisfy the huge demands of the textile manufacturers so that from 1800 to 1850 the production of hard soap in Britain rose from almost nothing to nearly 21 million tonnes per year. Use of such soap for personal hygiene was at first on a relatively minor scale, though eventually it doubtless contributed much to the health of the general population. Important early manufacturers included Pears of London (1789), Crosfield of Warrington (1815) and Gossage of Widnes (1857).

Fabrics needed not only to be clean. Most of them needed to be white, even if they were later coloured by dyestuffs. The bleaching of cloth by traditional means was to moisten it with sour milk, spread it out in the open air ('bleach fields') and leave it there for several months to be slowly bleached by whatever direct sunlight there might be

(Figure 3). Again, at the end of the century, chemistry produced the first economic alternative. This was to use chlorine, a bleaching agent recognized by Berthollet that reduced the duration of the process from months to minutes and released for other purposes acres of land in the countryside around textile factories. A modification, introduced by Charles Tennant, was to make bleaching powder from the chlorine, thus allowing the material to be transported easily from its place of manufacture.

It is thus apparent that chemistry played an absolutely critical role in the early phase of the Industrial Revolution. Some 60 years later synthetic dyes were introduced, thereby displacing such natural raw materials as madder, the indican plant, woad and so on and increasing still further the dependence of the textile industry on chemistry.

Soda was also used to make glass (by fusion with sand). A series of fiscal measures to raise revenue from the more opulent owners of buildings led to the Window Taxes, the last of which was not repealed until 1845. So the chief outlet for soda-based glass

Figure 3
A Scottish bleachfield.

was in clear glass bottles, but, as the nineteenth century advanced, the large northern mills were able to glaze their windows, thus allowing production to continue in all weathers. As the Clows observed:

It is difficult to decide whether manufacture of glass for windows or for glass containers had the greater effect on the trend of western civilization.

To make soda and chlorine one needed sulfuric acid on a large scale and, as we have seen, this was available at precisely the time when the Industrial Revolution was getting into full swing. From the 1840s it became needed additionally for the manufacture of superphosphate fertiliser, a calcium acid phosphate. The effect of this on agriculture was comparable to that of chlorine on the production of textiles. Above all, sulfuric acid, or its derivative hydrochloric acid, became necessary for the 'pickling' of iron before it was fabricated into a multitude of engineering artefacts and structures. This was essentially the removal of covering layers of oxide.

Merely to take these three chemicals, it is not hard to grasp that the immense technical advances of the early nineteenth century would have been utterly impossible without them. As the century advanced the role of chemistry became ever more important and obvious. Yet it is truly remarkable that so many histories of this period make no mention of the fundamental fact that, *had there been no prior Chemical Revolution, the Industrial Revolution could not have taken place* and with it the ensuing social and political changes that have transformed our world.

This book has recorded abundant instances of recent and important chemical discoveries, profoundly affecting the way we live today. It would be easy – but pointless – to provide further examples. Instead it may be more useful simply to indicate the main areas of life in which chemistry has played, and is still playing, a crucial role.

First there is the obvious area of warfare, though it would be very unfair to blame chemistry for the enormous loss of life through military conflict down the centuries; in the Thirty Years War (1618–48), for instance, one third of the population of Europe perished without any help from the chemists at all. If we exclude the employment of gunpowder (first at Crécy in 1346) and the manufacture of nitroglycerine in the Franco-Prussian War of 1870–71, the earliest large-scale utilization of chemistry for military purposes was in World War I. In Britain, cordite (containing nitroglycerine and nitrocellulose) had been employed for a few years as a propellant and was made in huge quantities at Gretna, where many of the workers were women. Picric acid had been employed as a high explosive for shells, but tends to be too sensitive to shock and was gradually replaced by the safer trinitrotoluene (TNT). As the war progressed, however, it soon became apparent that Britain was at a great disadvantage to Germany. There was an acute shortage of toluene and of pure sulfuric acid, which were needed for nitrations leading to picric acid and TNT. The German chemical industry was far better equipped to produce such materials. For in fact the European leadership in synthetic organic chemistry had long before passed from Britain to Germany, so, by 1914, 80% of Britain's dyestuffs were being imported from Germany. When British troops first went into the trenches their new khaki uniforms owed their colour to dyes made in German factories! It has been argued that, if Britain had been as well-prepared *chemically* as her enemy, the conflict would have been terminated far more speedily and the course of world history altered.

The later use of chlorine gas in warfare was another indicator of the chemist's role in World War I. A British response to this new menace was to set up the Chemical Defence Establishment at Porton Down in 1916. Given that World War I involved the detonation of 238 000 tonnes of TNT, 378 000 tonnes of ammonium nitrate and vast quantities of other explosives and at one stage a weekly release into the atmosphere of 100 tonnes of chlorine, it has with some justification been termed 'The Chemists' War'.

Chemistry was also deeply involved in World War II, as were various branches of physics. Here again the differences in chemical capabilities between the two warring powers were of profound importance. This time the Germans were at a disadvantage because they had only limited access to foreign petroleum and its products. Toluene, still needed for TNT, was beginning to be made from petroleum in the USA, while the Germans had to fall back on coal tar for this and many other chemicals, together with motor and aviation fuel. As a result they developed an impressive new field of aliphatic chemistry based on ethylene and acetylene. As the war continued German chemists capitalized on their pre-war experience and were able to develop a coal-based synthetic rubber comparable to that produced in America from petroleum. They also produced synthetic detergents, though these were inferior to those made by the Western Allies from petrochemical sources. Although they were never used by the combatants in Europe, vast amounts of poison gases were made by both sides, most lethal being the organophosphorus nerve gases developed in Germany. There can be no doubt whatever that chemical research of great ingenuity was pressed into service by both sides in the conflict and in that sense chemistry can be said to have continued its major impact on human history.

Biologically active chemicals have also been employed in war. The insecticidal properties of DDT, recognized in 1942, were put to good use by the USA's forces in the Far East in delousing powders to combat the spread of typhus and against mosquito larvae to combat malaria. Many lives were saved thereby. Later, in the Vietnam War, a herbicide containing 2,4,5-T and known as 'Agent Orange' was used in a programme of mass defoliation by the USA's forces, in order to flush out the enemy from forest cover. Some 10% of the country was defoliated by 84 million litres of the chemical. Each of these military uses produced unwelcome and unexpected consequences. Again, the ability of chemistry to be a potent instrument in warfare was demonstrated.

How far these military results of chemical research were, in the long term, good for humanity may be endlessly debated. What is not in doubt is the beneficial effect of chemistry in immeasurably improving the health of society. To take one example in the area of public health, it is often not realized that the problem of water-borne disease in Britain especially was (a) of huge proportions and (b) first solved by chemical rather than microbiological methods.

During the course and in the aftermath of the Industrial Revolution Britain experienced a demographic drift from the countryside to the towns where factories and mills tended to be situated. In the nineteenth century the population of London increased from 1.1 to 6.6 million, for example. Given the chronic overcrowding, one could expect the occasional epidemic to ravage the community. Most serious was its vulnerability to diseases caused by pathogenic organisms resident in supplies of drinking water. The nationwide cholera epidemic of 1831 was the first indication of troubles to come. It was followed by successive waves of the same disease or typhus or typhoid. Gradually it became clear that contaminated water was the culprit, though no one knew for sure exactly why. Suspicion fell (rightly) on sewage somehow entering the supply of drinking water, though how this could be checked was a mystery. Fortunately a uniquely talented chemist, Edward Frankland, recognized that past contact with sewage would leave detectable traces of organic material, especially of nitrates, in the water (Figures 4 and 5). Acting on behalf of the government, he designed an analytical method that almost invariably identified samples so contaminated. It is a long story, but its essence is this: the application of Frankland's methodology led to a dramatic decrease in deaths caused by water-borne organisms and arguably saved the large urban centres from something like a major pandemic. When, later, it was recognized that bacteria were responsible, the tests were switched from the chemical to the microbiological, but by that time the large-scale danger had been averted. Chemical analysis is still used, of course, and the purification of the water owes as much to the addition of chlorine or ozone as it does to other, less chemical, measures such as filtration.

Water is not the only commodity under constant examination by chemists. To inspect the records of the Society of Public Analysts, founded in Britain in 1874, is to gain some impression of the immense range of work performed behind the scenes to protect the public from contaminated food, milk, drugs and other hazards. Similar work was performed in the USA and elsewhere.

Of course, when it comes to drugs and medicines the chemist's role is more obvious. The armoury available to physicians today is a result of sustained research in synthetic organic chemistry. The discovery of aspirin in 1893 made available a drug that was not only a far less dangerous febrifuge than acetanilide, which had previously been used, but also had other, unexpected, therapeutic effects, not least those of an anticoagulant. A much more modern drug with the same function is warfarin, which is now used on an enormous scale in preventive medicine for treating all kinds of circulatory and other conditions. Endless examples of such chemical innovations can be listed and the public has little or no idea of the effort that goes into the synthesis and large-scale production of just one drug. A classic example is the sulfapyridine effective against the pneumococcus bacteria, which was discovered in the laboratories of May & Baker in 1938 after no less than 692 unsuccessful trials. It was known as 'M&B 693'.

Figure 4
Frankland's water-analysis
apparatus.

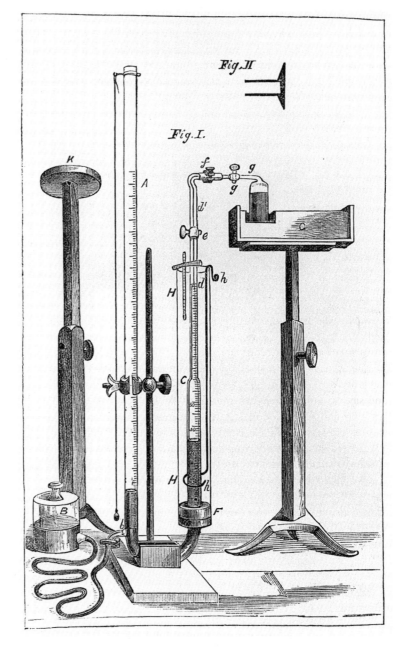

Neither have the chemists' contributions been restricted to research into drugs. They have made possible the relief of pain through anaesthesia, from the chloroform first used in 1847 by James Simpson for midwifery to the latest cocktail of local anaesthetics employed in modern surgery. Antiseptics, from phenol onwards, have transformed the practice of medicine, as have the new biocompatible substances in devices like artificial valves, contact lenses, catheters, blood vessel substitutes etc., rejection of which by the host organism has to be avoided.

The population explosion at the end of the eighteenth century led to gloomy predictions of a planet on which the population increased geometrically but food supplies increased only arithmetically. Possibly for this reason, the production of crops began at about the same time to receive some kind of scientific study. Chemists such as Davy and Liebig examined the correlation between the soil's composition and the growth of crops and, in 1843, an agricultural experimental station was established at Rothamsted by J. B. Lawes and J. H. Gilbert. Realization of the importance of phosphorus for the growth of plants led to the manufacture of superphosphate fertilizers in the same year. The immediate results were gratifying and chemistry became firmly linked with agriculture. Nitrates, from Chilean deposits, were also used to maintain the level of nitrogen in the soil. Warnings that these deposits were finite led to an intensive search for alternative sources of nitrogen. There was enough of it in the air but the problem was its 'fixation', namely combination in a form that could be chemically active. It was eventually solved in the early twentieth century by the Haber synthesis of ammonia (in which nitrogen and hydrogen react to form ammonia under high pressure). Despite known objections to artificial fertilizers (deterioration of the soil's structure, seepage of nitrate into water-courses and a general preference for 'organic' farming), recent work has shown that future demands from underdeveloped countries cannot possibly be met without major use of fertilizers produced chemically.

This group of underprivileged nations constitutes the Third World (or better the Two-Thirds World). It poses a challenge to the chemist that no amount of 'nature-first' ideology can evade. The fact is that, quite apart from the need to fertilize poor soil in those areas, their crops, like those in the First World, are highly susceptible to attack by organisms that will destroy whatever they can grow: hence the need for chemical insecticides. Today the worldwide destruction by pests of annual crops is around 50%. In the Two-Thirds World that figure is nearer 70% and, since the production of crops for those countries needs to be doubled in any case, they need a *sixfold* increase in the near future. The role of the chemist is therefore critically important. Provision of effective agrochemicals and their careful application to the land could significantly increase productivity and save countless lives. Neither is it directly a question of food alone. Many poor countries depend on cotton exports to run their economies. The FAO has recently shown that, without the use of pesticides to control pests, yields will drop by 50%. Cocoa is the largest export from Ghana, but crops are ravaged by the capsid bug; control of this pest by chemical means has trebled the country's annual crop. The sugar crop of Pakistan has increased by 30% following similar treatment and so on.

A great many crop failures are attributable to fungal diseases, which have presented a greater challenge to the chemist than those due to ravaging by insects. The Irish 'potato famine' in the 1840s, with a million deaths, was due to the fungus *Phytophthora infestans*. Efforts to control it by chemical means were unavailing. 'Bordeaux mixture' (copper sulfate and lime) was discovered in 1885 to have fungicidal properties and since then a few other substances have joined its ranks, but only as preventative treatments. No systemic fungicide existed until various benzimidazoles appeared in the 1960s, followed during the next decade by a number of 1,2,4-triazoles. More recently still, Zeneca conducted a series of exhaustic experiments on various naturally occurring fungicides (β-methoxy acrylates), hoping both to imitate and to modify nature. The result appears to be a new systemic fungicide (ICIA5504) of wide activity and low toxicity.

In terms of food supplies, agriculture has not been the only beneficiary of chemical research. Chemistry has long been employed in the business of detecting adulterants (which in Victorian times included iron filings in tea-leaves, lead ethanoate in sugar and red lead in cheese, to say nothing of deliberately added poisons). Today manufacturers are obliged by law to state the contents of their products and, when quantitative figures are quoted, extensive chemical analysis is implied. Then again, chemistry is deeply involved in the preservation of food and its monitoring, in the extensive use of additives to enhance the texture, colour and flavour as well as shelf-life of food and even in the pre-treatment of fruit and other vegetable products to inhibit early deterioration.

Figure 5
Edward Frankland.

Much more could be said of the way in which chemistry has a vital influence on these three areas of human life: war, health and the production of food. This is not propaganda for an obscure branch of science; it is simply the way things are. The strange thing is that, very largely, the public is quite unaware. So the question of chemistry's public image raises interesting and important questions.

The Public Image of Chemistry

It seems that chemistry, like many other things in the modern world, has endured something of a see-saw in its ratings with the general public. At present it is fairly low down in the popularity stakes, though the contention of this book is that this is entirely unjustified and, moreover, far from being in the best interests of society. Therefore we shall chart briefly the main changes in its public image, noting no less than four major slumps in esteem that may justifiably be called crises.

Members of the chemical profession who ruefully reflect on the way the public sometimes esteems their work could do worse than recall the reputation of two alchemists satirized in Chaucer's *Canon's Yeoman's Tale*. The one, a simple-minded cleric of 'sluttish' appearance and ruined complexion, was known by the 'smell of brimstone' and 'stinking as a goat'. The other, an unscrupulous con-man, earned the description 'this fiendly wretch, this false canon – the foul fiend him fetch!' Since then alchemists have been reviled repeatedly on the same grounds: pollution and pretence and, though the language is usually more moderate, today's chemists may face accusations of a similar kind. The issues are fundamentally the same as in Chaucer's day. Any chemical operation has the potential to harm the environment, even though this may be limited to unsavoury odours clinging to one's clothes. Furthermore, the profit motive in any manufacturing process – chemical or otherwise – inevitably carries with it the risk that producers will be tempted into deceiving the public in order to minimize costs and maximize profits.

In fact the image of chemistry recovered considerably with the decline of alchemy and the public had little to complain of until the early nineteenth century. The thirteenth, fourteenth and fifteenth centuries, when alchemy was most dominant, may be regarded as the first peak in the variable unpopularity of chemistry in Britain. For several hundred years following the Middle Ages chemistry was largely ignored because its effects on society were minimal. Even Isaac Newton was permitted to conduct alchemical experiments in his room at Trinity College, Cambridge. The more esoteric traditions of alchemy lingered on and, much more importantly, so did many of its techniques (such as distillation, sublimation, the construction of furnaces and use of the three mineral acids). The last alchemist (Peter Woulfe) died in the nineteenth century. Because of alchemy's association with fraud and self-deception, however, most chemists are all too happy to forget those origins of their craft that lay in the murky alchemical laboratories of the Middle Ages and earlier.

During this quiescent period of course some chemists had a bad press, though it was rarely for chemical reasons. Priestley was attacked by a mob in Birmingham and forced into exile in North America for largely political reasons: he backed the ideals of the French Revolution. Equally, Lavoisier was executed in the course of that revolution not because of his views on combustion but on account of his aristocratic connections and profession as a tax-collector.

However, the age of tolerance drew to an end with the coming of the Industrial Revolution and by about 1800 public antipathy to chemistry was starting to climb to a new and second peak. The most obvious cause was the proliferation of alkali works spewing into the atmosphere torrents of unpleasant gases, of which hydrogen chloride caused the most visible damage.

At that time there was no major use for the hydrogen chloride, so it was let into

Figure 6
'Tennant's Stalk' and the St Rollox Chemical Works, Glasgow, at one time the largest chemical works in Europe.

the atmosphere with predictable consequences. There was immense deterioration in the quality of air breathed by those unfortunate enough to live near the chemical works, destruction of vegetable matter including crops and corrosion of buildings, iron railings and so on. One of the earliest offenders to be brought to book for this anti-social behaviour was an Irishman, James Muspratt, who established an alkali works on Merseyside in the 1820s and within a few years was being vigorously prosecuted by local landowners and others. His own solution was to move his works from one parish to another and let the lengthy legal process take its course once more before moving on again. Others, like Charles Tennant on the Clyde, sought to minimize the damage by building ever higher chimneys, culminating in 'Tennant's Stalk' of 126 m, about as high as the Great Pyramid and 5 m taller than the spire of Salisbury Cathedral (Figure 6). It was a striking symbol of a new and unwelcome image for industrial chemistry.

Then, in 1836, a manufacturer in Worcestershire, William Gossage, showed how a counter-current of water streaming down brushwood in a disused windmill could wash out most of the HCl from rising gases introduced at the base. The Gossage tower gradually became a standard part of the multitude of alkali works being erected in Britain (chiefly on the banks of the Mersey, Tyne and Clyde).

By now, however, the public was becoming much more aware of the potential harm caused by the application of chemistry. Not only was there a ceaseless stream of polluted effluent entering the rivers near most chemical works, even the Gossage towers

Figure 7
Pollution from the Leblanc Factories, Widnes. (Reproduced from D. W. Broad, *Centennial History of Liverpool Section of SCI.*)

were not always effective, through poor construction, lack of repair or general mis-management. So great was the public pressure that, in 1863, the Government passed the first Alkali Act, which directed that alkali works should condense at least 95% of the hydrogen chloride from all gases emitted into the atmosphere. An Alkali Inspectorate was established with the almost unprecedented right to enter and inspect industrial premises in order to protect private property. It was also one of the earliest pieces of legislation to recognize the value of quantitative chemical analysis, which was needed to ensure that the law was being enforced. A further act of 1874 extended legislation to other kinds of chemical works and to other noxious gases.

Of course these Acts did not solve all the problems. They did not deal with all the causes of pollution, most notably the mountains of semi-solid calcium sulfide ('galligu') which, if they did not seep into the water supply, slowly but inexorably gave off hydrogen sulfide. Some are still there on Tyneside. Neither could legislation eliminate the mercenary instincts of some of the most successful proprietors. The modern catch-phrase 'economic gain by environmental loss' has a sad appropriateness to the Victorian chemical industry. A local guidebook to Merseyside compared the intense pollution around Runcorn and Widnes to the horrors of the doomed Biblical cities of Sodom and Gomorrah (Figure 7). Nonetheless, the author could not help adding that 'they have, however, this advantage, that they are by no means *dead* cities of the plain. Great and busy industries are carried on day and night in both, and in each large fortunes have been and are still being made'. Neither was the distress limited to the areas round the factories. On Tyneside, for example, alkali workers lost all their teeth at an early age through breathing acid gases and could be recognized by the crustless sandwiches they habitually consumed.

Modern supporters of the anti-science lobby frequently imagine the rise of the modern chemical industry as the beginning of an almost continuous process of pollution of the atmosphere and rivers, with the vested interests of a few wealthy landowners offering the only deterrent to wholesale plunder of the environment by chemistry. They are wrong. This period, about halfway through Victoria's reign, corresponded to a turning point in the public appreciation of chemistry; its image as a continuing archpolluter is an invention of the 1960s. There are several reasons for the transformation of its image into something much more acceptable.

In the first place, chemistry was putting its own house in order. The Alkali Inspectors were highly skilled analysts and so were many of the chemists recruited to the service of industry in the newly formed technical laboratories. Of course it could be argued that they were forced to comply with legislation, but many of them did so with an enthusiasm that went well beyond the strict call of duty. In any case that objection cannot apply to people like John Glover, a chemist from Tyneside, who devised a means of removing oxides of nitrogen from the effluent gases in the lead-chamber process for manufacturing sulfuric acid, with immense benefit to the environment as well as an economic advantage for the manufacturer (Figure 8). Yet Glover declined to patent his process and shared it openly with rivals. Then again, just as the Alkali Acts were beginning to bite, another process was being developed that would gradually replace the old method of Leblanc. The Solvay process made soda by the double-decomposition reaction

$$NaCl + NH_4HCO_3 = NH_4Cl + NaHCO_3$$

followed by ignition of the resultant sodium bicarbonate. This method involves no pollution of the environment with HCl or CaS, though it did not completely displace the Leblanc process until the 1920s.

Secondly, there are several cases on record of *manufacturers* taking the initiative and demanding more, not less, stringent controls. The alkali manufacturer and author of a standard text on the subject, John Lomas, was a realist:

The injudicious attempt is often made to prove that the alkali trade, as at present carried on, is no nuisance at all, and this in face of all the devastation in the neighbourhood of the works. A far wiser plan would be for manufacturers to acknowledge the danger, and reduce the evil to a minimum by the adoption of every practical means of condensation.

Still more radical in its pleas for 'further and enlightened legislation' was a clarion call addressed by the President of the Tyne Chemical Society to his fellow manufacturers:

To an ordinarily simple-minded individual, it seems neither unreasonable nor unfair that the manufacturer should be pressed to keep his smells and his miscellaneous nuisances to himself. And most must agree that however little else some of us may be heir to, we have all at least a divine inheritance – if not a prescriptive right – in air and water undefiled. If then the manufacturer – as the embodiment of power – rob the poor man – as the embodiment of weakness – of his oxygen, or dilute that oxygen with deleterious gases and vapours; or if he practically force him to make use of water which he has knowingly contaminated, the injury applies to what is of more value than silver – health.

Thus at very least there is evidence for rejecting the conventional view that chemical manufacturers are a uniformly exploitative class of irresponsible polluters. There is also a third reason for adopting a revisionist view of Victorian chemistry. This is the simple fact that, from around the 1860s, chemistry began not only to slough off its

Figure 8
Platform for analysis of chimney
effluent gases.

image of polluting monster but also to acquire a glamour and popularity amongst
working men that led to a dominant position amongst all the sciences and indeed other
subjects studied at Mechanics' Institutes and for the Department of Science and Art
examinations. It is unlikely that even the growing 'usefulness' of chemistry could have
given it such a dramatic appeal to the masses if it had self-evidently been poisoning
the very air they breathed and the water they drank.

Another reason for the growing public tolerance for chemistry was that, from 1877,
chemists constituted themselves into a distinct profession. The Institute of Chemistry
was the first professional institute for scientists anywhere in the world (though engineers had enjoyed professionalization several decades earlier). One important aspect
of being a professional is that one accepts certain responsibilities to society at large;
another is the need to go out and get public approval and support. So we find the emergence of a new kind of literature, books written not as textbooks but as expositions of
chemical progress for a lay public. For example in 1918 the Registrar and Secretary of
the Institute of Chemistry, R. B. Pilcher, produced a book of this kind, *What Industry
Owes to Chemical Science*. A slightly revised second edition in 1923 was rapidly sold
out, a much altered version appearing in 1945. Two years before Pilcher's first edition

another volume had appeared, *Chemistry in the Service of Man*, by Alexander Findlay (who was destined one day to become President of the Institute but then Professor at Aberystwyth). His avowed intention was to give 'some account of what the science of chemistry, both in its general principles and in its industrial applications, has accomplished for the material well-being and uplifting of mankind'. Writing in the middle of World War I, he added

The crisis through which this and other European countries are passing has brought home to us how greatly we, as a nation, have hitherto failed to recognise the intimate and vital dependence of our social and national prosperity on a knowledge and appreciation of the facts and principles of science, and not least of chemistry, and on their application in industry . . . The people as a whole, being ignorant of science, have mistrusted and looked askance at those who alone could enlarge the scope of their industries and increase the efficiency of their labours.

Findlay's book was an outstanding success, a seventh edition appearing as late as 1947. He took encouragement from the promotion and application of science in the previous 31 years. Yet, at the beginning of the nuclear age, he seemed to anticipate a new anti-science movement, for he warned that

Great as have been the suffering and destruction of life brought about by the misuse of the discoveries of chemists, very much greater have been the relief from suffering and the saving of life which their discoveries have made possible.

The immediate post-war years were overshadowed by the beginning of the nuclear arms race. After the atomic bombs had been dropped on Japan the USA declined to share any atomic secrets, even with Britain, and in the UK research into an 'atomic deterrent' was pursued both by Conservative and by Labour Governments. Such was the secrecy involved that public opposition was minimal. It was to be another 10 years before the Campaign for Nuclear Disarmament (CND) emerged and with it a wider disenchantment with science generally and nuclear chemistry in particular. Some (not necessarily in CND) found it difficult to distinguish between civilian and military uses of nuclear power.

On the whole the image of chemistry was not too badly dented. Partly this arose from the (false) impression that nuclear science was chiefly physics. There is also little doubt that the post-war advent of the petrochemicals industry in the UK may have helped to counteract some of the anti-chemical sentiments associated with nuclear science. After all, petrochemical products were now ubiquitously conspicuous, whether as plastics, motor fuels or new fibres for textiles, and the petroleum-based firms were expending vast sums on literature for schools and a general enhancement of their image. What the opposition to nuclear science did achieve was a wider disenchantment with science in general in the 1950s and, of course, chemistry participated in the 'swing from science' that owed more to ideology than to any specific hazard associated with chemical research. Meanwhile almost no one mentioned 'the environment'. In this sense the 1950s were only to be the calm before the storm.

An assault of unparalleled ferocity was launched against chemistry and the chemical industry in the early 1960s. Its effects are with us to this day. Wave after wave of almost virulent hostility to science broke as one crisis after another exposed weakness in the scientific establishment and suggested that enough was enough. The causes were complex and will continue to be debated, but some things are clear. There was, in Britain at least, a new realization of the economic cost of science as the UK opted out of the space race, experienced financial shortages in universities that had, until then, enjoyed generous provision and came genuinely face to face with its technological and financial limitations. Science began to appear an expensive luxury. However, it was

Figure 9
Rachel Carson. (Photograph courtesy of Brooks Studio, reproduced with permission from the Rachel Carson History Project.)

not from economics that the big challenge came. Far more serious, especially for chemistry, was the challenge over the *environment*. Today environmental issues are the subject of countless books, articles, speeches and even sermons and appear to constitute by far the biggest social challenge chemistry has ever had to face. We can never understand the nature of this challenge, let alone discover how to meet it, without comprehending how it came about in the first place.

In 1962 an American marine biologist (Figure 9) who had been working with the USA's Fish and Wildlife Service produced a book. She was an accomplished author who had written several delightful works on natural history. This one was different. Rachel Carson's *Silent Spring* proved to be a veritable time-bomb, a catalyst for a number of hugely important cultural changes, among them a re-appraisal of the role of science and technology in our modern society. Her book represented a watershed in the public appreciation of chemistry and brought to the fore the concept of the *environment*, together with a respectful recognition of ecology. The latter was not, of course, a new science; neither was 'the environment' invented (or discovered) in the 1960s. Concepts of what we might call 'environmental protection' were, as we have seen, circulating 100 years before, though not by that name.

The main thrust of *Silent Spring* was the danger to wildlife from chemicals deployed in agriculture, particularly insecticides like DDT. To these materials song-birds are especially sensitive, because they are not able to metabolize most of those ingested with food, so giving rise to the phenomenon of bio-accumulation. So many birds were found dying or dead with high levels of organochlorine residues, or were producing eggs with unacceptably thin shells, that a day when no more bird-song would be heard might come. There would be a 'silent spring'.

Such a phrase has high rhetorical value and the prospect aroused deep emotions in nature-lovers the world over. An environmentalist lobby orchestrated demands for action (actually believing that a victory on the question of DDT would give them, for future debates, 'a level of authority they [had] never had before'). In due course the USA's government responded, by first restricting and then banning the use of DDT. Environmentalists were jubilant and crusades for still more far-reaching action were mounted.

Not all responses to *Silent Spring* were as favourable. Not surprisingly, the chemical industry reacted strongly, with conferences, brochures and press releases, although not always in a tone of calm rational argument. A more temperate response came later from the American Chemical Society. Some of the arguments rehearsed have wider application than to insecticides and will shortly be discussed. Meanwhile it is important to recognize that the controversy over *Silent Spring* was only the beginning of a long dispute between environmentalists and chemists on the merits or otherwise of the presence of 'chemicals' in the wide world generally. Chemists, no less than their opponents, have had much to say on the subject in recent years. It is possible to identify three fundamental causes for the adverse reactions to chemistry. These can be summarized as follows.

(i) *Uncontrolled pollution.* This was particularly the case with early alkali works in Britain, though, after a series of legislative measures (Alkali Acts, Clean Air Acts, etc.), it tends now to be limited to large-scale emissions of carbon dioxide, to which, of course, the industry is only one of many contributors. In less developed countries the position is far less satisfactory, as may be seen in centres of industry in several parts of Eastern Europe today.

(ii) *Undesirable side-effects.* The case of DDT and other organochlorine or organophosphorus insecticides was followed by weed-killers such as 2,4-D. Later, PCBs from electrical equipment were shown to accumulate in human fatty tissue, potentially endangering health, and ozone-destroying CFCs from refrigeration equipment appeared in the upper atmosphere. One must, of course, also include unwelcome side-effects that have been attributed to numerous additives to foodstuffs (such as certain colourants, cyclamates, *trans* acids in margarine, etc.), to say nothing of drugs, of which thalidomide is probably the most notorious example.

(iii) *Unexpected accidents*. It may be said with some justification that the chemical industry has been accident-prone. In the light of the inflammable and/or explosive raw materials employed and products made, this is hardly surprising. What is remarkable is the rarity with which serious accidents occur today. In recent years major accidental spills have included Flixborough (1974), Mexico City (1976), Seveso (1976) and Bhopal (1984) (not to speak of nuclear spills such as Windscale (1957), Three Mile Island (1979) and above all Chernobyl (1986)). There have also been countless instances of less dramatic events including the seepage of stored chemicals into water supplies, overturned road tankers, minor explosions and inadvertent leakage of noxious gases into the atmosphere. More serious have been incidents of large-scale damage inflicted on the environment by a succession of accidents at sea involving oil-tankers and other carriers of toxic chemical products. The loss of the *Torrey Canyon* in 1967 marked a spectacular rise in public awareness of environmental issues and sensitivity to consequences of all accidents involving products of the chemical industry.

All these cases have been fully documented. Together they offer a formidable armoury to those engaged in denigrating the image of chemistry. At face value they appear to be a massive indictment. Whether this is really so, and how chemistry could respond to persistent attack, will be examined in our final section.

The Way Forward

As the new chemistry faces the new millennium it will encounter unprecedented opportunities as well as unwelcome opposition. The latter will spring chiefly from what may be called the 'environmental lobby' or the 'green movement'. It is up to chemical writers and speakers to face their challenge in a spirit of tolerance and realism.

Confronted with a long list of indictments, the first thing the chemical industry must do is surely to *admit past mistakes*. These will include all attempts at cover-up, be they by governments (such as Chernobyl), corporations (such as at Bhopal) or individuals such as the numerous polluters who befouled air and water and those (like Frankland) who advised them to sink alkali waste into the Severn and 'make the surrounding neighbourhood believe' that all the HCl was being washed away (whether it was or not). With hindsight there can be little excuse for this kind of attitude and it is better to offer a frank admission of guilt. The same must also go for certain actions of large corporations, then and now, when economic gain over-rode the necessity not to foul the environment.

It is also necessary to *expose the underlying fallacies of some anti-chemical propaganda*. One of the most persistent of these is the belief that anything 'natural' is good and anything artificial or 'chemical' is bad. It is a pervasive idea, as may be seen in the tendency of the media to call noxious substances 'chemicals' irrespective of whether or not they originated through the chemist's art. It is implied in a recent appeal letter from Greenpeace, in which a picture of a mother breast-feeding her baby is balanced, on the other side of the page, by a hideous plant belching out black smoke into the air and pouring torrents of a dark liquid over the earth. The point is an entirely proper one to make, but it does enshrine the huge contrast felt between 'natural' and

Figure 10
Result of a volcanic eruption (a rival polluter?).

'chemical'. Those who would exclude all chemicals from agriculture in favour of 'organic' farming may be heard to utter sentiments like 'only chemical products cause cancer', which have been discredited totally in the last few years (some ferns furnish particularly nasty examples of natural carcinogens). There are many other cases of plant toxins. When it comes to more cosmic issues it is now doubtful whether the partial destruction of the ozone layer is solely the work of CFCs, for volcanic emissions may well have a similar effect. Even to speak of 'the greenhouse effect' is an inaccuracy, for volcanic eruptions (Figure 10), forest fires and other natural phenomena have long ensured that there is sufficient carbon dioxide to stop us cooling down; technically we ought to speak of the 'excess greenhouse effect'.

A final fallacy concerns the so-called 'chlorine crisis'. Because that element enters into so many organic compounds (including many insecticides), because these degrade into smaller organochlorine molecules, which may have great longevitites, and because some of these may have an adverse effect on human health, some environmentalists have demanded a ban on all chlorinated products, some calling for the total elimination of PVC. It has even been said that 'God created 91 elements, mankind about a dozen, and the devil just one – chlorine'. Such wild sentiments do little good for a responsible 'green' look at the world and are easily countered by drawing attention to the facts that such assumptions about toxicity remain unproved and that organochlorine compounds occur widely in nature.

Then it becomes also necessary to *achieve a balanced view about chemicals in the environment.* It does not follow that one set of disadvantages inevitably outweighs the advantages that led to their use in the first place. We can take only one example: the banning of DDT consequent to *Silent Spring.* Perhaps one of the most extreme views of Carson's book was that of Norman Borlaug, Nobel Peace Prize winner, who attacked it as 'vicious, hysterical propaganda'. Such strong language was justified in part by the following fact: so effective is DDT in the control of the malaria-bearing parasite that in 25 years its banning had resulted in the exposure to the mosquitoes of a thousand million human beings. It has been argued that DDT has actually saved more lives than penicillin, yet we have virtually eliminated it from our environment. That is not a balanced approach to the problem, especially insofar as it is only recently that any real evidence that definitively links DDT to human maladies (in this case as a weak imitator of oestrogen) has appeared.

Future proposals include integrated pest management, in which biological *and* chemical measures are blended, together with the introduction of pest-resistant strains of plants and constant monitoring of the process. Similarly, insofar as 'organic' agriculture alone cannot supply the world's needs in the foreseeable future, it seems best to envisage a more controlled use of chemical fertilizers, in conjunction with the best techniques from 'alternative agriculture'.

A further strategy would be to *stress the historic course of chemical pollution*. That, indeed, is one reason why good historical analysis can actually serve the science today. It is very relevant to recall cases of responsible action by the chemical community, to avoid attribution of most atmospheric pollution to chemical works when, in fact, most was a result of burning coal in domestic fires and industrial furnaces and to note the successful measures actually taken by the industry to reduce pollution.

A much-needed strategy today is that of publicising the successes of the chemical industry. Ignored by many ordinary historians, misrepresented by its opponents and unknown to the general public, the industry can be enormously proud of its achievements. Not only is it a major earner of revenue for the state but also it has, as we have seen, conferred highly beneficial effects on society, from the Industrial Revolution onwards. For all its mistakes, to say nothing of its often lack-lustre response to criticism, the chemical industry is a prime example of science employed in the service of humanity.

It is always pertinent to draw attention to these things in connection with 'green' issues. In chemical education it is a great pity if opportunities to stress the benefits flowing from the application of chemistry are missed. For politicians an awareness of what chemistry has done in the wider world can never come amiss. One of the achievements of the Royal Society of Chemistry in the UK has been the establishment of a Parliamentary Links Scheme through which government is kept continually aware of recent developments in the application of chemistry.

There is, however, another aspect of chemistry that has nothing to do with its application in industry. This concerns its worth as a subject of academic study and research. Once again some knowledge of how chemistry arrived at its present state can only confirm the complexity and size of the daunting intellectual task confronting it at all stages. The majesty of such concepts as the chemical element, atomism, valency, structural theory, stereoisomerism, the periodic law, radioactive decay, quantum theory and thermodynamic and kinetic control can only be matched by the sheer ingenuity with which chemical analysis has been automated and the results have been interpreted. In a wider context it has helped to lead to the double helix and genetic finger-printing, to astrochemistry and an understanding of chemical evolution, to chemical archaeology and the dating of artefacts.

This book has offered abundant examples of recent research. We hope that something of the excitement of the chase has reached you, together with a sense of the worthwhileness of this kind of human enquiry. Indeed, the inherent satisfaction of progress in research is its own reward. Those who have discovered new regularities in nature, have prepared substances never seen before, have discovered the inner architecture of a complex molecule, have correctly predicted the course of a new chemical reaction or even obtained white needle-like crystals from the most treacly black tar all know in their different ways something of the joys of chemical research. However, there are other reasons for maintaining the momentum of research.

One of these is in order to obtain sufficient knowledge to prevent a repetition of past mistakes in the environment. If the lethal effects of hydrogen chloride on human tissues had been known its emission would surely have been proscribed much earlier; if the beneficial effects of DDT had been recognized it might not have been subject to such a stringent ban; if the importance of chirality had been perceived soon enough the tragedy of thalidomide might have been avoided; if the effects of CFCs on the ozone layer had been understood their emission would surely have been controlled more strictly and so on. The answer to the 'green' challenge is therefore not less science but better science.

A further reason for urging chemical research is of course in the pursuit of specified product objectives. That is the rationale behind the vast expenditure on the design and synthesis of drugs. It applies to all manner of other industries including dyestuffs, pigments, paints, textiles, photographic chemicals, explosives and plastics. The importance of R & D for such industries cannot easily be overstated.

Finally, in pure academic chemical research one can never be sure that something wholly unexpected may not emerge. Amongst many cases of serendipity (unexpected outcomes from research) have been the following: Perkin was trying to synthesize quinine when he accidentally prepared the first ever synthetic dyestuff (mauve); Frankland was trying to prepare 'ethyl' when he stumbled upon a new series of compounds (organometallics) and the fundamental chemical doctrine of valency; Alexander Todd discovered a synthesis for anthocyanins when a flask containing reagents to make a glucoside accidentally fell into a water-bath; research on certain substituted aromatic acids led to the production of liquid crystals that proved to be a major industrial development of the 1980s. We could go on. The very possibility of surprises round the corner is an effective inhibitor of confident predictions about the direction in which the new chemistry will go. That it will be very different from the old is likely; that it will be as rewarding and exciting is certain.

Further Reading

1. W. H. Brock, *The Fontana History of Chemistry* (Fontana, London, 1992).
2. A. Clow and N. Clow, *The Chemical Revolution* (Gordon & Breach Science Publishers, Philadelphia, 1992) (reprint of 1952 edition).
3. W. A. Campbell, *The Chemical Industry* (Longman, London, 1971).
4. J. A. G. Drake (editor), *The Chemical Industry – Friend to the Environment?* (Royal Society of Chemistry, Cambridge, 1992).
5. S. O. Duke, J. J. Menn and J. R. Plimmer, *Pest Control with Enhanced Environmental Safety* (American Chemical Society, Washington, 1993).
6. A. Findlay, *Chemistry in the Service of Man* (Longman, London, 1916 and many subsequent editions).
7. R. M. Harrison (editor), *Understanding Our Environment: An Introduction to Environmental Chemistry and Pollution*, second edition (Royal Society of Chemistry, London, 1992).
8. C. A. Russell, *The Earth, Humanity and God* (University College London Press, London, 1993).
9. C. A. Russell (editor), *Chemistry, Society and Environment: A New History of the British Chemical Industry* (Royal Society of Chemistry, Cambridge, 2000).
10. C. A. Russell, N. G. Coley and G. K. Roberts, *Chemists by Profession* (Royal Institute of Chemistry/Open University Press, Milton Keynes, 1977).

Index